VOLUME ELEVEN

Advances in Food Security and Sustainability

VOLUME ELEVEN

ADVANCES IN FOOD SECURITY AND SUSTAINABILITY

Edited by

DON ELISEO LUCERO-PRISNO III

Department of Global Health and Development, London School of Hygiene and Tropical Medicine, London, United Kingdom

MARC J. COHEN

Lead Researcher, Aid, Development Finance, and Food Systems, Oxfam America (retired)

ELSEVIER

ACADEMIC PRESS

An imprint of Elsevier

Academic Press is an imprint of Elsevier
125 London Wall, London, EC2Y 5AS, United Kingdom
50 Hampshire Street, 5th Floor, Cambridge, MA 02139, United States

First edition 2026

ISBN: 978-0-443-41421-3
ISSN: 2452-2635

For information on all Academic Press publications
visit our website at https://www.elsevier.com/books-and-journals

Publisher: Zoe Kruze
Acquisition Editor: Leticia Lima
Editorial Project Manager: Palash Sharma
Production Project Manager: Greetal Carolyn
Cover Designer: Muthukumaran Thangaraj

Typeset by MPS Limited, India

Contents

Contributors

Abdullateef Abdulsalam
Department of Biomedical Sciences, Nazarbayev University School of Medicine, Astana, Kazakhstan

Mubarak Oladimeji Abdulsalam
Department of Medicine and Surgery, University of Ilorin, Ilorin, Kwara State, Nigeria

Kolade Oluwakunmi Adekanye
Department of Agricultural Economics, Landmark University, Omu Aran, Kwara State, Nigeria

Gary G. Ador Dionisio
De La Salle College of Saint Benilde, Philippines

Michelle Agustin
Science Department, College of Science, Bulacan State University, Malolos, Bulacan, Philippines

Mohamed Mustaf Ahmed
Faculty of Medicine and Health Sciences, SIMAD University, Mogadishu, Somalia

Oluwatosin Seyide Alabi
Department of Child, Youth & Family Studies, College of Education and Human Sciences, University of Nebraska-Lincoln, Nebraska, NE, United States

Oliver R. Alaijos
Science Department, College of Science, Bulacan State University, Malolos, Bulacan, Philippines

Chris Jordan G. Aliac
Cebu Institute of Technology – University,Cebu City, Philippines

Abdulmajid Abdullahi Aliyu
Department of Microbiology and Biotechnology, Federal University Dutse, Dutse, Nigeria

Rose A. Arceño
Research Services Office, Palompon Institute of Technology, Palompon, Leyte, Philippines

Dauda Ayuba
Department of Local Government and Development Studies, Ahmadu Bello University, Zaria, Nigeria

John Rodel A. Balasabas
Graduate School of Agriculture and Related Sciences, University of Southeastern Philippines, Tagum City, Davao del Norte; Research and Creative Works Office, Agusan del Sur State University, Bunawan, Agusan del Sur; College of Agriculture and Forestry, Northeastern Mindanao State University, Tagbina Campus, Surigao del Sur, Philippines

Kevin Smith P. Cabuhat
Basic Education Department, La Consolacion University Philippines, Malolos, Bulacan, Philippines

N.E. Chei
Nway Eint Chei, Department of Management, Faculty of Humanities and Management, Myanmar Creative International College, Myanmar

Oshibe Joseph Daberechi
Department of Medical Laboratory Science, Ebonyi State University, Abakaliki, Nigeria

Rich Milton R. Dulay
Department of Biology, College of Science, De La Salle University, Manila; Center for Tropical Mushroom Research and Development, Central Luzon State University, Nueva Ecija, Philippines

Olabode Ekerin
Department of Child, Youth & Family Studies, College of Education and Human Sciences, University of Nebraska-Lincoln, Nebraska, NE, United States

Clement Ogar Elam
Department of Medicine and Surgery, Faculty of Clinical Sciences, College of Medicine, University of Ibadan, Ibadan, Nigeria

Alexander Idu Entonu
Department of Medicine and Surgery, University of Ilorin, Ilorin, Kwara State, Nigeria

Kehinde Precious Fadele
Department of Medicine and Surgery, College of Medicine, University of Nigeria, Enugu, Nigeria

Precious Kehinde Fadele
Department of Medicine and Surgery, Faculty of Clinical Sciences, College of Medicine, University of Nigeria, Enugu Campus, Enugu State, Nigeria

Larmie S. Feliscuzo
Cebu Institute of Technology – University,Cebu City, Philippines

Sonia Fontanilla-Pimentel
University of Makati, Philippines

Greta Gallo Gabinete
West Visayas State University, La Paz, Iloilo City, Philippines

Nsikakabasi Samuel George
Faculty of Health Sciences, Institute of Public Health, Jagiellonian University Medical College, Krakow, Poland

Ahmed Adamu Gololo
Faculty of Pharmaceutical Sciences, Chulalongkorn University, Bangkok, Thailand

Mohammed Raihanatu Hamid
Department of mathematics and Computer Science, College of Education, Billiri, Nigeria

Usman Abubakar Haruna
Faculty of Pharmaceutical Sciences, Ahmadu Bello University, Zaria, Nigeria; Department of Biomedical Sciences, Nazarbayev University School of Medicine, Astana, Kazakhstan

Adamu Muhammad Ibrahim
Department of Immunology, School of Medical Laboratory Science, Usmanu Danfodiyo University, Sokoto, Nigeria

Stephen Chukwuemeka Igwe
Department of Medicine and Surgery, University of Ilorin, Ilorin, Kwara State, Nigeria

Ejay Imperial
Advanced Technology Center in the Philippines (ATCP), Tech Architecture Capability, Accenture, Taguig City, Philippines; Research Unit, Imperialle AI Tech OU, Tallinn, Estonia

Oluwadarasimi Babafemi Jaiyeoba
Department of Medicine and Surgery, University of Ilorin, Ilorin, Kwara State, Nigeria

Roselle A. Jardin
Cebu Normal University, Philippines

Raymundo F. Javier
Science Department, College of Science, Bulacan State University, Malolos, Bulacan, Philippines

Muhammad Zainab Lawal
School of Global Health, Faculty of Medicine, Chulalongkorn University, Bangkok, Thailand

Don Eliseo Lucero-Prisno III
Department of Global Health and Development, Faculty of Public Health and Policy, London School of Hygiene and Tropical Medicine, London, United Kingdom; Research Office, Palompon Institute of Technology, Palompon, Leyte; Office for Research, Extension and Innovations, Bukidnon State University, Malaybalay City, Bukidnon; Research and Development Office, Biliran Province State University, Naval, Biliran; Research Planning and Development Office, WCC Aeronautical and Technological College Inc., Binalonan, Pangasinan, Philippines

Adriana Viola Miranda
Global Health Focus Asia, Bandung, Indonesia

Muhammad Kabir Musa
University of Michigan, Ann Arbor, United States

Shuaibu Saidu Musa
School of Global Health, Faculty of Medicine, Chulalongkorn University, Bangkok, Thailand; Department of Nursing Science, Ahmadu Bello University, Zaria, Nigeria

Jerome Jack O. Napala
Southern Leyte State University, Sogod, Southern Leyte, Philippines

Rahmatullah Nazari
Faculty of Dentistry, Cheragh University of Medical Sciences, Kabul, Afghanistan

Richard Ian Mark T. Necosia
Research Office, Bukidnon State University, Malaybalay, Bukidnon, Philippines

Joana May Nepomuceno
Science Department, College of Science, Bulacan State University, Malolos, Bulacan, Philippines

Jerico Bautista Ogaya
Department of Medical Technology, Institute of Health Sciences and Nursing, Far Eastern University Manila; Center for University Research, University of Makati, Makati; University Research and Development Office, Guimaras State University, Buenavista, Guimaras, Philippines

Molly Unoh Ogbodum
Department of Public Health, University of Calabar, Calabar, Nigeria

Sedoten Oluwaseun Ogun
Department of Food Science and Technology, University of Nebraska-Lincoln, Nebraska, United States

Isaac Olushola Ogunkola
Nuffield Department of Population Health, University of Oxford, Oxford, United Kingdom

Olalekan John Okesanya
Department of Public Health and Maritime Transport, University of Thessaly, Volos, Greece; Department of Medical Laboratory Science, Neuropsychiatric Hospital, Aro, Abeokuta, Ogun State, Nigeria

Kolawole Emmanuel Olayinka
Faculty of Engineering, University of Nigeria, Nsukka, Enugu State, Nigeria

Christian Joseph N. Ong
Department of Biology, College of Science, De La Salle University, Manila, Philippines

Jannie Fleur V. Oraño
Cebu Institute of Technology – University,Cebu City, Philippines

Zhinya Kawa Othman
Department of Pharmacy, Kurdistan Technical Institute, Sulaymaniyah, Kurdistan Region, Iraq

Virginia P. Paitan
Science Department, College of Science, Bulacan State University, Malolos, Bulacan, Philippines

Harvie P. Portugaliza
Department of Veterinary Clinical Sciences, Visayas State University, Baybay City, Leyte, Philippines

Fadele Kehinde Precious
College of Medicine, University of Nigeria, Enugu state, Nsukka, Nigeria

Ludito V. Ramirez
Faculty of Arts and Sciences, Southern Leyte State University – Main Campus, Sogod, Southern Leyte, Philippines

Mary Ann Jilly R. Ramirez
Faculty of Arts and Sciences, Southern Leyte State University – Main Campus, Sogod, Southern Leyte, Philippines

Joan M. Recente
Research Office, Bukidnon State University, Malaybalay, Bukidnon, Philippines

Razaele G. Resultay
Research Office, Pangasinan State University, Lingayen, Pangasinan, Philippines

Muktar Musa Shallangwa
Achieving Health Nigeria Initiative (AHNI), Biu, Nigeria

Deborah Oluwaseun Shomuyiwa
Department of Health Promotion & Behavior, College of Public Health, University of Georgia, Athens, GA, United States

Rhene C. Tabajen
University of Makati, Philippines

Ederson de los Trino Tapia
University of Makati; University of the Philippines, Diliman, Philippines

Debbie Ann Tuazon
Science Department, College of Science, Bulacan State University, Malolos, Bulacan, Philippines

Cherry I. Ultra
Office of the President, University of Eastern Philippines, Catarman, Northern Samar, Philippines

Wendell L. Ultra
College of Agriculture and Technology, Northwest Samar State University, Calbayog City, Samar; College of Agriculture and Food Science, Visayas State University, Visca, Baybay City, Leyte, Philippines

Preface

Don Eliseo Lucero-Prisno III[a,b,c,*], Marc J. Cohen[d], Deborah Oluwaseun Shomuyiwa[e], Jerico Bautista Ogaya[f,g,h], and M.B.N. Kouwenhoven[i]

[a]Department of Global Health and Development, London School of Hygiene and Tropical Medicine, London, United Kingdom
[b]Research and Extension Office, John B. Lacson Foundation Maritime University, Molo, Iloilo, Philippines
[c]Research and Development Office, Palompon Institute of Technology, Palompon, Leyte, Philippines
[d]Lead Researcher, Aid, Development Finance, and Food Systems, Oxfam America, Washington, DC, United States (retired)
[e]Department of Health Promotion & Behavior, College of Public Health, University of Georgia, Athens, Georgia
[f]Department of Medical Technology, Institute of Health Sciences and Nursing, Far Eastern University, Manila, Philippines
[g]Center for University Research, University of Makati, Makati City, Philippines
[h]Office of the Research Services, Kalinga State University, Tabuk City, Kalinga, Philippines
[i]Department of Physics, Xi'an Jiaotong-Liverpool University, Suzhou, China
*Corresponding author. e-mail address: don-eliseo.lucero-prisno@lshtm.ac.uk

INTRODUCTION

The challenge of feeding a growing, aging, and increasingly climate-vulnerable world population remains one of the defining imperatives of our time. The 2025 State of Food Security and Nutrition in the World report estimates that about 8.2% of the global population, approximately 673 million people, was chronically undernourished in 2024 (FAO, IFAD, UNICEF, WFP, & WHO, 2025). This reflects a slight improvement from 2023 (about 8.5% of the global population), but still leaves hunger levels well above those observed before the COVID-19 pandemic. According to the report, more than 295 million people across 53 countries and territories faced high levels of acute food insecurity (Integrated Phase Classification Phase 3 or above) (FAO, IFAD, UNICEF, WFP, & WHO, 2025). This represents roughly a 13.7-million increase from 2023 and marks yet another consecutive year in which acute food insecurity remains extremely elevated, driven primarily by conflict, climate extremes, and economic shocks. These figures are not abstractions. They reflect the accumulated effects of long-standing structural failures: fragmented and siloed food-system governance, accelerating ecological degradation, inequitable distributions of resources and power, and the persistent underprioritization of nutrition across health and agricultural systems (FAO, IFAD, UNICEF,

WFP, & WHO, 2025; Odoms-Young, et al, 2024). The contributors to this volume collectively set out to examine and address these profound challenges.

What the aggregate data tend to obscure is the systemic character of these conditions. Food insecurity is not reducible to insufficient agricultural output; it emerges from the interaction of multiple, mutually reinforcing failures across governance, health, ecology, technology, and political economy. This volume of Advances in Food Security and Sustainability recognizes this complexity and responds to it with a deliberately broad and intersectional analytical frame. Volume 11 brings together twelve chapters that collectively map the food security landscape across multiple scales, from global policy frameworks and continental agricultural histories to community pharmacies, coastal reef fisheries, and the governance of national food systems. In doing so, the volume advances a core argument: that sustainable food security cannot be achieved through any single sectoral lens, but only through coordinated, evidence-based action that crosses disciplines, borders, and systems.

The volume is divided into two parts, the first of which focuses on Food Security and Health. This opens with two chapters that establish its broadest conceptual frames. In Chapter 1, Greta Gallo Gabinete advances a governance-centered argument: that food security must be institutionalized as a core preventive function within health systems, embedded in routine screening, nutrition-sensitive service delivery, cross-sectoral policy coordination, and dedicated financing, rather than treated as a residual concern of social welfare or agricultural ministries. Drawing on the Social Determinants of Health framework and comparative evidence from across the globe, Gabinete demonstrates that the separation of food governance from health system planning is itself a structural driver of the preventable disease burdens associated with food insecurity, from maternal and child mortality to noncommunicable diseases and mental health disorders. The chapter makes a compelling case that achieving Sustainable Development Goal 2, Zero Hunger, requires not more food production alone, but deliberate institutional integration.

A complementary perspective is offered through the One Health framework in Chapter 2, which argues that the interdependence of human, animal, and environmental health is not incidental to food security but constitutive of it. Jerico Bautista Ogaya and colleagues examine how zoonotic disease outbreaks, the proliferation of antimicrobial resistance through agricultural antibiotic overuse, and climate-driven ecological disruption each expose critical fragilities within contemporary food systems. Drawing on case studies from Kenya, China, Vietnam, and the Netherlands, they demonstrate that integrated One Health responses can generate measurable improvements in food-system resilience and should be understood as central, rather than peripheral to food-security planning.

Chapters 3 and 4 address the role of emerging technologies and structured analytical methods in food safety governance. Mohammed Raihanatu Hamid and colleagues (Chapter 3) review the application of artificial intelligence (AI) and machine learning (ML) across the food safety landscape, from sensor-based real-time contamination monitoring and predictive pathogen modeling to digital public opinion surveillance and supply-chain traceability. The chapter is candid about constraints: data-quality deficits, regulatory gaps, limited generalizability of context-specific models, and the risk that AI-enabled risk profiling may discourage data sharing among food-industry actors, yet ultimately argues that AI and ML represent a genuinely transformative set of tools for food-safety management when embedded within accountable governance frameworks and designed for diverse global contexts. Chapter 4, by Mary Ann Jilly R. Ramirez and Ludito V. Ramirez, then shows how Multi-Criteria Decision Analysis (MCDA) can function as a decision-support mechanism that bridges complex food-safety science and practical policy in resource-constrained environments. Through empirical examples, MCDA is used to prioritize interventions such as consumer awareness campaigns and mandatory vendor hygiene training, offering a model for how structured, participatory tools can generate transparent, politically credible recommendations where governance capacity and resources are limited.

Nutritional vulnerability is foregrounded in Chapters 5 and 6, which examine population groups whose needs are often inadequately addressed by mainstream policy. Nway Eint Chei and Don Eliseo Lucero-Prisno III focus on older adults living with noncommunicable diseases, demonstrating how physiological ageing, including declining appetite, altered metabolism, reduced absorption capacity, and diminished mobility, intersects with structural food insecurity to produce a distinctive risk profile. With the global population aged 60 years and above projected to reach 2.1 billion by 2050, age-sensitive nutrition policies, community-based feeding programs, and integrated geriatric–nutritional care frameworks will become essential pillars of responsive food-security systems (WHO, 2025). Zhinya Kawa Othman and colleagues explore a related gap by positioning the community pharmacist as an underutilized but strategically placed actor in the nutrition-security response, particularly in low- and middle-income countries where dietitian services are scarce and community pharmacies often serve as the most accessible point of health contact for food-insecure households. Their chapter outlines practical pharmacy roles in risk identification, counselling, supplementation, drug–nutrient interaction review, and referral, and proposes short-, medium-, and long-term recommendations for integrating pharmacy practice into national nutrition strategies.

Stephen Chukwuemeka Igwe and colleagues (Chapter 7) examine biofortification as a sustainable strategy for addressing the "hidden hunger" of micronutrient deficiency that affects two billion people worldwide, with the burden concentrated in Africa. Tracing the history of biofortification from early breeding experiments to contemporary technologies, they analyse the scientific, regulatory, sociocultural, and logistical challenges that limit adoption and frame biofortification as a practical, farmer-centred pathway to improved nutrition that engages with the food systems on which poor households already rely.

The chapters in Part II of the volume focus on Resilient Food Systems, situating contemporary debates within longer agricultural and ecological trajectories. Wendell L. Ultra and Cherry I. Ultra (Chapter 8) provide a critical historical anchor through a reassessment of the Green Revolution from 1965 to 2000. Acknowledging the substantial increases in global cereal production during this period but moving beyond the dominant narrative of unqualified success, their analysis documents uneven regional consequences of high-yielding variety adoption, synthetic-input dependency, and irrigation expansion; transformative productivity gains in South and Southeast Asia; more limited and conditional benefits in Latin America; and a largely bypassed Africa, where agroecological heterogeneity, infrastructure deficits, and institutional constraints made the Green Revolution's technological package poorly suited to local conditions. Environmental legacies raise fundamental questions about the sustainability of input-intensive farming systems and must inform next-generation food-system transformations.

Christian Joseph N. Ong and colleagues (Chapter 9) extend this ecological analysis by examining biodiversity loss as a structural threat to food-system sustainability. Accelerating erosion of genetic, species, and ecosystem diversity is identified as one of the most under-appreciated long-term threats to food-system resilience, one that is significantly exacerbated by climate change. Their discussion explores how agrobiodiversity enhances crop productivity, pest regulation, soil fertility, pollination, and climate adaptation, while documenting how industrial monocultures, land-use change, pollution, and intensive agricultural practices are dismantling these foundations. Emerging frameworks such as agroecology, climate-smart agriculture, conservation agriculture, and landscape-level management are presented as integrated pathways for reconciling food production with biodiversity conservation, contingent on supportive governance and inclusive policies.

Chapter 10, by Jannie Fleur V. Oraño and colleagues, presents an AI-powered framework for coral-reef assessment in service of coastal food security in the Pacific blue economy. Coral reef–associated coastal fisheries provide an estimated 50–90% of total animal protein intake for many Pacific Island communities, yet these ecosystems face accelerating threats from climate change, ocean warming, destructive fishing, and coastal development (Bell, et al, 2018). The chapter describes a framework integrating YOLO (you only look once)-based instance segmentation, automated image preprocessing, and GIS mapping to produce spatially explicit coral-cover estimates, illustrated through a case study from Sogod Bay in Southern Leyte, Philippines. By linking reef-monitoring outputs to fisheries management, marine protected-area planning, and community-based governance, the framework positions AI-driven ecological assessment as a practical tool for sustaining the nutritional foundations of coastal populations.

Chapters 11 and 12 examine food-system governance in middle- and high-income contexts. Ederson de los Trino Tapia and colleagues (Chapter 11) analyse food-security governance in the Philippines, focusing on the Rice Tariffication Law (RA 11203), the Mandanas–Garcia ruling on local-government devolution, and the Philippine Development Plan. Through the lenses of institutional theory, policy-process analysis, and systems thinking, they trace how dependence on imported staples, climate vulnerability, land-use conflicts, inequitable food access, and fragmented inter-agency coordination have collectively undermined food-security outcomes despite significant policy-reform activity. The authors propose digital agriculture, smallholder empowerment, and coordinated multisectoral governance as a practical roadmap for future food security. Turning to the United States, Deborah Oluwaseun Shomuyiwa and colleagues (Chapter 12) examine the paradox of a food system that simultaneously leads the world in agricultural productivity and leaves tens of millions of Americans food insecure. Their analysis shows how innovations in biofortification, precision agriculture, and controlled-environment farming have reshaped crop production while often compromising nutritional quality and equity, and highlights community-supported agriculture, urban food systems, and food-sovereignty movements as emerging models for a more just food future.

What unites these diverse contributions is a shared commitment to resisting simplification. Food security is not solved by increasing yields alone, nor by any single technology, policy instrument, or institutional reform. It requires the patient and rigorous work of understanding how systems interact; how health governance shapes nutritional outcomes; how ecological integrity sustains agricultural productivity; how historical choices constrain present options; and how the most vulnerable populations experience food insecurity in ways that aggregate statistics cannot fully capture. Advancing food security and sustainability demands both urgency and nuance. This volume strives to equip researchers, policymakers, practitioners, and advocates with the analytical depth and empirical breadth needed to meet that challenge.

References

Bell, J. D., Cisneros-Montemayor, A., Hanich, Q., Johnson, J. E., Lehodey, P., Moore, B. R., Pratchett, M. S., Reygondeau, G., Senina, I., Virdin, J., & Wabnitz, C. C. C. (2018). Adaptations to maintain the contributions of small-scale fisheries to food security in the Pacific Islands. *Marine Policy, 88*, 303–314. https://doi.org/10.1016/j.marpol.2017.05.019

FAO, IFAD, UNICEF, WFP, & WHO. (2025). *The State of Food Security and Nutrition in the World 2025: Addressing high food price inflation for food security and nutrition*. Rome: FAO. https://doi.org/10.4060/cd6008en.

Odoms-Young, A., Brown, A. G. M., Agurs-Collins, T., & Glanz, K. (2024). Food Insecurity, Neighborhood Food Environment, and Health Disparities: State of the Science, Research Gaps and Opportunities. *The American Journal of Clinical Nutrition, 119*(3), 850–861. https://doi.org/10.1016/j.ajcnut.2023.12.019

WHO. (2025). *Ageing and Health*. World Health Organization. https://www.who.int/news-room/fact-sheets/detail/ageing-and-health Accessed 13 May 25.

PART 1

Food Security and Health

CHAPTER ONE

Food security and health systems

Greta Gallo Gabinete*
West Visayas State University, La Paz, Iloilo City, Philippines
*Corresponding author. e-mail address: gabinetegreta@yahoo.com

Contents

Advances in Food Security and Sustainability, Volume 11
ISSN 2452-2635, https://doi.org/10.1016/bs.af2s.2026.06.006

Abstract

Food security is a fundamental component of global health, economic stability, and sustainable development. This chapter examines the relationship between food security and health systems through the core dimensions of availability, access, utilization, and stability, while recognizing emerging dimensions of agency and sustainability. It explores how food insecurity contributes to malnutrition, non-communicable diseases, infectious disease vulnerability, and adverse mental health outcomes across the life course.

The chapter argues that food insecurity should be institutionalized within health systems as a preventive public health function rather than addressed solely through social assistance or agricultural policy. It highlights the role of healthcare systems in addressing food insecurity through nutrition-sensitive service delivery, integrated food assistance, and policy coordination across health, agriculture, trade, and social protection sectors. Emerging technologies—including artificial intelligence, blockchain, and precision agriculture—are examined for their potential to enhance food system transparency and health system responsiveness when embedded within accountable governance frameworks.

Policy implications emphasize the need for multisectoral strategies that align food security with health system planning, climate resilience, and social protection. Strengthening local food systems, improving governance coherence, and leveraging digital innovations are essential for building resilient food and health systems that support equitable and sustainable population health outcomes.

1. Introduction

Food insecurity remains one of the most persistent structural determinants of global health. Despite decades of progress in agricultural productivity and health service expansion, undernutrition, diet-related non-communicable diseases, and infection-related morbidity continue to coexist across regions and income levels. Food insecurity contributes to adverse maternal and child health outcomes, weakens immune responses, exacerbates mental health stressors, and undermines treatment adherence in chronic infectious diseases (Weiser et al., 2009; Yadav et al., 2022). Yet while the relationship between food security and health outcomes is widely recognized, institutional responses remain fragmented.

Food systems are typically governed through agricultural, trade, and economic policies, while health systems concentrate on curative services rather than upstream determinants. This separation limits the ability of governments to address multidimensional risks emerging from climate change, nutrition transitions, economic inequality, and supply chain instability. As a result, food insecurity is often treated as a social welfare concern rather than as a structural health system challenge. Achieving Sustainable Development Goal 2 (Zero Hunger) therefore requires more than increased food production; it demands deliberate governance alignment between food systems and health systems (Calder, 2013; Bello et al., 2024).

This chapter is anchored in the Social Determinants of Health (SDH) framework, which recognizes food insecurity as a structural driver of health inequities rather than merely an outcome of poverty. It also draws on systems theory perspectives that emphasize interdependence between food systems, health governance, and socio-political institutions.

Food security encompasses availability, access, utilization, and stability, as well as more recent dimensions of agency and sustainability (Makombe, 2023; Clapp et al., 2022). These expanded dimensions underscore that nutrition outcomes depend not only on production, but also on power relations, equity, ecological resilience, and institutional accountability.

This chapter advances a governance-centered argument: food insecurity must be institutionalized as a core preventive function within health systems. Institutionalization, in this context, refers to the systematic integration of food insecurity screening, nutrition-sensitive service delivery, cross-sectoral policy coordination, and dedicated financing mechanisms within routine health system operations. Without such integration, health systems remain reactive—managing disease outcomes rather than addressing upstream drivers of vulnerability. Through conceptual analysis and comparative global examples, the chapter demonstrates that integration failures—not food scarcity alone—sustain preventable burdens in maternal and child health, infectious diseases, non-communicable diseases, and climate-related health risks. It concludes by outlining a prioritized governance agenda for embedding food security within health system reform.

2. Conceptual framework and theoretical perspectives

2.1 Evolution of food security concepts

The concept of food security emerged in the 1970s in response to global food price crises, initially emphasizing national and global food availability (Makombe, 2023; Opara, 2013). By the 1980s, attention shifted toward food access, recognizing that adequate national supply does not ensure household-level security (Opara, 2013; Shakeel, 2018). This marked a transition from aggregate production concerns to distributional equity.

The 1996 World Food Summit formalized three dimensions—availability, access, and utilization—with stability later added to capture temporal consistency in food access (FAO, 1996; Berry et al., 2015). In the 2000s, the concept of nutrition security further emphasized diet quality and micronutrient adequacy, linking food systems to disease burden and poverty dynamics (Shetty, 2015).

More recently, agency and sustainability have expanded food security into governance and intergenerational equity domains (Clapp et al., 2022; Makombe, 2023). Agency refers to the capacity of individuals and communities—particularly women and marginalized groups—to make autonomous decisions regarding food production and consumption. Sustainability concerns the ecological and institutional resilience of food systems, including climate adaptation and long-term stability.

These developments shift food security from a production-oriented concept to a governance-centered framework. For health systems, compromised agency limits effective utilization of nutrition services, while weak sustainability exposes systems to recurrent shocks from climate variability and price instability (Habacon, 2025). Contemporary definitions therefore require stronger alignment between food governance and health system planning.

2.2 The health–Food security nexus

Food insecurity operates as a structural determinant of health through biological, behavioral, and institutional pathways. Biologically, undernutrition weakens immune function and increases susceptibility to infectious diseases, while micronutrient deficiencies impair cognitive and maternal outcomes. Behaviorally, constrained food budgets often shift consumption toward calorie-dense, nutrient-poor foods, contributing to obesity, diabetes, and cardiovascular disease (Yadav et al., 2022; Bello et al., 2024). These dynamics demonstrate that food insecurity underlies both undernutrition and diet-related non-communicable diseases.

The relationship is multidirectional. Illness-related income loss and high out-of-pocket health expenditures can deepen household food insecurity, while chronic infectious diseases such as HIV and tuberculosis are associated with reduced treatment adherence and poorer outcomes in food-insecure populations (Seligman et al., 2010). This cyclical interaction underscores the need for institutional integration rather than siloed interventions.

Health systems can respond through nutrition-sensitive primary care, food insecurity screening, referral mechanisms, social protection coordination, and preventive policies such as food fortification and school feeding (FAO, 2020). However, these efforts frequently operate as parallel initiatives rather than embedded components of routine care. Climate change and supply chain disruptions further threaten food availability, intensifying pressures on health systems (IPCC, 2019; Lipper et al., 2014). Without coordinated governance across agriculture, climate adaptation, and health sectors, structural vulnerability persists. Together, these interconnections reinforce the need to institutionalize food security within governance, financing, and service delivery frameworks.

3. Dimensions of food security and their implications for health systems

Food security is a fundamental determinant of global health, ensuring access to sufficient, safe, and nutritious food (FAO, 2011). While traditionally defined by availability, access, utilization, and stability, contemporary frameworks also incorporate agency and sustainability (Clapp et al., 2022). As illustrated in Fig. 1, these six dimensions represent not only agricultural or economic concerns but institutional responsibilities for health systems. Fig. 1

3.1 Availability

Availability refers to the physical presence of food through domestic agricultural production, imports, food reserves, market distribution systems, or humanitarian assistance (Boutrif, 2014). Traditionally, discussions of food availability focused primarily on agricultural output and trade flows. However, contemporary food systems demonstrate that availability is also shaped by infrastructure, transportation networks, storage capacity, labor conditions, and geopolitical stability. For health systems, food availability intersects directly with the reliability of supply chains for therapeutic foods, infant formula, vaccines requiring nutritional support, micronutrient supplementation, and essential medicines linked to nutrition-sensitive care. Hospitals and community health facilities depend upon stable procurement systems to manage acute malnutrition, maternal nutrition, and chronic disease treatment.

Fig. 1 The six pillars of food security.

Disruptions in food availability can rapidly translate into increased morbidity and mortality. Climate-related disasters such as droughts, floods, typhoons, and heatwaves reduce agricultural productivity and destabilize food distribution systems, particularly in vulnerable low- and middle-income countries. Similarly, armed conflicts, trade restrictions, and global commodity price volatility can interrupt import-dependent food systems and exacerbate hunger crises. The COVID-19 pandemic further revealed how fragile global supply chains can compromise both food systems and healthcare delivery simultaneously. These disruptions increase malnutrition caseloads, weaken immune resilience, and heighten susceptibility to infectious diseases, thereby generating surge pressures on already constrained health facilities (Gereffi and Lee, 2016; Lipper et al., 2014).

Consequently, strengthening food availability requires anticipatory governance and multisectoral coordination. Health systems must collaborate closely with agricultural agencies, disaster management institutions, and supply chain actors to develop early warning systems, maintain emergency nutrition stockpiles, and ensure continuity of food-related medical supplies during crises. Investments in climate-smart agriculture, local food production, and decentralized procurement mechanisms can also reduce dependence on unstable global markets. Embedding food availability considerations into health planning enables systems to transition from reactive emergency response toward proactive resilience building.

3.2 Accessibility

Accessibility encompasses the economic, social, and physical capacity of individuals and households to obtain sufficient and nutritious food. Even when food is physically available within markets, many populations remain food insecure because they lack the financial resources, transportation, social inclusion, or legal protections necessary to access it. Poverty, unemployment, disability, displacement, and geographic isolation all shape unequal patterns of food access. In urban settings, rising food prices and housing costs disproportionately affect low-income households, while in rural regions, weak infrastructure and limited market connectivity may constrain access to diverse and nutritious foods.

Health systems are deeply implicated in these dynamics because illness and poverty frequently reinforce one another. High out-of-pocket medical expenditures often compel households to reduce food consumption, purchase lower-quality foods, or forgo treatment entirely in order to meet competing financial demands (WFP et al., 2022). This trade-off is particularly severe among households managing chronic illnesses such as diabetes, cancer, or cardiovascular disease, where ongoing treatment costs coincide with the need for specialized diets. Food insecurity, in turn, contributes to poorer treatment adherence, delayed recovery, weakened immunity, and higher hospitalization rates, perpetuating a cycle of illness and deprivation.

Integrating food insecurity screening into primary healthcare settings therefore represents an important intervention point. Routine assessment of household food access during clinical consultations can help identify vulnerable populations before severe nutritional deterioration occurs. Linking patients to social protection mechanisms, food subsidy programs, school feeding initiatives, and income-support schemes such as the Supplemental Nutrition Assistance Program (SNAP) can mitigate the cyclical relationship between poverty and poor health outcomes (Schanzenbach, 2019). Furthermore, policies that reduce catastrophic healthcare spending through universal health coverage may indirectly strengthen food access by protecting household purchasing power. Addressing accessibility thus requires recognizing food insecurity not merely as a social welfare issue, but as a structural determinant of health.

3.3 Utilization

Utilization concerns the ability of individuals to derive adequate nutritional benefit from food through proper dietary quality, food safety, preparation practices, and nutrient absorption. This dimension extends beyond caloric intake to encompass the biological and social conditions necessary for healthy nourishment, including access to clean water, sanitation, healthcare, nutrition education, and culturally appropriate diets. Poor utilization may arise even when food is available and accessible if diets lack diversity, food preparation is unsafe, or underlying diseases impair nutrient absorption.

Health systems increasingly confront the "double burden" of malnutrition characterized by the simultaneous presence of undernutrition and diet-related non-communicable diseases (Walls et al., 2019). In many countries, stunting, micronutrient deficiencies, and maternal undernutrition coexist with rising rates of obesity, hypertension, diabetes, and cardiovascular disease linked to highly processed diets and sedentary lifestyles. This epidemiological transition places substantial pressure on healthcare systems that must simultaneously manage infectious diseases associated with undernutrition and chronic illnesses associated with dietary excess.

Embedding routine nutrition assessment within clinical workflows is therefore critical to preventing fragmented care. Nutritional screening should become a standard component of maternal health services, pediatric consultations, chronic disease management, and hospital admissions. Early identification of food insecurity, anemia, wasting, obesity, or micronutrient deficiencies enables timely intervention through dietary counseling, supplementation, therapeutic feeding, or referral to community support services. Multidisciplinary collaboration among physicians, nurses, dietitians, and community health workers is essential for delivering integrated nutrition-sensitive healthcare. In addition, strengthening food safety surveillance and public nutrition education can reduce foodborne disease burdens and improve long-term dietary behaviors. Without systematic integration of nutrition into healthcare delivery, health systems risk addressing symptoms of food insecurity while neglecting its underlying determinants.

3.4 Stability

Stability refers to the capacity of individuals and societies to maintain consistent access to food over time despite economic downturns, environmental disruptions, political instability, or public health emergencies (IPCC, 2019).

Unlike temporary shortages, instability reflects persistent vulnerability to shocks that interrupt food systems and undermine long-term nutritional security. Food stability is increasingly threatened by climate change, armed conflict, inflation, migration crises, and global market volatility, all of which can rapidly destabilize livelihoods and healthcare systems.

Protracted humanitarian crises illustrate the devastating health consequences of food instability. Situations in countries such as Venezuela and Syria demonstrate how political conflict, economic collapse, and displacement can precipitate severe malnutrition, infectious disease outbreaks, deteriorating maternal health, and the collapse of healthcare infrastructure (Doocy et al., 2019). Food instability often drives forced migration, overcrowding in urban settlements or refugee camps, and heightened competition for scarce resources, further exacerbating public health risks. Children, older adults, and individuals with chronic illnesses are particularly vulnerable to disruptions in food access during emergencies.

Integrating food stability indicators into emergency preparedness and health surveillance frameworks can strengthen system resilience. Health systems should incorporate food price monitoring, nutrition surveillance, climate forecasting, and supply chain vulnerability assessments into national preparedness strategies. Early warning systems can help governments anticipate surges in malnutrition or disease burdens before crises escalate. Establishing emergency nutrition reserves, decentralized food distribution networks, and adaptive financing mechanisms can further improve response capacity during shocks. Strengthening social protection systems and community resilience initiatives is equally important to buffering households against recurring instability. Ultimately, resilient health systems require resilient food systems.

3.5 Agency

Agency refers to the capacity of individuals and communities to participate meaningfully in decisions concerning food production, distribution, consumption, and governance (Malapit et al., 2015). The inclusion of agency as a core dimension of food security reflects growing recognition that food insecurity is not solely a technical or economic issue, but also a matter of power, representation, and human rights. Agency emphasizes the ability of people to exercise voice, make informed choices, and shape the food systems that affect their health and livelihoods.

Women's empowerment is particularly central to this dimension. Evidence consistently demonstrates that women's decision-making power within households and communities is strongly associated with improved child nutrition, maternal health, dietary diversity, and educational outcomes (FAO, 2023). However, structural inequalities—including gender discrimination, land insecurity, unequal wages, and limited political participation—often restrict women's agency within food systems. Marginalized populations such as Indigenous communities, migrants, informal workers, and persons with disabilities may likewise experience exclusion from food governance processes despite being disproportionately affected by food insecurity.

Institutionalizing participatory governance mechanisms can strengthen both accountability and intervention effectiveness. Community representation in nutrition planning, local agricultural policy, and public health decision-making enables interventions to better reflect cultural preferences, lived experiences, and local priorities. Incorporating measurable empowerment indicators into food security and health monitoring systems also helps shift policy attention from short-term relief toward structural transformation. A rights-based approach to food security recognizes individuals not merely as passive recipients of assistance, but as active agents entitled to dignity, participation, and equitable access to resources.

3.6 Sustainability

Sustainability ensures that food systems protect the nutritional security of present populations without compromising the ability of future generations to meet their own food and health needs (HLPE, 2020). Sustainable food systems must balance environmental integrity, economic viability, and social equity while remaining resilient to long-term ecological and climatic pressures. This dimension has become increasingly important as climate change, biodiversity loss, soil degradation, water scarcity, and unsustainable consumption patterns threaten global food production systems.

Unsustainable agricultural practices—including excessive pesticide use, deforestation, overfishing, monocropping, and intensive livestock production—contribute to ecosystem degradation and heighten risks of foodborne disease, zoonotic spillover, and nutritional instability (Pretty et al., 2018). Environmental degradation also disproportionately affects vulnerable populations that depend directly on natural resources for food and livelihoods. Climate change intensifies these risks by reducing crop yields, altering disease vectors, and increasing the frequency of extreme weather events that disrupt both food systems and healthcare infrastructure.

Health systems therefore have an important role in advancing sustainable food governance. Climate-resilient healthcare infrastructure, sustainable procurement policies, and nutrition-sensitive public purchasing can reduce environmental impacts while supporting healthier dietary patterns (Owens et al., 2023). Hospitals and public institutions can promote sustainable consumption through locally sourced foods, reduced food waste, and environmentally responsible supply chains. Integrating sustainability into food and health policy also requires cross-sectoral collaboration involving agriculture, environmental governance, urban planning, and public health institutions. Long-term food security depends not only on producing sufficient food today, but on preserving the ecological foundations necessary for future population health.

Together, these six dimensions underscore that food security is not merely an agricultural concern, but a cross-sectoral governance imperative closely tied to health equity, social protection, economic development, and environmental sustainability. Embedding these dimensions within health governance structures enables a transition from fragmented and reactive interventions toward preventive, integrated, and system-level management of food insecurity and its associated disease burdens.

4. Food insecurity and public health outcomes

Food insecurity generates measurable and preventable disease burdens across the life course. Its impacts extend beyond isolated nutritional deficiencies to encompass maternal and child mortality, non-communicable diseases, infectious disease progression, mental health disorders, and environmental health risks (Schlundt et al., 2020; Benton, 2016; Jaron and Galal, 2009). These outcomes are not incidental consequences of food system instability but predictable results of fragmented governance between food systems and health systems. Table 1 summarizes key health pathways through which food insecurity translates into clinical and public health pressures, reinforcing the need to institutionalize food insecurity as a preventive function within health systems.

4.1 Impact on maternal and child health

Food insecurity (FI) severely impacts child health, leading to undernutrition, stunted growth, and developmental deficiencies (Anjum et al., 2016). Undernutrition is estimated to contribute to approximately 45 % of deaths among children under five globally, particularly in low- and middle-income countries (World Health Organization, UNICEF, and World Bank Group, 2023).

Table 1 Key challenges in food security and public health.

Challenge	Description	Source
Chronic Foodborne Diseases	Prolonged exposure to contaminated food and harmful chemicals can cause long-term health issues. Strengthening health surveillance and regulatory systems is essential.	Schlundt et al., (2020)
Non-Communicable Diseases (NCDs)	Modern food production has led to increased food availability but also contributed to obesity, diabetes, and cardiovascular diseases. Sustainable and nutritious food systems are needed.	Benton (2016)
Food Supply and Health Inequality	Disruptions in food supply can worsen malnutrition and economic disparities, increasing health inequalities. Integrated food and health policies are required.	Jaron and Galal (2009)

Maternal food insecurity has been associated with adverse maternal outcomes, including increased risks of maternal morbidity and mortality, adverse pregnancy outcomes, mental health disorders, gestational diabetes, and feto-maternal complications, particularly in low-income settings (Iqbal and Ali, 2021; Bell et al., 2024). In South Asia, FI reduces women's access to high-quality food, affecting preconception health and increasing maternal depression and poor healthcare utilization (Diamond-Smith et al., 2020; Bishwajit and Yaya, 2017; Biete et al., 2025). These patterns indicate that maternal and child health services cannot treat nutrition as an ancillary concern. Institutionalizing routine food insecurity screening within antenatal, postnatal, and child health services is essential to interrupt intergenerational cycles of vulnerability.

4.2 Non-communicable diseases (NCDs)

Food insecurity is strongly linked to non-communicable diseases (NCDs) like diabetes, cardiovascular diseases, and obesity. Limited access to nutritious food increases the risk of obesity, hypertension, and diabetes, particularly in high-income countries (Dolin et al., 2021; Salinas-Roca et al., 2022; Laurentino et al., 2024; Carvajal-Aldaz et al., 2022). The consumption of ultra-processed foods, which are more accessible to food-insecure households, contributes to poor dietary quality and increased risks of NCDs, including cardiovascular disease and cancer (Odoms-Young et al., 2024). Addressing these risks requires policies that improve access to nutritious foods (Seligman et al., 2010).

Addressing the food insecurity–NCD nexus therefore requires not only clinical management of chronic disease but also regulatory engagement with food environments. Health systems must collaborate with fiscal and trade authorities to reshape dietary exposures that drive long-term disease burdens.

4.3 Infectious diseases and malnutrition

4.3.1 Nutritional deficiencies, immune impairment, and the infection–Malnutrition cycle

Food insecurity leads to inadequate intake of energy, protein, and essential micronutrients such as iron, zinc, vitamin A, and vitamin D, all of which are critical for maintaining immune competence. Nutritional deficiencies impair the body's physical barriers (e.g., skin and mucosal surfaces), reduce the production and function of immune cells, and weaken inflammatory and antibody-mediated responses. As a result, malnourished individuals exhibit diminished resistance to infectious agents and are more susceptible to both common and opportunistic infections (Calder, 2013; Yadav et al., 2022; Bello et al., 2024).

The relationship between malnutrition and infectious diseases is cyclical and self-reinforcing. Infection increases metabolic demands, induces inflammation, and reduces appetite, while gastrointestinal infections impair nutrient absorption and increase nutrient losses. These processes further worsen nutritional status, deepening immune suppression and prolonging illness. Recurrent or chronic infections—particularly in food-insecure settings—therefore perpetuate a vicious cycle in which undernutrition and disease reinforce one another, leading to increased morbidity, delayed recovery, and elevated mortality risk. This cycle is especially pronounced among children, pregnant women, and individuals with chronic infections, underscoring the need for health systems to integrate nutrition support as a core component of infectious disease prevention and treatment.

4.3.2 HIV/AIDS: Mechanisms linking food insecurity, treatment outcomes, and immune suppression

Food insecurity plays a critical role in shaping HIV-related health outcomes through multiple, interconnected mechanisms. At the behavioral level, food insecurity has been associated with increased engagement in high-risk behaviors, including transactional sex and reduced ability to negotiate safer sexual practices, thereby increasing HIV transmission risk (Weiser et al., 2009; Aibibula et al., 2017). At the clinical level, inadequate food intake undermines adherence to antiretroviral therapy (ART), as medications may exacerbate hunger or cause gastrointestinal side effects when taken without sufficient food. Economic constraints further force individuals to prioritize food acquisition over clinic visits, transportation, or medication refills, reducing continuity of care.

Biologically, malnutrition compromises immune recovery even when ART is available. Inadequate protein and micronutrient intake limits CD4 cell regeneration, weakens mucosal immunity, and increases susceptibility to opportunistic infections. Poor nutritional status has been associated with incomplete viral suppression, lower CD4 counts, and higher rates of disease progression and mortality (Weiser et al., 2009; Wang et al., 2011). These mechanisms illustrate that effective HIV management requires not only access to ART but also integrated food security and nutrition interventions within HIV care services.

4.3.3 Tuberculosis: Nutrition, treatment adherence, and health service access

Tuberculosis (TB) is both a cause and a consequence of malnutrition, making it a paradigmatic example of the infection–nutrition nexus. Undernutrition impairs cell-mediated immunity, particularly T-cell responses that are essential for controlling *Mycobacterium tuberculosis* infection. Food-insecure individuals are therefore more likely to progress from latent TB infection to active disease and to experience more severe clinical manifestations (Balinda et al., 2019; Benzekri et al., 2019).

Food insecurity also adversely affects TB treatment outcomes. Long treatment durations, medication side effects, and the physical demands of illness increase nutritional needs, yet food-insecure patients often lack adequate dietary intake to support recovery. This mismatch contributes to poor treatment adherence, higher default rates, and increased mortality (Ojo et al., 2022). Additionally, indirect costs such as transportation to health facilities and lost income further limit access to TB services among food-insecure populations. Evidence indicates that integrating food or cash assistance into TB treatment programs improves adherence, treatment completion, and survival, reinforcing the importance of linking nutrition support with infectious disease control within health systems.

These examples demonstrate that infectious disease control programs cannot succeed in isolation from food security interventions, reinforcing the need for integrated financing and service delivery models within health systems.

4.4 Mental health consequences

Food insecurity is consistently associated with anxiety, depression, and psychological distress across populations (Reeder et al., 2022; Myers, 2020). Among children, chronic food insecurity contributes to stress, impaired cognitive development, and behavioral difficulties, with long-term implications for educational attainment and social functioning (Gallegos et al., 2021; Kotchick et al., 2021). Integrating food insecurity assessment into mental health screening protocols can enhance early identification of psychosocial stressors and reduce long-term psychiatric morbidity associated with chronic deprivation.

4.5 Gender, power, and the political economy of food systems

4.5.1 Corporate power and ultra-processed food environments

Food insecurity is shaped by concentrated corporate power across seeds, agrochemicals, commodity trading, manufacturing, and retail. A small number of transnational corporations influence pricing, dietary patterns, and national policies—especially in low- and middle-income countries undergoing nutrition transitions.

The affordability and ubiquity of ultra-processed foods—high in refined carbohydrates, unhealthy fats, sodium, and added sugars—are driven by aggressive marketing, global supply chains, and trade liberalization. These products are often more accessible than fresh, nutrient-dense foods in food-insecure communities, reinforcing the coexistence of food insecurity and obesity (Odoms-Young et al., 2024; Walls et al., 2019).

Health systems bear the consequences through rising diabetes, cardiovascular disease, and cancer burdens. Addressing food insecurity as prevention therefore requires regulation of industry practices, including front-of-pack labeling, child-directed marketing restrictions, sugar-sweetened beverage taxes, and nutritious public procurement. Without confronting corporate influence, health systems remain focused on downstream disease management rather than upstream drivers.

4.5.2 Political economy of food systems and structural inequality

Food insecurity reflects political and economic structures governing land access, wages, subsidies, and social protection. Trade regimes often prioritize export monocultures over diversified local production, increasing dependence on imported processed foods.

Subsidies in high-income countries disproportionately support maize, wheat, and soy—key inputs for ultra-processed foods—distorting markets and shifting consumption toward calorie-dense, nutrient-poor products. Food insecurity thus intersects with labor precarity, rural dispossession, and uneven development.

Because health systems operate within this broader political economy, they absorb the consequences in the form of malnutrition and chronic disease. Institutionalizing food insecurity as a preventive function requires cross-sector governance that evaluates the health impacts of agricultural subsidies, trade agreements, and corporate consolidation. A health-informed political economy reframes food insecurity as an outcome of policy and power distribution.

4.5.3 Indigenous food sovereignty and self-determination

Indigenous food sovereignty centers self-determination, cultural continuity, land stewardship, and ecological sustainability. Food insecurity in Indigenous communities is often rooted in land dispossession, environmental degradation, and erosion of traditional food systems. Loss of access to ancestral lands disrupts fishing, hunting, and small-scale agriculture, increasing reliance on nutritionally inferior market foods.

This framework affirms communities' rights to define production methods, seed preservation, biodiversity, and dietary practices. Health systems must therefore move beyond standard nutrition programs toward culturally grounded, community-led approaches. Supporting land rights, traditional harvesting, and Indigenous-led initiatives enhances dietary diversity, cultural resilience, and equity. Recognizing Indigenous food sovereignty expands analysis from service delivery to justice and rights.

4.5.4 Trade–Nutrition conflicts and policy coherence

Trade agreements shape tariffs, imports, labeling standards, and regulatory autonomy. While liberalization may increase food availability, it can also accelerate ultra-processed food penetration. Public health regulations—such as sugar taxes and warning labels—have at times faced challenges under trade and investment agreements.

Policy coherence between trade and health is therefore essential. Ministries of health are often underrepresented in trade negotiations despite long-term dietary impacts. Institutional mechanisms—health impact assessments, nutrition-sensitive trade clauses, and inter-ministerial coordination—can align economic and public health goals.

Embedding food insecurity within health system planning, financing, and accountability transforms it from a reactive safety net issue into a preventive strategy. Across domains, food insecurity operates as a structural driver of disease. Without governance alignment across agriculture, trade, and health, systems will continue managing preventable burdens rather than addressing their root causes.

5. The role of health systems in addressing food security challenges

If food insecurity is to be treated as a core preventive health function, then health systems must move beyond episodic interventions toward institutionalized integration. This requires embedding food insecurity screening, nutrition services, policy engagement, and cross-sector coordination within routine health system functions rather than relying on short-term or donor-driven initiatives.

5.1 Primary health care and nutrition programs

5.1.1 Strengthening healthcare infrastructure to support food security

Strengthening healthcare infrastructure is vital for addressing food insecurity. Primary healthcare systems can help by implementing screening and referral programs, such as those by Kaiser Permanente Colorado, which connect patients to community food resources (Stenmark et al., 2018; Taher et al., 2022; Depuccio et al., 2022). Collaborations between healthcare providers and community organizations, like the Food as Medicine (FAM) program at Grady Health System, enhance food assistance efforts and improve patient health (Owens et al., 2023; Reinoso et al., 2022).

5.1.2 The role of primary healthcare centers in food distribution programs

Primary healthcare centers can serve as hubs for food distribution programs. Integrating food pantries within healthcare settings, as demonstrated by Eskenazi Health Center Pecar, can provide patients with access to healthy food and nutrition education, thereby addressing food insecurity and supporting chronic disease management (Reinoso et al., 2022). Implementing food insecurity screening tools within primary care practices helps identify patients in need and connect them to food assistance programs. This approach has been effective in various health systems, improving food security and health outcomes (Taher et al., 2022; Depuccio et al., 2022).

5.2 Integrating food and nutrition policies into health systems

5.2.1 *Policies for food fortification, nutrition labeling, and dietary interventions*

Food fortification is a cost-effective way to reduce micronutrient deficiencies. Mandating fortification of staple foods has significantly improved public health, especially in low- and middle-income countries (Fiedler and Macdonald, 2009; Osendarp et al., 2018; Garrett et al., 2018; Mkambula et al., 2020). Nutrition labeling helps consumers make informed dietary choices, with 95 countries implementing mandatory labeling policies, including nutrient declarations and front-of-pack labels to promote healthy eating (Afroza et al., 2024; Campbell et al., 2020).

5.2.2 *Global food fortification initiatives and their effectiveness*

Global food fortification initiatives have significantly reduced micronutrient deficiencies. The Global Alliance for Improved Nutrition (GAIN) has reached over 600 million people, highlighting the large-scale impact of fortification programs (Moench-Pfanner & Van Ameringen, 2012). However, challenges like quality control, monitoring, and compliance persist. Innovations such as fortified bouillon cubes and new fortificants are being explored to improve effectiveness (Osendarp et al., 2018; Mkambula et al., 2020; Dijkhuizen et al., 2013).

5.3 The role of healthcare professionals

5.3.1 *Engagement of doctors, nutritionists, and public health experts in addressing malnutrition*

Early identification and management of malnutrition by primary healthcare teams, led by general practitioners, significantly improve patient outcomes (Isenring et al., 2012). Dietitians play a key role, particularly for the elderly, though their impact can be limited by time constraints and insufficient screening knowledge among other healthcare professionals (Fleurke et al., 2020). Non-dietetic professionals contribute to malnutrition management, but their role in nutritional interventions requires further evaluation (Dabbous et al., 2021). Public health professionals help shape food system policies, integrating food and agricultural strategies to enhance public health (Muller et al., 2009).

5.3.2 *Importance of training healthcare professionals in food security policies*

Training healthcare professionals in evidence-based management of malnutrition, including the use of tools like the Malnutrition Universal Screening Tool (MUST), has been shown to improve their nutritional knowledge and management practices (Kennelly et al., 2010). The identification and treatment of severe malnutrition should be a core competency for pediatricians and related health professionals, highlighting the need for improved training curricula (Schofield et al., 2012). Continuous training and capacity building in food security are essential, particularly for resource-poor developing countries, to prevent foodborne illnesses and ensure food safety (Patil, 2024; Muit et al., 2025).

5.4 Innovations in healthcare for improving nutritional outcomes

Innovation in healthcare systems provides important opportunities to strengthen preventive responses to food insecurity. Integrating food insecurity screening into primary care settings has been shown to improve early identification of vulnerable households and facilitate timely referrals to nutrition assistance and social protection programs (Gundersen & Ziliak, 2015; Odoms-Young et al., 2024). When embedded within electronic health records, such screening tools enable systematic monitoring of nutrition-related risks and support continuity of care, reinforcing food security as a measurable health indicator rather than an external social issue.

Digital innovations—including AI-based dietary surveillance, mobile nutrition tracking, and supply-chain monitoring systems—offer opportunities to improve early detection of food insecurity and enhance precision in public health responses (Yevtushenko et al., 2024; Mortazavi and Gutierrez-Osuna, 2023). However, these tools must be implemented with safeguards addressing data privacy, algorithmic bias, and unequal digital access to avoid reinforcing health inequities (Bond et al., 2023).

Importantly, technological innovation cannot substitute for structural reform. Addressing food insecurity requires broader interventions targeting poverty, labor precarity, unhealthy food environments, and policy incoherence across agriculture and trade sectors (Walls et al., 2019; Clapp et al., 2022). Digital tools are most effective when embedded within multisectoral strategies that prioritize equity, prevention, and governance coherence. When aligned with accountable policy frameworks, healthcare innovations can strengthen early detection, coordination, and preventive capacity, supporting the institutionalization of food insecurity as a core function of health systems.

6. Case studies and global perspectives

6.1 Case study 1: Food security in low-income countries

6.1.1 Challenges in Sub-Saharan Africa

Sub-Saharan Africa continues to experience some of the highest levels of food insecurity globally, driven by a combination of structural poverty, climate vulnerability, political instability, and weak health and food system integration. Despite significant agricultural engagement, low productivity, limited access to inputs, and dependence on rain-fed farming systems constrain food availability, while high food prices and inadequate infrastructure restrict economic and physical access to nutritious food (Wudil et al., 2022).

Health systems in the region are disproportionately burdened by undernutrition-related morbidity, particularly among women and children. Stunting, wasting, and micronutrient deficiencies remain prevalent and are closely linked to preventable maternal and child mortality. However, food security interventions and health services are frequently implemented in parallel rather than through coordinated strategies. Nutrition programs are often donor-driven, short-term, and insufficiently embedded within primary health care systems, limiting their sustainability and impact.

Community-based interventions have demonstrated the value of integration. For example, unconditional cash transfer programs and nutrition-sensitive agriculture initiatives linked to early childhood development centers in countries such as Malawi have improved dietary diversity, reduced stunting, and enhanced household resilience (Durão et al., 2020; Gelli et al., 2022). These programs succeed largely because they connect agricultural production, income support, and health and nutrition services, rather than treating food insecurity as a purely agricultural problem.

Nonetheless, significant barriers persist. Weak governance, limited fiscal capacity, fragmented data systems, and inadequate monitoring mechanisms hinder scale-up and long-term impact. Climate shocks and conflict further disrupt both food supply chains and health service delivery, reinforcing cycles of malnutrition and disease. These challenges highlight the need for health systems in Sub-Saharan Africa to function not only as treatment platforms but also as entry points for food security screening, nutrition support, and climate-resilient social protection measures.

6.2 Case study 2: Food assistance programs in high-income countries

6.2.1 SNAP in the U.S. food banks in Europe

Food assistance programs in high-income countries play a critical role in mitigating food insecurity, particularly during economic downturns, public health emergencies, and periods of rising food prices. In the United States, the Supplemental Nutrition Assistance Program (SNAP) represents the cornerstone of the national food safety net. By increasing the food purchasing power of low-income households, SNAP has been consistently associated with reductions in food insecurity and improvements in dietary intake, particularly among children and older adults (Fernald & Gosliner, 2019; Schanzenbach, 2019).

Despite its broad reach, SNAP alone is often insufficient to meet household food needs. Benefit adequacy, fluctuating eligibility, and limited incentives for purchasing nutrient-dense foods contribute to continued reliance on charitable food assistance, including food pantries and emergency food banks (Morton, 2016). Moreover, SNAP operates largely outside the formal health system, with limited integration into primary care, nutrition counseling, or chronic disease management. This separation constrains its potential to function as a preventive public health intervention, particularly for diet-related non-communicable diseases.

In contrast, food assistance in Europe relies more heavily on charitable food banks coordinated through national and regional networks, such as the European Food Banks Federation. These organizations provide essential short-term relief to food-insecure populations, especially migrants, unemployed individuals, and the working poor. However, their effectiveness is constrained by dependence on surplus food donations, volunteer labor, and inconsistent supply chains (Loopstra, 2018; Daponte, 2017). As a result, food banks often struggle to ensure nutritional adequacy, cultural appropriateness, and stable access for beneficiaries.

Across both contexts, the reliance on food assistance highlights a common structural challenge: food insecurity is treated primarily as a social welfare issue rather than a determinant of health. Limited coordination between food assistance programs and health services reduces opportunities for early identification of food insecurity, targeted nutrition support, and prevention of diet-related health conditions. This fragmentation underscores the need for stronger institutional linkages between food assistance mechanisms and health systems, even in high-income settings. While SNAP and European food banks reduce immediate food hardship, their impact on long-term nutrition and health outcomes remains limited without systematic integration into health system functions.

6.2.2 Urban gardening and local food initiatives as complementary strategies

Beyond formal food assistance programs, urban food production initiatives have emerged as complementary strategies to improve food access, dietary quality, and community well-being in high-income countries. Urban gardening initiatives—including community gardens and allotment schemes—have been associated with increased fruit and vegetable consumption, improved dietary diversity, and higher levels of physical activity (Heise et al., 2017). These initiatives also contribute to mental well-being, social cohesion, and environmental awareness, reinforcing their multidimensional benefits.

Innovative interventions such as community-supported agriculture (CSA) programs and mobile produce markets further enhance access to fresh, nutritious foods in underserved urban neighborhoods. Evidence suggests that these approaches can improve food affordability and availability, particularly when paired with nutrition education or subsidy schemes (Marjanovic et al., 2023; Verfuerth et al., 2023). Importantly, such initiatives move beyond emergency food provision by strengthening local food systems and promoting healthier dietary behaviors.

However, urban gardening and local food initiatives face limitations related to land availability, seasonal production, and scalability. Without integration into broader food and health policies, their reach remains uneven and often limited to motivated or better-resourced communities. When linked with health systems—through prescriptions for fruits and vegetables, community health worker engagement, or local government support—these initiatives show greater potential to address food insecurity as a preventive health strategy rather than a temporary solution. Urban food initiatives are most effective when embedded within coordinated food–health system frameworks that align local food access with nutrition education and public health goals.

6.3 Case study 3: The role of international organizations in food security and health systems

6.3.1 Coordinated interventions by global institutions

International organizations play a central role in shaping global food security and health system responses, particularly in low- and middle-income countries facing complex humanitarian, nutritional, and public health challenges. Collectively, the World Health Organization, Food and Agriculture Organization, World Food Programme, and United Nations Children's Fund address food insecurity through complementary yet distinct mandates that span emergency response, policy guidance, system strengthening, and nutrition-focused interventions.

The FAO focuses on improving food security and nutrition by strengthening agricultural systems, food safety governance, and national food control mechanisms. Its work includes providing scientific and technical advice on food safety, supporting the development of international food standards, and enhancing national capacities for surveillance and regulation (FAO et al., 2023; Boutrif, 2014; Wennberg, 2014). A key initiative is the Emergency Prevention System (EMPRES), which addresses transboundary animal and plant diseases that threaten food supply stability, livelihoods, and public health (Welte and Terán, 2004). Through such initiatives, FAO contributes to preventive food security by reducing systemic risks before they escalate into humanitarian crises.

The WFP, recognized as the world's largest humanitarian organization addressing hunger, plays a critical role in emergency food assistance and crisis response. Traditionally focused on in-kind food aid, the organization has progressively shifted toward cash-based transfers, voucher systems, and local food procurement to enhance efficiency, stimulate local markets, and improve dietary diversity (Johnson, 2021; Lele and Goswami, 2021). In addition to emergency relief, WFP increasingly supports national governments in developing early warning systems, logistics capacity, and nutrition-sensitive safety nets—positioning itself as both a humanitarian and development actor.

The WHO's primary contribution lies in protecting public health by ensuring food safety and nutrition standards. In collaboration with FAO, WHO co-manages the Codex Alimentarius Commission, which establishes international food standards, guidelines, and codes of practice to safeguard consumer health and facilitate fair food trade (DeWaal and Brito, 2005; Wennberg, 2014). WHO also supports member states in strengthening national food safety systems, surveillance, and regulatory frameworks, thereby linking food security directly to health system capacity and disease prevention.

UNICEF complements these efforts by focusing on maternal, infant, and child nutrition within broader health and social protection systems. Its programs address acute malnutrition, promote optimal infant and young child feeding practices, and integrate nutrition services into primary health care, education, and emergency response settings. While often operating in partnership with FAO, WFP, and WHO, UNICEF's child-centered approach ensures that food security interventions explicitly address the needs of the most nutritionally vulnerable populations.

These organizations collectively demonstrate that food security is not solely an agricultural or humanitarian issue but a multisectoral challenge requiring coordinated governance across food and health systems and social protection mechanisms.

6.3.2 Successes and system-level impacts of global food assistance

Global food assistance programs led by international organizations have achieved notable successes in reducing hunger, improving food safety, and strengthening resilience during crises. The awarding of the Nobel Peace Prize in 2020 to the WFP highlights the organization's critical role in addressing acute food insecurity during conflicts and global emergencies, including the COVID-19 pandemic (Johnson, 2021; Lele and Goswami, 2021). This recognition underscores the importance of timely, large-scale food assistance in stabilizing vulnerable populations and preventing humanitarian catastrophes.

Similarly, the development and global adoption of Codex Alimentarius standards represent a major achievement in safeguarding food safety and public health. These standards have enhanced consumer protection, reduced foodborne illness risks, and facilitated international food trade by harmonizing regulatory requirements across countries (Boutrif, 2014; DeWaal and Brito, 2005). By linking food quality and safety directly to health outcomes, Codex exemplifies effective food–health system integration at the global level.

Despite these successes, challenges remain. Coordination across agencies can be uneven, funding is often reactive rather than preventive, and national-level implementation varies widely depending on political commitment and institutional capacity. Moreover, emergency-focused interventions may struggle to translate into long-term improvements in food

Table 2 Interventions of international organizations.

Organization	Primary mandate	Key food security interventions	Health system linkages	System-level contribution
Food and Agriculture Organization (FAO)	Food security, agriculture, nutrition, food safety	• Development of international food standards • Strengthening national food control systems • Technical assistance on food safety and nutrition • Emergency Prevention System (EMPRES) for transboundary animal and plant diseases	• Prevention of foodborne diseases • Reduced public health risks from unsafe food • Support to national surveillance and regulatory systems	Preventive food security through risk reduction, resilience building, and governance strengthening
World Food Programme (WFP)	Emergency food assistance and hunger response	• In-kind food aid • Cash-based transfers and vouchers • Local food procurement • Early warning and logistics systems	• Nutrition support during health emergencies • Improved treatment adherence through food/cash assistance • Support to vulnerable populations during crises	Stabilization during shocks; transition from humanitarian relief to resilience and systems support

(continued)

Table 2 Interventions of international organizations. (*cont'd*)

Organization	Primary mandate	Key food security interventions	Health system linkages	System-level contribution
World Health Organization (WHO)	Public health and disease prevention	• Food safety guidance • Nutrition standards and policy advice • Co-management of Codex Alimentarius	• Strengthened national food safety systems • Reduced foodborne illness burden • Integration of nutrition into public health policy	Linking food quality and safety directly to population health outcomes
United Nations Children's Fund (UNICEF)	Child health, nutrition, and protection	• Management of acute malnutrition • Infant and young child feeding (IYCF) programs • Nutrition in emergencies	• Integration of nutrition into primary health care • Maternal and child health services • Community-based nutrition interventions	Targeting the most nutritionally vulnerable; strengthening human capital through early-life interventions

system resilience and population health without sustained integration into national health and development strategies. The experiences of international organizations demonstrate that global food security initiatives are most effective when emergency response, food system governance, and health system strengthening are pursued simultaneously rather than in isolation. Table 2.

Taken together, the evidence presented in this section demonstrates that food security and health systems are deeply interconnected and mutually reinforcing. Across low-, middle-, and high-income contexts, food insecurity consistently undermines health outcomes through nutritional deficiencies, impaired immunity, reduced treatment adherence, and constrained access to services. Case studies from diverse settings illustrate that fragmented approaches where food assistance, nutrition, and health care operate in parallel limit long-term effectiveness. In contrast, integrated strategies that embed food security interventions within health systems, social protection, and climate-resilient food policies offer more sustainable and equitable outcomes. Addressing food insecurity as a core determinant of health is therefore essential for strengthening health systems, improving population resilience, and achieving global development and nutrition goals.

7. Policy recommendations and future directions

The preceding analysis demonstrates that food insecurity functions as a structural determinant of health across clinical, social, and political domains. Translating this evidence into meaningful action requires a sequenced, accountable, and multisectoral reform agenda capable of addressing both the immediate and long-term drivers of nutritional inequities. The following priorities are organized according to feasibility, institutional leverage, and their potential impact on reducing preventable disease burden. Collectively, they provide a governance roadmap for institutionalizing food insecurity within health systems and embedding nutrition security into broader development planning.

7.1 Priority 1: Integrate food security into health system functions

The first priority is the integration of food security into core health system functions. Governments and health institutions must institutionalize routine food insecurity screening within primary health care settings using standardized assessment tools that can identify vulnerable households early and accurately. Nutrition counseling, supplementation, and referral mechanisms should be systematically embedded within maternal and child health programs to ensure that nutritional vulnerabilities are addressed alongside routine clinical care. Importantly, food insecurity must no longer be treated solely as a social welfare issue but recognized as a preventive public health concern with direct implications for morbidity, mortality, and health system sustainability. Ministries of Health and primary health care providers are central actors in implementing these reforms, particularly in the short to medium term. Progress should be monitored through nutrition indicators, service utilization patterns, and maternal and child health outcomes.

7.2 Priority 2: Strengthen governance and policy coherence

The second priority involves strengthening governance and policy coherence across sectors. Food systems, health outcomes, agricultural production, trade regulations, and social protection mechanisms are deeply interconnected, yet they are often governed through fragmented institutional arrangements. National governments and planning agencies must therefore align food, health, trade, and social protection policies through sustained inter-ministerial coordination. Health impact assessments should become a routine component of trade and agricultural policymaking to prevent unintended consequences that undermine local food access, dietary diversity, and nutrition quality. In parallel, accountability mechanisms must be strengthened to track the nutritional and health impacts of food-related policies and investments. These reforms are expected to unfold over the medium term and can be evaluated through cross-sector policy reviews and nutrition-sensitive budgeting frameworks.

7.3 Priority 3: Use technology to enhance equity, not just efficiency

The third priority focuses on harnessing technology to advance equity rather than merely improve efficiency. Digital innovations such as artificial intelligence, nutrition surveillance platforms, and supply-chain monitoring systems hold significant potential for strengthening public health nutrition responses. However, technological deployment must be accompanied by safeguards that address risks of data bias, exclusion, and unequal access, particularly among marginalized populations. Governments, health systems, and research institutions should prioritize technologies that enhance early warning systems, improve nutrition monitoring, and support evidence-informed decision-making. Medium- to long-term implementation strategies should incorporate continuous assessment of coverage, equity outcomes, and cost-effectiveness to ensure that technological advancements contribute to inclusive and equitable health gains.

7.4 Priority 4: Build climate-resilient nutrition and health systems

The fourth priority is the development of climate-resilient nutrition and health systems. Climate change increasingly threatens food production, supply stability, and nutritional outcomes, particularly among populations already experiencing socioeconomic vulnerabilities. Agriculture and environment ministries, together with local governments, should integrate climate-resilient agricultural practices into community nutrition and public health programs. Agroforestry and regenerative agriculture approaches can serve not only environmental objectives but also nutrition-sensitive development goals by improving food diversity and long-term sustainability. At the same time, adaptive social safety nets must be strengthened to protect vulnerable populations from climate-induced food shocks and nutritional crises. These long-term reforms should be monitored through indicators related to food stability, resilience capacity, and climate adaptation outcomes.

These priorities are mutually reinforcing and should be implemented in a coordinated sequence. Immediate integration of food security within health systems establishes the operational foundation for broader governance reforms, which can subsequently be strengthened through technological innovation and climate resilience planning. Sequencing reforms in this manner enables governments to move progressively from service-level interventions toward structural transformation. Ultimately, such an approach ensures that food insecurity is addressed not as a reactive social assistance issue, but as a core preventive public health function essential to achieving equitable and sustainable health outcomes.

8. Conclusion

This chapter argues that food insecurity is not merely a social or agricultural concern but a structural determinant of health that must be institutionalized within health systems as a preventive function. Across the life course, food insecurity generates predictable and preventable burdens, including malnutrition, diet-related non-communicable diseases, infectious disease vulnerability, and mental health stressors. These outcomes are amplified by economic inequality, climate change, and political instability, underscoring the need for systemic rather than fragmented responses. Institutionalizing food insecurity within health systems requires integrating routine screening, nutrition-sensitive service delivery, and coordinated governance mechanisms that align health, agriculture, trade, and social protection policies.

Technological innovations—including digital surveillance, artificial intelligence, and supply-chain monitoring—can enhance equity and efficiency when embedded within accountable public health frameworks. However, technology alone cannot substitute for governance coherence and political commitment. Moving forward, research and policy must focus on evaluating implementation pathways, strengthening cross-sector accountability, and measuring health system responsiveness to food-related risks. By embedding food security within core health system functions, societies can shift from reactive crisis management toward preventive, resilience-oriented public health systems that advance equity, stability, and sustainable development.

Declaration of generative AI and AI-assisted technologies in the writing process

During the preparation of this work, the author used Scopus AI to assist in identifying and searching relevant literature and ChatGPT to aid in drafting and refining text. The author critically evaluated all AI-assisted outputs for accuracy, relevance, and coherence, ensuring alignment with the objectives. The final manuscript was reviewed, edited, and approved by the author, who takes full responsibility for its content.

References

Afroza, U., Abrar, A.K., Nowar, A., Choudhury, S.R., 2024. Global overview of government-endorsed nutrition labeling policies of packaged foods: a document review. Front. Public. Health.

Aibibula, W., Cox, J., Hamelin, A.-M., Brassard, P., 2017. Association between food insecurity and HIV viral suppression: a systematic review and meta-analysis. AIDS Behav.

Anjum, F., Asim, M., Awan, K.A., Nawaz, Y., 2016. Household food insecurity and its impact on children health in rural Faisalabad, Pakistan. Rawal Med. J.

Balinda, I.G., Sugrue, D.D., Ivers, L.C., 2019. More than malnutrition: a review of the relationship between food insecurity and tuberculosis. Open. Forum Infect. Dis.

Bell, Z., Nguyen, G., Andreae, G., Scott, S., Sermin-Reed, L., Lake, A.A., et al., 2024. Associations between food insecurity in pregnancy and maternal and infant health outcomes: a systematic review and meta-analysis. PLOS Med. 21 (9), e1004450. https://doi.org/10.1371/journal.pmed.1004450.

Bello, T.O., Ibrahim, F.A., Oladele, J.O., Oladiji, A.T., 2024. Molecular mechanism underlying nutritional influence on the onset of infectious diseases. Nutr. Diet. Health: Princ. Appl.

Benton, T.G., 2016. Food security. Book Chapter In Encyclopedia of Applied Plant Sciences. DOI:10.1016/B978-0-12-394807-6.00039-3.

Benzekri, N.A., Sambou, J.F., Tamba, I.T., Gottlieb, G.S., 2019. Nutrition support for HIV-TB co-infected adults in Senegal, West Africa: a randomized pilot implementation study. PLoS ONE.

Berry, E.M., Dernini, S., Burlingame, B., Conforti, P., 2015. Food security and sustainability: can one exist without the other? Public. Health Nutr.

Biete, C., Biete, A., Patriota, E.S.O., Pizato, N., 2025. Household food insecurity and symptoms of anxiety and depression during pregnancy: systematic review and meta-analysis. Matern. Child. Nutr.

Bishwajit, G., Yaya, S., 2017. Household food insecurity is independently associated with poor utilization of maternal healthcare services in Bangladesh. Facets.

Bond, A., McCay, K., Lal, S., 2023. Artificial intelligence & clinical nutrition: what the future might have in store. Clin. Nutr. ESPEN.

Boutrif, E., 2014. Institutions involved in food safety: food and agriculture organization of the United Nations (FAO). Encycl. Food Saf.

Calder, P.C., 2013. Feeding the immune system. Proc. Nutr. Soc.

Campbell, E.J., Hoadley, J.E., Post, R.C., 2020. Nutrition in labeling. Present. Knowl. Nutr.: Clin. Appl. Top. Nutrition.

Carvajal-Aldaz, D., Cucalon, G., Ordonez, C., 2022. Food insecurity as a risk factor for obesity: a review. Front. Nutr.

Clapp, J., Moseley, W.G., Burlingame, B., Termine, P., 2022. Viewpoint: the case for a six-dimensional food security framework. Food Policy.

Dabbous, M., Hastings, R., Weekes, C.E., Baldwin, C., 2021. The role of non-dietetic healthcare professionals in managing interventions among adults at risk of malnutrition: a systematic review. Clin. Nutr.

Daponte, B.O., 2017. Linking SNAP to food security: exploring reinstating a purchase requirement and tying benefits to the low-cost food plan. Economists' Voice.

Depuccio, M.J., Garner, J.A., Hefner, J.L., Walker, D.M., 2022. Multi-stakeholder perspectives on the implementation of a clinic-based food referral program for patients with chronic conditions: a qualitative examination. Transl. Behav. Med.

DeWaal, C.S., Brito, G.R.G., 2005. Safe food international: a blueprint for better global food safety. Food Drug. Law J.

Diamond-Smith, N., Shieh, J., Puri, M., Weiser, S., 2020. Food insecurity and low access to high-quality food for preconception women in Nepal: the importance of household relationships. Public. Health Nutr.

Dijkhuizen, M.A., Wieringa, F.T., Soekarjo, D., Laillou, A., 2013. Legal framework for food fortification: examples from Vietnam and Indonesia. Food Nutr. Bull.

Dolin, C.D., Compher, C.C., Oh, J.K., Durnwald, C.P., 2021. Pregnant and hungry: addressing food insecurity in pregnant women during the COVID-19 pandemic in the United States. Am. J. Obstet. Gynecol. MFM.

Doocy, S., Ververs, M.-T., Spiegel, P., Beyrer, C., 2019. The food security and nutrition crisis in Venezuela. Soc. Sci. & Med. 226, 63–68. https://doi.org/10.1016/j.socscimed.2019.02.007.

Durão, S., Visser, M.E., Ramokolo, V., Schoonees, A., 2020. Community-level interventions for improving access to food in low- and middle-income countries. Cochrane Database Syst. Rev.

Fernald, L.C.H., Gosliner, W., 2019. Alternatives to SNAP: global approaches to addressing childhood poverty and food insecurity. Am. J. Public. Health.

Fiedler, J.L., Macdonald, B., 2009. A strategic approach to the unfinished fortification agenda: feasibility, costs, and cost-effectiveness analysis of fortification programs in 48 countries. Food Nutr. Bull.

Fleurke, M., Voskuil, D.W., Beneken genaamd Kolmer, D.M., 2020. The role of the dietitian in the management of malnutrition in the elderly: a systematic review of current practices. Nutr. Dietetics.

Food and Agriculture Organization of the United Nations, 1996. Rome Declaration on World Food Security and World Food Summit Plan of Action. FAO.

Food and Agriculture Organization of the United Nations, 2011. An Introduction to the Basic Concepts of Food SecurityFAO. http://www.fao.org/docrep/013/al936e/al936e00.pdf.

Food and Agriculture Organization of the United Nations, 2020. The State of Food Security and Nutrition in the World. FAO.

Food and Agriculture Organization, 2023. The State of Food Security and Nutrition in the World 2023. FAO.

Food and Agriculture Organization of the United Nations (FAO) Lipp, M., Cannavan, A., Dupouy, E., Bessy, C., 2023. Encyclopedia of food safety, 2nd ed, Vol. 1-4.

Gallegos, D., Eivers, A., Sondergeld, P., Pattinson, C., 2021. Food insecurity and child development: a state-of-the-art review. Int. J. Environ. Res. Public. Health.

Garrett, G.S., Luthringer, C.L., Yetley, E.A., Neufeld, L.M., 2018. Food fortification policy. Encycl. Food Sec. Sustain.

Gelli, A., Kemp, C.G., Margolies, A., Levin, C., 2022. Economic evaluation of an early childhood development center–based agriculture and nutrition intervention in Malawi. Food Sec.

Gereffi, G., Lee, J., 2016. Economic and social upgrading in global value chains and industrial clusters. J. Bus. Ethics.

Gundersen, C., Ziliak, J.P., 2015. Food insecurity and health outcomes. Health Aff. 34 (11), 1830–1839. https://doi.org/10.1377/hlthaff.2015.0645.

Habacon, J.P.P., 2025. Exploring the lived experiences of filipino smallholder farmers under contract rice farming within a neoliberal context. Recoletos Multidiscip. Res. J. 13 (1), 83–94.

Heise, T.L., Romppel, M., Molnar, S., Lhachimi, S.K., 2017. Community gardening, community farming and other local community-based gardening interventions to prevent overweight and obesity in high-income and middle-income countries: protocol for a systematic review. BMJ Open.

High Level Panel of Experts on Food Security and Nutrition, 2020. *Food security and nutrition: building a global narrative towards 2030* (HLPE Report 15). Committee on World Food Security, FAO. https://www.fao.org/3/ca9731en/ca9731en.pdf.

Intergovernmental Panel on Climate Change, 2019. Climate change and land: An IPCC special report on climate change, desertification, land degradation, sustainable land management, food security, and greenhouse gas fluxes in terrestrial ecosystems. https://www.ipcc.ch/srccl/.

Iqbal, S., Ali, I., 2021. Maternal food insecurity in low-income countries: revisiting its causes and consequences for maternal and neonatal health. J. Agric. Food Res.

Isenring, E., Bauer, J., Banks, M., Miller, M., 2012. Managing malnutrition: identifying and treating unintentional weight loss in adults. Med. Today.

Jaron, D., Galal, O., 2009. Food security and population health and well-being. Asia Pac. J. Clin. Nutr.

Johnson, B., 2021. Why do people starve? Technology review. MIT Technology.

Kennelly, S., Kennedy, N.P., Rughoobur, G.F., Sugrue, S., 2010. An evaluation of a community dietetics intervention on the management of malnutrition for healthcare professionals. J. Hum. Nutr. Dietetics.

Kotchick, B.A., Whitsett, D., Sherman, M.F., 2021. Food insecurity and adolescent psychosocial adjustment: indirect pathways through caregiver adjustment and caregiver–adolescent relationship quality. J. Youth Adolescence.

Laurentino, J.S.L., Brito, R.C.D.S., Oliveira-Silva, R.T., Palmeira, P.A., 2024. Association between food insecurity and chronic noncommunicable diseases in Brazil: a systematic review. Rev. Brasileira de. Epidemiologia.

Lele, U., Goswami, S., 2021. The world food programme. Food All: Int. Organ. Transform. Agric.

Lipper, L., Thornton, P., Campbell, B., et al., 2014. Climate-smart agriculture for food security. Nat. Clim. Change 4, 1068–1072. https://doi.org/10.1038/nclimate2437.

Loopstra, R., 2018. Interventions to address household food insecurity in high-income countries. Proc. Nutr. Soc.

Makombe, G., 2023. The food security concept: definition, conceptual frameworks, measurement and operationalisation. Afr. Dev.

Malapit, H.J.L., Kovarik, C., Sproule, K., Meinzen-Dick, R., Quisumbing, A., 2015. The Women's empowerment in agriculture index. World Dev. 72, 71–91.

Marjanovic, S., Kent, K., Morrison, N., MacMillan, F., 2023. The impact of community-based food access strategies in high-income countries: a systematic review of randomised controlled trials. Public. Health Res. Pract.

Mkambula, P., Mbuya, M.N.N., Rowe, L.A., Gorstein, J., 2020. The unfinished agenda for food fortification in low- and middle-income countries: quantifying progress, gaps and potential opportunities. Nutrients.

Moench-Pfanner, R., Van Ameringen, M., 2012. The global alliance for improved nutrition (GAIN): a decade of partnerships to increase access to and affordability of nutritious foods for the poor. Food Nutr. Bull.

Mortazavi, B.J., Gutierrez-Osuna, R., 2023. A review of digital innovations for diet monitoring and precision nutrition. J. Diabetes Sci. Technol.

Morton, E., 2016. Alleviating food insecurity with SNAP: overview and impacts of the supplemental nutrition assistance program. Alleviating Food Insecurity SNAP: Overv. Impacts Supplemental Nutr. Assist. Program.

Muit, M.C.B., Manaog, A.T., Masagca, J.L.S., Montañez, J.J.F., 2025. FreshTrack: enhancing household food safety with RFID-enabled food expiry monitoring system. Recoletos Multidiscip. Res. J. 13 (2), 25–34.

Muller, M., Tagtow, A., Roberts, S.L., MacDougall, E., 2009. Aligning Food Systems Policies to Advance Public Health. J. Hunger Environ. Nutr. 4 (3–4), 225–240. https://doi.org/10.1080/19320240903321193.

Myers, C.A., 2020. Food insecurity and psychological distress: a review of the recent literature. Curr. Nutr. Rep.

Odoms-Young, A., Brown, A.G.M., Agurs-Collins, T., Glanz, K., 2024. Food insecurity, neighborhood food environment, and health disparities: state of the science, research gaps, and opportunities. Am. J. Clin. Nutr.

Ojo, T., Ruan, C., Hameed, T., Peprah, E., 2022. HIV, tuberculosis, and food insecurity in Africa—A syndemics-based scoping review. Int. J. Environ. Res. Public. Health.

Opara, U.L., 2013. Perspective: the evolving dimensions and perspectives on food security—What are the implications for postharvest technology research, policy, and practice? Int. J. Postharvest Technol. Innov.

Osendarp, S.J.M., Martinez, H., Garrett, G.S., Darnton-Hill, I., 2018. Large-scale food fortification and biofortification in low- and middle-income countries: a review of programs, trends, challenges, and evidence gaps. Food Nutr. Bull.

Owens, C., Cook, M., Goetz, J., Chakkalakal, R.J., 2023. Food is medicine intervention shows promise for engaging patients attending a safety-net hospital in the Southeast United States. Front. Public. Health.

Patil, S.S., 2024. Future of food safety: way Forw. Clim. Change Food Secur.
Pretty, J., et al., 2018. Global assessment of agricultural system redesign for sustainable intensification. Nat. Sustain.
Reeder, N., Tolar-Peterson, T., Bailey, R.H., Evans, M.W., 2022. Food insecurity and depression among US adults: NHANES 2005–2016. Nutrients.
Reinoso, D., Haut, D., Claffey, S., Carter, A., 2022. Addressing food insecurity: lessons learned from co-locating a food pantry with a federally qualified health center. Int. J. Integr. Care.
Salinas-Roca, B., Rubió-Piqué, L., Carrillo-Álvarez, E., Franco-Alcaine, G., 2022. Impact of health and social factors on the cardiometabolic risk in people with food insecurity: a systematic review. Int. J. Environ. Res. Public. Health.
Schanzenbach, D.W., 2019. Exploring options to improve the supplemental nutrition assistance program (SNAP). Ann. Am. Acad. Political Soc. Sci.
Schofield, C., Ashworth, A., Annan, R., Jackson, A.A., 2012. Malnutrition treatment to become a core competency. Arch. Dis. Child.
Schlundt, J., Tay, M.Y.F., Chengcheng, H., Liwei, C., 2020. Food security: microbiological and chemical risks. In: Masys, A., Izurieta, R., Reina Ortiz, M. (Eds.), Global Health Security: Advanced Sciences and Technologies for Security Applications. Springer, pp. 231–274. https://doi.org/10.1007/978-3-030-23491-1_11.
Seligman, H.K., Laraia, B.A., Kushel, M.B., 2010. Food insecurity is associated with chronic disease among low-income NHANES participants. J. Nutr.
Shakeel, A., 2018. Food security: theorizing the evolution and involution of the concept. Arab. World Geographer.
Shetty, P., 2015. From food security to food and nutrition security: role of agriculture and farming systems for nutrition. Curr. Sci.
Stenmark, S.H., Steiner, J.F., Marpadga, S., Seligman, H., 2018. Lessons learned from implementation of the food insecurity screening and referral program at Kaiser Permanente Colorado. Permanente J.
Taher, S., Muramatsu, N., Odoms-Young, A., Courtney, K.S., 2022. An embedded multiple case study: using CFIR to map clinical food security screening constructs for the development of primary care practice guidelines. BMC Public. Health.
Verfuerth, C., Sanderson Bellamy, A., Adlerova, B., Dutton, A., 2023. Building relationships back into the food system: addressing food insecurity and food well-being. Front. Sustain. Food Syst.
Walls, H., Baker, P., Chirwa, E., Hawkins, B., 2019. Food security, food safety & healthy nutrition: are they compatible? Glob. Food Secur.
Wang, E.A., McGinnis, K.A., Fiellin, D.A., Justice, A.C., 2011. Food insecurity is associated with poor virologic response among HIV-infected patients receiving antiretroviral medications. J. Gen. Intern. Med.
Weiser, S.D., Anema, A., Vogenthaler, N., Kadiyala, S., 2009. Food insecurity and HIV/AIDS: current knowledge, gaps, and research priorities. Curr. HIV/AIDS Rep.
Welte, V.R., Terán, M.V., 2004. Emergency prevention system (EMPRES) for transboundary animal and plant pests and diseases - The EMPRES-livestock: an FAO initiative. Ann. N. Y. Acad. Sci.
Wennberg, A., 2014. Food and Agriculture Organization of the United Nations, 3rd ed. Encyclopedia of Toxicology.
World Food Programme, Food and Agriculture Organization, & Global Network Against Food Crises, 2022. Global report on food crises 2022. World Food Programme. https://www.fsinplatform.org/global-report-food-crises-2022.
World Health Organization, United Nations Children's Fund (UNICEF), & World Bank Group, 2023. Levels and Trends in Child Malnutrition: UNICEF/WHO/World Bank Group Joint Child Malnutrition Estimates—Key Findings of the 2023 Edition. World Health Organization.

Wudil, A.H., Usman, M., Rosak-Szyrocka, J., Boye, M., 2022. Reversing years for global food security: a review of the food security situation in Sub-Saharan Africa (SSA). Int. J. Environ. Res. Public. Health.
Yadav, N., Shah, A., George, R., Miraj, S.S., 2022. Role of nutrients in combating infection. Viral, Parasitic, Bacterial, and Fungal Infections: Antimicrobial, Host Defense, and Therapeutic Strategies.
Yevtushenko, V., Rukkas, K., Chumachenko, T., 2024. Development of intelligent nutrition surveillance system with OpenAI API. CEUR Workshop Proceedings.

Further reading

Alaimo, K., Chilton, M., Jones, S.J., 2020. Food insecurity, hunger, and malnutrition. Present. Knowl. Nutr.: Clin. Appl. Top. Nutrition.
AlTheyab, N.A., Housawi, A.A., Lytras, M.D., 2023. Redesigning the global healthcare system through digital transformation: insights from Saudi Arabia. Digital Transform. Healthc. Post-COVID-19 Times.
Behzadifar, M., Martini, M., Behzadifar, M., Bragazzi, N.L., 2020. The barriers to the full implementation of strategic purchasing and the role of health policy and decision-makers: Past, current status, ethical aspects and future challenges. J. Preventive Med. Hyg.
Belans, R., Odom, J., Kolm, P., Rethy, J.A., 2024. The association between food insecurity, parental stress, and child development. Fam. Commun. Health.
Bertolini, A.M., Jaime, P.C., Di Giulio, G.M., 2023. The role of urban and peri-urban agriculture in global food security from post-war to the COVID-19 crisis: New perspectives on food justice, global health, and sustainability. Saude e Sociedade.
Beshir, H.A., Maystadt, J.-F., 2020. utero seasonal food insecurity and cognitive development: evidence on gender imbalances from Ethiopia. J. Afr. Economies.
Clapp, J., 2017. Food self-sufficiency: making sense of it, and when it makes sense. Food Policy 66, 88–96. https://doi.org/10.1016/j.foodpol.2016.12.001.
Côté , M., Lamarche, B., 2023. Artificial intelligence in nutrition research. Artificial Intelligence in Clinical Practice: How AI Technologies Impact Medical Research and Clinics.
Cruz Casados, J., Cervantes López, M.J., Gil Herrera, R.D.J., 2024. Promoting healthy eating habits via intelligent virtual assistants, improving monitoring by nutritional specialists: State of the art. Lecture Notes in Computer Science (including subseries Lecture Notes in Artificial Intelligence and Lecture Notes in Bioinformatics).
Dankan Gowda, V., Sharma, A., Suneel, S., Anil Kumar, N., 2024. Design and implementation of an IoT-based control system for precision food manufacturing. Lecture Notes Netw. Syst.
Doustmohammadian, A., Fadavi, G., Alibeyk, S., Mohammadi-Nasrabadi, F., 2023. An overview of food insecurity during the global COVID-19 outbreak: transformative change and priorities for the Middle East. Agric. Food Sec.
Durão, S., Ramokolo, V., Schoonees, A., Drimie, S., 2022. Cochrane corner: community-level interventions to increase access to food in low- and middle-income countries. Pan Afr. Med. J.
Food and Agriculture Organization of the United Nations, 2021. The State of Food and Agriculture 2021: Making Agri-Food Systems More Resilient to Shocks and Stresses. FAO.
Frongillo, E.A., Bernal, J., 2014. Understanding the coexistence of food insecurity and obesity. Curr. Pediatrics Rep.
Gami, S.J., Sharma, M., Bhatia, A.B., Whig, P., 2024. Artificial intelligence for dietary management: transforming nutrition through intelligent systems. Nutr. Controversies Adv. Autoimmune Dis.

Hadley, C., Patil, C.L., 2006. Food insecurity in rural Tanzania is associated with maternal anxiety and depression. Am. J. Hum. Biol.

Hopma, J., Woods, M., 2014. Political geographies of 'food security' and 'food sovereignty. Geogr. Compass.

Israni, M., 2024. A gender lens on the political economy of food and nutrition insecurity in the Global South. Food Secur. Dev. World: Status, Challen. Oppor.

Karan, R., Mowlds, S., McKeown, P.C., Spillane, C., 2023. Data for decision-making for sustainable food systems transformation in the Eastern cape of South Africa: what is needed? Front. Sustain. Food Syst.

Leung, C.W., Stewart, A.L., Portela-Parra, E.T., Epel, E.S., 2020. Understanding the psychological distress of food insecurity: a qualitative study of children's experiences and related coping strategies. J. Acad. Nutr. Dietetics.

Li, D., Wahlqvist, M.L., 2011. Securing food from field to table: what can we do? Asia Pac. J. Clin. Nutr.

Murray, S., Gale, F., Adams, D., Dalton, L., 2024. Evolution of food and nutrition policy: a Tasmanian case study from 1994 to 2023. Nutrients.

Nikoonia, M., Khosravi, M., Islam, S.M.S., Senobari, M.-A., 2023. Is household food insecurity related to mothers' stress, anxiety, and depression in Iran? Preventive Med. Rep.

Pryor, L., Lioret, S., van der Waerden, J., Melchior, M., 2016. Food insecurity and mental health problems among a community sample of young adults. Soc. Psychiatry Psychiatr. Epidemiol.

Robertson, A., Tirado, C., Lobstein, T., James, W.P., 2004. Food and health in Europe: a new basis for action. WHO Regional Publications, Eur. Ser.

Theodore Armand, T.P., Nfor, K.A., Kim, J.-I., Kim, H.-C., 2024. Applications of artificial intelligence, machine learning, and deep learning in nutrition: a systematic review. Nutrients.

Tilman, D., et al., 2017. Future threats to biodiversity and pathways to their prevention. Nature.

Vermeulen, S.J., et al., 2012. Climate change and food systems. Annu. Rev. Environ. Resour.

World Health Organization, 2015. The Global View of Foodborne Disease: Food Safety and Public HealthWorld Health Organization. https://www.who.int/publications/i/item/9789241565165.

World Health Organization, 2020. Micronutrients: Vitamin and Mineral Nutrition Information SystemWorld Health Organization. https://www.who.int/health-topics/micronutrients.

World Health Organization, 2022. Strengthening Gender-Responsive Health Systems. WHO.

CHAPTER TWO

Food security and One Health

Jerico Bautista Ogaya[a,b,*], Dauda Ayuba[c], Olalekan John Okesanya[d], Mohamed Mustaf Ahmed[e], Fadele Kehinde Precious[f], Zhinya Kawa Othman[g], Shuaibu Saidu Musa[h,i], Adriana Viola Miranda[j], Usman Abubakar Haruna[k], Harvie P. Portugaliza[l], John Rodel A. Balasabas[m,n,o], and Don Eliseo Lucero-Prisno III[p,q,r]

[a]Department of Medical Technology, Institute of Health Sciences and Nursing, Far Eastern University Manila, Manila, Philippines
[b]Center for University Research, University of Makati, Makati, Philippines
[c]Department of Local Government and Development Studies, Ahmadu Bello University, Zaria, Nigeria
[d]Department of Public Health and Maritime Transport, University of Thessaly, Volos, Greece
[e]Faculty of Medicine and Health Sciences, SIMAD University, Mogadishu, Somalia
[f]College of Medicine, University of Nigeria, Enugu state, Nsukka, Nigeria
[g]Department of Pharmacy, Kurdistan Technical Institute, Sulaymaniyah, Kurdistan Region, Iraq
[h]School of Global Health, Faculty of Medicine, Chulalongkorn University, Bangkok, Thailand
[i]Department of Nursing Science, Ahmadu Bello University, Zaria, Nigeria
[j]Global Health Focus Asia, Bandung, Indonesia
[k]Faculty of Pharmaceutical Sciences, Ahmadu Bello University, Zaria, Nigeria
[l]Department of Veterinary Clinical Sciences, Visayas State University, Baybay City, Leyte, Philippines
[m]Graduate School of Agriculture and Related Sciences, University of Southeastern Philippines, Tagum City, Davao del Norte, Philippines
[n]Research and Creative Works Office, Agusan del Sur State University, Bunawan, Agusan del Sur, Philippines
[o]College of Agriculture and Forestry, Northeastern Mindanao State University, Tagbina Campus, Surigao del Sur, Philippines
[p]Department of Global Health and Development, London School of Hygiene and Tropical Medicine, London, United Kingdom
[q]Office for Research, Extension and Innovations, Bukidnon State University, Malaybalay City, Bukidnon, Philippines
[r]Research and Development Office, Biliran Province State University, Naval, Biliran, Philippines
*Corresponding author. e-mail address: jericoogaya13@gmail.com

Contents

Advances in Food Security and Sustainability, Volume 11
ISSN 2452-2635, https://doi.org/10.1016/bs.af2s.2026.03.001

Abstract

Food security and One Health are inextricably connected through the human-animal-environment relationship, which emphasizes sustainable solutions to global health and nutrition concerns. Food security refers to the availability, access, utilization, and stability of food, whereas One Health highlights the interconnection between human, animal, and environmental health. This chapter investigates the components of food security, specifically availability, access, utilization, and stability, within the One Health framework, focusing on zoonotic diseases, antibiotic resistance, and the implications of climate change. Additionally, this chapter analyzes drivers of food security from One Health perspective, focusing on population growth, environmental degradation, pandemics, and political instability. Case studies, such as Rift Valley fever and antibiotic resistance, demonstrate integrated approaches to enhancing food system resilience. Incorporating One Health principles into food security strategies requires cross-sector collaboration and transdisciplinary research.

1. Introduction

Food security and public health are two interrelated pillars that are necessary for global well-being. The incorporation of these issues within the concept of One Health, as a collaborative, multi-sectoral, and transdisciplinary approach, is widely acknowledged as essential for tackling challenges in human, animal, and environmental health (Prata et al., 2022). The Food Agriculture Organization of the United Nations

(FAO, 2001) described food security as when all people, at all times, have physical, social, and economic access to sufficient, safe, and nutritious food that meets their dietary needs and food preferences for an active and healthy life. Food security has four important dimensions: availability, access, utilization, and stability, all of which are influenced by socio-economic and environmental factors. Food availability refers to the sufficient production and supply of food through agricultural techniques, trade, and local distribution systems. Globally, acute food insecurity affects 318 million people across 68 countries, with Sub-Saharan Africa, the Middle East, and South Asia particularly vulnerable (WFP, 2025; Bahar et al., 2020) state that global agrarian production should increase by 70 % by 2050 to meet the demands of a projected 9.7 billion people. However, problems such as soil degradation, water scarcity, and climate change endanger agricultural productivity, disproportionately affecting vulnerable regions. Factors including household income, market mechanisms, and equitable resource distribution play a crucial role in determining access to food. According to the World Bank (2024), over 702 million people live in extreme poverty worldwide, making food inaccessible for a significant proportion of the population. Economic disparities and fluctuations in food prices exacerbate inequality, particularly in developing countries.

Food utilization considers the nutritional content, safety, and cultural acceptability of food, as well as an individual's ability to metabolize nutrients. The Food and Agriculture Organization (FAO, 2015) emphasized that poor dietary diversity and the incidence of foodborne infections, which affect nearly 600 million people annually, impede optimal nutrition. In addition, economic shocks, natural disasters, and conflicts threaten food security by undermining its stability. In 2026, an estimated 41.1 million people are projected to be in the emergency phase of acute food insecurity or worse (WFP, 2025). Despite progress, ensuring global food security remains a significant challenge due to environmental, socioeconomic, and policy-driven constraints. The FAO, (2015) noted that climate change poses a significant risk, as escalating temperatures and erratic weather patterns can lead to reductions in crop yields of up to 30 % in some regions. For livestock, a study conducted in Laos indicated that by 2085, rising global temperatures could result in decreases of up to 22 % in meat production and 21 % in milk production (Spiller et al., 2023; Maqbool et al. (2024)) emphasized that biodiversity loss and unsustainable farming practices weaken the ecosystems necessary for food production. The

COVID-19 pandemic exposed new vulnerabilities in global supply chains, interrupting food access for millions and worsening existing imbalances (Bonadio et al., 2021).

Nzietchueng et al. (2023) defined One Health as an approach to designing and implementing programs, policies, legislation, and research where multiple sectors collaborate to achieve optimal health outcomes by recognizing the interconnectedness of people, animals, plants, and their shared environment. One Health focuses on three fundamental principles: engaging human medicine, veterinary science, and environmental management; implementing unified actions to address complex issues, such as zoonotic diseases i.e., those transferred between animals and humans; see (Rahman et al., 2020) and antimicrobial resistance (AMR); and prioritizing sustainable solutions that balance environmental protection with socio-economic needs. The One Health concept is founded on the recognition that human, animal, and environmental health are interconnected and interdependent. Marie and Gordon (2023) argued that emerging infectious diseases, 75 % of which are zoonotic, highlight species-wide vulnerabilities. For example, the H5N1 avian influenza and the COVID-19 pandemic demonstrate how diseases can cross species boundaries, exploiting gaps in detection and containment systems. (Manyi-Loh et al. (2018)) noted that indiscriminate antibiotic use in agriculture promotes antimicrobial resistance, threatening human and veterinary health. Environmental health is an integral part of the One Health triangle that supports food production systems. Deforestation, soil erosion, and water contamination reduce agricultural output and accelerate the spread of vector-borne diseases. Integrating environmental conservation with health interventions is essential for reducing risks and increasing resilience in food systems.

Utilizing a comprehensive One Health approach to enhance food security is in harmony with the United Nations Sustainable Development Goals (SDGs), particularly SDG 2 (Zero Hunger), SDG 3 (Good Health and Well-being), and SDG 15 (Life on Land). This strategy, which addresses the health and environmental determinants of food security, can help reduce inequities, strengthen resilience to climate change, and promote sustainable agricultural practices. In addition, the One Health paradigm enables the identification of structural weaknesses in food systems, supporting fairer and more sustainable practices. Relevant stakeholders can effectively manage the challenges of an interconnected world by addressing the relationship between food security and human, animal, and environmental health.

2. Food security and one health for a holistic approach to sustaining life and well-being

The concept of One Health emphasizes the interdependence of human, animal, and environmental health. When applied to food security, this framework is a comprehensive approach to addressing the multifaceted challenges of ensuring adequate and safe food supplies while preserving ecological balance and protecting public health.

2.1 Unraveling how food systems shape global health

Livestock, crop production, forestry, and fishing are necessary components of food systems, delivering essential nutrients through animal- and plant-based diets. Globally, livestock accounts for approximately 33 % of dietary protein and provides a significant source of income for over 1.3 billion people. For human health, livestock production ensures food and nutrition security, particularly in low-income countries where animal products supplement diets that are otherwise deficient in key micronutrients such as iron, zinc, and vitamin B12 (Homann-Kee Tui et al., 2020; Kappes et al., 2023) emphasized the significant role of livestock in supporting both local and global economies, particularly concerning supply chains, market pricing, and trade dynamics. Therefore, outbreaks of animal diseases can lead not only to declines in livestock populations but also to serious disruptions in people's livelihoods. A notable example is foot-and-mouth disease (FMD), which can result in a staggering 80 % drop in milk production from affected animals, thereby heightening food insecurity in vulnerable regions. Additionally, FMD can greatly hinder exports of milk and dairy products, mainly due to the restrictions and regulations enforced by impacted countries and their trading partners. Although implementing effective management strategies can mitigate these risks, the costs associated with managing FMD—such as biosecurity measures and vaccination programs—can further strain profitability (Lewis et al., 2023).

Food production systems also have far-reaching implications for environmental health. Modern agricultural practices usually involve intensive land use, contributing to deforestation, biodiversity loss, and soil degradation (Wang et al., 2025). According to Ingrao et al. (2023), agriculture accounts for 70 % of global freshwater withdrawals, and inefficient water usage exacerbates drought conditions in already arid regions. Moreover, the excessive use of fertilizers and pesticides in intensive farming pollutes water sources and degrades soil quality (Ali and Bijay-Singh., 2025). Agriculture is responsible for 19–29 % of global greenhouse gas emissions, which drive climate change (Li et al., 2024). These environmental changes directly and indirectly impact human and animal health by altering ecosystems, reducing crop yields, and increasing the prevalence of vector-borne diseases.

2.2 Emerging challenges between food security and one health

Zoonotic diseases pose significant challenges to food security and the One Health framework. According to Rahman et al. (2020), over 60 % of new infectious diseases in humans have zoonotic origins and food systems significantly contribute to their spread. Rift Valley fever (RVF) and Highly Pathogenic Avian Influenza (HPAI) are two significant zoonotic diseases, and each has its own unique transmission pathways and public health and economic implications. RVF is primarily transmitted to humans through contact with infected livestock or by mosquito bites. Humans may also contract RVF by handling the blood or organs of infected animals. RVF outbreaks, which are usually associated with excessive rainfall and livestock rearing, have devastated pastoralist communities in Sub-Saharan Africa. During the 1997–1998 outbreak in Kenya, nearly 50,000 human infections and thousands of livestock deaths led to widespread food shortages and economic losses exceeding US$32 million. Outbreaks of HPAl in Asia during 2003 and 2004 resulted in significant declines in bird populations. In Vietnam, approximately 44 million birds were lost, accounting for about 17.5 % of its poultry population, costing approximately US$76–450 million while Thailand experienced losses of around 29 million birds, which represented 14.5 % of its poultry population (McLeod et al., 2005). Regarding human health, from 2003 to 2023, the World Health Organization (WHO) recorded 878 cases of HPAI (H5N1) infections, leading to 458 fatalities and resulting in a mortality rate of 52.16 % across 23 countries (Charostad et al., 2023).

The use of antibiotics in agriculture significantly contributes to the global burden of antimicrobial resistance (AMR; see National Academies of Sciences, Engineering, and Medicine, 2021; Tiseo et al., 2020) noted that approximately 73 % of all antimicrobials sold worldwide are used in animal husbandry to stimulate growth and prevent sickness. This indiscriminate use accelerates the development of resistant infections, which can spread to humans through the food chain, direct contact with animals, or environmental contamination. For instance, multidrug-resistant *Salmonella* infections associated with chicken production have created a major public health concern. Tang et al. (2023) projected that AMR could cause 10 million deaths per year by 2050, with foodborne infections accounting for a large portion.

Climate change exacerbates food insecurity and disease burdens by reducing agricultural production and altering disease patterns. Studies have shown that rising global temperatures and extreme weather events negatively impact crop yields, with maize and wheat production expected to decline by 7 % and 6 %, respectively, for each degree of warming (Bouteska et al., 2024; Gebeyehu et al., 2025). These environmental changes also affect vector-borne diseases such as malaria and dengue fever, which are becoming more prevalent as ecological niches shift. Heat stress reduces fertility and productivity in animals while shifting rainfall patterns influence the spread of zoonotic diseases such as Rift Valley fever.

2.3 Insights on food security and one health

The outbreaks of RVF highlight the importance of a One Health approach in mitigating the impact of zoonotic diseases on food security. RVF is transmitted by mosquitoes and impacts both humans and livestock. During outbreaks, infected livestock suffer reduced productivity, while humans face significant risks of severe disease and death. Kenya implemented an integrated One Health approach that included veterinary vaccination campaigns, mosquito control, and public health education programs to reduce the burden of RVF. Mansfield et al. (2015) reported that early warning systems using satellite data to predict heavy rainfall, a precursor to increased mosquito reproduction, nable timely interventions. As a result, the 2006–2007 RVF outbreak resulted in fewer human cases than earlier outbreaks. Avian influenza, particularly the highly virulent H5N1 variant, poses a severe threat to food security and human health. The virus can devastate poultry populations, which serve as a primary protein source in many countries.

According to Li and Chen (2021), China experienced a significant H7N9 outbreak in 2013, leading to over 600 human infections and substantial economic losses in the poultry industry. Chinese officials employed a One Health framework to enforce stricter biosecurity measures, enhance surveillance of live bird markets, and implement vaccination campaigns. These actions not only curbed the virus's spread but also protected livelihoods and improved food security. The Netherlands adopted a One Health policy prioritizing responsible antibiotic use in agriculture to combat AMR. The country achieved a 64 % reduction in veterinary antibiotic use between 2007 and 2015, which significantly decreased resistant bacteria levels in food products (Manyi-Loh et al., 2018). This success resulted from stringent regulation and coordinated efforts among legislators, veterinarians, and farmers, demonstrating that cross-sector collaboration can effectively address complex issues. Livestock and crop production form necessary components of food systems, but improper management poses significant threats to public health and the environment. In low- and middle-income countries, where surveillance capacity is weaker and agricultural livelihoods often rely on antibiotic-driven productivity gains, similar policies falter. Studies on AMR interventions confirm that without integrated governance and economic support, antibiotic stewardship remains aspirational (Costa et al., 2023; Pinho et al., 2025). Thus, the critical tension lies in balancing short-term agricultural productivity with long-term public health imperatives.

3. Drivers of food insecurity from a one health perspective

Food insecurity is a complex global concern closely related to human, animal, and environmental health, and it is effectively addressed within the One Health paradigm (Fig. 1). This approach emphasizes ecological interdependence and focuses on collaborative solutions to food insecurity and other global challenges.

3.1 Population growth and urbanization

Population growth and urbanization pose significant challenges and opportunities for ensuring food security within the One Health framework. Gu et al. (2021) observed that the world population reached 8 billion in 2023, and forecasts suggest it will rise to 9.7 billion by 2050, with metropolitan areas accounting for 68 % of this growth. This demographic transition increases food demand in countries such as Africa and Asia, where cities are rapidly expanding. Bricas (2019) noted that urbanization causes substantial modifications in food systems, affecting production, distribution, and consumption patterns. Ahmed et al. (2019) stated that urban populations usually adopt diets high in animal protein and processed foods, which place more strain on agricultural systems and contribute to environmental degradation. Intensive animal husbandry, which is usually required to meet urban food demand, can increase zoonotic disease risks, as

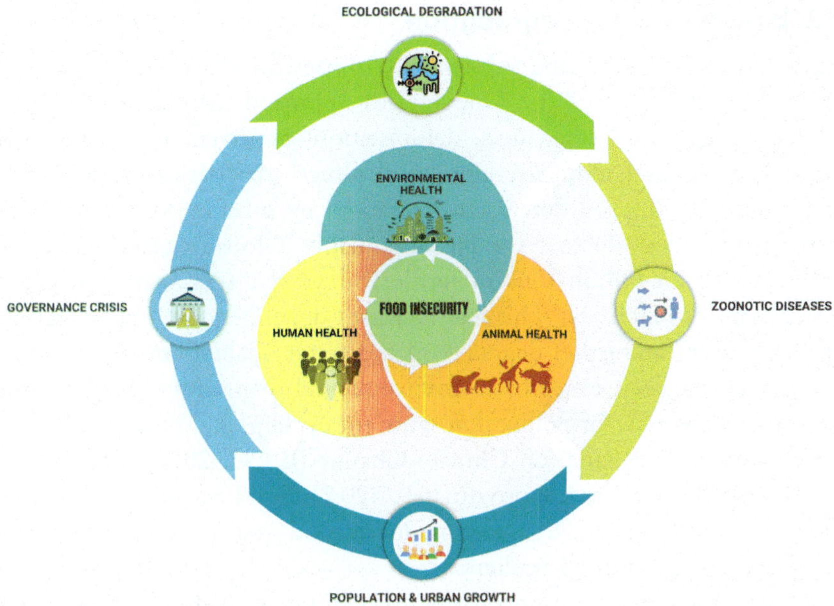

Fig. 1 One health framework of drivers of food insecurity. *Note: This figure is created by the authors as synthesized from various literature sources.*

demonstrated by the emergence of diseases such as avian influenza and COVID-19. Furthermore, urban development frequently encroaches on fertile land, reducing agricultural capacity and exacerbating food shortages (Gržinić et al., 2023).

Balancing urban growth with sustainable food systems is necessary within the One Health approach. Urban agriculture and vertical farming have emerged as innovative strategies to address land constraints and provide fresh produce in urban areas. Devendra (2006) noted that Singapore's 30 by 30 strategy aims to meet 30 % of its nutritional needs locally by 2030 through urban farming technologies. Simultaneously, the environmental consequences of increased agricultural activities, such as deforestation, biodiversity loss, and greenhouse gas emissions, threaten ecosystem stability and human health. Viana et al. (2022) pointed out that policies should promote sustainable farming practices, reduce food waste, and strengthen rural-urban food networks to address these challenges. The challenges of population growth and urbanization underscore the importance of implementing a One Health strategy to ensure food security. Integrating innovative farming strategies, sustainable policies, and robust health surveillance systems can foster resilient food systems that protect human, animal, and environmental health.

3.2 Environmental degradation

Environmental degradation substantially impacts food security and undermines the principles of the One Health approach. Müller et al. (2022) observed that land degradation, deforestation, biodiversity loss, and pollution directly and indirectly affect agricultural production, food supply, and public health. Soil deterioration, caused by erosion, salinization, and nutrient depletion, decreases arable land quality. Globally, erosion removes an estimated 24 billion tons of fertile soil yearly, significantly reducing agricultural productivity. This loss increases dependency on chemical fertilizers, which can further degrade soil and water quality, creating a vicious cycle. Deforestation exacerbates environmental degradation by disrupting ecosystems essential for pollination, nutrient cycling, and pest control. The Intergovernmental Panel on Climate Change (IPCC, 2023) reported that between 1990 and 2020, approximately 420 million hectares of forest were lost worldwide, severely impacting local food systems and biodiversity necessary for agricultural resilience.

Water contamination, primarily caused by agricultural runoff containing nitrates and phosphates, reduces the availability of clean water for irrigation and consumption. The (WHO Regional Office for Europe, 2022) noted that contaminated water sources facilitate the spread of zoonotic diseases, endangering food safety and public health. Environmental degradation also increases the risk of zoonotic infections by destroying wildlife habitats. The Food and Agriculture Organization (FAO) and United Nations Environment Programme (UNEP, 2020) reported that the conversion of forests to agricultural land has been associated with disease outbreaks such as Ebola and the Nipah virus. Furthermore, biodiversity loss undermines ecosystem services that regulate pests and diseases, threatening crop yields and livestock health. According to van der (Vander Sluijs and Vaage, 2016), in some regions, a 70 % decline in pollinators, such as bees, has jeopardized the yields of pollination-dependent crops, posing a significant risk to global food security.

3.3 Pandemics and disease outbreaks

The interconnectivity of pandemics, food security, and the One Health concept has become increasingly evident in recent years. Pandemics and zoonotic disease outbreaks have significant impacts on global food security, human health, and environmental sustainability. Pandemics disrupt food production, distribution, and access, exacerbating food insecurity. Aday and Aday (2020) noted that the COVID-19 pandemic significantly disrupted global food supply chains, increasing prices and limiting food accessibility for vulnerable populations. The World Food Programme (WFP, 2024) reported that food insecurity in Sub-Saharan Africa rose by 30 %, affecting nearly 250 million people. Labor shortages and limited access to markets and inputs have disproportionately impacted agriculture-dependent livelihoods, such as smallholder farming (Van et al., 2024). Furthermore, zoonotic diseases such as avian influenza and African swine fever directly impact food security because they decrease livestock productivity and increase costs. Ma et al. (2021) reported that the 2018 African swine fever outbreak reduced China's pork production by 20 %, driving global pork price surges.

According to Yang et al. (2024), human encroachment on wildlife habitats, deforestation, and intensive farming strategies elevate the risk of zoonotic spillovers, evidenced by viruses such as Ebola and Nipah. Integrating disease surveillance across the human, animal, and environmental sectors can effectively mitigate these risks. For example, following the H5N1 avian influenza outbreak in Vietnam, the One Health approach involved coordinated efforts by veterinary and public health agencies to curb disease transmission and protect poultry-dependent livelihoods. Governments and international organizations should invest in resilient food systems, sustainable agricultural practices, and robust One Health frameworks to ensure food security during pandemics. Precision farming and early warning systems can help mitigate the risks of future outbreaks (Roy et al., 2025; Sharan et al., 2023) argued that equitable vaccine distribution for zoonotic diseases and capacity building in disease surveillance are necessary for reducing pandemic impacts. Addressing pandemics and disease outbreaks from a One Health perspective is essential to ensuring global food security. Coordinated and proactive measures are necessary to strengthen resilience in interconnected systems and ensure a sustainable future.

3.4 Economic and political instability

Economic and political instability jeopardize efforts to attain food security and achieve the One Health approach. Le et al. (2022) argued that instability, whether driven by economic crises, civil conflict, or ineffective governance, disrupts food production, distribution, and access, increasing health risks and obstructing long-term growth. Economic instability usually reduces agricultural production by restricting farmers' access to financial resources and diminishing investment in rural development. Factors such as high inflation, exchange rate fluctuations, inadequate infrastructure and technological innovation contribute to inefficiencies in food delivery systems. According to Odoh et al. (2023), the World Bank projected that economic disruptions could reduce agricultural output by 30 % in some countries, intensifying Sub-Saharan Africa's food insecurity. Beyene, (2023) examined the impact of food insecurity on the availability of nutritious food, noting an increased reliance on less diverse, lower-quality diets that, over time, worsen malnutrition and amplify foodborne infections.

Political instability, characterized by poor governance, corruption, and conflict, also undermines food security. Armed conflicts disrupt supply chains, devastate agricultural land, and displace populations, leading to food shortages and heightened vulnerability to famine. Salama et al. (2012) reported that violence in Somalia in 2011 caused a 50 % decline in food production, leaving 2.3 million people acutely food insecure. The presence of armed conflicts may also lead to reduced resilience in facing other shocks, such as pandemics: in Yemen, the ongoing civil war led to 80–90 % of food being imported pre-COVID-19 pandemic. During the pandemic, food import was substantially restricted, leaving Yemen with even more dire food insecurity (Hashim et al., 2021). Furthermore, unpredictable political environments hinder the implementation of effective public health programs necessary for managing diseases in both human and animal populations. Sharan et al. (2023) noted that in politically unstable regions, zoonotic diseases such as Ebola and Rift Valley fever can spread unchecked, posing severe threats to public health and food security. Lee and Brumme (2013) contended that political instability hampers the coordination essential for a One Health strategy, as governance failures prevent cross-sector collaboration among agriculture, healthcare, and environmental management. This lack of coordination impedes the formulation and implementation of policies needed to address food insecurity, public health, and environmental degradation, which are key pillars of the One Health concept. Economic and political instability intensify food insecurity and obstruct the effective implementation of One Health policies.

4. Role of sustainable agriculture in food security and one health

Sustainable agriculture is essential for ensuring food security while promoting the principles of One Health. Integrating agroecological practices, responsible livestock and aquaculture management, and technological advancements is important for building resilient food systems that can support the growing global population while preserving ecological balance.

4.1 Agroecological practices

Agroecology applies ecological principles to agricultural systems, advancing sustainable food production while protecting human, animal, and environmental health. This approach addresses food security challenges by enhancing biodiversity, reducing dependence on synthetic inputs, and strengthening resilience to climate change. Bezner Kerr et al. (2021) highlighted that agroecological approaches enhance food security by improving crop productivity and promoting sustainable agricultural systems. Studies show that agroecological farms can produce 80 % more food than conventional systems under similar conditions, particularly in developing countries (Hilson et al., 2021; Mouratiadou et al., 2024), Kaur et al. (2023) found that crop diversification, agroforestry, and organic farming techniques increase yields and improve soil health and water retention capacity, thereby reducing vulnerability to droughts and floods. Amadu et al. (2020) demonstrated that agroforestry practices in Malawi increased maize yields by 50 %, benefiting over 200,000 households. Oyediran (2023) reported that integrated pest management (IPM) reduced pesticide use by 30 %, resulting in safer food for consumers and healthier ecosystems. Agroecology aligns with One Health principles by mitigating the risks associated with zoonotic infections and antibiotic resistance. Deguine et al. (2023) emphasized that prioritizing natural pest control and reducing antibiotic use in agroecological systems lowers the risk of disease transmission between humans and animals. Maintaining various environments reduces habitat fragmentation, a major driver of zoonotic spillovers. Glaze-Corcoran et al. (2020) illustrated that agroecological methods, such as intercropping, decreased malaria prevalence by 30 %, highlighting their potential to address agricultural health challenges in Madagascar. Furthermore, integrating livestock with crop production enhances nutrient recycling and ecosystem services while promoting animal welfare (Sekaran et al., 2021).

4.2 Livestock and aquaculture management

Livestock and aquaculture are essential for achieving food security and advancing One Health principles. Fukase and Martin (2020) predicted that food demand will increase by 70 % as the global population reaches

9.7 billion by 2050. Livestock contributes over 40 % of global agricultural output, while aquaculture supplies over half the fish consumed worldwide (World Bank, 2021). However, sustainable management practices are essential for mitigating health and environmental risks. Akash et al. (2022) explained that livestock production supports food security by providing meat, milk, and eggs, which are key sources of protein and vitamins. For instance, cattle feed 1.3 billion people globally but contribute 14.5 % of greenhouse gas (GHG) emissions. Getiso and Mijena, (2021) opined that sustainable practices, such as improving feed efficiency and managing manure effectively, could reduce these emissions by 30 %.

Aquaculture is the world's fastest-growing food production sector and has expanded significantly, with global production expected to reach 122.6 million tonnes in 2020, valued at $281.5 billion. Fish provide an essential source of omega-3 fatty acids, accounting for approximately 20 % of the animal protein consumed by 3.3 billion people (FAO, 2024). Despite its benefits, aquaculture faces challenges such as antibiotic resistance, overexploitation of natural resources, and water pollution. Bostock et al. (2020) argued that innovations in recirculating aquaculture systems (RAS) and plant-based fish feed can reduce environmental impacts.

The One Health framework underscores the interconnectedness of livestock, aquaculture, and human health. The presence of zoonotic diseases like avian influenza and the issue of antimicrobial resistance demonstrate the necessity for a cohesive management approach. According to Pitt and Gunn (2024), around 60 % of viral diseases affecting humans are zoonotic, with livestock playing a substantial role in the rise of antimicrobial resistance due to the overuse of antibiotics. To tackle these issues, it is essential to strengthen veterinary services, encourage vaccination, and enforce biosecurity measures. Sustainable management practices for livestock and aquaculture are crucial for ensuring food security and promoting the One Health agenda. Adopting practices that address environmental, animal, and human health risks is essential to achieving global health and nutrition goals.

4.3 Technological innovations

Technological advancements in agriculture have transformed production and sustainability. Precision agriculture integrates sensors, drones, and internet of things devices to monitor crop health, manage irrigation, and reduce waste. Chandra et al. (2024) reported that farmers using these technologies achieved yield increases of 20–30 % while reducing water and fertilizer usage by 40 %.

Artificial intelligence (AI) assists farmers in predicting insect outbreaks and weather patterns, enabling timely preventive measures. Jurgens and Kaushik (2024) stated that AI-powered tools helped Indian farmers reduce post-harvest losses by 30 % in 2021 through accurate weather forecasting. These technologies advance One Health's objectives by mitigating environmental risk and enhancing agricultural efficiency. Biotechnology significantly contributes to developing crops and livestock that resist diseases and adapt to climate stress. Peshin et al. (2021) documented that India's adoption of Bt cotton between 2002 and 2020 led to a 56 % reduction in pesticide use and a 30 % increase in yields. Furthermore, Shrock and Güell, (2017) highlighted the use of gene editing to develop disease-resistant animals, such as pigs resistant to porcine reproductive and respiratory syndrome (PRRS), potentially saving billions of dollars annually. These innovations strengthen food security while reducing agriculture's environmental impact. Sustainable agriculture, including agroecological practices, ethical livestock and aquaculture management, and technological advancements, is necessary to ensure food security and support One Health principles. These measures address global challenges such as climate change and population growth by improving productivity and resilience while protecting the environment and public health. Integrating these approaches is essential for achieving a sustainable future.

5. Policy and governance frameworks supporting one health and food security

The United Nations' SDGs establish a comprehensive global framework for achieving sustainable development by 2030. SDG 2 (Zero Hunger) and SDG 3 (Good Health and Well-being) emphasize food security and health. SDG 13 (Climate Action) indirectly supports these goals by addressing the impacts of climate change on agriculture and health systems (United Nations, 2023). The SDGs promote sustainable agricultural practices to ensure a stable food supply and nutrition-sensitive agriculture, thereby reducing malnutrition and related health concerns. Their synergistic approach fosters stakeholder engagement to achieve the One Health goal. The WHO, FAO, World Organisation for Animal Health (WOAH), and the United Nations Environment Programme (UNEP) established the One Health High-Level Expert Panel (OHHLEP) to provide strategic guidance on One Health challenges. The OHHLEP promotes evidence-based policymaking by focusing on zoonotic diseases, AMR, and environmental degradation (WHO, FAO, WOAH, UNEP, 2022). It advocates for integrating One Health principles into global food security strategies. Its recommendations have shaped global policies, such as the Global Action Plan on Antimicrobial Resistance, which addresses antibiotic misuse in agriculture (WHO, FAO, WOAH, UNEP, 2022).

5.1 National and regional policy initiatives

Food security and One Health are inherently interconnected, emphasizing the relationship between human, animal, and environmental health. Policymakers worldwide increasingly address these intersections through national and regional frameworks. Many countries have prioritized food security and One Health in their policy agendas. For example, Nigeria's National Agricultural Transformation Agenda (ATA) prioritizes food security by increasing agricultural productivity and incorporating One Health principles to reduce zoonotic disease risks (Obiora, 2014). Similarly, Shrivastava (2019) noted that India's National Action Plan for Climate Change and Human Health (NAPCCHH) addresses the health impacts of agricultural practices while promoting sustainable farming to enhance food safety and security. In the United States, the Food Safety Modernization Act (FSMA), implemented jointly by the Food and Drug Administration (FDA) and the United States Department of Agriculture (USDA), integrates One Health principles to reduce foodborne diseases and ensure sustainable food production systems (U.S. Food and Drug Administration, 2024).

Considering the global nature of the food supply chain, various regional policies, frameworks, and initiatives have been implemented. The Comprehensive Africa Agriculture Development Programme (CAADP) aims to eliminate hunger and promote food security by 2025. CAADP emphasizes a One Health approach by addressing ecosystem health and reducing animal diseases (CAADP, 2015). The European Green Deal integrates sustainable food systems into its Farm to Fork Strategy, aiming to reduce antimicrobial usage by 50 % by 2030 to improve animal and human health (European Parliament, 2020).

In Asia, the Association of Southeast Asian Nations (ASEAN) established the ASEAN Health Cluster to address cross-border food security challenges by integrating food safety, zoonotic disease prevention, and environmental sustainability. ASEAN also published a strategic framework to combat AMR through One Health approach, with food safety being one of its core activity focuses. As a part of regional data harmonisation initiatives, ASEAN established the ASEAN Rapid Alert System for Food and Feed (ARASFF), enabling rapid notification of safety concerns of imported food, both from within and outside the region (ASEAN Secretariat, 2024).

5.2 Challenges in policy implementation

The convergence of food security and One Health policy is essential for addressing the interconnectedness of human, animal, and environmental health. However, implementing these policies presents significant challenges, reducing their effectiveness and impact. One important challenge is fragmented governance, which occurs when different entities with overlapping mandates fail to interact effectively. Intersectoral collaboration in nutrition faces significant challenges, including weak political commitment, competing sectoral priorities, and institutional isolation that hinder coordinated action across the agriculture, health, and education sectors. Benson (2008) emphasizes that undernutrition persists in Sub-Saharan Africa due to fragmented policy processes where nutrition lacks dedicated institutional homes, leading to diffused responsibilities and inadequate budgeting. Additional barriers include poor data sharing, differing professional cultures, and short-term project funding that undermines sustained partnerships, necessitating stronger leadership and incentive structures to foster joint accountability. Garbero and Jäckering (2021) observed that food security programs usually incorporate agriculture ministries, whereas One Health initiatives involve the public health and veterinary sectors. This lack of coordination causes inefficiencies and policy gaps. According to Bakiika et al. (2023), only 30 % of countries have successfully implemented integrated One Health frameworks due to poor intersectoral collaboration. Limited financial and technical resources impede the successful implementation of food security and One Health policies. Globally, poor countries confront a $15 billion yearly budget gap for agricultural and health system integration (World Bank, 2024). The World Organisation for Animal Health (2024) noted that in Sub-Saharan Africa, low investment in veterinary services exacerbates zoonotic disease outbreaks, threatening food production and public health. Cultural customs and societal resistance usually inhibit policy implementation. Traditional farming methods in rural regions and a reluctance to adopt sustainable practices impede food security goals. O'Mara-Eves et al. (2015) stated that skepticism towards health interventions hampered community engagement in One Health projects, as demonstrated during the Ebola outbreak in West Africa. Owino et al. (2022)

maintained that climate change exacerbates these problems by causing food instability and increasing the incidence of zoonotic illnesses. Extreme weather events, for example, diminish food yields in vulnerable areas by 30–40 %, while habitat destruction hastens pathogens spread from animals to humans. According to Aborode et al. (2021), unreliable data and surveillance systems hamper evidence-based decision-making. Approximately 40 % of low-income countries lack comprehensive systems for monitoring food production and zoonotic disease outbreaks, which limits timely policy responses.

6. Community-level interventions in food security and One Health

Integrating community-level initiatives into food security programs and advancing One Health principles is essential. One Health provides a comprehensive strategy for addressing food security concerns sustainably. Durao et al. (2020) emphasized empowering local communities by incorporating cultural and indigenous knowledge and leveraging public-private partnerships through evidence-based strategies and practical examples. Education and capacity-building initiatives are essential for equipping communities to achieve food security and address One Health challenges. Community education projects can fill knowledge gaps by teaching individuals how to manage resources sustainably, boost agricultural productivity, and mitigate health risks. For instance, training programs for smallholder farmers in Sub-Saharan Africa have significantly impacted agrarian practices. Farmer Field Schools (FFS) have enhanced farmers^{+} technical knowledge of sustainable farming techniques, pest management, and soil health. Najjar et al. (2013) reported that FFS participants in Kenya increased crop yields by 35 % more compared to non-participants.

Furthermore, targeted interventions in community health education can reduce the incidence of zoonotic diseases. Programs focusing on animal safety and early disease detection have successfully minimized outbreaks in zoonotic-prone regions. The EcoHealth Alliance recorded a 25 % decrease in zoonotic disease cases in Southeast Asia after implementing community-centred disease surveillance training (Sharan et al., 2023; Baffour-Ata et al., 2023) highlighted the importance of building resilience among smallholder farmers, who contribute 80 % of global food production in developing countries. These farmers are disproportionately affected by climate variability, market disruptions, and pandemics, jeopardising food security. Climate-smart agriculture (CSA) refers to techniques that enable farmers to adapt to climate change while maintaining productivity. Mango et al. (2018) noted that CSA initiatives in Malawi, such as crop diversification and agroforestry, improved food security for 70 % of participating households during an extended drought. Cole and Xiong (2017) observed that microloans and crop insurance programs help farmers recover from economic and environmental shocks. For example, index-based insurance schemes in Ethiopia have supported 30,000 farmers in managing drought-related risks.

In addition, consumer empowerment should be prioritized, as their choices can drive changes in the whole industry. For example, increased calls for reduced animal protein consumption—commonly associated with high carbon footprints, land use change, and climate change— have fueled the growth of the plant-based diet industry (McInnes, Carstairs, and Cecil, 2021). With regards to antibiotic use in farming, Zhou et al. (2024) have demonstrated that, when provided adequate information on its health and AMR risks, the public has stronger support for a ban on preventive antibiotic use.

In ASEAN countries, workshops and awareness programs have been conducted by various groups, from schoolchildren, teachers, and university students, to housewives. The driving force is the need for consumers to understand food safety, including safety risks across the food chain, guidance in selecting safe foods, and safe cooking measures. The development of programs such as Food Safety Cadres in ASEAN countries helps train community members to be adept at standard food safety practices. As these cadres come from within the community, they are able to share information in a culturally sensitive manner, setting positive examples for their respective communities (Asadullah et al., 2025).

6.1 Cultural and indigenous perspectives

Indigenous knowledge and cultural traditions provide relevant insights for developing sustainable food systems aligned with the One Health approach. Messer and Cohen (2023) emphasized that incorporating these perspectives is essential to address global food insecurity, which affected approximately 29.6 % of the global population in 2021. Indigenous communities have historically employed sustainable agricultural practices that ensure food security while preserving ecosystems. Traditional methods for enhancing soil fertility, managing pests, and promoting biodiversity include agroforestry systems, crop rotation, and polyculture. Virtanen (2019) reported that Amazonian Indigenous people in Brazil have used "terra preta" (black soil) for generations to enhance soil fertility and agricultural productivity. Moreover, indigenous diets are diverse and nutrient-rich, providing resilience against malnutrition. Orr et al. (2022) highlighted the importance of traditional grains such as millet and sorghum, which are drought-resistant and culturally significant. Incorporating these grains into modern diets can improve nutritional outcomes while reducing reliance on monoculture crops such as maize and wheat.

Kaoga et al. (2021) observed that the Maasai community in Kenya integrates cattle management with land conservation, promoting food security and ecosystem health. Their practices also mitigate zoonotic disease risks by fostering harmonious relationships between humans and animals. Cultural attitudes significantly influence disease prevention and food safety. For example, indigenous food preservation techniques in Southeast Asia, such as fermentation, reduce food waste while improving microbiological safety. These methods align with One Health principles by addressing public health challenges and minimizing environmental impact (Baiyeri et al., 2019; Khupe, 2020) argued that indigenous knowledge systems are usually overlooked in policymaking despite their value. Land dispossession, cultural erosion, and climate change exacerbate food insecurity among indigenous communities. Policymakers should integrate indigenous knowledge into food security strategies while prioritizing land rights and cultural preservation. Indigenous and cultural perspectives on food security and One Health underscore the importance of sustainable practices rooted in local knowledge. Incorporating these perspectives can bridge gaps in modern agricultural systems and enhance global resilience to food insecurity.

6.2 Public-private partnerships

Public-Private Partnerships (PPPs) are increasingly recognized as necessary for addressing food security and supporting the One Health concept. PPPs provide innovative and sustainable solutions to global challenges by leveraging resources and expertise from governments, the private sector, and civil society. Food security remains a pressing global issue. According to the (FAO, IFAD, UNICEF, WFP, & WHO. 2024a), 2.4 billion people experienced moderate to severe food insecurity in 2023, with 735 million suffering from chronic hunger. This underscores the need for effective interventions, such as PPPs. The World Food Programme collaborates with commercial companies such as PepsiCo to promote sustainable agricultural practices and supply chain innovations, benefiting over 250,000 smallholder farmers globally (WFP, 2021). The One Health concept recognizes the mutual interdependence of food security, control of zoonotic diseases, and environmental sustainability. El Amri et al. (2020) reported that nearly 60 % of viral diseases in humans are zoonotic, highlighting the importance of integrated health systems. PPPs enable cross-sector collaboration to mitigate such risks. Nel et al. (2017) illustrated how the Global Alliance for Rabies Control partners with governments and private sector entities to control rabies, saving an estimated $8.6 billion annually in economic losses and healthcare expenditures. PPPs serve as an important link between food security and One Health. For instance, the Livelihoods and Food Security Fund (LIFT) partners with commercial agribusinesses to improve cattle health in Sub-Saharan Africa. Capper and Williams (2023) asserted that LIFT improved cattle productivity by 40 % and reduced zoonotic disease outbreaks by 20 % through the integration of veterinary services and agricultural advancements. Despite their significance, PPPs face challenges such as unequal power dynamics, financial disparities, and regulatory challenges. Transparent governance, equitable benefit-sharing, and robust monitoring procedures are essential for effective partnerships. Additionally, investments in digital technologies, such as blockchain for traceability and artificial intelligence for predictive analytics, can amplify their impact (Chiswa, 2024).

7. Challenges and opportunities for integration

Realizing the synergy between food security and the One health concept presents challenges and opportunities. King (2012) identified that a major challenge to incorporating food security into the One Health paradigm is the limited awareness among stakeholders about the interconnection between human, animal, and environmental health. Many policymakers, health professionals, and agricultural specialists continue to work in isolation, unaware of the systemic benefits of an interdisciplinary approach. Mulder (2017) highlighted that bridging these fields requires enhanced technical expertise, which remains scarce. Traditional medical, veterinary, and agricultural education programs rarely emphasize the interdependence of their respective domains. Fasina et al. (2021) reported that a survey of African colleges revealed fewer than 15 % offered curricula that included One Health principles, perpetuating knowledge gaps. Tinto et al. (2023) argued that this knowledge gap restricts the development of holistic solutions to food security challenges, particularly in regions experiencing zoonotic disease outbreaks, such as Rift Valley Fever in East Africa. Institutional isolation is a major barrier to implementing One Health. Public health, veterinary, and environmental sectors usually operate independently, leading to fragmented regulations and inefficient resource allocation. Balana and Oyeyemi (2022) observed that in low-income countries, separate financial structures for agricultural and public health interventions reduce the effectiveness of integrated solutions. Furthermore, resources for One Health programs are usually limited, as funding agencies prioritize immediate crises over long-term interdisciplinary initiatives. According to United Nations and FAO estimates, fewer than 5 % of global agricultural aid in 2021 was allocated to initiatives explicitly aligned with One Health objectives (WHO, 2023). This underfunding contributes to conflicting stakeholder interests, hindering goal alignment across sectors.

7.1 Opportunities for holistic integration

Digital platforms provide a transformative opportunity to integrate food security with One Health. These systems facilitate real-time data sharing and collaboration among stakeholders from various disciplines. The FAO, WHO, and World Organisation for Animal Health (OIE) have collaborated to develop the Global Early Warning System (GLEWS), which leverages digital tools to monitor and respond to zoonotic disease outbreaks, directly linking animal health to food security outcomes (Hassan and Winkler, 2023). Advances in data analytics and artificial intelligence have significantly enhanced the predictive capabilities of these platforms. Gómez et al. (2021) demonstrated that machine learning algorithms could anticipate locust outbreaks, a major threat to food security, with 90 % accuracy using meteorological and vegetation data. Integrating such tools into One Health frameworks empowers stakeholders to foresee hazards and implement proactive solutions. Furthermore, digital platforms facilitate capacity building through online training modules, webinars, and e-learning courses, making multidisciplinary knowledge more accessible. The U.S. Agency for International Development (USAID) supported programs such as the One Health Workforce, which has trained over 10,000 professionals worldwide, equipping them with the skills necessary for cross-sectoral problem-solving (U.N, 2022).

Global advocacy networks significantly contribute to advancing the integration of food security and One Health. These networks unite governments, non-governmental organizations (NGOs), academic institutions, and the private sector to align their goals and resources. Pungartnik et al. (2023) observed that the One Health Global Network has been instrumental in fostering policy discussions and facilitating the adoption of One Health principles at national and international levels. Yang et al. (2020) argued that international frameworks such as the SDGs provide compelling justification for integration. SDGs 2 and 3 align closely with One Health's objectives, creating a favorable policy environment that supports collaborative initiatives. According to Naomi (2023), countries such as Bangladesh have effectively leveraged this alignment by developing integrated livestock and public health programs, which reduced zoonotic disease rates by 30 % while improving food security. Global financial organizations like the World Bank have increasingly prioritized funding for One Health initiatives. In 2020, the World Bank launched a $500 million project to enhance pandemic preparedness using One Health frameworks, highlighting the growing recognition of this approach's importance (World Bank, 2022). While limited awareness, inadequate interdisciplinary skills, institutional isolation, and resource constraints hinder the integration of food security and One Health, several opportunities exist to address these challenges. Stakeholders can strengthen food systems, mitigate zoonotic risks, and promote sustainable health outcomes for humans, animals, and ecosystems by addressing these challenges and leveraging these opportunities.

8. Future directions and research priorities on food security and One Health

Food security and One Health are becoming increasingly important global concerns that demand novel approaches and extensive research to address complex challenges. Climate change seriously threatens food security and One Health by influencing crop yields, livestock health, and disease patterns. Rising temperatures, unpredictable precipitation, and extreme weather events disrupt agricultural production, resulting in food insecurity for millions. Clark et al. (2023) observed that, under high-emission scenarios, global cereal production could decline by 10 % by 2050, disproportionately affecting disadvantaged populations. According to Rupasinghe et al. (2022), climate change impacts the distribution of zoonotic diseases, which are central to the One Health paradigm. For instance, malaria and dengue fever are expanding into new regions as temperatures rise.

The interconnectivity of food security and the One Health approach highlights the need for comprehensive policy actions. Research should prioritize developing prediction models to assess the impact of climate change on food systems and health by integrating data from meteorology, epidemiology, and agriculture. Rodrigues et al. (2023) highlighted that genetic engineering and synthetic biology advancements can enhance food security and sustainability. Genetically modified organisms (GMOs) can improve agricultural tolerance to drought, pests, and diseases, potentially increasing yields by 15–20 % in stress-prone areas. Synthetic biology also enables the production of lab-grown meat and other proteins, reducing reliance on resource-intensive livestock rearing. However, these technologies raise ethical, safety, and regulatory concerns. Research should focus on refining genetic engineering techniques to ensure environmental protection and address public skepticism. Moreover, evaluating synthetic biology's scalability and economic feasibility in food systems is necessary for its widespread adoption.

8.1 Policy recommendations

Policymakers should implement methods that promote sustainable food systems while reducing health and environmental risks. They should prioritize effective surveillance systems for zoonotic diseases that endanger food security. Marrana, (2022) reported that roughly 75 % of new infectious diseases in humans are zoonotic. Investing in early detection mechanisms can help prevent epidemics that disrupt food supply chains. Adopting sustainable agriculture strategies such as crop diversification, agroforestry, and integrated pest management can enhance food security while reducing environmental damage. Nsabiyeze et al. (2024) stated that reducing agricultural greenhouse gas emissions by 30 % by 2030 can help offset the effects of climate change on food systems. A One Health strategy requires coordination between sectors such as agriculture, health, and the environment. Policymakers should form multisectoral task forces to coordinate responses to challenges such as antimicrobial resistance (AMR), and provide civil servants with appropriate incentives to successfully participate in these bodies (FAO/IFAD/WFP, 2023) estimated that treating AMR in livestock could save up to $1 trillion in global GDP by 2050.

Investment in research and development (R&D) is necessary for developing new solutions to food security issues. Aljabali et al. (2024)

noted that genome editing technologies such as CRISPR-Cas9 have been identified as necessary tools for generating climate-resilient crops. Policymakers should provide financing to research institutions focused on these innovations. Empowering local communities with education and resources can boost their resilience in the face of food insecurity. Berrian et al. (2018) noted that community-based One Health programs in East Africa have reduced livestock mortality by 25 %, boosting household nutrition. The United Nations Environment Programme (UNEP, 2013) noted that global food waste accounts for one-third of total food production, or 1.3 billion tons per year. Policymakers should create frameworks to reduce waste and redistribute surplus food to vulnerable groups. Policymakers can address the gap between food security and One Health by implementing these recommendations to promote resilient and sustainable systems that benefit all sectors.

8.2 Advancing monitoring and evaluation systems

Monitoring and evaluation (M&E) systems are necessary for analyzing and improving food security measures as well as the One Health concept. In 2024, the gender gap in the global prevalence of moderate or severe food insecurity increased to 1.9 % points compared to 1.3 in 2023; women experienced higher rates (26.1 per cent) than men (24.2 per cent) (FAO, IFAD, UNICEF, WFP, & WHO, 2024b). Enhancing M&E systems allows policymakers to assess the effectiveness of agricultural practices, disease control measures, and environmental conservation programs in building sustainable food systems. Incorporating real-time data collection and analysis technologies is essential for improving these systems. Nefros et al. (2022) emphasized that remote sensing techniques and geographic information systems (GIS) can monitor land use changes, crop yields, and disease outbreaks. Studies demonstrate that using such technologies improves food security assessments by 30 %. Ghai et al. (2022) argued that One Health M&E systems should monitor zoonotic disease transmission. Cross-sector collaboration is indispensable. Effective M&E frameworks require partnerships among governments, non-governmental organizations, and research institutions to share data and harmonize measurement standards. Indicators such as food availability, nutritional status, and disease incidence must align with global frameworks such as the SDGs. However, significant challenges persist, such as data fragmentation and inadequate funding. Miebach (2024) emphasized that investing in capacity building for local stakeholders is essential to bridge these gaps because it leverages AI to analyze complex datasets. These advancements can strengthen early warning systems for food crises and mitigate risks to human and animal health. Improving M&E systems for food security and One Health necessitates adopting technology, fostering collaboration, and addressing funding disparities. Such initiatives will contribute to a resilient, sustainable, and health-integrated global food system.

8.3 Promoting interdisciplinary research

Food security and One Health are interconnected challenges that require interdisciplinary research to address complex global issues such as public health, animal health, and environmental sustainability. In 2025, an estimated 318 million people across 68 countries are experiencing acute food insecurity, with 41.1 million in the emergency phase or worse, while 60 % of human infectious diseases are zoonotic (WFP, 2025; Pfenning-Butterworth et al., 2024) argued that climate change, biodiversity loss, and unsustainable agricultural practices exacerbate food shortages and increase health risks. Extreme weather events reduce food production by 30 % in vulnerable regions, while deforestation facilitates the spread of zoonotic diseases such as Ebola. An interdisciplinary approach is essential to integrate biotechnology, veterinary medicine, and public health advancements to develop sustainable food systems with minimal disease risks and environmental impact.

Research priorities include evaluating AMR in food production, strengthening zoonotic disease surveillance systems, and developing climate-resilient crops. The University of Oxford (2024) reported that current projections indicate AMR could cause 10 million deaths annually by 2050, with agricultural misuse accounting for 73 % of global antibiotic consumption.

Designing holistic solutions requires collaboration among researchers, policymakers, and industry stakeholders. Investing in transdisciplinary research provides transformative opportunities. Initiatives such as the CGIAR Research Program on Agriculture for Nutrition and Health have demonstrated how integrated approaches can improve food security and health outcomes (CGIAR, 2023).

9. Conclusion

The relationship between food security and the One Health approach underscores the complex interplay of human, animal, and environmental health. Climate change, zoonotic diseases, AMR and unsustainable farming practices are necessary concerns that should be addressed to ensure food security. These challenges are further compounded by rapid population growth, with the United Nations projecting the global population will exceed 9.7 billion by 2050, necessitating a 70 % increase in food production to meet demand. Simultaneously, approximately 2.4 billion people experience moderate to severe food insecurity, emphasizing the urgency of a coordinated response. The One Health paradigm provides a relevant framework for achieving sustainable food security. It fosters interdisciplinary collaboration among human medicine, veterinary science, and environmental management. For instance, investments in climate-smart agriculture, including resilient crop varieties and precision farming, can reduce greenhouse gas emissions from the agricultural sector, which contributes 24 % of global emissions. In addition, strengthening zoonotic disease surveillance and control systems is necessary because 75 % of emerging infectious diseases are zoonotic, posing significant threats to food systems. Governments should prioritize policies that promote sustainable farming practices and improve infrastructure to reduce post-harvest losses, which account for 14 % of global food waste (Shukla et al., 2025). International organizations should advocate for equitable resource allocation and provide technical assistance to low-income regions disproportionately affected by food insecurity. Academic and research institutions can advance biotechnology and data analytics, while the private sector can drive scalable solutions through long-term agricultural investments. Civil society and local communities also play essential roles in ensuring equitable access to nutritious food and adopting culturally appropriate interventions. The One Health approach is desirable and indispensable for ensuring the long-term sustainability of food systems. Stakeholders can effectively address the multifaceted challenges of food insecurity by recognizing the interdependence of health domains and fostering coordinated efforts. Achieving global food security requires a commitment to transdisciplinary strategies, innovative practices, and inclusive policies that uphold the principles of equity, sustainability, and resilience.

References

Aborode, A.T., Hasan, M.M., Jain, S., Okereke, M., Adedeji, O.J., Karra-Aly, A., et al., 2021. Impact of poor disease surveillance system on COVID-19 response in Africa: Time to rethink and rebuild. Clin Epidemiol Glob. Health 12, 100841. https://doi.org/10.1016/j.cegh.2021.100841.

Aday, S., Aday, M.S., 2020. Impact of COVID-19 on the food supply chain. Food Qual. Saf. 4 (4), 167–180. https://doi.org/10.1093/fqsafe/fyaa024.

Ahmed, S., Dávila, J.D., Allen, A., Tacoli, C., Fèvre, E.M., 2019. Does urbanization make emergence of zoonosis more likely? Evidence, myths and gaps. Environ Urbanization 31 (2), 443. https://doi.org/10.1177/0956247819866124.

Ali, A.M., Bijay-Singh, 2025. Environmental pollution and climate change implications of agricultural fertilizer use. In: Hussain, N., Hung, C.Y., Wang, L. (Eds.), Agricultural Nutrient Pollution and Climate Change. Springer. https://doi.org/10.1007/978-3-031-80912-5_1.

Aljabali, A.A.A., El-Tanani, M., Tambuwala, M.M., 2024. Principles of CRISPR-Cas9 technology: advancements in genome editing and emerging trends in drug delivery. J Drug Delivery Sci. Technol. 92, 105338. https://doi.org/10.1016/j.jddst.2024.105338.

Akash, M.H., Mondal, S., Adusumilli, S., 2022. Chapter Four - Sustainable livestock production and food security. In: Mondal, S., Singh, R.L. (Eds.), Emerging Issues in Climate Smart Livestock Production. Academic Press, pp. 71–90. https://doi.org/10.1016/B978-0-12-822265-2.00011-9.

Amadu, F.O., Miller, D.C., McNamara, P.E., 2020. Agroforestry as a pathway to agricultural yield impacts in climate-smart agriculture investments: evidence from southern Malawi. Ecol. Econ. 167, 106443. https://doi.org/10.1016/j.ecolecon.2019.106443.

Asadullah, M.N., Jilani, A.H., Negara, S.D., Suryadarma, D., 2025. Improving the quality of basic education in ASEAN—Emerging challenges and reforms. Int. J. Educ. Dev. 116, 103292. https://doi.org/10.1016/j.ijedudev.2025.103292. (Article).

ASEAN Secretariat, 2024. ASEAN socio-cultural community trend report No. 4: enhancing and integrating regional food safety to face the changing landscape of food system and health threats (Vol. 4). https://www.asean.org.

Baffour-Ata, F., Atta-Aidoo, J., Said, R.O., Nkrumah, V., Atuyigi, S., Analima, S.M., 2023. Building the resilience of smallholder farmers to climate variability: Using climate-smart agriculture in Bono East Region, Ghana. Heliyon 9 (11), e21815. https://doi.org/10.1016/j.heliyon.2023.e21815.

Bahar, N.H.A., Lo, M., Sanjaya, M., Van Vianen, J., Alexander, P., Ickowitz, A., et al., 2020. Meeting the food security challenge for nine billion people in 2050: What impact on forests? Glob. Environ. Change 62, 102056. https://doi.org/10.1016/j.gloenvcha.2020.102056.

Baiyeri, P.K., Foleng, H.N., Machebe, N.S., Nwobodo, C.E., 2019. Crop-livestock interaction for sustainable agriculture. In: Farooq, M., Pisante, M. (Eds.), Innovations in Sustainable Agriculture. Springer. https://doi.org/10.1007/978-3-030-23169-9_18.

Balana, B.B., Oyeyemi, M.A., 2022. Agricultural credit constraints in smallholder farming in developing countries: evidence from Nigeria. World Dev Sustainability 1, 100012. https://doi.org/10.1016/j.wds.2022.100012.

Benson, T., 2008. Improving Nutrition as a Development Priority: Addressing Undernutrition within National Policy Processes in Sub-Saharan Africa (IFPRI Research Report 156). International Food Policy Research Institute. https://ageconsearch.umn.edu/record/37875/files/rr156.pdf.

Beyene, S.D., 2023. The impact of food insecurity on health outcomes: empirical evidence from sub-Saharan African countries. BMC Public Health 23, 338. https://doi.org/10.1186/s12889-023-15244-3.

Bezner Kerr, R., Madsen, S., Stüber, M., Liebert, J., Enloe, S., Borghino, N., et al., 2021. Can agroecology improve food security and nutrition? A review. Glob. Food Security 29, 100540. https://doi.org/10.1016/j.gfs.2021.100540.

Bonadio, B., Huo, Z., Levchenko, A.A., Pandalai-Nayar, N., 2021. Global supply chains in the pandemic. J. Int. Econ. 133, 103534. https://doi.org/10.1016/j.jinteco.2021.103534.

Bostock, E.C.S., Kirkby, K.C., Taylor, B.V., Hawrelak, J.A., 2020. Consumer Reports of "Keto Flu" Associated With the Ketogenic Diet. Front. nutr. 7, 20. https://doi.org/10.3389/fnut.2020.00020.

Bouteska, A., Sharif, T., Bhuiyan, F., Abedin, M.Z., 2024. Impacts of the changing climate on agricultural productivity and food security: evidence from Ethiopia. J. Clean Prod. 449, 141793. https://doi.org/10.1016/j.jclepro.2024.141793.

Bricas, N., 2019. Urbanization issues affecting food system sustainability. Designing Urban Food Policies. Springer. https://doi.org/10.1007/978-3-030-13958-2_1. (xx–xx).

CAADP, 2015. *The Comprehensive Africa Agriculture Development Programme (CAADP)*. New partnership for Africa's development (NEPAD). Retrieved 13 December 2024 from https://www.nepad.org/caadp.

CGIAR, 2023. *2023 annual report: advancing nutrition, health, and food security*. Retrieved from https://www.cgiar.org.

Capper, J.L., Williams, P., 2023. Investing in health to improve the sustainability of cattle production in the United Kingdom: A narrative review. Vet. J. 296–297, 105988. https://doi.org/10.1016/j.tvjl.2023.105988.

Chandra, S.S.V., Hareendran, S.A., Albaaji, G.F., 2024. Precision farming for sustainability: an agricultural intelligence model. Comput. Elect. Agric. 226, 109386. https://doi.org/10.1016/j.compag.2024.109386.

Charostad, J., Rukerd, M.R.Z., Mahmoudvand, S., Bashash, D., Hashemi, S.M.A., Nakhaie, M., et al., 2023. A comprehensive review of highly pathogenic avian influenza (HPAI) H5N1: an imminent threat at doorstep. Travel Med. Infect. Dis. 55, 102638. https://doi.org/10.1016/j.tmaid.2023.102638.

Chiswa, N., 2024. Unlocking Africa's economic potential: the role of PPPs in fuelling infrastructure development and investment–South Africa's perspective. Jindal Glob. Law Rev. 15 (2), 157–185. https://doi.org/10.1007/s41020-024-00221-7.

Clark, B., Xia, L., Robock, A., et al., 2023. Optimal climate intervention scenarios for crop production vary by nation. Nat. Food 4 (9), 902–911. https://doi.org/10.1038/s43016-023-00853-3.

Cole, S.A., Xiong, W., 2017. Agricultural insurance and economic development. Annu. Rev. Econ. 9 (1), 235–262. https://doi.org/10.1146/annurev-economics-080315-015225.

Costa, M.M., Cardo, M., Ruano, Z., Alho, A.M., Dinis-Teixeira, J., Aguiar, P., et al., 2023. Effectiveness of antimicrobial interventions directed at tackling antimicrobial resistance in animal production: a systematic review and meta analysis. Preventive Veterinary Med. 218, 106002. https://doi.org/10.1016/j.prevetmed.2023.106002.

Deguine, J.-P., Aubertot, J.-N., Bellon, S., CÔte, F., Lauri, P.-E., Lescourret, F., et al., 2023. Agroecological crop protection for sustainable agriculture. In: Sparks, D.L. (Ed.), Advances in Agronomy 178. Academic Press, pp. 1–59. https://doi.org/10.1016/bs.agron.2022.11.002.

Devendra, C., 2006. Perspectives on animal production systems in Asia. Livest Sci. 106 (1), 1–18. https://doi.org/10.1016/j.livsci.2006.05.005.

Durao, S., Visser, M.E., Ramokolo, V., Oliveira, J.M., Schmidt, B.M., Balakrishna, Y., et al., 2020. Community-level interventions for improving access to food in low- and middle-income countries. Cochrane Database Syst. Rev. 7 (7), CD011504. https://doi.org/10.1002/14651858.CD011504.pub2.

El Amri, H., Boukharta, M., Zakham, F., Ennaji, M.M., 2020. Emergence and reemergence of viral zoonotic diseases: concepts and factors of emerging and reemerging globalization of health threats. Emerg reemerging viral Pathog. 619–634. https://doi.org/10.1016/B978-0-12-819400-3.00027-2.

European Parliament, 2020. *The Farm to Fork Strategy: for a Fair, Healthy and Environmentally-Friendly Food System*. https://www.europarl.europa.eu/news/en/press-room/20200520IPR77421/the-farm-to-fork-strategy-for-a-fair-healthy-and-environmentally-friendly-food-system.

FAO (Food and Agriculture Organization of the United Nations), 2001. *The State of Food Insecurity in the World*. Rome, Italy. https://www.fao.org/publications.

FAO (Food and Agriculture Organization of the United Nations), 2015. *Climate Change and Food Security: Risks and Responses*. FAO. https://www.fao.org/publications.

FAO, 2024. *Global Fisheries and Aquaculture Production Reaches a New Record High*. FAO. https://www.fao.org.

FAO, IFAD, UNICEF, WFP, & WHO, 2024a. *Hunger Numbers Stubbornly High foR Three Consecutive Years as Global Crises Deepen: UN Report. United Nations*. https://www.un.org/news/item/Hunger-numbers-stubbornly-high-for-three-consecutive-years-as-global-crises-deepen.

FAO, IFAD, UNICEF, WFP, & WHO, 2024b. *The State of Food Security and Nutrition in the World 2024 – Financing to End Hunger, Food Insecurity and Malnutrition in all Its Forms*. https://doi.org/10.4060/cd1254en.

FAO/IFAD/WFP, 2023. *Foodborne Antimicrobial Resistance (AMR): An Economic Concern. Food and Agriculture Organization of the United Nations*. https://doi.org/10.4060/cc6014en.

FAO & United Nations Environment Programme (UNEP), 2020. *The State of the World's Forests 2020: Forests, Biodiversity and People*. Rome. https://doi.org/10.4060/ca8642en.

Fasina, F.O., Fasanmi, O.G., Makonnen, Y.J., Bebay, C., Bett, B., Roesel, K., 2021. The one health landscape in Sub-Saharan African countries. One Health 13, 100325. https://doi.org/10.1016/j.onehlt.2021.100325.

Fukase, E., Martin, W., 2020. Economic growth, convergence, and world food demand and supply. World Dev 132, 104954. https://doi.org/10.1016/j.worlddev.2020.104954.

Garbero, A., Jäckering, L., 2021. The potential of agricultural programs for improving food security: a multi-country perspective. Glob. Food Security 29, 100529. https://doi.org/10.1016/j.gfs.2021.100529.

Gebeyehu, L., Gemechu, A., Bedemo, A., 2025. Effects of climate change on food security in Africa: meta-analysis. J Agric. Food Res. 23, 102214. https://doi.org/10.1016/j.jafr.2025.102214.

Getiso, A., Mijena, D., 2021. Feeding and nutritional strategies to reduce methane emission from large ruminants: review. J. Anim. Livest Production 2 (2). https://doi.org/10.47363/JALP/2021(2)109. Article 109.

Ghai, R.R., Wallace, R.M., Kile, J.C., et al., 2022. A generalizable one health framework for the control of zoonotic diseases. Sci. Rep. 12 (1), 8588. https://doi.org/10.1038/s41598-022-12619-1.

Glaze-Corcoran, S., Hashemi, M., Sadeghpour, A., Jahanzad, E., Keshavarz Afshar, R., Liu, X., et al., 2020. Understanding intercropping to improve agricultural resiliency and environmental sustainability. In: Sparks, D.L. (Ed.), Advances in Agronomy 162. Academic Press, pp. 199–256. https://doi.org/10.1016/bs.agron.2020.02.004.

Gómez, D., Salvador, P., Sanz, J., Rodrigo, J.F., Gil, J., Casanova, J.L., 2021. Prediction of desert locust breeding areas using machine learning methods and SMOS (MIR_SMNRT2) near real time product. J. Arid. Environ. 194, 104599. https://doi.org/10.1016/j.jaridenv.2021.104599.

Gržinić, G., Piotrowicz-Cieślak, A., Klimkowicz-Pawlas, A., Górny, R.L., Ławniczek-Wałczyk, A., Piechowicz, L., et al., 2023. Intensive poultry farming: a review of the impact on the environment and human health. Sci. Total Environ. 858 (Pt. 3). https://doi.org/10.1016/j.scitotenv.2022.160014. Article 160014.

Gu, D., Andreev, K., Dupre, M.E., 2021. Major trends in population growth around the world. China CDC Wkly 3 (28), 604. https://doi.org/10.46234/ccdcw2021.160.

Hassan, O.A., Winkler, A.S., 2023. One Health early warning and response system for zoonotic diseases outbreaks: emphasis on the involvement of grassroots actors. Veterinary Med. Sci. 9 (4), 1881. https://doi.org/10.1002/vms3.1135.

Hashim, H.T., Miranda, A.V., Babar, M.S., Essar, M.Y., Hussain, H., Ahmad, S., et al., 2021. Yemen's triple emergency: food crisis amid a civil war and COVID-19 pandemic. Public health Pract. (Oxford, Engl) 2, 100082. https://doi.org/10.1016/j.puhip.2021.100082.

Hilson, G., Sauerwein, T., Cardoso, M.E., 2021. Small-scale mining, rural resilience and the sustainable development goals in sub-Saharan Africa. Handbook of Sustainable Politics and Economics of Natural Resources. Edward Elgar Publishing, Cheltenham. https://doi.org/10.4337/9781789908770.00021.

Homann-Kee Tui, S., Valdivia, R.O., Descheemaeker, K., Senda, T., Masikati, P., Makumbe, M.T., et al., 2020. Crop–livestock integration to enhance ecosystem services in sustainable food systems. In: Rusinamhodzi, L. (Ed.), The Role of Ecosystem Services in Sustainable Food Systems. Academic Press, pp. 141–169. https://doi.org/10.1016/B978-0-12-816436-5.00008-1.

Ingrao, C., Strippoli, R., Lagioia, G., Huisingh, D., 2023. Water scarcity in agriculture: an overview of causes, impacts and approaches for reducing the risks. Heliyon 9 (8), e18507. https://doi.org/10.1016/j.heliyon.2023.e18507.

Intergovernmental Panel on Climate Change, 2023. *IPCC Sixth Assessment Report: Impacts, Adaptation and Vulnerability*. Cross-chapter paper 7: Tropical forests. Retrieved from https://www.ipcc.ch/report/ar6/wg2/.

Jurgens, J., Kaushik, P., 2024. *Farmers in India Are Using AI for Agriculture – Here's How They Could Inspire the World*. World Economic Forum. https://www.weforum.org/agenda/2024/01/ai-agriculture-india.

Kaoga, J., Olago, D., Ouma, G., Ouma, G., Onono, J., 2021. The evolving cultural values and their implications on the Maasai pastoralists, Kajiado County, Kenya. Sci. Afr. 13, e00881. https://doi.org/10.1016/j.sciaf.2021.e00881.

Kappes, A., Tozooneyi, T., Shakil, G., Railey, A.F., McIntyre, K.M., Mayberry, D.E., et al., 2023. Livestock health and disease economics: a scoping review of selected literature. Front Veterinary Sci. 10, 1168649. https://doi.org/10.3389/fvets.2023.1168649.

Kaur, A., Paruchuri, R.G., Nayak, P., Devi, K.B., 2023. The role of agroforestry in soil conservation and sustainable crop production: a comprehensive review. Int. J. Environ. Clim. Change 13 (11), 3089–3095. https://doi.org/10.9734/IJECC/2023/v13i113478.

Khupe, C., 2020. Indigenous knowledge systems. In: Akpan, B., Kennedy, T.J. (Eds.), Science Education in Theory and Practice. Springer Texts in Education. Springer, Cham. https://doi.org/10.1007/978-3-030-43620-9_30.

King, L.J., 2012. One health and food safety. In Institute of Medicine (US), Improving food safety through a One Health approach: workshop summary (A8). National Academies Press (US). Available from https://www.ncbi.nlm.nih.gov/books/NBK114498/.

Le, T.-H., Bui, M.-T., Uddin, G.S., 2022. Economic and social impacts of conflict: a cross-country analysis. Economic Model 115, 105980. https://doi.org/10.1016/j.econmod.2022.105980.

Lee, K., Brumme, Z.L., 2013. Operationalizing the One Health approach: the global governance challenges. Health Policy Plan 28 (7), 778–785. https://doi.org/10.1093/heapol/czs127.

Lewis, R.A., Kashongwe, O.B., Bebe, B.O., 2023. Quantifying production losses associated with foot and mouth disease outbreaks on large-scale dairy farms in Rift Valley, Kenya. Tropical Anim Health Prod 55 (5), 293. https://doi.org/10.1007/s11250-023-03707-z. (Article).

Li, C., Chen, H., 2021. H7N9 influenza virus in China. Cold Spring Harb Perspect Med. 11 (8), a038349. https://doi.org/10.1101/cshperspect.a038349.

Li, L., Awada, T., Shi, Y., Jin, V.L., Kaiser, M., 2024. Global greenhouse gas emissions from agriculture: pathways to sustainable reductions. Glob. Change Biol. 31 (1), e70015. https://doi.org/10.1111/gcb.70015.

Ma, M., Wang, H.H., Hua, Y., Qin, F., Yang, J., 2021. African swine fever in China: impacts, responses, and policy implications. Food Policy 102, 102065. https://doi.org/10.1016/j.foodpol.2021.102065.

Mango, N., Makate, C., Mapemba, L., others, 2018. The role of crop diversification in improving household food security in central Malawi. Agr. Food Secur. 7 (1). https://doi.org/10.1186/s40066-018-0160-x. Article 7.

Mansfield, K.L., Banyard, A.C., McElhinney, L., Johnson, N., Horton, D.L., Hernández-Triana, L.M., et al., 2015. Rift Valley fever virus: a review of diagnosis and vaccination, and implications for emergence in Europe. Vaccine 33 (42), 5520–5531. https://doi.org/10.1016/j.vaccine.2015.08.020.

Manyi-Loh, C., Mamphweli, S., Meyer, E., Okoh, A., 2018. Antibiotic use in agriculture and its consequential resistance in environmental sources: potential public health implications. Molecules: A J. Synth. Chem. Nat. Product Chem. 23 (4), 795. https://doi.org/10.3390/molecules23040795.

Maqbool, M.E., Farhan, A., Qamar, M.A., 2024. Global impact of COVID-19 on food safety and environmental sustainability: pathways to face the pandemic crisis. Heliyon 10 (15), e35154. https://doi.org/10.1016/j.heliyon.2024.e3515.

Marie, V., Gordon, M.L., 2023. The (re-)emergence and spread of viral zoonotic disease: a perfect storm of human ingenuity and stupidity. Viruses 15 (8), 1638. https://doi.org/10.3390/v15081638.

Marrana, M., 2022. Epidemiology of disease through the interactions between humans, domestic animals, and wildlife. In: Prata, J.C., Ribeiro, A.I., Rocha-Santos, T. (Eds.), One Health. Academic Press, pp. 73–111. https://doi.org/10.1016/B978-0-12-822794-7.00001-0.

McLeod, A., Morgan, N., Prakash, A., Hinrichs, J., 2005. Economic and social impacts of avian influenza. In Proceedings of the joint FAO/OMS/OIE/World Bank conference on avian influenza and human pandemic influenza, November (pp. 7–9).

McInnes, G., Lavertu, A., Sangkuhl, K., Klein, T.E., Whirl-Carrillo, M., Altman, R.B., 2021. Pharmacogenetics at Scale: An Analysis of the UK Biobank. Clin. pharmacol. ther. 109 (6), 1528–1537. https://doi.org/10.1002/cpt.2122.

Messer, E., Cohen, M.J., 2023. Chapter ten - Conflict, food insecurity, and globalization: an update 20 years on. In: Cohen, M.J. (Ed.), Advances in Food Security and Sustainability 8. Elsevier, pp. 287–345. https://doi.org/10.1016/bs.af2s.2023.08.001.

Miebach, M., 2024. *The Pace of Innovation Is Speeding Up. Digital Fragmentation Threatens That.* World Economic Forum. https://www.weforum.org/agenda/2024/01/the-pace-of-innovation-is-speeding-up-digital-fragmentation-threatens-that.

Mouratiadou, I., Wezel, A., Kamilia, K., Marchetti, A., Paracchini, M.L., Bàrberi, P., 2024. The socio-economic performance of agroecology. A review. Agron. Sustain. Dev. 44 (2). https://doi.org/10.1007/s13593-024-00945-9.

Müller, K., Oliver, M.A., Siebe, C., 2022. Overview chapter on soil degradation. Encycl. Soils Environ. (Second Ed) 165–171. https://doi.org/10.1016/B978-0-12-822974-3.00270-6.

Mulder, M., Ed., 2017. *Competence-Based Vocational and Professional Education: Bridging the Worlds of Work and Education.* Springer. https://doi.org/10.1007/978–3-319–41713-4.

Najjar, D., Spaling, H., Sinclair, A.J., 2013. Learning about sustainability and gender through farmer field schools in the Taita Hills, Kenya. Int. J. Educ. Dev. 33 (5), 466–475. https://doi.org/10.1016/j.ijedudev.2012.06.004.

Naomi, S.S., 2023. Bangladesh Foreign Policy Approach Should Shift to Multi-Alignment. Institute for Security and Development Policy. https://www.isdp.eu/bangladesh-foreign-policy-approach-should-shift-to-multi-alignment/.

National Academies of Sciences, Engineering, and Medicine, Health and Medicine Division, Board on Population Health and Public Health Practice, & Committee on the Long-Term Health and Economic Effects of Antimicrobial Resistance in the United Stateso, 2021. Combating antimicrobial resistance and protecting the miracle of modern medicine. In G. H. Palmer & G. J. Buckley (Eds.), The Health and Economic Burden of Resistance (Chapter 3). National Academies Press (US). https://www.ncbi.nlm.nih.gov/books/NBK577288/.

Nefros, C., Kitsara, G., Loupasakis, C., 2022. Geographical information systems and remote sensing techniques to reduce the impact of natural disasters in smart cities. IFAC-PapersOnLine 55 (11), 72–77. https://doi.org/10.1016/j.ifacol.2022.08.051.

Nel, L.H., Taylor, L.H., Balaram, D., Doyle, K.A.S., 2017. Global partnerships are critical to advance the control of neglected zoonotic diseases: the case of the global alliance for rabies control. Acta Tropica 165, 274–279. https://doi.org/10.1016/j.actatropica.2015.10.014.

Nsabiyeze, A., Ma, R., Li, J., Luo, H., Zhao, Q., Tomka, J., et al., 2024. Tackling climate change in agriculture: a global evaluation of the effectiveness of carbon emission reduction policies. J. Clean Prod. 468, 142973. https://doi.org/10.1016/j.jclepro.2024.142973.

Nzietchueng, S., Kitua, A., Nyatanyi, T., Rwego, I.B., 2023. Facilitating implementation of the one health approach: a definition of a one health intervention. One Health 16, 100491. https://doi.org/10.1016/j.onehlt.2023.100491.

Obiora, C.J., 2014. Agricultural transformation agenda in Nigeria: how prepared is the technology transfer-sub system? J. Biol. Agric. Healthc 4 (2), 1–6. Retrieved from. http://www.iiste.org.

Odoh, C., Okwuchukwu, Onyeanu, E.O., 2023. Exchange rate fluctuations and inflation on economic development in Nigeria: the economic heart of a nation. Univ Niger. https://doi.org/10.5110/77.1082.

O'Mara-Eves, A., Brunton, G., Oliver, S., et al., 2015. The effectiveness of community engagement in public health interventions for disadvantaged groups: a meta-analysis. BMC Public Health 15, 129. https://doi.org/10.1186/s12889-015-1352-y.

Orr, A., Weltzien, E., Rattunde, F., 2022. Research and development for sorghum and millets in Sub-Saharan Africa: what have we learned? Outlook Agric. 51 (1), 1–13. https://doi.org/10.1177/00307270221133127.

Owino, V., Kumwenda, C., Ekesa, B., Parker, M.E., Ewoldt, L., Roos, N., et al., 2022. The impact of climate change on food systems, diet quality, nutrition, and health outcomes: a narrative review. Front Clim. 4. https://doi.org/10.3389/fclim.2022.941842. Article 941842.

Oyediran, I.O., 2023. Chapter three - Integrated pest management in Africa: the necessary foundation for insect resistance management. In: Onstad, D.W., Knolhoff, L.M. (Eds.), Insect Resistance Management, 3rd ed. Academic Press, pp. 61–93. https://doi.org/10.1016/B978-0-12-823787-8.00017-9.

Peshin, R., Hansra, B.S., Singh, K., Nanda, R., Sharma, R., Yangsdon, S., et al., 2021. Long-term impact of Bt cotton: an empirical evidence from North India. J. Clean Prod. 312, 127575. https://doi.org/10.1016/j.jclepro.2021.127575.

Pfenning-Butterworth, A., Buckley, L.B., Drake, J.M., Farner, J.E., Farrell, M.J., Gehman, A.-L.M., et al., 2024. Interconnecting global threats: climate change, biodiversity loss, and infectious diseases. Lancet Planet Health 8 (4), e270–e283. https://doi.org/10.1016/S2542-5196(24)00021-4.

Pinho, J.O., Plácido, A.I., Monteiro, A., Nogueira, R., Oliveira, P.A., Coelho, A.C., et al., 2025. Evaluation of guidelines on antimicrobials use in food-producing animals: a systematic review. One Health Outlook 7 (1), 39. https://doi.org/10.1186/s42522-025-00160-w.

Pitt, S.J., Gunn, A., 2024. The One Health concept. Br. J. Biomed. Sci. 81, 12366. https://doi.org/10.3389/bjbs.2024.12366.

Pungartnik, P.C., Abreu, A., Brito dos Santos, C.V., Cavalcante, J.R., Faerstein, E., Werneck, G.L., 2023. The interfaces between One Health and global health: a scoping review. One Health 16, 100573. https://doi.org/10.1016/j.onehlt.2023.100573.

Prata, J.C., Ribeiro, A.I., Rocha-Santos, T., 2022. An introduction to the concept of One Health. One Health 1–31. https://doi.org/10.1016/b978-0-12-822794-7.00004-6.

Rahman, M.T., Sobur, M.A., Islam, M.S., Ievy, S., Hossain, M.J., El Zowalaty, M.E., et al., 2020. Zoonotic diseases: etiology, impact, and control. Microorganisms 8 (9), 1405. https://doi.org/10.3390/microorganisms8091405.

Rodrigues, R.C., Pereira, H.S., Senra, R.L., Ribon, A. de O.B., Mendes, T.A.D.O., 2023. Understanding the emerging potential of synthetic biology for food science: achievements, applications and safety considerations. Food Chem. Adv. 3, 100476. https://doi.org/10.1016/j.focha.2023.100476.

Roy, S., Malla, S., Dekari, D., Choudhury, T.G., 2025. One Health approach: human, environment, and animal health. In: Mallik, S.K., Shahi, N., Pandey, P.K. (Eds.), Management of Fish Diseases. Springer. https://doi.org/10.1007/978-981-96-0270-4_12.

Rupasinghe, R., Chomel, B.B., MartÍnez-López, B., 2022. Climate change and zoonoses: a review of the current status, knowledge gaps, and future trends. Acta. Tropica. 226, 106225. https://doi.org/10.1016/j.actatropica.2021.106225.

Salama, P., Moloney, G., Bilukha, O.O., Talley, L., Maxwell, D., Hailey, P., et al., 2012. Famine in somalia: evidence for a declaration. Glob Food Security 1 (1), 13–19. https://doi.org/10.1016/j.gfs.2012.08.002.

Sekaran, U., Lai, L., Ussiri, D.A.N., Kumar, S., Clay, S., 2021. Role of integrated crop-livestock systems in improving agriculture production and addressing food security – A review. J. Agric. Food Res. 5, 100190. https://doi.org/10.1016/j.jafr.2021.100190. (Article).

Sharan, M., Vijay, D., Yadav, J.P., Bedi, J.S., Dhaka, P., 2023. Surveillance and response strategies for zoonotic diseases: a comprehensive review. Sci One Health 2, 100050. https://doi.org/10.1016/j.soh.2023.100050.

Shrivastava, A., 2019. National programme on climate change and human health-India. J. Communicable Dis. 52 (03), 43–48. https://doi.org/10.24321/0019.5138.202029.

Shrock, E., Güell, M., 2017. CRISPR in animals and animal models. In: Torres-Ruiz, R., Rodriguez-Perales, S. (Eds.), Progress in Molecular Biology and Translational Science 152. Academic Press, pp. 95–114. https://doi.org/10.1016/bs.pmbts.2017.07.010.

Shukla, S., Chaudhary, K., Aahana, Phutela, S., Bhutani, R., Shukla, S.K., 2025. Smart crop varieties and precision agriculture: a way ahead for climate-resilient sustainable agriculture. In: Kumar, A., Babalola, O.O., Panwar, J.S., Santoyo, G. (Eds.), Climate Change and Agricultural Ecosystems, 2nd ed. Woodhead Publishing, pp. 435–466. https://doi.org/10.1016/B978-0-443-26520-4.00027-5.

Spiller, D., Franceschini, G., Henry, M., Cinardi, G., Falcucci, A., Wisser, D., et al., 2023. A case study in the Lao people's democratic republic. An Analysis of the Effects of Climate Change on Livestock. FAO, Rome. https://doi.org/10.4060/cc7320en.

University of Oxford, 2024. *Antibiotic Resistance Could Cause 39 Million Deaths Between Now and 2050*. Nuffield Department of Medicine. https://www.ox.ac.uk/news/2024-09-17-antibiotic-resistance-could-cause-39-million-deaths-between-now-and-2050.

U.S. Food and Drug Administration, 2024. *Federal/State Integration and the FSMA. U.S. Department of Health and Human Services*. https://www.fda.gov/food/guidance-regulation-food-and-dietary-supplements/food-safety-modernization-act-fsma/federalstate-integration-and-fsma.

Tang, K.W., Millar, B.C., Moore, J.E., 2023. Antimicrobial resistance (AMR). Br. J. Biomed. Sci. 80, 11387. https://doi.org/10.3389/bjbs.2023.11387.

Tinto, B., Quellec, J., Cêtre-Sossah, C., Dicko, A., Salinas, S., Simonin, Y., 2023. Rift valley fever in West Africa: a zoonotic disease with multiple socio-economic consequences. One Health 17, 100583. https://doi.org/10.1016/j.onehlt.2023.100583.

Tiseo, K., Huber, L., Gilbert, M., Robinson, T.P., Van Boeckel, T.P., 2020. Global trends in antimicrobial use in food animals from 2017 to 2030. Antibiotics 9 (12), 918. https://doi.org/10.3390/antibiotics9120918.

United Nations, 2022. *UN Report: Global Hunger Numbers Rose to as many as 828 Million in 2021*. World Health Organization. https://www.who.int/news/item/06–07-2022-un-report-global-hunger-numbers-rose-to-as-many-as-828-million-in-2021.

United Nations, 2023. *Goal 2: Zero Hunger*. The Sustainable Development Goals Report 2023. United Nations. https://www.un.org/sustainabledevelopment/hunger/.

United Nations Environment Programme, 2013. *UN Report: One-Third of World's Food Wasted Annually, at Great Economic, Environmental Cost*. United Nations. https://www.un.org/News/News/FullStory.asp?newsID=45525.

Van, T., Tan, D.K.Y., Cook, B.R., Liu, D.L., Cross, R., Tran, T.A., et al., 2024. Smallholder farmers' challenges and opportunities: implications for agricultural production, environment and food security. J. Environ. Manag. 370. https://doi.org/10.1016/j.jenvman.2024.122536. Article 122536.

Vander Sluijs, J.P., Vaage, N.S., 2016. Pollinators and global food security: the need for holistic global stewardship. Food Ethics 1 (1), 75–91. https://doi.org/10.1007/s41055-016-0003-z.

Viana, C.M., Freire, D., Abrantes, P., Rocha, J., Pereira, P., 2022. Agricultural land systems importance for supporting food security and sustainable development goals: A systematic review. Sci. Total Environ. 806, 150718. https://doi.org/10.1016/j.scitotenv.2021.150718.

Virtanen, P.K., 2019. Ancestors' times and protection of Amazonian Indigenous biocultural heritage. AlterNative: An Int. J. Indigenous Peoples 15 (4), 330–339. https://doi.org/10.1177/1177180119893132.

Wang, Y., Wu, L., Zhao, J., Yu, Y., Huang, K., 2025. Navigating the sustainability challenges of food systems: a review of life cycle-based environmental footprints and future directions. J. Environ. Manag. 392, 126794. https://doi.org/10.1016/j.jenvman.2025.126794.

WFP (World Food Programme), 2021. *PepsiCo Partners with the UN World Food Programme in Iraq to Support Farmers Affected by Conflict, Displacement, and Climate Change*. United Nations World Food Programme. https://www.wfp.org/news/pepsico-partners-un-world-food-programme-iraq-support-farmers-affected-conflict-displacement-and-climate-change.

WFP, 2025. *WFP 2026 Global Outlook: Hunger and Hope: Innovative Solutions to Address Food Insecurity*. https://doi.org/10.71958/wfp131186.

World Bank, 2021. *Moving Towards Sustainability: The Livestock Sector and the World Bank*. The World Bank. https://www.worldbank.org.

World Bank, 2022. *New Fund for Pandemic Prevention, Preparedness And Response Formally Established*. https://www.worldbank.org/pandemicfund.

World Bank, 2024. *Poverty, Prosperity, and Planet Report 2024: Pathways out of the Polycrisis*. The World Bank Group. https://www.worldbank.org.

World Health Organization, 2023. *One Health. World Health Organization*. Retrieved December 14, 2024, from https://www.who.int/news-room/fact-sheets/detail/one-health.

WHO Regional Office for Europe, 2022. A Health Perspective on the Role of the Environment in One HealthWHO Regional Office for Europe, Copenhagen. https://creativecommons.org/licenses/by-nc-sa/3.0/igo/.

World Organisation for Animal Health, 2024. *Strengthening of Veterinary Services in Africa*. Retrieved from https://www.oie.int/en/strengthening-veterinary-services-africa.

Yang, L., Xu, H., Pan, S., Chen, W., Zeng, J., 2024. Identifying the impact of global human activities expansion on natural habitats. J. Clean Prod. 434, 140247. https://doi.org/10.1016/j.jclepro.2023.140247.

Yang, S., Zhao, W., Liu, Y., Cherubini, F., Fu, B., Pereira, P., 2020. Prioritizing sustainable development goals and linking them to ecosystem services: a global expert's knowledge evaluation. Geogr. Sustain. 1 (4), 321–330. https://doi.org/10.1016/j.geosus.2020.09.004.

Zhou, Y., Zhang, A., Klinken, van, Wang, J., 2024. The effect of information provision on consumers' risk perceptions of, support for a ban, and behavioral intention towards the preventive use of antibiotics in food animals. BMC Public Health 24 (1). https://doi.org/10.1186/s12889-024-18859-2.

WFP, 2024. Annual country reports 2024. https://www.wfp.org/publications/annual-country-reports-2024.

WHO, FAO, WOAH, UNEP, 2022. One Health joint plan of action (2022–2026): Working together for the health of humans, animals, plants and the environment*. United Nations Environment Programme. https://wedocs.unep.org/handle/20.500.11822/40843.

CHAPTER THREE

Artificial intelligence and machine learning in food safety: Challenges, innovations, and future perspectives

Mohammed Raihanatu Hamid[a], Shuaibu Saidu Musa[b,c], Molly Unoh Ogbodum[d], Mohamed Mustaf Ahmed[e], Zhinya Kawa Othman[f], Usman Abubakar Haruna[g], Adamu Muhammad Ibrahim[h,*], Muhammad Zainab Lawal[b], Isaac Olushola Ogunkola[i], Abdullateef Abdulsalam[j], Muhammad Kabir Musa[k], Ahmed Adamu Gololo[l], Muktar Musa Shallangwa[m], Abdulmajid Abdullahi Aliyu[n], Ejay Imperial[o,p], and Don Eliseo Lucero-Prisno III[q,r,s]

[a]Department of mathematics and Computer Science, College of Education, Billiri, Nigeria
[b]School of Global Health, Faculty of Medicine, Chulalongkorn University, Bangkok, Thailand
[c]Department of Nursing Science, Ahmadu Bello University, Zaria, Nigeria
[d]Department of Public Health, University of Calabar, Calabar, Nigeria
[e]Faculty of Medicine and Health Sciences, SIMAD University, Mogadishu, Somalia
[f]Departmen of Pharmacy, Kurdistan Technical Institute, Sulaimani, Kurdistan Region, Iraq
[g]Department of Biomedical Sciences, Nazarbayev University School of Medicine, Astana, Kazakhstan
[h]Department of Immunology, School of Medical Laboratory Science, Usmanu Danfodiyo University, Sokoto, Nigeria
[i]Nuffield Department of Population Health, University of Oxford, Oxford, United Kingdom
[j]Department of Biomedical Sciences, Nazarbayev University School of Medicine, Astana, Kazakhstan
[k]University of Michigan, Ann Arbor, United States
[l]Faculty of Pharmaceutical Sciences, Chulalongkorn University, Bangkok, Thailand
[m]Achieving Health Nigeria Initiative (AHNI), Biu, Nigeria
[n]Department of Microbiology and Biotechnology, Federal University Dutse, Dutse, Nigeria
[o]Advanced Technology Center in the Philippines (ATCP), Tech Architecture Capability, Accenture, Taguig City, Philippines
[p]Research Unit, Imperialle AI Tech OU, Tallinn, Estonia
[q]Department of Global Health and Development, Faculty of Public Health and Policy, London School of Hygiene and Tropical Medicine, London, United Kingdom
[r]Research Office, Palompon Institute of Technology, Palompon, Leyte, Philippines
[s]Office for Research, Extension and Innovations, Bukidnon State University, Malaybalay City, Bukidnon, Philippines
*Corresponding author: e-mail address: amuhammadibrahim37@gmail.com

Contents

Advances in Food Security and Sustainability, Volume 11
ISSN 2452-2635, https://doi.org/10.1016/bs.af2s.2026.06.004

Abstract

The integration of artificial intelligence (AI) and machine learning (ML) into food safety practices has revolutionized the food industry by enhancing the detection, prevention, and management of food safety hazards. In this article, we reviewed the application of AI and ML in monitoring and predicting food safety issues, highlighting technological advancements and challenges in the field. AI-driven tools enable the real-time monitoring of food production processes, early detection of contamination, and improved precision in food preparation, thus reducing human errors and contamination risks. ML algorithms analyze large volumes of data to identify potential food safety threats and develop predictive models for proactive risk management. Several applications of AI and ML have been explored, including real-time monitoring of food manufacturing processes, which allows for the early detection of quality issues, and AI-powered robots, which enhance precision and uniformity in food handling. These technologies improve supply chain traceability and transparency, facilitating faster recall and accountability. Despite their promising potential, the implementation of AI and ML for food safety faces significant challenges. These include data quality and access issues, resource constraints, and the need for customized solutions tailored to specific agricultural and food supply chain contexts. To address these challenges, we recommend fostering collaboration between industry stakeholders, policymakers, and technologists by focusing on the importance of developing robust AI and ML algorithms tailored to the food industry, enhancing data infrastructure, and ensuring ethical considerations in technology deployment. The future of AI and ML in food safety appears promising with the potential to significantly improve public health outcomes, economic stability, and consumer trust in the food sector.

Abbreviations

AI	Artificial Intelligence.
ANNs	Artificial Neural Networks.
BDA	Big Data Analytics.
CNN	Convolutional Neural Network.
FTIR	Fourier Transform Infrared.
HCA	Hierarchical Cluster Analysis (HCA).
ML	Machine Learning.
NAFI	Nanozyme-activated Food Immunosensors.
PCA	Principal Components Analysis.
PLS	Partial Least Squares regression analysis.
RF	Random Forest.
SVM	Support Vector Machine.

1. Introduction

Food safety is the practice of handling, preparing, and storing food in a manner that minimizes the possibility of contracting a food-borne illness. Food safety has become a top priority in the modern world because of the worldwide nature of food production, distribution, and consumption (Aljasir, 2023). For centuries, numerous conventional ways of guaranteeing food safety have been in use, relying mostly on the basic concepts of storage, management, and cleanliness. These procedures often include canning, salting, and drying, which have been used to increase the shelf life of food and avoid spoiling (Amit et al., 2017). Traditional cooking and food preparation practices such as boiling, roasting, and fermentation have eradicated unwanted germs and parasites from food. Furthermore, the establishment of local food markets with stringent quality standards and watchful merchants has long been a customary method of ensuring food safety through community regulations. While these approaches have been effective for millennia, the changing agricultural scene and greater globalization have increased the complexity of supply chains and the risk of food-borne diseases.

In order to maintain strict food safety standards, it is essential to use advanced technology and practices to address new challenges (Amit et al., 2017). Sustaining the safety of food supplies is critical for public health, economic stability, international trade, protecting customers' well-being, and retaining consumer faith in the food sector (Food safety Internet, 2026). An estimated 600 million people suffer from foodborne illnesses each year worldwide, resulting in approximately 420,000 deaths and a loss of about 33 million disability-adjusted life years (DALYs) (Food safety Internet, 2026). The DALY is a critical metric used to measure the total burden of disease by combining years of life lost (YLL) due to premature mortality and years lived with disability (YLD) (Leading causes of DALYs Internet, 2026).

Artificial intelligence (AI) is a branch of computer science that mimics human thinking processes, learning ability, and knowledge storage (Kudashkina et al., 2022). The agriculture, food, and processing industries are among the sectors that make use of AI methods. AI's inviting performance has made it a highly appealing tool to use in industries, particularly for decision making and process estimation, with the objective of reducing overall costs, increasing efficiency, and increasing profitability (Kudashkina et al., 2022). AI is revolutionizing food safety by enabling predictive analytics, real-time monitoring, and rapid pathogen detection across the supply chain from farm to fork These tools address key challenges like contamination risks and outbreaks, particularly relevant for public health surveillance in resource-limited settings (Onyeaka et al., 2024). Machine learning (ML), as applied to food safety, is a branch of AI that studies and analyzes large volumes of data using algorithms to identify underlying patterns and trends that may indicate potential food contamination (Wang et al., 2022). Furthermore, ML methods facilitate the construction of predictive frameworks designed to identify and mitigate safety hazards before they manifest in the supply chain. This technological integration is becoming increasingly vital as the global population continues to expand, with food demand projected to surge by 50 % to 60 % over 2019 levels by the year 2050 (Falcon et al., 2026).

To address the substantial need to increase food output, AI-driven solutions are being implemented across various sectors, including supply chain logistics, automated food sorting, production scaling, quality enhancement, and the maintenance of rigorous industrial hygiene standards. ML models like random forests, neural networks, and deep learning analyze microbial genomics and environmental data to predict contamination risks and resistance patterns in pathogens. Supervised learning detects anomalies in food quality via imaging, while unsupervised methods identify emerging outbreaks from surveillance data. Reinforcement learning optimizes supply chain decisions to minimize spoilage (Onyeaka et al., 2024). The integration of ML and AI more generally has the potential for improving food safety. Real-time monitoring of food manufacturing processes enabled by ML and AI technology allows for early detection of contamination or quality concerns (Qian et al., 2023). These tools can scan large databases to discover patterns and trends, allowing authorities to forecast food-borne illness outbreaks and respond quickly. AI-powered robots and automation can improve precision and uniformity in food preparation, thereby lowering the risk of human mistakes and contamination. ML and AI can increase supply chain traceability and transparency, allowing for faster recall and greater accountability. Additionally, in an ever-changing food scene, these technologies have the potential to transform risk assessments, making food safety measures more proactive and successful (Wang et al., 2022). Hence, AI and ML can be used to ensure food safety by developing forecasting algorithms and automated agriculture and farming systems that can increase crop yields, minimize waste, and adapt to changing environmental circumstances for more sustainable and resilient food safety (Qian et al., 2023).

The key issue for AI and ML in guaranteeing food safety is related to data quality and access, shortage of resources, and the requirement for tailored solutions in various agricultural and food supply chain scenarios. In comparison with other food-related fields such as food security and marketing, the application of AI in food safety remains relatively limited due to a combination of technical, regulatory, and stakeholder-related constraints. A major technical barrier is the scarcity of high-quality, standardized data necessary for developing and implementing AI systems (Qian et al., 2023). Although food safety data may be shared, they are often derived from multiple sources with differing formats and structures, making harmonization, standardization, and validation challenging (Qian et al., 2023). Critical details such as testing methodologies and sample sizes are frequently inconsistent or unavailable. Moreover, most AI approaches in food safety are designed for specific products, microorganisms, or environmental conditions, thereby limiting generalizability and requiring multiple systems to achieve comprehensive benefits.

Regulatory and governance challenges further hinder progress. The absence of a clearly defined legal and regulatory framework for AI applications, along with concerns regarding the protection and security of sensitive data, exacerbates uncertainty and slows adoption. In addition, stakeholder concerns play a significant role. Some food industry actors fear that AI technologies, if widely accessible, could be used to predict or expose a firm's food safety risk profile, potentially undermining commercial interests. As a result, hesitancy in data sharing and technology implementation persists, collectively contributing to the slower integration of AI in food safety compared to other food-related domains (Qian et al., 2023).

This chapter proceeds as follows: first, we examine the applications of ML in food safety, and go on to discuss the use of more general AI techniques and Big Data Analytics. We then discuss challenges and emerging trends. Finally, we offer conclusions and recommendations.

2. Applications of machine learning in food safety

AI has become a cornerstone of the modern food sector, serving diverse functions including modeling, prediction, control tools, food drying, sensory assessment, quality control, and tackling complicated challenges in food processing. Beyond technical execution, AI enhances corporate strategy by providing high-accuracy forecasts for sales and yield increases (Kudashkina et al., 2022). In particular, ML can be used to solve challenging tasks involving a large amount of data, as well as variables that do not already have existing equations or formulas, as they adapt dynamically to complex, shifting parameters. In such circumstances, ML has been used to predict and classify food samples (Wang et al., 2022).

2.1 Systemic processing and strategic optimization

Research by Kyaw et al. (Kyaw et al., 2024). underscores the potential of ML in food safety by proposing automated, technology-aided processing lines. For instance, a conceptual design for automated liquid food processing specifically for milk and beverages targets enhanced microbial safety through integrated operational layers.

2.2 Predictive modeling for pathogen control and spoilage

While global progress in reducing food-borne infections has remained stagnant evidenced by the United States' difficulty in meeting "Healthy People" public health goals (Farber et al., 2021), ML offers a proactive alternative to traditional monitoring. A primary example is "MeatReg," a web-based application developed by Estelles-Lopez et al. (Estelles-Lopez et al., 2017), that utilizes seven different regression approaches to predict microorganism counts in minced beef. By providing specific recommendations for analytical platforms, such models transition food safety from reactive testing to predictive prevention.

2.3 Sensor-based real-time monitoring and IoT

Building on these predictive capabilities, the integration of the Internet of Things (IoT) allows for real-time environmental oversight. Varun et al (Sonwani et al., 2026), demonstrated a system using NodeMCU controllers and the ThingSpeak cloud platform to monitor air temperature and humidity in real time. These data, processed via random forest regressors, enable the precise control of moisture in vegetables and maintain the purity of dairy products.

Further innovations focus on hardware-software synergy to overcome the subjectivity of traditional freshness checks:

* Integrated Microcontrollers: Bhuiyan et al. (Bhuiyan et al., 2023). utilized Arduino and ESP32-CAM systems combined with gas sensors (MQ135 and MQ4) to extract pertinent freshness data from meat samples.
* Electronic Noses (e-noses): Similarly, Binson et al. (Binson and Thomas, 2023). developed a portable e-nose using eight metal-oxide-semiconductor sensors. When paired with a Support Vector Machine (SVM) algorithm, the device achieved over 90 % accuracy in detecting complex smells associated with beef spoilage and adulteration.

3. Synthesis of findings

The evidence from these various studies suggests that the strength of AI and ML lies in their multimodal application. While regression models like "MeatReg" provide the analytical "brain" for spoilage prediction, IoT sensors and e-noses act as the "nervous system," providing the real-time data necessary for those models to function. Collectively, these technologies mitigate human error and bypass the limitations of data scarcity, offering a high-precision, automated framework for the next generation of food safety management.

Table 1 summarizes case studies of successful applications of ML to food safety across various components of the food industry.

4. Artificial intelligence techniques in food safety

Beyond the real and potential applications of ML to food safety, AI offers additional ways to combine physical processing precision with digital

Table 1 Case studies showcasing successful ML applications in food safety across various industries.

Food type	Machine learning technique	Results
Pineapple	Artificial Neural Networks (ANNs) learn to recognize hidden patterns in food data by imitating how human neurons connect, K-Nearest Neighbours (KNN) works by labelling a new item based on whichever known group it looks most like. Support Vector Machine (SVM) is a divider that separates good and bad samples, and Random Forest (RF) works by selecting from samples of different decision-making trees to reach a more reliable and stable final result (Hasan et al., 2020)	The KNN approach produced the greatest results when it came to figuring out how sweet the pineapple was, with an accuracy of 0.820. The created model will be helpful, particularly if there is a huge assortment of pineapples.
Apple	Convolutional Neural Networks (CNN) and (SVM) (Fan et al., 2020).	To separate the good apples from the bad ones
Olive Oil	SVM (Xu et al., 2020).	SVM was successful in identifying the contaminated olive oils by fusion of near infrared and Raman spectroscopy and the SVM model was able to forecast the level of adulteration in olive oil with accuracy.
Rice	RF and Principal Components Analysis (PCA) work by shrinking the massive datasets down to the most important to spot patterns, with Linear Discriminant Analysis (LDA) distinguishing between different groups to	With good performance, all machine learning approaches were able to recognize and categorize the rice based on its composition (glycemic index, amylose-based) and the intensity of the hydrothermal treatment.

	make them as distinct as possible, and Partial Least Squares regression analysis (PLS) connecting raw sensor data like smells directly to a specific predicted value like age (Rizwana and Hazarika, 2020).	
Beef	SVM (Estelles-Lopez et al., 2017).	Two parts of the e-nose model are used to assess the freshness of different beef samples. Of these, the SVM technique yields higher results in terms of accuracy, sensitivity, and specificity—91.5 %, 91.08 %, and 90.90 %, respectively.

surveillance and targeted agricultural automation. These offer the food industry crucial tools to reduce manual errors and respond more rapidly to emerging food safety threats.

4.1 Advanced food processing and quality control

AI directly and indirectly draws the interest of organizations from a wide range of disciplines and sectors. For effective production and management of food, a considerable number of innovative and new techniques are needed for post-harvest food processing. AI-driven food processing enhances the food-creation model through automated inspection and learning techniques. In addition to digital data analysis, it precisely substitutes for most human actions. Utilizing dependable AI models permit intricate evaluation and administration of food quality and quantity. Numerous examples and connections exist between AI and the processing of dairy products (detecting adulteration, standardizing, and keeping an eye on cleaning in place systems), cereals (grading and sorting), meat and poultry products (analyzing the quality of carcasses and eggs), beverages (simulating the mechanical breakdown of produce while monitoring antioxidant levels (polyphenols), filtration flow rates (permeate flux), and liquid thickness (viscosity) and brand discrimination to ensure consistent beverage quality), fruits and vegetables (estimating physical properties, internal defects, total soluble solids, pH, and texture quality of processed fruits and vegetables), tea (evaluating quality using E-nose), baked goods, and confections (surface color changes for chocolate) (Tiwari et al., 2023). Deep learning approaches also reduce the risk of dangerous goods, minimize manual inspections, improve accuracy and efficiency, and automate food safety checks by identifying pollutants and quality concerns.

4.2 Digital surveillance and public opinion monitoring

A study conducted by Goldberg et al. (Goldberg et al., 2022) found that internet media were swiftly searched for references to potential threats to food safety using text mining and ML. In addition to evaluating performance on more than four million online reviews, this research found distinctive terms and phrases in user posts and compared them to conventional sentiment-based text mining. This study offers useful and affordable methods to support real-time food safety monitoring. To systematically analyze real-time public opinion and regional disparities in food safety, Fan et al. (Fan et al., 2020) created the IFoodCloud platform in China, which gathers data from more than 3100 public sources. These data can be used to investigate trends in public opinion, public sentiments, and regional attention differences in food safety incidents. For example, it gathers opinions on food adulteration, food-borne diseases, agricultural pollution, irregular food distribution, and food production issues.

4.3 Agricultural automation and workforce dynamics

The strategy suggested by Kate et al. (Kate et al., 2026) to government agencies was to apply cutting-edge text mining and ML techniques that autonomously identify and analyze user posts that discuss food safety offenses in online forums, such as blogs, user forums, and citizen journalism websites. Reinforcement learning can be used to optimize food safety regulations and reaction mechanisms by creating adaptive systems that continually learn and improve in real time. By streamlining response plans and improving the early identification of possible safety risks in the food supply chain, these technologies help lessen the effects of food-borne illness outbreaks (Kate et al., 2026). Prabhukhot analyzes regulatory border actions related to food safety, using data from the US Food and Drug Administration (FDA) and the EU Rapid Alert System for Food and Feed (RASFF) between 2008 and 2025. It finds significant differences in enforcement priorities: the EU focuses on mycotoxins in nuts and seeds, whereas the FDA emphasizes visual violations without lab analysis. Additionally, geopolitical patterns emerge, with localized enforcement affecting trading partners like Mexico and Turkey, and contrasting risk profiles for China. Indian exporters face varied compliance demands from the US and EU, highlighting the existence of divergent regulatory enforcement profiles that create parallel compliance requirements for exporting countries (Prabhukhot, 2026).

Because the food and agricultural industry is dynamic, AI and robots cannot be generalized to solve all challenges; instead, they can only address specific, complicated issues. Numerous solutions have been created globally to address various issues and to offer improved strategies. The lengthy history of labor-intensive work in the agricultural sector may have an impact on farmers' attitudes toward automation (Sparrow et al., 2021). A few early trials with drone delivery of seed-related supplies and automated irrigation revealed that these are highly mechanized and that farmers will continue to choose human management over machines in uncertain conditions to maintain control. The ways in which industrialized and developing nations use machinery and technology for agricultural production differ. In emerging nations, production per unit area is lower than in industrialized ones because of technical diversity, implying that a developing nation needs many more laborers than a developed nation does to generate a unit of agricultural production (Sparrow et al., 2021).

4.4 Synthesis of findings

The integration of these AI techniques reveals a shift toward proactive, data-driven safety management. While technical diversity and workforce concerns remain, these technologies offer a comprehensive framework that can contribute to stabilizing the global food supply chain.

5. Big data analytics and food safety

Big data refers to large amounts of information collected from multiple sources. It is often characterized by three main features: high volume, high velocity, and high variety. In the food and agriculture sector, data can come from farms, processing plants, laboratories, transport systems, and even retail outlets. When analyzed properly, it can support the use of AI and ML in managing food systems. At the same time, technology does not work in isolation. Yu et al. (Kate et al., 2026) note that human involvement remains important in applying AI to food safety. Professional judgment, experience, and collaboration help ensure that the results produced by AI systems are understood and used appropriately. ML has already been applied in different stages of the agricultural supply chain, including preproduction, manufacturing, processing, and distribution. When combined with big data in a supply chain performance assessment system, it can help monitor operations and support steady and sustainable industry performance (Zhang et al., 2023; Yu et al., 2022).

Incorporation of automation using AI and Big Data Analytics (BDA) will result in many positive outcomes, including: less time spent, less chance for mistakes, improved cleanliness, more secured food, lower costs, better decision-making, more efficient product sorting, enhanced inventory management, waste reduction, toxicity prediction, better supply chain and logistics, improved product development, and prompter food delivery. Small spectral cameras are imaging technology that capture large information beyond the use of the naked eye. These cameras are becoming increasingly common in the food industry for evaluating food authenticity and quality, and efforts are being made to provide consumers with this feature via their cellphones. Big data can also be applied by a variety of businesses and farmers to analyze and extract valuable information that will increase their productivity and pave the way for smart farming (Yu et al., 2022).

For big data problems to be successfully processed and analyzed in a timely manner, a sizable computer infrastructure is required. Research on food safety is now made possible by the implementation of supercomputing centers, high-performance computing infrastructure, and cloud computers. Big data and cloud computing technology research in food safety is still in its early stages, although other research disciplines have accepted a similar infrastructure as a solution. Ultimately, numerous findings have demonstrated the technology's enormous potential, and there are reports of its successful use in anticipating, tracking, and managing food safety issues (Sharma et al., 2021). Big data is mainly applied in observing concerns governing global problems, such as food security, safety, feasibility, and progress in efficiency. These uses significantly broaden the application of big data, creating an integrated data ecosystem that spans the entire food supply chain 20.

Hassoun et al. (Hassoun et al., 2023) highlighted the economic benefits of predicting milk production yields through BDA, particularly its importance in future dairy industry decision-making due to population growth. The study reveals that modern technologies in milk procurement, product composition, packaging, supply chain integration, and traceability have improved data implementation and reduced time and cost. A system has been established to predict the shelf life of processed cheese through linear layer design for specifying input and output dimensions and multiple regression models. The latter are used to control the date of expiry of the processed cheese, taking into account various parameters such as pH, percentage of soluble nitrogen, and numerous counts of bacteria, yeasts, and molds (Al-Salman and Haddad, 2026). The project 'Dairy Brain' focuses on utilizing big data for decision-making on dairy farms for the nutritional fulfillment of lactating cows and predicting clinical mastitis through decision-support tools. Further, it facilitates improved service quality, enhanced traceability, and reduced costs to attain optimum profits (Cabrera et al., 2020).

There are a number of additional examples of the use of BDA in food safety. Intelligent packaging technologies detect seafood spoilage using techniques such as hierarchical cluster analysis (HCA), a statistical procedure used to merge similar materials into clusters, as well as principal components analysis (PCA), a statistical technique that reduces the dimensionality of large data, and partial least squares (PLS) regression analysis, which measures the relationship between two variables. Microorganism growth and enzymatic activity cause spoilage of meat and its products, thereby posing health and economic threats. An adaptive fuzzy logic system and FTIR (Fourier Transform Infrared) spectroscopy with the fusion of principal microbiological analysis techniques were used to determine the various parameters responsible for beef spoilage (Kurćubić et al., 2025). A paper surveyed the grading system of various agricultural products using AI. This utilized a machine vision strategy to ensure food quality, detect defects, and ensure safety compliance throughout the supply chain from digital images and videos, and included separate hardware and software components for efficiency. This study highlights AI's potential to improve fresh food logistics, reduce waste, enhance transportation and distribution efficiency, and provide recommendations for analytical platforms and ML techniques. Up scaled automation in food processing uses food quality sensors to convert food properties into electrical signals, thereby enabling the determination of various quality parameters. These sensors, grouped online and offline, are assembled in a smart oven for baking, generating accurate product data, and improving performance (Yared, 2016).

Researchers have found that Artificial Neural Networks (ANNs) can balance various aspects of the food sector, including medicine, beverages, and agriculture. In the brewing industry, ANNs are used to evaluate product quality and consumer acceptance. RoboBEER, a computer-vision-enabled automatic pourer, was used to assess foam generation, color combinations, and consumer acceptability (Gonzalez Viejo et al., 2019). The e-nose system enables the identification of odor and gas composition in the beverage industry and enhances food preservation methods. It quantifies volatile gas mixtures, identifies chemical constituents, addresses bio-deterioration, and predicts bacterial types and culture growth using a multi-layer perception network to allow for precise classification of food safety risks (Binson and Thomas, 2023). Researchers have suggested employing Metal Oxide Semiconductor sensors and a variety of ANN algorithms to evaluate data and prevent data manipulation to assess the scent and flavor of tea and spices using temperature and humidity drift compensation techniques. These are mathematical algorithms that filter out "noise" caused by shifting ambient conditions. By adjusting the raw data to account for these variables, the system ensures that the resulting big data patterns reflect actual food quality such as chemical spoilage rather than mere changes in the storage environment (Sharma et al., 2021). Automation facilitates food processing lines, and tremendous quantities of data are gathered, archived, scrutinized for further enhancements in the food supply chain, and employed for risk evaluations. There are currently some significant limitations to this technology, however. It attracts hackers, adding another layer of risk to the sector. Also, current students in the food safety field tend to have low levels of computer science proficiency, exposure to the subject of food science, and acquaintance with cybersecurity concerns (Feye et al., 2020).

6. Challenges

Concerns about privacy are among the ethical ramifications of ML and AI more generally in food safety, as gathering and examining data pertaining to food may violate people's private information and consumption patterns. Unfair access to wholesome food might be caused by bias in the AI models employed to ensure food safety, which would especially harm underprivileged populations. Such bias occurs when AI training data reflect existing social or economic inequalities rather than objective biology. This shifts food safety from a universal right to a technical privilege where the most accurate protection is reserved for specific demographics or regions (Feye et al., 2020). To foster confidence among customers, regulators, and industry stakeholders, it is imperative to guarantee transparency and accountability of AI-powered food safety systems. Everyone concerned needs to know how choices are made and who is accountable for any problem that may occur.

Furthermore, the implementation of ML and AI in food safety is significantly hindered by fragmented regulations and a lack of standardised data collection practices. Without a universal framework for data handling, integrating diverse analytical tools across the global food supply chain remains difficult. Also, while several niche AI approaches could theoretically improve supply chain cleanliness and safety, their adoption is stalled by the absence of dependable decision-support instruments and clear legal guidelines for their application.

From a technical perspective, the physical deployment of AI-driven robotics faces harsh operational realities, particularly in remote agricultural areas. Robotic systems often struggle with the dexterity required for high-labor-intensity tasks and frequently require downtime for repairs and system resets. These reliability issues are compounded by infrastructure gaps; many rural farming regions lack the continuous, dependable electricity required to power autonomous systems, making widespread robotic adoption currently unfeasible for many farmers.

Finally, the transition toward AI-driven food systems raises significant social and economic concerns regarding the workforce. The automation of labor-intensive roles risks increasing unemployment and creating resistance toward technological adoption among traditional workers. Successfully executing an AI strategy requires not only significant financial investment but also the development of a highly competent team capable of managing complex digital systems. Consequently, the "human element" – including workforce displacement and the need for specialized technical training – remains one of the most critical challenges to the digital transformation of food safety and quality (Sharma et al., 2021).

7. Future perspectives and emerging trends

AI and robotics provide high efficiency and better production quality, which can potentially lead to improved crop productivity. These innovations would have the greatest impact on large, relatively low productivity sectors of the economy, such as agri-food (Paul, 2020). The agricultural sector, a massive consumer of freshwater, accounts for about 70% of global water use. Simultaneously, large-scale farm equipment does excessive compaction harm to soils, while several species of British wildlife are at risk of extinction due to current agricultural practices, including extensive use of herbicides and pesticides and the industrialization of machinery. These problems can be overcome by using AI and robotics-based solutions (Amit et al., 2017).

8. Conclusion and recommendations

Agriculture and the food industry must recognize the shift in people's wants and mindsets if AI and robotics-based technologies should have an impact on agriculture. Primarily, it is necessary to assemble a cohort of physical engineers who possess a sufficient understanding of biological principles and agriculture and, conversely, to cultivate a corresponding cohort of biologists who comprehend the opportunities presented by associated engineering and AI components. As a result, society may collaborate to forge links that will allow the next generation of engineers to facilitate the development of biological sciences, plant breeding, and soil health (Onyeaka et al., 2024).

To improve the results, governments should provide a systematic approach to creating a suitable ML and AI algorithm for the food business, which will motivate academics and industry participants to use tried-and-true technology. AI is crucial in food safety for modeling, prediction, control, food drying, sensory evaluation, quality control, and processing problems. It can improve business strategies, predict output and sales, and increase yields. However, the future depends on robust and equitable collaboration among industries, policymakers, technologists, and consumers.

Ethical approval

Not required.

Contribution of authors

Mohammed Raihanatu Hamid and Shuaibu Saidu Musa conceived the idea. Mohammed Raihanatu Hamid, Shuaibu Saidu Musa, Molly Unoh Ogbodum, Mohamed Mustaf Ahmed, Zhinya Kawa Othman, Usman Abubakar Haruna, Adamu Muhammad Ibrahim,

Muhammad Zainab Lawal, Isaac Olushola Ogunkola, Abdullateef Abdulsalam, Muhammad Kabir Musa, Ahmed Adamu Gololo, and Muktar Musa Shallangwa collected and analysed the data and information and rotated in writing different versions of the drafts with intellectual additions by Don Eliseo Lucero-Prisno III. All authors read and approved the final manuscript.

Funding

This research did not receive any specific grant from funding agencies in the public, commercial, or not-for-profit sectors.

Declaration of competing interest

None to declare.

References

Aljasir, S.F., 2023. Food safety knowledge and practices among food handlers and consumers in Gulf countries: an integrative review. Glob. Public. Health 18, 2287584. https://doi.org/10.1080/17441692.2023.2287584. Taylor & Francis.

Al-Salman, S.M., Haddad, M.A., 2026. Impact of gamma irradiation on the shelf-life enhancement of refrigerated and desalted soft white cheese. Int. Dairy. J. 175, 106523. https://doi.org/10.1016/j.idairyj.2025.106523.

Amit, S.K., Uddin, M.M., Rahman, R., Islam, S.M.R., Khan, M.S., 2017. A review on mechanisms and commercial aspects of food preservation and processing. Agric. Food Secur. BioMed. Cent 6, 51. https://doi.org/10.1186/s40066-017-0130-8.

Bhuiyan, Z.W., Redwanul Haider, S.A., Haque, A., Hasan, M., Uddin, M.R., 2023. Meat freshness classifier with machine and AI. 2023 IEEE Reg 10 Symp TENSYMP [Internet]. 2023 [cited 2026 May 10]. p. 1–5. https://doi.org/10.1109/TENSYMP55890.2023.10223681.

Binson, V.A., Thomas, S., 2023. The development of a mobile e-nose system for real-time beef quality monitoring and spoilage detection. Eng. Proc. Multidiscip. Digital Publ. Inst 56, 256. https://doi.org/10.3390/ASEC2023-15960.

Qian, C., Murphy, S.I., Orsi, R.h., Wiedmann, M., 2023. How can AI help improve food safety? PubMed [Internet]. [cited 2026 May 10]; https://pubmed.ncbi.nlm.nih.gov/36542755/. Accessed 10 May 2026.

Cabrera, V.E., Barrientos-Blanco, J.A., Delgado, H., Fadul-Pacheco, L., 2020. *Symposium review:* real-time continuous decision making using big data on dairy farms. J. Dairy. Sci. 103, 3856–3866. https://doi.org/10.3168/jds.2019-17145.

Estelles-Lopez, L., Ropodi, A., Pavlidis, D., Fotopoulou, J., Gkousari, C., Peyrodie, A., et al., 2017. An automated ranking platform for machine learning regression models for meat spoilage prediction using multi-spectral imaging and metabolic profiling. Food Res. Int. 99, 206–215. https://doi.org/10.1016/j.foodres.2017.05.013.

Falcon, W.P., Naylor, R.L., Shankar, N.D., 2026. Population and Development Review | Population Council Journal | Wiley Online Library. [cited 10 May 2026]; https://doi.org/10.1111/padr.12508.

Fan, S., Li, J., Zhang, Y., Tian, X., Wang, Q., He, X., et al., 2020. On line detection of defective apples using computer vision system combined with deep learning methods. J. Food Eng. 286, 110102. https://doi.org/10.1016/j.jfoodeng.2020.110102.

Farber, J.M., Zwietering, M., Wiedmann, M., Schaffner, D., Hedberg, C.W., Harrison, M.A., et al., 2021. Alternative approaches to the risk management of *Listeria monocytogenes* in low risk foods. Food Control 123, 107601. https://doi.org/10.1016/j.foodcont.2020.107601.

Feye, K.M., Lekkala, H., Lee-Bartlett, J.A., Thompson, D.R., Ricke, S.C., 2020. Survey analysis of computer science, food science, and cybersecurity skills and coursework of undergraduate and graduate students interested in food safety. J. Food Sci. Educ 19, 240–249. https://doi.org/10.1111/1541-4329.12200.

Food safety [Internet]. [cited 9 May 2026]. https://www.who.int/news-room/fact-sheets/detail/food-safety. Accessed 9 May 2026.

Goldberg, D.M., Khan, S., Zaman, N., Gruss, R.J., Abrahams, A.S., 2022. Text mining approaches for postmarket food safety surveillance using online media. Risk Anal 42, 1749–1768. https://doi.org/10.1111/risa.13651.

Gonzalez Viejo, C., Torrico, D.D., Dunshea, F.R., Fuentes, S., 2019. Development of artificial neural network models to assess beer acceptability based on sensory properties using a robotic pourer: a comparative model approach to achieve an artificial intelligence system. Beverages 5, 33. https://doi.org/10.3390/beverages5020033. Multidisciplinary Digital Publishing Institute.

Hasan, M., Sarno, R., Sabila, S., 2020. Optimizing machine learning parameters for classifying the sweetness of pineapple aroma using electronic nose. Int. J. Intell. Eng. Syst 13, 122–132.

Hassoun, A, Garcia-Garcia, G., Trollman, H., Jagtap, S., Parra-López, C., Cropotova, J., et al., 2023. Birth of dairy 4.0: opportunities and challenges in adoption of fourth industrial revolution technologies in the production of milk and its derivatives. Curr. Res. Food Sci. 7, 100535. https://doi.org/10.1016/j.crfs.2023.100535.

Kate, K., Negi, S., Kalagnanam, J., 2026. Monitoring food safety violation reports from internet forums. PubMed [Internet] Accessed 10 May. https://pubmed.ncbi.nlm.nih.gov/25160357/.

Kudashkina, K., Corradini, M.G., Thirunathan, P., Yada, R.Y., Fraser, E.D.G., 2022. Artificial intelligence technology in food safety: a behavioral approach. Trends Food Sci. Technol 123, 376–381. https://doi.org/10.1016/j.tifs.2022.03.021.

Kurćubić, V.S., Munjić, M.D., Dmitrić, M.P., Živković, S., Stajić, S.B., Tomasevic, I., 2025. Bacterial antimicrobial resistance in meat products—Current concepts. Foods 14, 2792. https://doi.org/10.3390/foods14162792. (Multidisciplinary Digital Publishing Institute).

Kyaw, K.S., Adegoke, S.C., Ajani, C.K., Nwabor, O.F., Onyeaka, H., 2024. Toward in-process technology-aided automation for enhanced microbial food safety and quality assurance in milk and beverages processing. Crit. Rev. Food Sci. Nutr 64, 1715–1735. https://doi.org/10.1080/10408398.2022.2118660. Taylor & Francis.

Leading causes of DALYs [Internet]. [cited 2026 May 9]. https://www.who.int/data/gho/data/themes/mortality-and-global-health-estimates/global-health-estimates-leading-causes-of-dalys. Accessed 9 May 2026.

Onyeaka, H., Akinsemolu, A., Miri, T., Nnaji, N.D., Emeka, C., Tamasiga, P., et al., 2024. Advancing food security: the role of machine learning in pathogen detection. Appl. Food Res. 4, 100532. https://doi.org/10.1016/j.afres.2024.100532.

Paul, P.K., 2020. Artificial intelligence and robotics in agriculture and allied areas - A study. Int. J. Bioinforma. Biol. Sci. [Internet]. https://doi.org/10.30954/2319-5169.01.2020.2. [cited 2026 May 10];8.

Prabhukhot, G., 2026. Frontiers | regulatory responses to foodborne illness outbreaks in the United States and their implications for food safety. Front. Nutr. https://doi.org/10.3389/fnut.2025.1717980. (cited).

Rizwana, S., Hazarika, M.K., 2020. Application of near-infrared spectroscopy for rice characterization using machine learning. J. Inst. Eng. India Ser. A 101 (4), 579–587.

Sharma, S., Gahlawat, V.K., Rahul, K., Mor, R.S., Malik, M., 2021. Sustainable innovations in the food industry through artificial intelligence and big data analytics. Logistics 5, 66. https://doi.org/10.3390/logistics5040066. (Multidisciplinary Digital Publishing Institute).

Sonwani, E., Bansal, U., Alroobaea, R., Baqasah, A.M., Hedabou, M., 2026. Frontiers | An Artificial intelligence approach toward food spoilage detection and analysis. [cited 2026 May 10]; https://doi.org/10.3389/fpubh.2021.816226.

Sparrow, R., Howard, M., Degeling, C., 2021. Managing the risks of artificial intelligence in agriculture. NJAS Impact Agric. Life Sci. Taylor & Francis 93, 172–196. https://doi.org/10.1080/27685241.2021.2008777.

Tiwari, M., Pandey, H., Mukherjee, A., Sutar, R.F., 2023. Artificial intelligence in food processing. Novel Technology Food Science [Internet]. John Wiley & Sons, Ltd, pp. 511–550. https://doi.org/10.1002/9781119776376.ch14.

Wang, X., Bouzembrak, Y., Lansink, A.O., van der Fels-Klerx, H.J., 2022. Application of machine learning to the monitoring and prediction of food safety: a review. Compr. Rev. Food Sci. Food Saf. 21, 416–434. https://doi.org/10.1111/1541-4337.12868.

Xu, Q., Yang, X., Hu, Z., Li, Y., Zheng, X., 2020. Identification of adulterated olive oil by fusion of near infrared and Raman spectroscopy. J. Phys. Conf. Ser. 1592, 012041.

Yared, R., 2016. Sensor-based smart oven system to enhance cooking safety. Proceedings. Multidiscip. Digital Publ. Inst 1, 23. https://doi.org/10.3390/ecsa-3-S2003.

Yu, Z., Jung, D., Park, S., Hu, Y., Huang, K., Rasco, B.A., et al., 2022. Smart traceability for food safety. Crit. Rev. Food Sci. Nutr 62, 905–916. https://doi.org/10.1080/10408398.2020.1830262. Taylor & Francis.

Zhang, H., Zhang, D., Wei, Z., Li, Y., Wu, S., Mao, Z., et al., 2023. Analysis of public opinion on food safety in Greater China with big data and machine learning. Curr. Res. Food Sci. 6, 100468. https://doi.org/10.1016/j.crfs.2023.100468.

CHAPTER FOUR

The use of multi-criteria decision analysis for food safety endeavors: A global perspective with a case study from the Philippines

Mary Ann Jilly R. Ramirez* and **Ludito V. Ramirez**
Faculty of Arts and Sciences, Southern Leyte State University – Main Campus, Sogod, Southern Leyte, Philippines
*Corresponding author. e-mail address: mramirez@southernleytestateu.edu.ph

Contents

Abstract

Food safety remains a major global public health concern, particularly in low- and middle-income countries where weak infrastructure, fragmented regulatory systems, informal food markets, and limited surveillance capacity increase the risk of foodborne diseases. As food systems become increasingly complex due to urbanization, globalization, climate change, and changing consumer behaviors, there is growing demand for evidence-based and participatory approaches to food safety governance. Multi-Criteria Decision Analysis (MCDA) has emerged as an effective decision-support framework for addressing complex food safety problems involving competing considerations such as public health protection, economic feasibility, social acceptability, environmental sustainability, and regulatory practicality. This chapter examines the theoretical foundations, methodologies, and global applications of MCDA in food

Advances in Food Security and Sustainability, Volume 11
ISSN 2452-2635, https://doi.org/10.1016/bs.af2s.2026.06.007

safety governance and highlights its role in integrating scientific evidence, expert judgment, stakeholder perspectives, and policy priorities into transparent decision-making processes. The chapter also discusses food safety governance challenges in the Philippines, particularly within informal street food sectors. A case study applying Evidence-Based Decision-Making (EBDM), quantitative microbiological assessment, and MCDA using the PROMETHEE-GAIA method evaluated *Escherichia coli* contamination risks in street-vended pork barbecue in Baybay City, Philippines. Results revealed unsatisfactory microbiological conditions exceeding Philippine FDA standards. MCDA identified consumer awareness campaigns and mandatory Good Hygienic Practices (GHP) and Good Manufacturing Practices (GMP) training as the most effective interventions. Overall, the chapter demonstrates the potential of MCDA as a transparent and adaptable framework for strengthening food safety governance and public health protection.

1. Introduction

Food safety is a fundamental pillar of global public health, economic development, and sustainable food systems. Each year, an estimated 600 million people—nearly one in every ten individuals worldwide—suffer from foodborne illnesses caused by contaminated food, resulting in approximately 420,000 deaths, with children under five accounting for almost 30 % of these fatalities World Health Organization (WHO, 2015). Beyond its direct health consequences, foodborne disease imposes substantial economic burdens through healthcare costs, productivity losses, and disruptions in domestic and international trade (Jaffee et al., 2019). Developing countries bear a disproportionate share of this burden due to inadequate infrastructure, weak surveillance systems, limited regulatory enforcement, poor sanitation, and low public awareness regarding food safety practices Grace, 2015; Food and Agriculture Organization (FAO and WHO, 2019). These challenges are further intensified by rapid urbanization, the expansion of informal food markets, climate change, and increasingly complex food supply chains, all of which influence contamination patterns and food safety risks (Tirado et al., 2010; FAO, 2020).

The globalization of food trade has likewise increased the complexity of food safety governance by introducing transboundary risks that require harmonized standards, coordinated regulatory responses, and evidence-based policy interventions to protect consumer health (Narrod et al., 2009; Unnevehr and Ronchi, 2014). Food safety decisions today involve balancing multiple and often competing considerations, including public health protection, economic efficiency, environmental sustainability, social equity, and regulatory feasibility (Katsikopoulos et al., 2021; Thomopoulos et al., 2019). However, conventional decision-making approaches are often limited in their ability to integrate diverse stakeholder perspectives, address conflicting policy objectives, and manage uncertainty within complex food systems (Linkov et al., 2006; Cinelli et al., 2021). In response, evidence-based decision-making (EBDM) has emerged as a critical framework that systematically utilizes scientific evidence, surveillance data, risk assessments, and expert analysis to guide food safety policies, interventions, and regulatory actions (FAO and WHO, 2019). EBDM enhances the accuracy, transparency, accountability, and defensibility of food safety decisions, particularly when integrated with modern decision-support approaches such as Multi-Criteria Decision Analysis (MCDA) and the Preference Ranking Organization Method for Enrichment Evaluation–Geometrical Analysis for Interactive Aid (PROMETHEE-GAIA) (Belton and Stewart, 2002; Cinelli et al., 2021).

MCDA has emerged as a valuable decision-support framework capable of systematically evaluating policy alternatives using multiple and often conflicting criteria. By integrating scientific evidence, expert judgment, stakeholder preferences, and contextual realities, MCDA supports more transparent, participatory, and balanced food safety governance (Belton and Stewart, 2002; Linkov et al., 2006). Among MCDA approaches, PROMETHEE-GAIA has gained considerable attention due to its flexibility, transparency, and ability to visually interpret relationships among alternatives and evaluation criteria. The method generates comprehensive rankings through preference flows, while the GAIA plane facilitates clearer visualization and interpretation of decision outcomes.

PROMETHEE-GAIA has been widely applied in food safety, environmental management, agriculture, engineering, public health, and sustainability studies because of its ability to incorporate both quantitative and qualitative criteria into decision-making processes. Recent studies have demonstrated its effectiveness in prioritizing emerging food safety risks, evaluating antimicrobial interventions, and assessing agricultural and environmental management practices (Biswas et al., 2024; Hommels et al., 2025; Ozkaya et al., 2025). These applications highlight the value of PROMETHEE-GAIA as a transparent, evidence-based, and adaptable analytical tool for supporting policy development, risk prioritization, and strategic planning in food safety governance, particularly in resource-constrained settings and informal food sectors such as street food vending.

This chapter examines the application of Multi-Criteria Decision Analysis (MCDA) in addressing complex food safety challenges, emphasizing its theoretical foundations, methodological approaches, and practical applications in risk-based decision-making. Specifically, the chapter explores how MCDA and PROMETHEE-GAIA can strengthen evidence-based food safety governance within the Philippine context, particularly in relation to street-vended foods and informal food systems. The chapter is organized as follows: Section 2 discusses the theoretical foundations and major variants of MCDA; Section 3 reviews global applications of MCDA in food safety governance; Section 4 contextualizes food safety governance challenges in the Philippines; Section 5 presents the case study and methodology; Section 6 discusses the findings within national and global contexts; and Section 7 concludes with implications, recommendations, and future directions for research and policy.

2. Theoretical foundations of multi-criteria decision analysis (MCDA)

Multi-Criteria Decision Analysis (MCDA) is a structured decision-support methodology designed to address complex problems involving multiple and often conflicting criteria. It provides a systematic framework for evaluating alternatives in situations where decision-makers must balance competing objectives such as public health protection, economic efficiency, regulatory feasibility, environmental sustainability, and socio-cultural acceptability (Belton and Stewart, 2002; Bana e Costa et al., 2014). MCDA has become increasingly important in policy and governance contexts, particularly in food safety, where decisions frequently involve uncertainty, diverse stakeholder interests, and multidimensional risks. By organizing decision problems into transparent and logical processes, MCDA supports more consistent, evidence-based, and accountable decision-making.

The MCDA process generally begins with defining the decision context, objectives, and scope of analysis. Decision-makers then identify potential alternatives, establish evaluation criteria, assign weights reflecting the relative importance of each criterion, and assess the performance of alternatives based on available evidence. The results are subsequently aggregated using a decision model to generate rankings or identify preferred alternatives. Sensitivity analysis is commonly conducted to evaluate the robustness and stability of results under varying assumptions, criteria weights, or data conditions (Dodgson et al., 2009; Cinelli et al., 2021). One of the major strengths of MCDA is its ability to integrate both quantitative and qualitative information. Quantitative indicators may include microbial contamination levels, exposure rates, disease incidence, or economic costs, while qualitative factors may involve regulatory practicality, ethical concerns, stakeholder perceptions, and social acceptability (Linkov et al., 2006; Wang et al., 2021). This integrated approach enables a more comprehensive assessment of food safety interventions and policy alternatives. MCDA also enhances transparency and accountability by clearly defining evaluation criteria, weighting systems, assumptions, and trade-offs among alternatives, thereby strengthening public trust and institutional credibility in decision-making processes (Kiker et al., 2005).

Another important feature of MCDA is its capacity to facilitate stakeholder participation and collaborative governance. Policymakers, scientists, regulators, food industry representatives, vendors, and consumers can contribute to defining criteria, assigning priorities, and evaluating alternatives, thereby improving the relevance, legitimacy, and acceptance of decisions (Burgman, 2015; Linkov and Moberg, 2012). This participatory characteristic is especially valuable in food safety governance, where diverse interests, values, and risk perceptions must be carefully balanced. Various MCDA methods have been developed to address different decision contexts and analytical requirements.

The Analytic Hierarchy Process (AHP), introduced by Thomas L. Saaty, structures decision problems into hierarchical levels and uses pairwise comparisons to derive priority scales, making it useful for integrating qualitative judgments with quantitative data. The Technique for Order Preference by Similarity to Ideal Solution (TOPSIS), developed by Hwang and Yoon (1981), ranks alternatives according to their distance from ideal and negative-ideal solutions and is widely applied in resource prioritization problems. The Preference Ranking Organization Method for Enrichment Evaluation (PROMETHEE), formulated by Brans and Mareschal (2005), utilizes preference functions and pairwise comparisons to generate outranking relationships among alternatives and is particularly suited for complex decision environments involving uncertainty and multiple stakeholders. Similarly, the ELimination Et Choix Traduisant la REalité (ELECTRE) family of methods, proposed by Bernard Roy, employs concordance and discordance indices to determine dominance among alternatives and is useful when trade-offs among criteria are limited or when information is incomplete.

Given the multidimensional nature of food safety risks, evolving regulatory demands, and uncertainty associated with policy responses, MCDA has emerged as a practical and adaptable framework for supporting informed and evidence-based decision-making. By integrating scientific evidence, stakeholder values, expert judgment, and local knowledge, MCDA enhances the ability of policymakers and researchers to evaluate complex food safety issues, prioritize interventions, allocate resources effectively, and develop responsive governance strategies across diverse institutional and socio-economic contexts.

3. Global applications of multi-criteria decision analysis (MCDA) in evidence-based food safety governance

Multi-Criteria Decision Analysis (MCDA) has emerged as a significant evidence-based decision-support framework in modern food safety governance because of its ability to systematically evaluate complex food safety issues involving multiple and often competing criteria. As food safety challenges become increasingly multidimensional—encompassing public health risks, economic impacts, regulatory feasibility, social acceptability, and environmental sustainability—traditional single-criterion approaches are frequently insufficient for effective decision-making. MCDA addresses these limitations by integrating scientific evidence, expert judgment, stakeholder perspectives, and policy priorities within a transparent and structured analytical framework. Consequently, governments, international organizations, and research institutions increasingly utilize MCDA to support hazard prioritization, microbial risk assessment, food inspection planning, surveillance strategies, intervention selection, and resource allocation. One of the major strengths of MCDA is its capacity to combine quantitative and qualitative information in a unified decision-making process. Unlike conventional risk assessment models that primarily focus on scientific risk estimates, MCDA enables decision-makers to simultaneously consider epidemiological evidence, economic costs, intervention feasibility, social implications, and stakeholder preferences. This multidimensional capability is particularly important in food safety governance, where decisions often involve balancing consumer protection, economic sustainability, and regulatory practicality. Through systematic weighting and prioritization of criteria, MCDA enhances the transparency, consistency, accountability, and legitimacy of regulatory decisions while strengthening stakeholder confidence in food safety policies.

Globally, MCDA has been widely applied in microbial risk assessment, hazard prioritization, and regulatory planning. In the Netherlands, the Dutch National Institute for Public Health and the Environment (RIVM) applied MCDA to prioritize foodborne pathogens by integrating disease burden, exposure levels, illness severity, and intervention feasibility into a comprehensive ranking system (Nauta et al., 2009). Similarly, Canadian researchers utilized MCDA to evaluate risks associated with listeriosis in ready-to-eat foods by combining scientific risk assessments with social and economic criteria to support evidence-informed policymaking (Carter et al., 2012). The European Food Safety Authority (EFSA) also incorporated MCDA into the prioritization of zoonotic diseases using weighted criteria such as disease incidence, severity, outbreak potential, and socioeconomic impact (Efsa Biohaz Panel, 2011). In the United Kingdom, the Food Standards Agency employed MCDA to establish risk-based inspection priorities for food businesses by considering compliance history, food category, operational complexity, and public health significance (Maguire et al., 2016). These applications demonstrate the value of MCDA in strengthening risk-based food safety systems and improving evidence-driven regulatory planning and resource allocation.

In low- and middle-income countries (LMICs), where food safety resources are often limited, MCDA supports strategic decision-making by identifying interventions that maximize public health benefits while remaining economically feasible and socially acceptable. Studies conducted in Kenya and Vietnam applied MCDA to prioritize interventions in informal food markets, including vendor training and infrastructure improvements, using criteria related to cost-effectiveness, cultural acceptability, and risk reduction potential (Alarcon et al., 2017; Loc et al., 2020). Comparative reviews further reveal that high-income countries commonly employ data-intensive and model-driven MCDA approaches supported by extensive surveillance systems, whereas LMICs often adopt participatory and qualitative MCDA frameworks that emphasize stakeholder engagement and contextual relevance (Tompkins et al., 2020).

Despite these differences, common success factors include transparent criteria weighting, stakeholder participation, adaptability to evolving regulatory contexts, and integration of scientific evidence with local knowledge (Cinelli et al., 2021; Linkov and Moberg, 2012). International organizations such as the Food and Agriculture Organization (FAO) and the World Health Organization (WHO) likewise recognize MCDA as an important tool for evidence-based food safety policymaking, particularly within One Health and Sustainable Development Goal (SDG) frameworks (FAO and WHO, 2019). Overall, the growing global adoption of MCDA demonstrates its effectiveness as a flexible, transparent, and scientifically grounded framework for strengthening food safety governance and supporting informed, balanced, and context-sensitive decision-making across diverse regulatory environments.

4. Food safety governance and challenges on street-vended foods in the Philippines

Food safety remains a critical public health concern in the Philippines due to the complex interaction among rapid urbanization, population growth, informal food economies, and fragmented regulatory systems. The country's food safety governance framework is primarily anchored on the *Food Safety Act of 2013* (Republic Act No. 10611), which establishes the legal and institutional mechanisms for ensuring the safety of the national food supply through a farm-to-fork approach (Congress of the Philippines, 2013). The Philippines adopts a decentralized and multi-agency food safety governance structure involving the Food and Drug Administration (FDA) under the Department of Health (DOH), which regulates processed and prepackaged foods; the Department of Agriculture (DA), which oversees primary agricultural production, fisheries, and slaughterhouses; the Department of Science and Technology (DOST), which provides scientific and laboratory support through institutions such as the Food and Nutrition Research Institute (FNRI); and Local Government Units (LGUs), which implement food safety regulations at the community level, particularly in public markets and among street food vendors (FAO, 2019; FAO/WHO, 2020). Although comprehensive in scope, this decentralized structure often faces coordination, monitoring, and enforcement challenges, especially within informal food sectors.

Street-vended foods constitute a major component of the informal food economy in many developing countries, including the Philippines. Generally defined as ready-to-eat foods and beverages prepared or sold in public spaces or streets (FAO, 2013; WHO, 1996), street foods provide affordable, accessible, and culturally relevant meals to urban populations, particularly low-income consumers. Street food vending also contributes significantly to employment generation and household income, especially among women, who comprise a large proportion of the workforce in the informal food sector in countries such as the Philippines, Zimbabwe, and Nigeria (Graffham et al., 2005; Chukuezi, 2010). Beyond their socio-economic contributions, street foods also help preserve traditional food cultures and support food security in urban communities. However, despite these advantages, street-vended foods are frequently associated with food safety concerns due to inadequate sanitation, poor hygiene practices, unsafe food handling, limited access to clean water, improper waste disposal, and weak regulatory oversight (Roesel and Grace, 2014; Ada et al., 2020). In the Philippines, meat- and poultry-based street foods are particularly popular because they provide affordable protein sources using low-cost meat cuts and offal (Manguiat and Fang, 2013; Canini et al., 2013).

Foodborne pathogens associated with street-vended meat products remain a significant public health threat, particularly in low- and middle-income countries where informal food vending systems often operate outside formal inspection and certification systems. Coliform contamination resulting from unsanitary processing, cross-contamination, improper storage, and inadequate cooking practices is a major contributor to foodborne illness among street-vended foods (FAO/WHO, 2023). Among the most important microbial hazards linked to meat and poultry products is pathogenic *Escherichia coli*, which is frequently associated with fecal contamination and improper food handling practices. Enterotoxigenic *Escherichia coli* (ETEC), one of the major diarrheagenic pathotypes, remains a significant cause of foodborne disease and traveler's diarrhea in developing regions, including Southeast Asia Centers for Disease Control and Prevention (CDC, 2024). Several outbreaks associated with contaminated pork and poultry products have implicated pathogenic *Escherichia coli* as a major etiological agent, particularly in improperly cooked or poorly handled meat products World Health Organization (WHO, 2024). These challenges highlight the urgent need for stronger evidence-based food safety interventions, including improved hygiene practices, sanitation monitoring, microbial risk assessment, vendor education, and regulatory enforcement.

In response to these challenges, there is increasing recognition of the need for structured and evidence-based decision-support tools capable of assisting regulatory authorities in prioritizing food safety interventions under conditions of limited resources and uncertainty. Multi-Criteria Decision Analysis (MCDA) offers a promising framework for food safety governance because it integrates quantitative risk assessments with qualitative stakeholder perspectives to systematically evaluate and prioritize food safety risks and intervention strategies. MCDA's participatory and flexible framework is particularly relevant within the decentralized food safety governance structure of the Philippines because it supports localized and context-specific decision-making while maintaining broader policy coherence (Linkov and Moberg, 2012; Loc et al., 2020). As food safety governance becomes increasingly important for both domestic public health protection and international trade compliance, the adoption of MCDA and similar evidence-based approaches can strengthen transparency, accountability, efficiency, and stakeholder participation within the Philippine food safety system.

5. Case report: Assessment of the microbiological quality of meat barbecue for *Escherichia coli* contamination in the Philippines

5.1 Hazard identification of potential sources of microbial contamination

As part of the Evidence-Based Decision-Making (EBDM) framework, the microbiological quality of meat-based street foods must first be assessed through quantitative microbial analysis prior to conducting Multi-Criteria Decision Analysis (MCDA). The initial step involves hazard identification, which evaluates the microbiological condition of raw meat before it reaches street food vendors and consumers. In the Philippines, concerns regarding meat safety remain significant because many municipalities still lack adequately equipped and accredited slaughterhouses, contributing to illegal backyard slaughtering and poor meat handling practices. The National Meat Inspection Service (NMIS) and Local Government Units (LGUs) continue to encourage the establishment and upgrading of compliant slaughter facilities to improve meat quality and reduce food safety risks nationwide. However, only a limited number of government-run abattoirs currently meet accreditation standards, raising concerns regarding the safety and suitability of meat products distributed in local markets (Meniano, 2016).

Wet markets also represent important sources of microbial contamination. Many small-scale meat vendors and street food operators rely on wet markets because they often lack adequate cold storage facilities. Previous microbiological studies have demonstrated the presence of coliforms and *Escherichia coli* in raw meat sold in Philippine wet markets. Evaluation through microbiological assays remains essential for objectively assessing meat freshness and safety (Gracey and Church, 1999). In Central Visayas, it is reported that 55 % of pork meat samples obtained from wet markets tested positive for *E. coli* (Yandug et al., 2016). Significant levels of *E. coli* contamination were likewise observed in marinated raw pork barbecue and chicken intestine products, suggesting possible unhygienic handling practices during slaughtering, processing, transportation, and retail display (Feng et al., 2002; South Africa Department of Health Directorate, 1997).

The risk of contamination is further aggravated by unsanitary market conditions and improper storage practices. Raw meat is commonly displayed openly in wet markets, exposing products to environmental contaminants, insects, and repeated consumer handling. Combined with the Philippines' tropical climate and elevated ambient temperatures, these conditions create favorable environments for the rapid growth and proliferation of pathogenic microorganisms (Gregory, 2010; Guntner and Hautzinger, 2007; Muit et al., 2025). Another concerning practice is the freezing of unsold meat at the end of the trading day followed by re-display and thawing at ambient temperature the following day. The

Table 1 Coliform and *E. coli* count of raw and marinated chicken intestine and pork barbeque.

Meat sample	**Mean Coliform count ($\log_{10}$/g)**	**Limit ($\log_{10}$/g)**	**Mean E. coli count ($\log_{10}$/g)**	**Limit***
Raw Pork (Wet Market) (n = 75) (Yundag, 2016)	8.04 - >8.77	6.0 (NMIS standard)	<0.3−15	500 MPN (NMIS standard)
Marinated raw pork barbeque (n = 10)	3.91−6.06	6.0 (NMIS standard)	2.04−5.67	Not more than 100 cells
Boiled Chicken Intestine (n = 3)	4.81−4.98	≥2.0 (FSANZ, 2001)	3.27−4.43	Must be negative
Marinated chicken intestines (n = 10)	3.91−6.77	≥2.0 (FSANZ, 2001)	2.04−5.86	Must be negative

*Limits were based on the Revised Guidelines for the Assessment of Microbiological Quality of RTE Foods (FDA, 2013).

National Meat Inspection Service (NMIS, 2013) demonstrated progressive increases in *E. coli* growth in thawed meat sold in Philippine wet markets, indicating heightened microbial risks associated with improper thawing and prolonged exposure at room temperature. These findings emphasize the importance of proper sourcing, storage, sanitation, and thorough cooking practices among street food vendors to minimize foodborne disease risks (Table 1).

6. Microbiological analysis of meat-based street food in the Philippines

6.1 Microbiological analysis of meat-based street foods in the Philippines

6.1.1 Microbial risk-based sampling plan

The hazard identification findings suggested that initial microbial contamination may significantly influence the microbiological quality of meat-based street foods (Ramirez et al., 2024). Consequently, quantitative microbial analysis was conducted to determine the extent of contamination in the final ready-to-eat products. Prior to laboratory analysis, a risk-based sampling plan was established following the principles of microbiological food sampling systems. Sampling plans provide systematic approaches for assessing food safety based on statistically appropriate numbers of representative samples International Commission on Microbiological Specifications for Foods (ICMSF, 1986). The development of the sampling plan involved identifying relevant hazard parameters, establishing probability and effect scores, and assigning weights based on risk significance (Fig. 1).

In this study, *Escherichia coli* was identified as the primary microbial hazard associated with grilled meat products because coliform contamination resulting from poor sanitation and food handling practices has been recognized as a major contributor to foodborne disease outbreaks among street-vended foods. In the Philippines, the Department of Health reported 6701 cases of acute bloody diarrhea and 36 deaths in 2017, with *Escherichia coli* among the commonly identified etiological agents (DOH, 2018). The probability and effect scoring system used in the study was adapted from the Food and Agriculture Organization and World Health Organization framework for microbiological risk assessment (FAO/WHO, 2009).

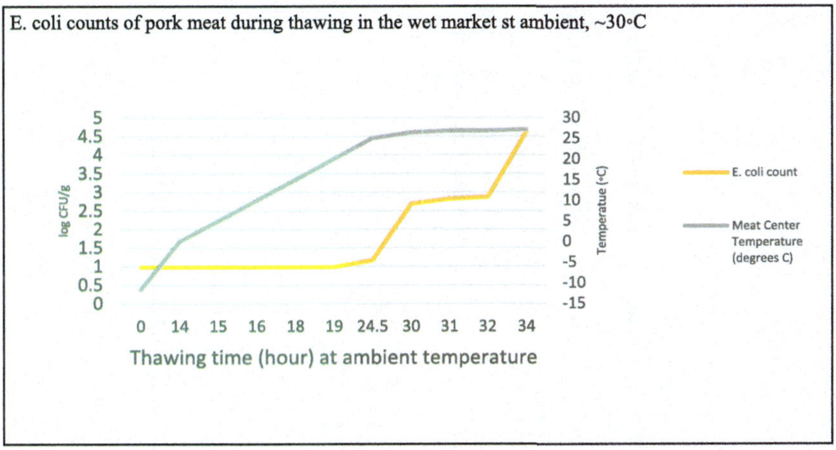

Fig. 1 Rise of E. coli counts with meat temperature at the Philippine wet market thawed at ambient temperature (NMIS, 2013).

Previous studies reporting *E. coli* contamination in Philippine street foods provided the basis for estimating contamination probability. *E. coli* contamination in grilled chicken intestine samples sold in Ozamiz City were found (Canini et al. 2013), while unsatisfactory *E. coli* levels in grilled chicken and pork products in Laguna, Philippines were reported (Manguiat and Fang, 2013). Additional risk profile parameters such as consumption volume and supplier risk profile were incorporated into the assessment, generating a cumulative risk score of 11 points. Based on the classification framework proposed by Lahou et al. (2014), this score corresponded to a medium-risk category. Using the binomial distribution approach recommended by Evers (2001) and ICMSF (1986), a target sample size of 59 samples per product category was initially determined using a 95 % confidence level and a maximum expected positive sample fraction of 5 %. However, only 42 pork barbecue samples were collected because these represented the total number of available street vendors selling pork barbecue in Baybay City during the sampling period (Table 2).

6.1.2 Microbial assessment of meat-based street foods

The microbiological assessment of grilled pork barbecue and chicken intestine revealed unsatisfactory food safety conditions among street-vended meat products in Baybay City, Philippines. The pH values of the samples were relatively low due to prior marination and post-grilling application of condiments such as ketchup and spiced vinegar (Table 3).

Table 2 Risk profile assessment and scoring framework for the sampling plan of grilled meat-based street foods.

Grilled product	Hazard group	Hazard	Probability (P) of the occurrence of E. coli in grilled meat: 4-High; 3-real; 2-small; 1-very small	Effect (E) severity of *E. coli* to human health: 1-limited; 2; moderate; 3-serios; 4-very serious	Risk (R_1) = PxE	Volume of consumption: 1-Minor<1 kg/ person/year; 3-major ≥1 kg/ person/year (R_2)	Risk profile supplier: 1- FSMS certified; 3: Not certified (R_3)	Peeling/ cooking before consumption (R_4)	Total scores ΣR_n	Risk level: Score 3–8 = low risk; 9–13 = medium risk; 14–17 = high risk	Samples/ year: Low risk - 29; medium risk – 59; high risk-149
Meat Barbeque/ chicken intestine barbecue	Micro-bial	*E. coli**	4	1	4	3	3	1	11	medium	**59**

Table 3 Aerobic plate count and physicochemical properties of the selected meat-based street food samples.

Meat sample	Mean Aerobic plate count (log_{10}/g); n = 42	Limit[a] (log_{10}/g)	Physicochemical properties		
			Temperature (°C)[b]	pH[b]	Water activity (A_w)[b]
Grilled chicken intestines	6.30	4.0	8.50	5.98	0.886
Pork barbeque	5.28	4.0	6.50	5.89	0.876

[a]Limits were based on the Revised Guidelines for the Assessment of Microbiological Quality of RTE Foods, (FDA, 2013).
[b]Values are means of three trials.

Water activity values were likewise reduced because grilling removed moisture from the products. Although low water activity generally limits microbial proliferation, aerobic plate count results still exceeded the microbiological quality limits established by the Philippine Food and Drug Administration (FDA, 2013).

The microbiological analysis further revealed that *Escherichia coli* levels exceeded the Philippine FDA standard for ready-to-eat foods (>3 MPN/g). Among the 42 pork barbecue samples analyzed, 23 tested positive for total coliforms and 14 were confirmed positive for *E. coli*. Comparable contamination levels were also observed among grilled chicken intestine samples. Previous studies have shown that cross-contamination from food handlers, utensils, food-contact surfaces, and handling practices contributes significantly to *E. coli* transmission in ready-to-eat foods (Taulo et al., 2009; Shahreza et al., 2017; Jensen et al., 2013; Bartz et al., 2016). Approximately 45 % of pork barbecue samples were contaminated with fecal coliforms, with several samples exceeding 1100 MPN/g, while 57 % of grilled chicken intestine samples surpassed tolerable microbiological limits established by the Philippine FDA. Similar findings were reported by Manguiat and Fang (2013) in grilled meat products collected in Laguna, Philippines as presented earlier.

Escherichia coli contamination was more prevalent in grilled pork barbecue samples (24 %) than in grilled chicken intestine samples (17 %), with both exceeding acceptable microbiological standards (Tables 4, 5, 6). Elevated bacterial counts suggest that grilling procedures may have been insufficient to

Table 4 Frequency of the occurrence of coliform bacteria on pork barbeque.

Total coliform		Fecal coliform		Limit[a] (MPN/g)
Count (MPN/g)	Frequency (n = 42)	Count (MPN/g)	Frequency (n = 42)	
<3.0	19	<3.0	22	3.0
3.0	1	3.0	1	3.0
3.6	6	3.6	4	3.0
7.4	1	6.2	1	3.0
11	3	9.2	2	3.0
15	1	11	1	3.0
23	1	16	1	3.0
28	1	23	2	3.0
43	4	36	1	3.0
290	1	43	1	3.0
>1100	4	64	1	3.0
—	—	290	1	3.0
—	—	>1100	4	3.0

[a]Limits were based on the Revised Guidelines for the Assessment of Microbiological Quality of RTE Foods, (FDA, 2013).

eliminate microbial contaminants completely. Previous studies have reported that meat products are sometimes undercooked to minimize shrinkage during grilling (Manguiat and Fang, 2013; Yan et al., 2010). Additional contamination may also occur through contact with uncooked ingredients such as spices, condiments, and bamboo skewers added after cooking. Studies conducted in Philippine wet markets further revealed the presence of *E. coli* contamination in fresh vegetables used as garnishes and spices, including tomatoes and bell peppers (Vital et al., 2014).

6.1.3 MCDA using PROMETHEE-GAIA software

Given that detected *E. coli* levels exceeded Philippine FDA microbiological standards for ready-to-eat meat products, the study applied Multi-Criteria

Table 5 Frequency of the occurrence of coliform bacteria on grilled chicken intestines.

Total coliform		Fecal coliform		Limit[a] (MPN/g)
Count (MPN/g)	Frequency (n = 42)	Count (MPN/g)	Frequency (n = 42)	
<3.0	13	<3.0	18	3.0
3.0	2	3.6	4	3.0
3.6	2	6.1	1	3.0
7.2	2	11	2	3.0
7.4	1	15	1	3.0
9.2	2	23	2	3.0
9.4	1	28	1	3.0
11	1	36	1	3.0
16	1	43	2	3.0
23	2	240	1	3.0
36	1	460	3	3.0
43	4	1100	3	3.0
240	1	>1100	3	3.0
1100	1	—	—	3.0
>1100	8	—	—	3.0

[a]Limits were based on the Revised Guidelines for the Assessment of Microbiological Quality of RTE Foods, (FDA, 2013).

Decision Analysis (MCDA) using Visual PROMETHEE-GAIA software to identify appropriate food safety interventions. Evaluated interventions included mandatory Good Hygienic Practices (GHP) and Good Manufacturing Practices (GMP) training for vendors, stricter licensing requirements, intensified inspections, penalties for violations, public awareness campaigns, and community-based vendor organizations. These interventions were assessed according to effectiveness, acceptability, implementability, and sustainability criteria. Results showed that when equal walking weights of 50 % were

Table 6 *E. coli* concentration on grilled chicken intestines and pork barbeque.

Grilled chicken intestine			Pork barbeque		
Count (MPN/g)	**Frequency (n = 42)**	**FDA Standard[a] (MPN/g)**	**Count (MPN/g)**	**Frequency (n = 42)**	**FDA Standard[a] (MPN/g)**
3.0	2	3.0	3.0	4	3.0
3.6	2	3.0	3.6	3	3.0
6.2	1	3.0	7.2	1	3.0
7.4	1	3.0	9.2	2	3.0
9.2	1	3.0	15	1	3.0
23	1	3.0	23	2	3.0
93	1	3.0	150	1	3.0
TOTAL # of Positive Samples	9/42			14/42	

[a]FDA Standard was based on the Revised Guidelines for the Assessment of Microbiological Quality of RTE Foods, (FDA, 2013).

applied across all criteria, consumer awareness campaigns and compulsory GHP/GMP training consistently emerged as the highest-ranked intervention strategies. These findings indicate that education-based interventions and capacity-building initiatives may represent the most practical and sustainable food safety measures for informal street food sectors.

6.1.4 MCDA for food safety governance and evidence-based interventions

This case study demonstrated the value of integrating Evidence-Based Decision-Making (EBDM), quantitative microbial analysis, and Multi-Criteria Decision Analysis (MCDA) in addressing food safety risks associated with informal street food systems. The combined framework enabled systematic identification of microbial hazards, estimation of contamination risks, and prioritization of evidence-based interventions suitable for resource-constrained settings. Findings revealed critical weaknesses in meat inspection practices, sanitation enforcement, surveillance systems, and regulatory oversight in slaughterhouses, wet markets, and street food operations. These observations are consistent with broader food safety governance challenges commonly reported in developing countries (Grace, 2015; FAO and WHO, 2019). Limited microbiological surveillance capacity, weak laboratory infrastructure, inadequate reporting systems, and fragmented governance further constrain effective food safety risk assessment and monitoring (Fig. 2).

The Philippine experience aligns with findings from other countries where MCDA has successfully integrated scientific evidence with economic, social, cultural, and operational considerations in food safety decision-making (Linkov et al., 2006; Cinelli et al., 2021). Within the Philippine context, MCDA provided a transparent and participatory framework for evaluating interventions based on effectiveness, feasibility, cost, acceptability, and sustainability. The study further emphasized the importance of political commitment, stakeholder participation, inter-agency collaboration, vendor education, and consumer awareness in strengthening food safety governance. Overall, the case study highlights the strong potential of MCDA as a scientifically grounded decision-support tool capable of improving evidence-based policymaking, prioritizing food safety interventions, and strengthening the resilience and effectiveness of food safety management systems in informal food sectors and resource-limited settings.

6.2 Future directions for MCDA in food safety governance

Addressing food safety challenges within informal food sectors requires more coordinated, adaptive, and science-driven approaches to governance and decision-making. In the Philippines, Multi-Criteria Decision Analysis (MCDA) has demonstrated significant potential for strengthening food safety management by supporting evidence-based, transparent, and risk-based interventions. To maximize its effectiveness, MCDA should be institutionalized within national food safety programs, local government initiatives, and inter-agency regulatory frameworks. Government agencies such as the Department of Health (DOH), Food and Drug Administration (FDA), Department of Agriculture (DA), and Local Government Units (LGUs) can integrate MCDA into food safety planning, inspection prioritization, surveillance systems, and resource allocation processes to improve coordination, consistency, and policy responsiveness across the food supply chain. Institutionalizing MCDA within regulatory systems can also enhance accountability and support more balanced decision-making by integrating scientific evidence, economic considerations, stakeholder perspectives, and public health priorities.

The study likewise underscores the importance of strengthening data systems, laboratory infrastructure, and digital technologies to improve the

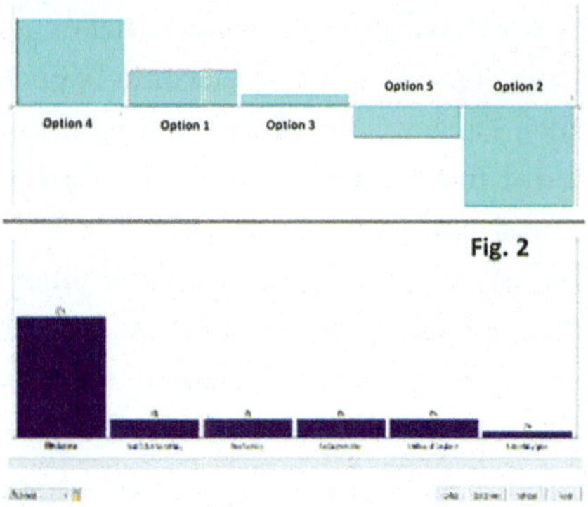

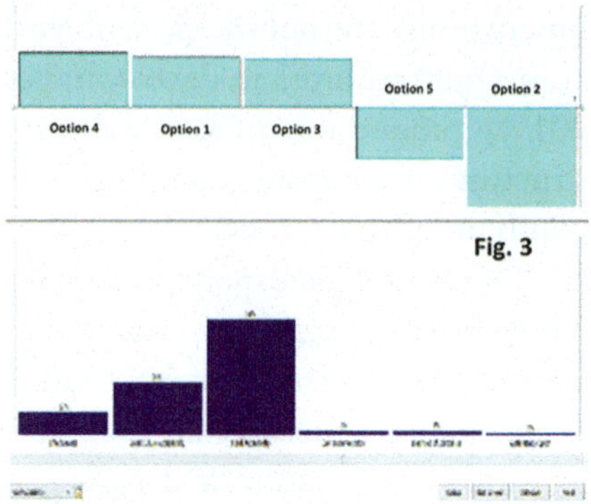

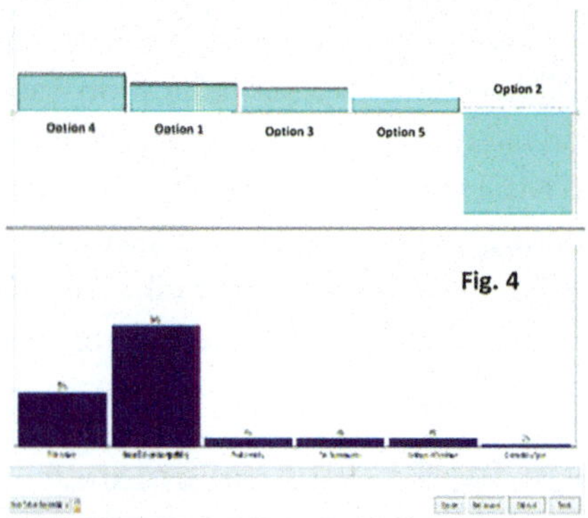

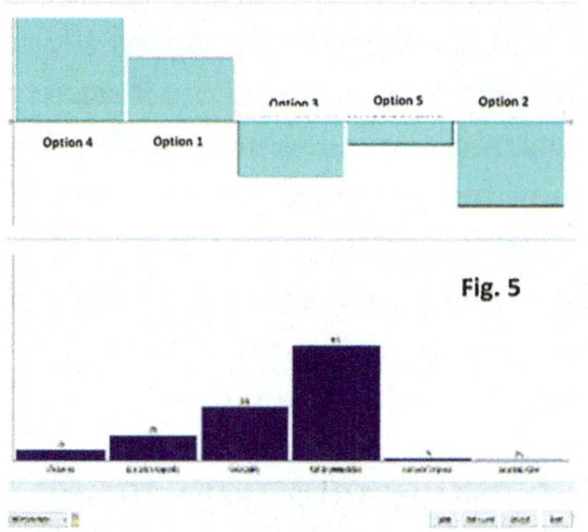

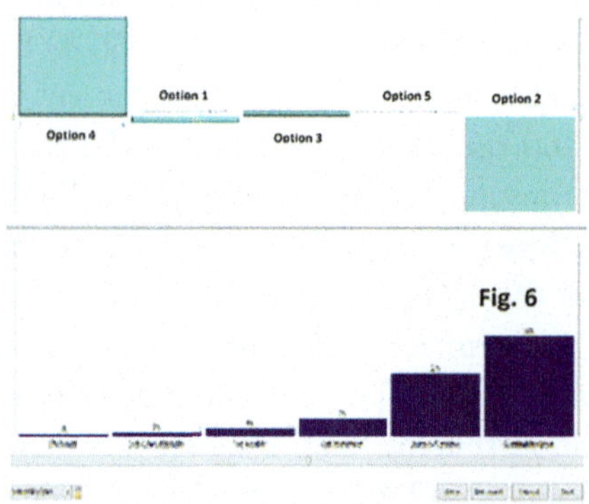

Legend:

Options

1. GHP and GMP training/seminars street vendors
2. Provide license/Permit to operate to street vendors
3. Increase local government inspection, enforcement
4. Consumer awareness Campaign
5. Establish a street food organization in the locality

Criteria

1. Effectiveness
2. Socio-Cultural Acceptability
3. Food Accessibility
4. Cost of Implementation
5. Likelihood of compliance
6. Sustainability option

Fig. 2 -6. The walking weight (sensible score analysis): top (light blue) shows the ranking of the options after adjusting the weight each criterion shown at the bottom (dark blue).

application of MCDA in food safety governance. The development of digital platforms for microbial surveillance, risk assessment, data reporting, and decision-support analytics can significantly improve the efficiency, accessibility, and timeliness of food safety interventions, particularly for LGUs and agencies with limited technical capacity. Transparent communication and public reporting of MCDA findings can further strengthen stakeholder engagement, public trust, and regulatory legitimacy. Regional collaboration within the Association of Southeast Asian Nations (ASEAN) is also essential because food safety threats increasingly transcend national borders through trade, migration, and climate-related risks.

Shared surveillance systems, harmonized risk assessment frameworks, and coordinated food safety policies can enhance regional preparedness and response to foodborne outbreaks and emerging hazards. At the local level, continuous capacity-building programs for food safety officers, laboratory personnel, policymakers, and food vendors remain necessary to strengthen competencies in MCDA, microbial risk assessment, sanitation management, and community engagement. Future applications of MCDA should also move beyond reactive responses toward anticipatory and resilience-based approaches capable of addressing emerging threats associated with climate change, antimicrobial resistance, globalization, evolving food systems, and changing consumer behaviors. Overall, the Philippine experience demonstrates that MCDA can serve as a valuable framework for building more resilient, participatory, and science-driven food safety governance systems.

7. Conclusion

Evidence-Based Decision-Making (EBDM) through Multi-Criteria Decision Analysis (MCDA) has emerged as a valuable framework for strengthening food safety governance by supporting systematic, transparent, participatory, and balanced decision-making processes. The Philippine case study on street-vended pork barbecue demonstrated that integrating MCDA with quantitative microbiological analysis can effectively identify food safety hazards, assess microbial risks, and prioritize practical and context-appropriate interventions such as consumer awareness campaigns and mandatory vendor training on food hygiene and sanitation practices. In informal food sectors where street-vended foods play an important role in food accessibility, cultural preservation, and local livelihoods, MCDA provides decision-makers with a structured approach for simultaneously considering scientific evidence, public health risks, socioeconomic realities, cultural practices, stakeholder perspectives, and implementation feasibility in the design and evaluation of food safety interventions.

Experiences from countries such as Canada, Netherlands, and New Zealand further demonstrate the adaptability and applicability of MCDA across diverse food safety systems, regulatory environments, and governance frameworks. MCDA also promotes stronger coordination among regulatory agencies, researchers, industry stakeholders, local governments, and communities by enhancing transparency, accountability, and stakeholder participation in decision-making processes. As global food systems continue to face growing challenges associated with globalization, urbanization, climate change, antimicrobial resistance, and changing consumer behaviors, the integration of MCDA into food safety policies, microbial risk assessment frameworks, surveillance systems, and local intervention programs can significantly contribute to the development of safer, more resilient, and adaptive food systems. Overall, the Philippine experience highlights the strong potential of MCDA as an evidence-based and scientifically grounded decision-support tool capable of improving food safety governance, strengthening risk-based policymaking, optimizing resource allocation, and enhancing public health protection, particularly within informal and resource-constrained food systems.

References

Ada, C.A., Gabriel, A.A., Alonzo, A.A., 2020. Food safety practices among street food vendors in the Philippines. Food Res. 4 (5), 1502–1510.

Alarcon, P., Fèvre, E.M., Muinde, P., Murungi, M.K., 2017. Urban livestock keeping in the city of Nairobi: diversity of production systems, supply chains, and their disease management and risks. Front. Veterinary Sci. 4, 171. https://doi.org/10.3389/fvets.2017.00171.

Bana e Costa, C.A., De Corte, J.M., Vansnick, J.C., 2014. MACBETH. In: Greco, S., Ehrgott, M., Figueira, J. (Eds.), Multiple Criteria Decision Analysis: State of the Art Surveys. Springer, pp. 245–286.

Bartz, F.E., de Oliveira, T.C.R.M., Moreira, A.N., 2016. Cross-contamination of Escherichia coli in food handling environments. Food Control 68, 267–274. https://doi.org/10.1016/j.foodcont.2016.04.009.

Belton, V., Stewart, T.J., 2002. Multiple Criteria Decision Analysis: An Integrated Approach. Kluwer Academic Publishers.

Biswas, T., et al., 2024. The PROMETHEE-GAIA: a multi-criteria decision-making method for identifying best conservation agricultural practices. Soil. Tillage Res. 245, 106315. https://doi.org/10.1016/j.still.2024.106315.

Brans, J.P., Mareschal, B., 2005. PROMETHEE methods. In: Figueira, J., Greco, S., Ehrgott, M. (Eds.), Multiple Criteria Decision Analysis: State of the Art Surveys. Springer, pp. 163–186.

Burgman, M., 2015. Trusting Judgements: How to Get the Best Out of Experts. Cambridge University Press.
Canini, M.H.N., et al., 2013. Microbiological quality assessment of street-vended grilled chicken intestine in Ozamiz City, Philippines. Asian J. Health 3 (1), 45–54.
Carter, A., et al., 2012. Application of multi-criteria decision analysis in Listeria monocytogenes risk management. Risk Anal. 32 (4), 617–631. https://doi.org/10.1111/j.1539-6924.2011.01679.x.
Centers for Disease Control and Prevention, 2024. Foodborne germs and illnesses: Escherichia coli (E. coli). U.S. Department of Health and Human Services. https://www.cdc.gov/ecoli/index.html.
Chukuezi, C.O., 2010. Food safety and hygienic practices of street food vendors in Owerri, Nigeria. Stud. Sociol. Sci. 1 (1), 50–57.
Cinelli, M., Kadziński, M., Gonzalez, M., Słowiński, R., 2021. How to support the application of multiple criteria decision analysis? Let us start with a comprehensive taxonomy. Omega 96, 102261. https://doi.org/10.1016/j.omega.2019.102261.
Congress of the Philippines, 2013. Republic Act No. 10611: Food Safety Act of 2013. Official Gazette of the Republic of the Philippines.
Department of Health, 2018. Philippine health statistics report. Department of Health Philippines.
Dodgson, J., Spackman, M., Pearman, A., Phillips, L., 2009. Multi-criteria analysis: A manual. Department for Communities and Local Government, United Kingdom.
Efsa Biohaz Panel, 2011. Scientific opinion on the public health hazards to be covered by inspection of meat from swine. EFSA J. 9 (10), 2351. https://doi.org/10.2903/j.efsa.2011.2351.
Evers, E.G., 2001. A rationale for microbiological criteria for food products. Int. J. Food Microbiol. 67 (3), 203–213. https://doi.org/10.1016/S0168-1605(01)00468-6.
Feng, P., Weagant, S.D., Grant, M.A., 2002. Enumeration of Escherichia coli and the coliform bacteria. Bacteriological Analytical Manual, 8th ed. U.S. Food and Drug Administration.
Food and Agriculture Organization of the United Nations, 2013. Street Foods. FAO. https://www.fao.org/.
Food and Agriculture Organization of the United Nations, 2019. Food Safety Technical Guidance for the Philippines. FAO.
Food and Agriculture Organization of the United Nations, 2020. Climate Change and Food Safety Risks. FAO.
Food and Agriculture Organization of the United Nations, & World Health Organization, 2009. Risk Characterization of Microbiological Hazards in Food: Guidelines. FAO/WHO.
Food and Agriculture Organization of the United Nations, & World Health Organization, 2019. Strengthening Food Control Systems: Guidelines for Developing Countries. FAO/WHO.
Food and Agriculture Organization of the United Nations, & World Health Organization, 2020. Food Safety Risk Governance in Southeast Asia. FAO/WHO.
Food and Agriculture Organization of the United Nations, & World Health Organization, 2023. Food Safety Risks Associated with Street-Vended Foods. FAO/WHO.
Food and Drug Administration Philippines, 2013. Revised Guidelines for the Assessment of Microbiological Quality of Processed Foods. Department of Health Philippines.
Grace, D., 2015. Food safety in low- and middle-income countries. Int. J. Environ. Res. Public. Health 12 (9), 10490–10507. https://doi.org/10.3390/ijerph120910490.
Gracey, J.F., Church, P.N., 1999. Meat Hygiene, 10th ed. W.B. Saunders.
Graffham, A., Karehu, E., MacGregor, J., 2005. Impact of EurepGAP on Small-Scale Vegetable Growers in Kenya (Fresh Insights No. 6). Natural Resources Institute.

Gregory, N.G., 2010. How climatic changes could affect meat quality. Food Res. Int. 43 (7), 1866–1873. https://doi.org/10.1016/j.foodres.2009.05.018.
Guntner, M., Hautzinger, P., 2007. Hygiene management in wet markets. Food Control 18 (4), 327–332. https://doi.org/10.1016/j.foodcont.2005.10.012.
Hommels, N.M.C., Mourits, M.C.M., Focker, M., van der Fels-Klerx, H.J., 2025. Categorization of drivers of change for emerging food safety risks. Curr. Res. Food Sci. 10, 101098. https://doi.org/10.1016/j.crfs.2025.101098.
Hwang, C.L., Yoon, K., 1981. Multiple Attribute Decision Making: Methods and Applications. Springer.
International Commission on Microbiological Specifications for Foods, 1986. Microorganisms in Foods 2: Sampling for Microbiological Analysis. University of Toronto Press.
Jaffee, S., Henson, S., Unnevehr, L., Grace, D., Cassou, E., 2019. The Safe Food Imperative: Accelerating Progress in Low- and Middle-Income Countries. World Bank.
Jensen, D.A., Friedrich, L.M., Harris, L.J., Danyluk, M.D., Schaffner, D.W., 2013. Cross contamination of Escherichia coli in food service environments. J. Food Prot. 76 (4), 641–648. https://doi.org/10.4315/0362-028X.JFP-12-404.
Katsikopoulos, K.V., et al., 2021. Decision-making in food safety governance under uncertainty. Food Policy 99, 101982. https://doi.org/10.1016/j.foodpol.2020.101982.
Kiker, G.A., Bridges, T.S., Varghese, A., Seager, T.P., Linkov, I., 2005. Application of multicriteria decision analysis in environmental decision making. Integr. Environ. Assess. Manag. 1 (2), 95–108. https://doi.org/10.1897/IEAM_2004a-015.1.
Lahou, E., Jacxsens, L., Daelman, J., Van Landeghem, F., Uyttendaele, M., 2014. Microbiological performance indicators for food safety management systems. Food Control 40, 150–161. https://doi.org/10.1016/j.foodcont.2013.11.031.
Linkov, I., Moberg, E., 2012. Multi-Criteria Decision Analysis: Environmental Applications and Case Studies. CRC Press.
Linkov, I., Satterstrom, F.K., Kiker, G., et al., 2006. From comparative risk assessment to multi-criteria decision analysis and adaptive management. Environ. Int. 32 (8), 1072–1093. https://doi.org/10.1016/j.envint.2006.06.013.
Loc, H.T.T., et al., 2020. Application of multi-criteria decision analysis in informal food markets in Vietnam. Food Control 108, 106842. https://doi.org/10.1016/j.foodcont.2019.106842.
Maguire, K.B., et al., 2016. Risk-based inspection systems for food establishments using multi-criteria decision analysis. Food Control 73, 114–121. https://doi.org/10.1016/j.foodcont.2016.07.032.
Manguiat, L.S., Fang, T.J., 2013. Microbiological quality of grilled street foods in Laguna, Philippines. Food Control 31 (2), 621–626. https://doi.org/10.1016/j.foodcont.2012.10.039.
Meniano, S., 2016. Poor slaughterhouse conditions raise food safety concerns in the Philippines. Philippine N. Agency.
Muit, M.C.B., Manaog, A.T., Masagca, J.L.S., Montañez, J.J.F., 2025. FreshTrack: enhancing household food safety with RFID-enabled food expiry monitoring system. Recoletos Multidiscip. Res. J. 13 (2), 25–34.
Narrod, C., et al., 2009. Public-private partnerships and collective action in high value fruit and vegetable supply chains. Food Policy 34 (1), 8–15. https://doi.org/10.1016/j.foodpol.2008.10.005.
National Meat Inspection Service, 2013. Growth of Escherichia coli in Thawed Meat Sold in Philippine Wet Markets. NMIS.
Nauta, M.J., et al., 2009. A comparison of risk ranking frameworks for foodborne pathogens. Risk Anal. 29 (5), 659–674. https://doi.org/10.1111/j.1539-6924.2008.01194.x.

Ozkaya, G.U., et al., 2025. Combined experimental design based-PROMETHEE analysis for antimicrobial activity optimization. J. Food Meas. Charact. https://doi.org/10.1007/s11694-025-02891-0.
Ramirez, M.A.J.R., Mesias, I.C.P., Ramirez, L.V., 2024. Quantitative Microbial Risk Assessment and Multi-Criteria Decision Analysis of Street-Vended Pork Barbeque Contaminated with *Escherichia coli*. J. Sci. Eng. Technol. 12, 44–59.
Roesel, K., Grace, D., 2014. Food Safety and Informal Markets: Animal Products in sub-Saharan Africa. Routledge.
Shahreza, M.S., et al., 2017. Food-contact surface contamination and Escherichia coli transmission in street-vended foods. Foodborne Pathog. Dis. 14 (6), 341–347. https://doi.org/10.1089/fpd.2016.2245.
South Africa Department of Health Directorate, 1997. Guidelines for environmental health officers on the interpretation of microbiological analysis data of food. Department of Health, South Africa.
Taulo, S., Wetlesen, A., Abrahamsen, R., et al., 2009. Microbiological hazard identification and exposure assessment of street food in developing countries. Food Control 20 (3), 186–190. https://doi.org/10.1016/j.foodcont.2008.04.002.
Thomopoulos, R., et al., 2019. Food safety governance and sustainability transitions. Trends Food Sci. & Technol. 84, 14–20. https://doi.org/10.1016/j.tifs.2018.11.019.
Tirado, M.C., Clarke, R., Jaykus, L.A., McQuatters-Gollop, A., Frank, J.M., 2010. Climate change and food safety: a review. Food Res. Int. 43 (7), 1745–1765. https://doi.org/10.1016/j.foodres.2010.07.003.
Tompkins, K., et al., 2020. Participatory multi-criteria decision analysis in food safety governance. Food Policy 93, 101839. https://doi.org/10.1016/j.foodpol.2020.101839.
Unnevehr, L., Ronchi, L., 2014. Food safety and developing markets. Agric. Econ. 45 (S1), 131–141. https://doi.org/10.1111/agec.12132.
Vital, P.G., et al., 2014. Microbial contamination of fresh produce sold in open-air markets in the Philippines. Food Control 37, 282–287. https://doi.org/10.1016/j.foodcont.2013.09.050.
Wang, Y., et al., 2021. Integrating quantitative and qualitative criteria in food safety decision-making using MCDA. Food Control 123, 107712. https://doi.org/10.1016/j.foodcont.2020.107712.
World Health Organization, 1996. Essential Safety Requirements for Street-Vended Foods. WHO.
World Health Organization, 2015. Estimates of the Global Burden of Foodborne Diseases. WHO Press.
World Health Organization, 2024. Escherichia coli. https://www.who.int/news-room/fact-sheets/detail/e-coli.
Yan, H., Li, M., Wang, J., 2010. Effects of grilling practices on microbial reduction in meat products. J. Food Saf. 30 (4), 789–798.
Yandug, R.T., et al., 2016. Microbiological quality of pork sold in wet markets in Central Visayas, Philippines. Philippine J. Veterinary Med. 53 (2), 45–53.

CHAPTER FIVE

Food and nutrition security in the context of ageing and non-communicable diseases (NCDs)

N.E. Chei[a,*] and Don Eliseo Lucero-Prisno, III[b,c,d]
[a]Nway Eint Chei, Department of Management, Faculty of Humanities and Management, Myanmar Creative International College, Myanmar
[b]Department of Global Health and Development, Faculty of Public Health and Policy, London School of Hygiene and Tropical Medicine, London, United Kingdom
[c]Research Office, Palompon Institute of Technology, Palompon, Leyte, Philippines
[d]Research Planning and Development Office, WCC Aeronautical and Technological College Inc., Binalonan, Pangasinan, Philippines
*Corresponding author. e-mail address: drnwayec.mcu@gmail.com

Contents

Advances in Food Security and Sustainability, Volume 11
ISSN 2452-2635, https://doi.org/10.1016/bs.af2s.2026.04.001

Abstract

Food and nutrition security are agents of necessity for health in older adults, particularly in the context of ageing and the rising prevalence of non-communicable diseases (NCDs). With the mounting in the ageing population globally, older adults face heightened vulnerability to food insecurity due to age-related physiological changes, social isolation, and limited access to nutritious food. Simultaneously, the growing burden of NCDs such as diabetes, cardiovascular diseases, and hypertension complicates the dietary needs of the ageing population, as these conditions often require specialized nutritional interventions, making it challenging for older adults to maintain proper nutrition. Poor dietary intake and food insecurity can exacerbate the risk malnutrition and accelerate the progression of NCDs, creating a detrimental cycle that undermines the health and quality of life of older adults. This essay explores the complex interaction between food and nutrition security, ageing, and NCDs, emphasizing how ageing and chronic diseases influence nutritional needs and the role of factors in shaping food access. It also underscores the importance of targeted interventions, including improved food environments and community-based programs, to address food insecurity among ageing populations. Additionally, it highlights the need for integrated public health policies to ensure that older people with NCDs have equitable access to adequate, nutritious food. Addressing food and nutrition security in this context is essential for promoting healthy ageing, mitigating the burden of NCDs, and ultimately improving health outcomes and quality of life for older individuals.

1. Introduction

Food and nutrition security are fundamental pillars of health and well-being, particularly for ageing populations and individuals living with non-communicable diseases (NCDs). The Food and Agriculture Organization of the United Nations (FAO) defines food security as "the condition where all individuals, at all times, have physical and economic access to sufficient, safe, and nutritious food that meets their dietary needs and preferences for an active and healthy life" (Food and Agriculture Organization FAO, 1996).

Food and nutrition security has emerged as a pivotal item on the UN's 21st Century development agenda, with the second Sustainable Development Goal (SDG) aiming to "End hunger, achieve food security and improved nutrition, and promote sustainable agriculture" (United Nations, n.d.). Despite global efforts to address hunger and malnutrition, food insecurity remains an unresolved dilemma, with an estimated 673 million people (8.2 % of global population) facing hunger in 2024, representing a decrease of approximately 22 million people compared to 2022 (FAO, 2025). Furthermore, beyond hunger, the prevalence of moderate or severe food insecurity continues to exceed pre-pandemic levels; 2.3 billion people experienced moderate to severe food insecurity in 2024, with 828 million facing severe food insecurity, highlighting the persistent challenges in ensuring equitable access to nutritious food and endangering their health and well-being. (FAO, 2025). Nutrition security is defined as "consistent and equitable access to healthy, safe, and affordable foods that promote optimal health and well-being" (Economic Research Service, U.S, 2023), which is equally critical, especially for older adults who require adequate nutrition to manage chronic conditions and maintain overall health.

The global population is swiftly expanding with older adults at an unprecedented rate,reaching 1 billion in 2019, and the number of individuals aged 60 and older is projected to grow to 1.4 billion by 2030 and 2.1 billion by 2050 (United Nations Department of Economic and Social Affairs, 2023). Ageing is associated with inevitable and irreversible processes of physiological, psychological, and social changes throughout the lifetime that lead to a decline in physical functions and structure, and increased vulnerability to malnutrition and non-communicable diseases (NCDs) such as diabetes, cardiovascular diseases, and hypertension (WHO, 2020a). Particularly in developing countries, this demographic shift, coupled with the increasing prevalence of NCDs and degenerative disorders, underscores the need for targeted interventions to ensure that older adults, particularly those living with NCDs, have equitable access to nutritious food. Therefore, these chronic conditions often require specialized dietary management, yet older adults frequently face barriers to accessing nutritious food due to socio-economic constraints, physical limitations, and social isolation.

With potential to significantly influence the accumulation and distribution of resources, people's capabilities, and, crucially, their autonomy (WHO, 2020b), the interplay between ageing, NCDs, and food insecurity presents one of the most compelling and complex problems that undermines the health and quality of life of older individuals (Beaglehole et al., 2011).

Food insecurity, which affects a significant portion of the global population, has wide-ranging implications for health and well-being, particularly among older individuals (FAO, 2025). These conditions exacerbate the vulnerabilities associated with ageing, leading to a higher risk of developing chronic diseases and impairing overall health outcomes (Zhou et al., 2023). Therefore access to nutritious food is essential for older adults not only to age healthily but also to maintain physical health, cognitive function, and immune resilience (Volkert et al., 2019). Moreover, the concept of nutrition security emerged to address the linkages between food insecurity and diet-related diseases (Mozaffarian et al., 2021).

Food and nutrition security are vital for maintaining the health and well-being of older adults, especially those living with NCDs. It is unquestionable that proper nutrition is integral not only for reducing the risk and progression of NCDs, which are highly prevalent among ageing populations, but also for ultimately enhancing the quality of life for seniors (Larrea-Killinger et al., 2024). Access to adequate and nutritious food boosts the immune function, supports cognitive health, and enhances physical well-being, all of which are critical for maintaining a high quality of life among older people (Gianfredi et al., 2024a). Ensuring food and nutrition security for older adults is therefore a key factor in promoting healthy ageing and preventing adverse health outcomes, as it helps mitigate of age-related vulnerabilities such as NCDs, cognitive decline, and weakened immunity. Furthermore, food security empowers older individuals to better manage the challenges associated with NCDs, reducing the risk of complications and supporting their ability to age healthily (Gianfredi et al., 2024a; Larrea-Killinger et al., 2024).

This essay explores the intricate relationship between food and nutrition security, ageing, and NCDs, focusing on how the ageing process and chronic diseases influence nutritional needs and access to food. It emphasizes the role of multiple factors in shaping food security and underscores the importance of targeted interventions, such as improved food environments and community-based programs, to improve food security among ageing populations. Furthermore, it highlights the importance of integrated public health policies that address food and nutrition insecurity, ensuring equitable access to nutritious food for older adults, promoting healthy ageing, and mitigating the burden of NCDs. In doing so, it underscores the need for targeted interventions for enhancing health outcomes and quality of life for ageing populations worldwide.

2. Overview of food systems

Food systems encompass the entire range of activities involved in the production, processing, distribution, consumption, and disposal of food. These systems are complex networks that integrate environmental, economic, social, and political dimensions, influencing food security, nutrition, and sustainability (FAO, 2021a). Despite the crucial role food systems play in human well-being and environmental sustainability, they face numerous challenges that require urgent actions to ensure their continued effectiveness and resilience.

2.1 Components of food systems

Food systems consist of several interconnected components that work in tandem to ensure the availability and accessibility of food. The main components are:

- **Production:** This involves the cultivation of crops, raising of livestock, and harvesting of seafood and food from forests. Agricultural practices, land use, and resource management play critical roles in determining the sustainability of food production. Sustainable practices, such as crop rotation and soil and water conservation, are essential for maintaining soil health and long-term agricultural productivity while minimizing environmental degradation (Altieri et al., 2017).
- **Processing:** After harvesting, food typically undergoes processing to enhance its shelf life, safety, and convenience, and to add value. This stage includes activities such as milling, canning, and packaging (FAO, 2021a). Food processing raises concerns about nutritional quality, the use of additives, and the environmental impacts of packaging materials (Popkin et al., 2020a).
- **Distribution:** The distribution component covers transportation, storage, and marketing of food products, including domestic and international trade. Efficient distribution systems are critical to ensure that food reaches consumers promptly in a cost-effective manner, and in good condition (Vermeulen et al., 2012). As long supply chains can contribute to food loss and environmental impacts (Gustavsson et al., 2011), these systems must address challenges such as food waste during transportation and inequities in access, particularly in underserved areas.
- **Consumption:** This refers to the acquisition (by purchase or through public and charitable transfers), preparation, and consumption of food by individuals and households. Dietary choices and consumption patterns

significantly impact health and nutrition outcomes (Popkin et al., 2020a). Societal trends, such as the rise in consumption of ultra-processed foods, are contributing to increased rates of diet-related diseases, including obesity and diabetes (Willett et al., 2019).

- **Disposal:** Food waste and loss occur at all stages of the food system. A significant portion of food is wasted due to inefficiencies in processing, distribution, and consumption. Proper disposal and recycling of food waste are critical for reducing environmental impacts, such as greenhouse gas emissions and landfills (Gustavsson et al., 2011). Moreover, strategies to reduce food waste include improved storage facilities, better inventory management, and consumer education (Gustavsson et al., 2011).

2.2 Challenges in food systems

Despite their importance, food systems face a range of challenges that threaten food security, environmental sustainability, and public health:

- **Food Insecurity:** Millions of people worldwide suffer from hunger and malnutrition due to inadequate access to nutritious food. Food insecurity is often exacerbated by factors such as poverty, conflict, and climate change (FAO, 2021a), leading to malnutrition and hunger, especially in low-income regions where resources for food production and distribution are limited.
- **Environmental Degradation:** Unsustainable agricultural practices, such as deforestation, overuse of fertilizers, over grazing and excessive water consumption, contribute to soil erosion, biodiversity loss, greenhouse gas emissions, and climate change (IPCC, 2019). The negative impact of food systems on the environment underscores the need for more sustainable agricultural practices that minimize ecological footprints.
- **Climate Change:** Food systems are both affected by and contribute to climate change. It disrupts food production by causing extreme weather events, such as droughts and floods, while agricultural activities account for 19 %–29 % of global greenhouse gas emissions (Vermeulen et al., 2012). Rising temperatures and shifting precipitation patterns threaten food security by affecting crop yields and livestock health.
- **Health and Nutrition:** The global rise in diet-related diseases, such as obesity, diabetes, and cardiovascular diseases, highlights the need for healthier and more sustainable diets. The consumption of ultra-processed foods high in sugar, fats, and salt is major contributor to these health issues (Popkin et al., 2020a). Shifting dietary patterns toward

plant-based and whole foods could improve public health outcomes while reducing the environmental impact of food production.

- **Economic Inequality:** Small-scale farmers and food workers often face economic hardships, including low incomes, lack of access to markets, and exploitation by intermediaries and market actors (such as middlemen, brokers, and traders). This exacerbates poverty and limits their ability to access resources for sustainable production (FAO, 2021a). Addressing these inequalities within food systems is crucial for building equitable and resilient food networks (Hinrichs, 2003).

2.3 Transforming food systems

Transformative changes are needed across food systems in order to address these challenges. Key strategies include:

- **Sustainable Agriculture:** Promoting agroecological practices, such as crop diversification, organic farming, and integrated pest management, can enhance the resilience of food production systems while minimizing their environmental impact (Altieri et al., 2017). Agroecology encourages the use of local knowledge and ecosystems to achieve food security in a sustainable manner.
- **Reducing Food Waste:** Implementing measures to minimize food loss and waste at all stages of the food system can improve efficiency and reduce pressure on natural resources (Gustavsson et al., 2011). Efforts to reduce food waste must be integrated into policy frameworks, agricultural practices, and consumer behavior to ensure that food is used more efficiently and sustainably.
- **Promoting Healthy Diets:** Encouraging the consumption of healthier diets that are plant-based and nutritionally balanced, can have both public health and environmental benefits. Reducing the consumption of ultra-processed foods and promoting food systems that prioritize sustainable food production can contribute to lower rates of chronic diseases and a smaller environmental footprint (Willett et al., 2019).
- **Strengthening Local Food Systems:** Supporting local food production and distribution networks can enhance food security, reduce transportation emissions, and reduce environmental impacts (Hinrichs, 2003). Local food systems reduce the need for long-distance transportation, decrease greenhouse gas emissions, and empower communities to control their food supply (Hinrichs, 2003). Strengthening local

economies can also lead to more equitable food access and opportunities for small-scale farmers.

- **Policy and Governance:** Governments and international organizations play a critical role in creating policies that support sustainable food systems, such as subsidies for sustainable farming practices, regulations on food waste, and investments in agricultural research that promotes innovation in food production and distribution (FAO, 2021a). Effective governance is necessary to coordinate the efforts of various stakeholders and align food system practices with global sustainability goals.

3. Ageing and its impact on food and nutrition security

Ageing is a significant global demographic shift, characterized by an increasing proportion of older people. According to the United Nations (2019a, 2019b), it is expected that the number of people aged 60 years and above will rise to reach 2.1 billion by 2050, from 962 million in 2017. This ageing trend poses considerable challenges to food and nutrition security, particularly as older people often face physiological, economic, and social changes that can affect dietary intake, nutrient absorption, and access to food.

3.1 Ageing: A natural, irreversible process

The biological aspects of ageing are marked by gradual changes in physical functions and structures in terms of decreased muscle mass, reduced bone density, and changes in skin elasticity (López-Otín et al., 2013). Ageing leads to various physiological and social changes that influence an individual's nutritional needs and access to food. One of the key changes is the decline in muscle mass and metabolic rate, which increases the need for certain nutrients, such as protein, vitamins, and minerals (Haddad et al., 2020). These physiological changes also contribute to an increased vulnerability to various poor health conditions and chronic diseases, such as cardiovascular diseases, arthritis, and diabetes, which are more prevalent among ageing populations, with significant detrimental effects on both their physical and psychological well-being (Lindau et al., 2020a, 2020b).

Psychologically, ageing is linked with changes in cognitive functions. While some cognitive abilities, such as memory and processing speed, may depreciate with age, other aspects like emotional regulation, tend to improve (Labouvie-Vief, 2003a, 2003b). However, the experience of ageing processes is not solely negative, and a majority of older adults surveyed reported high levels of life satisfaction and personal growth, especially when they have access to meaningful social connections and social engagement activities (Pinquart and Sörensen, 2001). Social changes also play a critical role in the ageing process. As people age, many older adults often experience changes in their social networks, including retirement, loss of loved ones, and fluctuations in living situations, which can negatively impact mental health and overall quality of life (Kahana et al., 2012). Nevertheless, social support and active engagement in community activities can help to mitigate the effects of these social changes, contributing to better health outcomes for older adults (Hawkley and Cacioppo, 2010a).

It is true that for ageing populations, food and nutrition security is significantly crucial, as older adults face unique nutritional challenges, including reduced appetite, altered metabolism, and difficulties in food preparation and access (Sharkey et al., 2012). Older adults often struggle to maintain adequate nutritional intake due to changes in taste, chewing, and swallowing abilities, leading to reduced food consumption (Nesbitt et al., 2021). Moreover, the prevalence of multiple chronic conditions, such as hypertension, diabetes, and arthritis, commonly associated with ageing, exacerbates these issues, making the management of both food security and nutritional needs more complex (Stelmach-Mardas et al., 2021).

In addition to the physiological factors, social determinants such as income, mobility, and social networks play a significant role in food and nutrition security among older adults. Many elderly individuals live in isolation, have limited access to nutritious foods, and lack the financial resources to make healthy food choices (Shankar et al., 2020). The intersection of these physical and socioeconomic factors contributes to food insecurity, which, in turn, negatively impacts the nutritional status of older adults, increasing their vulnerability to malnutrition and various NCDs (Ribeiro et al., 2022).

3.2 The ageing process and food security

Difficulties in chewing and swallowing can lead to nutritional deficiencies, which in turn can contribute to the onset or worsening of NCDs such as osteoporosis, diabetes, and cardiovascular disease (Nesbitt et al., 2021; Stelmach-Mardas et al., 2021). Moreover, older adults often undergo changes in their sense of taste and smell, which can trigger reduced appetite and decreased food intake. This decrease in appetite, coupled with changes in metabolic requirements, can leave older adults struggling to meet their caloric and nutritional needs (Nesbitt et al., 2021). The World Health Organization (WHO, 2015a; WHO, 2015b) reported that although older adults may require fewer calories, their need for certain nutrients—such as protein, calcium, and vitamins—remains the same or even increases. Consequently, malnutrition evolves into a serious matter among older populations, with food insecurity potentially aggravating pre-existing health conditions.

3.3 Ageing: A demographic trend and policy challenge

The global population is ageing at an unprecedented rate, with the number of individuals aged 60 and older projected to double from 962 million in 2017–2.1 billion by 2050 (United Nations, 2017). While this trend reflects positive improvements in human well-being, it also presents significant challenges for food and nutrition security. As worldwide populations age, the negative implications for both individual health and societal structures become increasingly evident, particularly with respect to the unique nutritional needs and vulnerabilities of older adults. Addressing these challenges requires targeted policies and interventions to ensure that ageing populations have access to adequate and nutritious food. In addition, strengthening elder care resources, expanding sustainable pension systems, and enhancing comprehensive social protection programs are essential to mitigate financial insecurity, social isolation, and functional decline among older adults. Adequate income support and accessible long-term care services not only improve food affordability and dietary quality but also contribute to health equity, dignity, and overall well-being in later life.

Unquestionably, malnutrition remains a critical issue for ageing populations, particularly those in low-income or rural areas. Older adults are at an increased risk of malnutrition, which can worsen health outcomes, impair immune function, and diminish quality of life (Gibson, 2006). When combined with the rise of NCDs, such as diabetes, osteoporosis, and cardiovascular diseases, malnutrition presents a complex challenge. These interrelated issues complicate healthcare treatment plans and resource allocation, necessitating multifaceted policy solutions that consider not only healthcare infrastructure but also the economic, social, and educational factors impacting older adults' well-being.

Population ageing is a global phenomenon; however, the pace of ageing varies across regions. Developing countries are experiencing more rapid demographic transitions than developed nations, and countries with significant rural populations are ageing even faster. In many rural areas, limited access to healthcare services and adequate nutrition resources further exacerbates the vulnerabilities of older adults (United Nations, 2022). The increasing number of older individuals is expected to drive growing demand for health services (United Nations, 2022). Alongside this demographic shift, the prevalence of NCDs is rising, with conditions such as obesity and hypertension becoming more prominent compared to communicable diseases, under-nutrition, tuberculosis, and HIV (WHO, 2021a; WHO, 2021b). While life expectancy at birth and at old age is steadily increasing, alongside healthier life expectancy (WHO, 2021a; WHO, 2021b), hunger, under-nutrition, and micronutrient deficiencies persist in certain global regions, and food insecurity remains particularly prevalent among individuals aged 50–70 years and those aged 70 and above (United Nations, 2022).

The demographic shift toward an older population, coupled with the rising burden of NCDs and malnutrition, represents a notable challenge for today's global policymakers (WHO, 2020a; WHO, 2020b). Managing this issue will require innovative strategies, international collaboration, and a focus on promoting health across the lifespan. Older adults face specific nutritional requirements due to physiological changes, such as reduced appetite, altered metabolism, and increased risk of chronic diseases, making them particularly vulnerable to malnutrition (WHO, 2021a; WHO, 2021b). Moreover, social and economic challenges, such as limited mobility, fixed incomes, lack of pensions, limited entitlement to health insurance, and social isolation, further hinder older adults' access to nutritious food. To ensure food and nutrition security for older adults, we recommend that policies address several challenges:

- **Age-Sensitive Nutrition Policies:** The development of age-sensitive nutrition policies, such as tailored dietary guidelines and specialized food programs, is crucial. Current food systems are often designed to meet the needs of younger populations, leaving older adults unserved or underserved. Policymakers must prioritize the creation of policies that promote healthy ageing and address these specific needs.
- **Economic Support Programs:** Many older adults face economic constraints, particularly those living on fixed incomes or in poverty, which limit their ability to afford nutritious food. Governments must implement social protection programs, such as food subsidies, voucher assistance, or cash transfers, to help older adults' access adequate nutrition.

- **Community-Based Initiatives:** Programs such as meal delivery services and senior nutrition programs can alleviate food insecurity for those with limited mobility or who live in remote areas.
- **Preventive Healthcare Approaches:** The ageing population puts additional pressure on healthcare systems as malnutrition and diet-related chronic diseases increase healthcare costs. A preventive approach, emphasizing nutrition education and healthy eating habits for older adults, is essential to reducing the burden of diet-related diseases and improving overall health outcomes.

4. Ageing and nutritional requirements

Ageing is a natural and inevitable process that leads to changes in nutritional needs due to alterations in physiology, metabolism, digestive capacity, and physical activity levels. Such factors as changes in body composition, decreased lean body mass, altered digestion and absorption, and the presence of chronic health conditions further influence the nutritional requirements of senior people (Morley, 2012a). Furthermore, age-related changes in the gastrointestinal system, such as reduced gastric acid secretion and slower intestinal motility, can impair nutrient absorption (Morley, 2012b), resulting in increased risk of malnutrition, micronutrient deficiencies, and chronic diseases such as osteoporosis and anemia (Ahmed and Haboubi, 2010a, 2010b).

In addition, older adults often experience a decline in sensory functions, such as taste and smell, which can reduce appetite and lead to inadequate food intake and inefficient nutrient absorption (Drewnowski and Schultz, 2001a, 2001b; Locher et al., 2005a, 2005b). These changes can lead to undernutrition, which is associated with a higher risk of chronic diseases, reduced immune function, and decreased quality of life (Keller, 2012a, 2012b). It is important to note that malnutrition in older adults is frequently underdiagnosed, as it may not always present obvious symptoms, making it essential to assess both food intake and nutrient absorption comprehensively (Keller et al., 2015a, 2015b).

Currently, the world faces two opposing but closely interrelated challenges: undernutrition and malnutrition on one hand, and obesity and its associated chronic diseases on the other. Food security, which ensures access to an adequate dietary intake for an active and healthy life, plays a central role in addressing these issues. It encompasses both the sufficient availability of basic foodstuffs, typically ensured by local agricultural resources at national or regional levels, and the population's purchasing power (FAO, 2006). The most widely used method for assessing food security involves investigating the family food budget to estimate the average daily per capita availability of food in terms of quantity and quality, relative to recommended dietary allowances. However, this approach often adopts a static perspective, focusing on average dietary needs without accounting for within-diet variability or the health profiles of different populations (Drewnowski and Specter, 2004).

The concept of food security is evolving to embrace a broader perspective, emphasizing the right of all individuals to an adequate diet. This shift, however, confronts the rapid expansion of unhealthy eating habits, which contribute to a wide spectrum of chronic disease risks across populations. As the global population ages, there is a pressing need to address the dual burden of malnutrition and diet-related chronic diseases by promoting balanced diets and healthier food choices (Popkin et al., 2020a, 2020b). Therefore, meeting nutritional needs is crucial for maintaining health, preventing chronic diseases, and promoting an optimal quality of life among an ageing population.

Malnutrition, encompassing both undernutrition and overnutrition, is prevalent among older adults and is closely linked to poor health outcomes. Undernutrition in older adults is associated with increased frailty, sarcopenia (progressive loss of skeletal muscle mass and function, which can impact physical ability and overall health; see Bauer et al., 2013a, 2013b), and mortality, while overnutrition contributes to obesity and related NCDs (Cederholm et al., 2017a, 2017b). Ensuring food and nutrition security for older adults requires targeted interventions that address their specific nutritional needs and barriers to accessing adequate food. Although older adults require specific nutrients, food systems often fail to address these unique dietary needs.

4.1 Macronutrients and ageing

The ageing process brings about several physiological changes in the body, particularly in muscle mass and fat distribution, which significantly influence nutritional needs of older adults. These changes make it essential to carefully consider macronutrient intake to promote health and prevent the decline in physical function often associated with ageing. Macronutrients—protein, carbohydrates, and fats—play a critical role in maintaining health and functional capacity of older individuals.

Understanding the specific requirements for these macronutrients is essential for promoting healthy ageing and preventing age-related health complications.

Protein is a primary macronutrient essential for maintaining muscle mass and strength, which are particularly important for older adults. As individuals age, they often experience sarcopenia (Bauer et al., 2013a, 2013b). Studies suggest that older adults require higher amounts of protein than younger populations to counteract muscle loss and preserve functional capacity (WHO, 2020c; WHO, 2020d). The recommended dietary allowance (RDA) for protein is typically 0.8 g per kilogram of body weight for adults, but some experts suggest that older adults may benefit from an intake of 1.0–1.2 g per kilogram to maintain muscle health (Volpi et al., 2013a, 2013b). Adequate protein intake is crucial in preventing muscle wasting and improving strength, which are vital for maintaining independence in older age.

Carbohydrates are another essential macronutrient that requires careful attention in ageing individuals. Older adults may experience slower glucose metabolism and increased insulin resistance, making it important to balance carbohydrate intake to manage blood sugar levels effectively. Consuming high fiber foods, such as fruits, vegetables, and whole grains, can improve digestion and prevent constipation, which is a common issue among older populations (Slavin, 2013a, 2013b). Furthermore, complex carbohydrates with a low glycemic index (which are slowly digested) are recommended to provide sustained energy throughout the day while avoiding blood sugar spikes that can exacerbate insulin resistance (Harvard Health Publishing, 2012).

For all age groups, fat intake is important for facilitating absorption of micronutrients. Fat intake also plays an important role in maintaining health during ageing. Healthy fats, particularly monounsaturated and polyunsaturated fats, contribute to cardiovascular health by reducing the risk of heart disease, which is more prevalent in older adults. Omega-3 fatty acids, found in fish, flaxseeds, and walnuts, are particularly beneficial in reducing inflammation and supporting brain health (Santos et al., 2020a, 2020b). Incorporating these fats into the diet can help mitigate the risk of cardiovascular disease and cognitive decline, conditions that are common in ageing populations. Thus, ensuring an adequate intake of these healthy fats can help mitigate age-related health risks and promote overall well-being.

4.2 Micronutrients and ageing

As individuals age, the ability to absorb certain vitamins and minerals can diminish due to changes in digestive function, decreased stomach acid production, and other age-related factors (Saltzman and Russell, 1998). Micronutrient deficiencies are a significant concern for older adults due to age-related changes in absorption and metabolism (Yilmaz et al., 2024). Consequently, micronutrient requirements may increase for older adults to maintain optimal health and prevent deficiencies. Key vitamins and minerals that require particular attention in ageing populations include vitamin D, calcium, vitamin B12, folate, and magnesium, which are essential for maintaining bone health, cognitive function, and overall well-being as individuals age (WHO, 2020c; WHO, 2020d). Thus, dietary adjustments, supplementation, and biofortification [Editors' note: see the related chapter on this topic in this volume], when necessary, can help older adults meet their micronutrient needs and reduce the risk of age-related health complications.

Vitamin D is essential for bone health, as it enhances calcium absorption and helps maintain bone density. However, older adults often experience a reduced capacity to synthesize vitamin D from sunlight, leading to an increased risk of deficiency. Deficient vitamin D levels can lead to osteoporosis, falls, and fractures, which are major concerns for older adults (Lips, 2006a, 2006b). To address this, older adults should aim for an intake of 800–1000 IU of vitamin D daily through either dietary sources such as fortified foods and fatty fish or supplementation (International Osteoporosis Foundation IOF, 2021).

Likewise, calcium is another critical nutrient for bone health and preventing osteoporosis and fractures, both of which are more prevalent in older adults (WHO, 2020c; WHO, 2020d). The recommended calcium intake for adults over the age of 50 is approximately 1200 mg per day (Institute of Medicine, 2011). Calcium-rich foods like dairy products, leafy greens, and fortified plant-based milks should be incorporated into the diet to meet these daily requirements.

Vitamin B12 is essential for nerve function and red blood cell production; however, its absorption decreases with age due to changes in stomach acid production and the use of medications that interfere with absorption. Vitamin B12 deficiency in older adults can lead to cognitive decline, anemia, and neurological issues (Smith et al., 2012a, 2012b). Therefore, to prevent the deficiency risk, supplementation or consumption of fortified foods is often recommended for older adults.

Folate and magnesium are other essential micronutrients that support heart and muscle health, respectively. As age-related deficiencies in these nutrients can lead to various health issues, including cardiovascular disease and muscle weakness, older adults are encouraged to consume adequate amounts of folate and magnesium through a balanced diet or supplements if needed. Ensuring sufficient intake of these nutrients is important for maintaining cardiovascular and muscular health in the ageing population (Mustafa Khalid et al., 2022).

4.3 Hydration and ageing

Hydration is a vital aspect of health for older adults, yet it is often overlooked due to age-related changes in thirst perception and kidney function. Maintaining proper hydration is critical for older adults, as ageing is often accompanied by physiological changes that increase the risk of dehydration. The sense of thirst tends to diminish with age, and the kidneys' ability to concentrate urine declines, making it more challenging for older adults to maintain adequate fluid balance. Dehydration in older adults can lead to a variety of health issues, including urinary tract infections, kidney stones, and cognitive impairment. Additionally, older adults are more prone to dehydration due to a reduced total body water content and the potential diuretic effects of medications they may be taking (Thomas et al., 2016a).

Ensuring adequate fluid intake through a combination of beverages and water-rich foods can help prevent dehydration and its associated health risks. Therefore, it is recommended that older adults consume sufficient fluids throughout the day, including water, herbal teas, and foods with high water content, such as fruits and vegetables (Thomas et al., 2016b; Thomas and Duthie, 2016c). Proactive hydration strategies are essential, as relying solely on thirst may not be sufficient to meet fluid needs in older adults. Thus, caregivers and healthcare providers should emphasize the importance of regular fluid consumption to maintain overall health, support kidney function, to prevent complications related to dehydration, and to support overall well-being in ageing populations.

4.4 Early life nutrition and long-term health

Early life nutrition is a critical determinant of long-term health outcomes, particularly for ageing populations. The relationship between early life nutrition and the health of older adults is another critical aspect of food and nutrition security. Early life, encompassing prenatal development, infancy, and childhood, is a foundational period for growth and development, with lasting implications for health in later life. Research suggests that the nutritional environment during childhood can have lasting effects on health outcomes in adulthood, including susceptibility to cognitive decline and chronic diseases such as diabetes and hypertension (Victora et al., 2008). This early-life influence, combined with dietary habits formed over the course of a lifetime, shapes the ability of older adults to maintain proper nutrition and manage NCDs in their later years (Ren et al., 2025). This connection underscores the importance of addressing nutritional disparities across the lifespan to promote healthy ageing. Therefore, ensuring adequate nutrition during prenatal development, infancy, and childhood can mitigate the risk of chronic diseases and promote healthy ageing. Addressing food and nutrition security across the lifespan, with a focus on early life, is essential for breaking the cycle of malnutrition and improving health outcomes for future generations.

Early life nutrition plays a pivotal role in shaping an individual's health trajectory. During this time, nutritional deficiencies or excesses can lead to epigenetic changes, which refers to heritable changes in gene expression that occur without alterations in the DNA sequence, often influenced by environmental and lifestyle factors; altered organ development; and an increased vulnerability to NCDs in adulthood (Barker, 2007). Maternal malnutrition during pregnancy can result in low birth weight, which is associated with a higher risk of cardiovascular diseases, diabetes, and hypertension in later life (Gluckman et al., 2008a, 2008b). Similarly, inadequate intake of essential nutrients such as iron, iodine, and vitamin A during early childhood can impair cognitive development and immune function, with long-term consequences for health and productivity (Peng et al., 2025).

The Developmental Origins of Health and Disease (DOHaD) hypothesis provides a framework for understanding how early life nutrition influences long-term health outcomes. This hypothesis posits that environmental factors, including nutrition, during critical developmental periods can program an individual's risk of chronic diseases in adulthood (Hanson and Gluckman, 2014). For instance, studies show that individuals exposed to famine during early life, such as those born during the Dutch Hunger Winter of 1944–1945, have a higher prevalence of obesity, diabetes, and cardiovascular diseases in later life (Roseboom et al., 2006a, 2006b). These findings underscore the importance of ensuring adequate nutrition during early life to mitigate the risk of NCDs among the ageing population and break the intergenerational cycle of malnutrition and poor health.

Moreover, disparities in food and nutrition security during early life can have lasting effects on long-term health outcomes. Children from food-insecure households are more likely to experience stunted growth, micronutrient deficiencies, and impaired cognitive development, which can increase their risk of chronic diseases in adulthood (Gundersen and Ziliak, 2015).

As individuals age, their physiological resilience declines, making them more susceptible to the cumulative effects of poor nutrition earlier in life, and the long-term effects of early life nutritional deficiencies. Older adults who experienced malnutrition during childhood may have reduced muscle mass, bone density, and cognitive function, increasing their risk of frailty, osteoporosis, and dementia (Sayer and Cooper, 2005a, 2005b). Moreover, the rising prevalence of NCDs among the ageing population further emphasizes the need for interventions targeting the root causes of these conditions, including early life nutrition.

5. Non-communicable Diseases and their impact on food and nutrition security

NCDs, including cardiovascular diseases, diabetes, cancer, and chronic respiratory diseases, are the leading causes of morbidity and mortality worldwide, particularly in older adults (WHO, 2025c; WHO, 2021d). These chronic conditions not only affect the quality of life but also complicate food and nutrition security. NCDs often require specialized dietary modifications, such as low-sodium, low-sugar, and low-fat foods, which can be difficult to manage for ageing individuals, especially when they have limited access to nutritious food or lack the resources to maintain a balanced diet (Beaglehole et al., 2011; FAO, 2021a; FAO, 2021b). Access to such particular diets remains a significant challenge, particularly in low-resource settings. Additionally, these dietary restrictions may lead to confusion, frustration, and difficulties in accessing foods that are both suitable for managing the disease and nutritionally adequate (Nishida et al., 2004a, 2004b).

NCDs have a profound impact on the nutritional status of older adults. These diseases not only reduce the quality of life but also impair the body's ability to utilize nutrients efficiently. For instance, individuals with diabetes may experience fluctuations in blood sugar levels that necessitate careful attention to diet, which can be challenging in the context of food insecurity (Nishida et al., 2004a). Hypertension likewise necessitates management of carbohydrates, salt, and fat intakes (Lee et al., 2021).

The relationship between food and nutrition security and NCDs is bidirectional. On the one hand, poor dietary habits and inadequate access to nutritious food are key risk factors for the development of NCDs (Global Burden of Disease Collaborative Network, 2020a, 2020b). On the other hand, the presence of NCDs often leads to altered dietary requirements, making it more challenging for individuals to maintain proper nutrition. Older adults with NCDs may struggle to adhere to these specialized diets due to physical and cognitive limitations, lack of access to appropriate foods, or financial constraints (Lee et al., 2021). Furthermore, the side effects of medications used to manage NCDs, such as nausea, loss of appetite, or changes in taste perception, can further reduce food intake, compounding the challenges of maintaining nutritional security (Müller et al., 2022). Moreover, NCDs often require long-term medications that may have side effects, such as appetite suppression, gastrointestinal disturbances, or increased nutritional requirements. Medications used to manage cardiovascular diseases, for instance, may interfere with the absorption of essential vitamins and minerals (Baum et al., 2013a, 2013b). These side effects, combined with the chronic health conditions themselves, create additional barriers to maintaining adequate food and nutrition security for older adults.

Many countries face the dual challenge of undernutrition and overnutrition, with older adults and NCD-affected individuals often caught in the middle (FAO, 2021a). Malnutrition in older adults with NCDs can worsen the progression of these diseases, leading to poorer health outcomes and higher healthcare costs (Keller, 2012a, 2012b). The ageing population presents significant challenges for food and nutrition security, as older adults are particularly vulnerable to both inadequate food intake and the increasing burden of NCDs. Understanding the impact of ageing on food and nutrition security requires a comprehensive examination of the physical, social, and economic factors that influence seniors' dietary habits and health outcomes.

Interventions targeting food and nutrition security in older populations with NCDs have been shown to significantly improve health outcomes. For instance, community-based programs that promote the availability of affordable, nutritious foods, along with health education and nutritional counseling, can reduce the incidence and severity of NCDs among older adults (Swinburn et al., 2019). Similarly, improving the food environment, enhancing food literacy, and offering support for meal preparation can help mitigate the risks of malnutrition and NCDs in ageing populations (Basiotis and Lino, 2020).

Food and nutrition security, as a policy aspiration, embodies ensuring that food is available, accessible, adequate, and culturally acceptable. It is closely linked to three of the four central challenges of population ageing: ensuring income security in old age, access to affordable and appropriate health services, and the need for suitable housing and supportive living environments (WHO, 2015a). Addressing the dynamic interaction between ageing, food and nutrition security, and NCDs is a strategic and critical policy approach. Given the significance of early childhood as a period of risk for many chronic NCD risk factors—coinciding with a period of rapid cognitive, social, and physical developmental flourishing—it is essential to adopt a preventive trajectory that supports food and nutrition security from an early age (Black et al., 2013).

Protecting and enhancing our capacity to age successfully, while aligning healthy lifestyles with increasing life expectancy, extends beyond individual dietary choices. It requires a comprehensive policy framework focused on achieving lifecycle health, which includes preventing and controlling childhood undernutrition, ensuring adequate nutrition, and providing nurturing care for early childhood growth and development (Victora et al., 2008). Adult health and well-being, including cognitive function, can be shaped in influential and potentially reversible ways, ultimately determining the pace and manner of our departure from late life. Nutrition is central to this lifecycle health, underscoring its critical role in promoting healthy ageing and reducing the burden of NCDs (Bloom et al., 2011a, 2011b).

5.1 The burden of non-communicable diseases in ageing populations

The global rise in NCDs, driven by unhealthy diets high in processed foods, sugar, salt, and unhealthy saturated fats, places additional strain on food systems to provide healthier options (WHO, 2020a; WHO, 2020b). Such diets contribute to the development of NCDs, while diets rich in fruits, vegetables, whole grains, and lean proteins are protective (Afshin et al., 2019).

Ageing is a significant risk factor for NCDs due to physiological changes such as reduced metabolic rate, altered body composition, and increased oxidative stress (which links age-related physiological decline to the molecular and cellular damage that underlies chronic disease development) (López-Otín et al., 2013). These age-related changes, combined with poor nutrition, create a vicious cycle where NCDs exacerbate nutritional deficiencies, and poor nutrition worsens NCD outcomes (Bloom et al., 2011a, 2011b). For example, older adults with diabetes are at higher risk of malnutrition due to dietary restrictions and complications such as neuropathy, which refers to nerve damage caused by prolonged high blood glucose levels, and gastroparesis, which results from damage to the vagus nerve (a type of autonomic neuropathy) that controls stomach muscle contractions (Volkert et al., 2019).

5.2 Food insecurity and health outcomes

Research shows that food insecurity among older adults is associated with adverse health outcomes, including poor mental health, increased hospitalization rates, and higher mortality (Seligman et al., 2010). A study by Ziliak and Gundersen (2016) found that food-insecure seniors are more likely to report chronic conditions such as diabetes and hypertension. Furthermore, food insecurity can lead to compromised dietary quality, with older adults prioritizing calorie-dense, nutrient-poor foods due to financial constraints (Tarasuk et al., 2016). Food insecurity in older adults is linked to NCDs and obesity, which are often associated with unhealthy lifestyle behaviors (Gianfredi et al., 2024b). Unhealthy food environments, which refer to the physical, economic, policy, and sociocultural conditions that contribute to food insecurity, increase the risk of developing NCDs, thereby further affecting food and nutrition security (Du Plooy et al., 2018a, 2018b).

Food insecurity has profound implications for the physical and mental health of older adults, exacerbating existing health conditions and increasing the risk of new ones.

- **Malnutrition:** Food insecurity often leads to inadequate intake of essential nutrients, resulting in malnutrition. This can weaken the immune system, increase susceptibility to infections, and delay recovery from illnesses (Seligman et al., 2010).
- **Functional Decline:** Poor nutrition can accelerate physical decline, reducing muscle mass, strength, and mobility. This increases the risk of falls, fractures, and loss of independence (Lee and Frongillo, 2001a, 2001b).
- **Depression and Anxiety:** Food insecurity is strongly associated with mental health issues, including depression and anxiety. The stress of uncertain food access and the stigma associated with food insecurity can exacerbate these conditions (Dean et al., 2015).

- **Cognitive Decline:** Inadequate nutrition, particularly deficiencies in vitamins and minerals such as B12 and folate, has been linked to cognitive impairment and an increased risk of dementia among older adults (Shatenstein et al., 2012).

Food insecurity among older adults is influenced by a combination of economic, social, and health-related factors:

- **Economic Challenges:** Fixed incomes, such as pensions or social security, often fail to keep pace with rising living costs, leaving older adults with limited resources to purchase nutritious food (Ziliak and Gundersen, 2020a; Ziliak and Gundersen, 2020b).
- **Health and Mobility Issues:** Chronic health conditions, physical disabilities, and reduced mobility can make it difficult for older adults to shop for groceries or prepare meals (Lee and Frongillo, 2001a, 2001b).
- **Social Isolation:** Older adults who live alone or lack social support networks are at higher risk of food insecurity due to limited access to assistance or community resources (Dean et al., 2015).
- **Food Deserts:** Many older adults reside in areas with limited access to affordable and nutritious food, particularly in rural or low-income urban neighborhoods (Sharkey et al., 2010).

5.3 Diet and non-communicable diseases (NCDs)

Good nutrition is essential for preventing and managing NCDs, making it a global health priority (Carrara and Schulz, 2018). For instance, the Mediterranean diet, which emphasizes plant-based foods, olive oil, and moderate fish consumption, along with daily exercise, has been associated with a lower risk of cardiovascular diseases and improved cognitive function in older adults (Estruch et al., 2018a, 2018b). Similarly, diets that are high in fiber and low in refined sugars can help manage diabetes and prevent obesity, both of which are prevalent among the ageing population (Reynolds et al., 2025).

Addressing food and nutrition insecurity requires a multifaceted approach, together with policies and programs that provide affordable, nutrient-dense food options, such as subsidized meal programs or community gardens, aiming to improve access for older adults. Additionally, nutrition education tailored to the needs of ageing individuals can empower them to make informed dietary choices and improve their health outcomes.

5.3.1 Challenges to food and nutrition security

The ageing population faces unique challenges related to diet, nutrition, and the prevention of NCDs. Food and nutrition security are vital for ageing populations to prevent and manage non-communicable diseases. Ensuring access to healthy, safe, and affordable food can significantly impact the health and well-being of older adults, reducing the prevalence and severity of NCDs (Gianfredi et al., 2024c; Odoms-Young et al., 2024; Carrara and Schulz, 2018). However, the intersection of food and nutrition security, ageing, and NCDs presents a complex public health challenge. As individuals age, their nutritional requirements change due to shifts in metabolism, muscle mass, and the absorption of nutrients.

5.4 Socioeconomic factors

Older adults face significant socioeconomic barriers to food security, which are further complicated by the prevalence of NCDs. A primary challenge is the economic vulnerability of older individuals, many of whom live on fixed incomes, such as pensions or social security, which limit their ability to afford nutritious food (Lloyd et al., 2017; HelpAge International, 2023, 2023b). This issue is particularly severe in low- and middle-income countries (LMICs), where social safety nets are often inadequate. In these settings, rising food costs combined with limited financial resources force older adults to rely on cheaper, nutrient-poor food options, which increases their risk of malnutrition and worsens the management of NCDs (Lüdecke et al., 2018). Furthermore, nutrient-dense foods like fruits, vegetables, and lean proteins tend to be more expensive than calorie-dense, less nutritious options, creating a financial barrier to healthy eating (FAO, 2021a). Older adults living in rural or food-desert areas are at a significant disadvantage. They often face challenges accessing fresh, nutritious food due to limited transportation options and a lack of nearby grocery stores (Kim and Binns, 2019).

Social isolation further compounds food insecurity among older adults. Many elderly individuals experience loss of family or friends, which reduces their access to shared meals and community support systems (Dean et al., 2014). This problem is particularly pronounced among older women, who are more likely to live alone and face poverty (Lee and Frongillo 2001a). Elderly people who live alone may find it more difficult to prepare meals or access social support networks that could assist with shopping and meal preparation. Studies have shown that older adults who are socially isolated are more likely to experience food insecurity and poorer nutritional outcomes (Liu et al., 2020). Without the help of family members, friends, or community programs, food access can become even more problematic, particularly for those who live in rural areas or lack transportation (Kim and Binns, 2019). This economic barrier disproportionately affects older adults and those managing NCDs (Lee and Kim, 2022).

The availability of healthy food outlets is also a critical factor in food security. In many low-income or rural communities, food deserts pose a substantial challenge. In these areas, older adults may be forced to rely on convenience stores or fast-food outlets that offer nutrient-poor options, making it harder to manage NCDs and increasing the risk of diet-related health problems (Ver Ploeg et al., 2009).

5.5 Physical and social barriers

Ageing populations often face physical limitations that restrict their ability to shop for food or prepare meals, leading to greater reliance on processed or convenience foods that are typically lower in nutritional value (HelpAge International, 2023, 2023b). Mobility issues, chronic health conditions, and disabilities can make it difficult for older adults to access grocery stores or prepare balanced meals, further compromising their nutritional status (FAO, 2021a). In addition to physical challenges, loneliness and social isolation among older adults can decrease motivation to cook or eat well, further compromising food security and nutritional status (FAO, 2021a).

5.6 Healthcare and food system integration

As populations age globally, the intersection of healthcare and food systems has become increasingly vital in ensuring the well-being of older adults. Food and nutrition security are essential components of public health, directly influencing health outcomes, chronic disease management, and quality of life. However, the ageing population faces unique challenges related to accessing nutritious food, maintaining adequate dietary intake, and managing health conditions through diet. Integrating healthcare services with food systems is a key strategy to address these challenges, promote better health outcomes, and improve food and nutrition security (Fanzo et al., 2020a, 2020b; WHO, 2020e; FAO, 2020).

Healthcare and food systems often operate independently, leading to missed opportunities for incorporating nutrition into the prevention and management of NCDs (Fanzo et al., 2020a, 2020b). Older adults and individuals with NCDs frequently lack access to information on how to manage their diets effectively, particularly in rural or underserved areas (WHO, 2020e; FAO, 2020). Additionally, age-related physiological changes, such as reduced appetite, slower metabolism, and altered taste and smell, can lead to inadequate food intake in seniors (Erickson et al., 2019a, 2019b). Moreover, mental health issues such as depression and cognitive decline, common among seniors, can further complicate food choices and access to proper nutrition (Harris et al., 2015a, 2015b).

It is essential to integrate healthcare services with food systems, particularly for ageing populations, to ensure they have the resources and support to maintain a healthy, balanced diet. One key challenge is the fragmentation of services. In many healthcare settings, food and nutrition are treated as separate issues, resulting in limited collaboration between healthcare providers, nutritionists, and community food programs (Meyers et al., 2020). This fragmentation means that older adults may not receive sufficient dietary support or education as part of their routine healthcare visits (Meyers et al., 2020). Moreover, healthcare providers may lack the resources to connect patients with local food assistance programs or community-based nutrition services, leaving many seniors without essential support (Meyers et al., 2020).

Financial barriers also contribute significantly to food insecurity among older adults, as many seniors live on fixed incomes, coupled by the rising cost of food, particularly fresh fruits, vegetables, and other healthy items, making it difficult to afford healthy and nutritious food, particularly in low-income or rural areas (Guthrie et al., 2016). Without integration between healthcare services and food systems, these financial barriers are unlikely to be addressed comprehensively, resulting in persistent food insecurity.

5.7 Climate change, environmental and agricultural challenges

The worldwide demographic shift toward an ageing population poses significant challenges to food and nutrition security, principally in the context of environmental degradation and unsustainable agricultural practices. Environmental challenges such as climate change, soil degradation, and water scarcity are intensifying the pressures on agricultural systems, threatening the availability and accessibility of nutritious food (IPCC, 2021a; IPCC, 2021b). Extreme weather events, soil degradation, and water scarcity threaten agricultural productivity, further diminishing access to fresh and nutritious foods (IPCC, 2021a; IPCC, 2021b). Additionally, industrial agriculture often prioritizes calorie production over nutrient density, which does not align with the dietary needs of ageing populations or those with NCDs (Fanzo et al., 2020a, 2020b).

Climate change is exacerbating environmental challenges, with rising temperatures, unpredictable weather patterns, and extreme events such as droughts and floods disrupting agricultural production worldwide (IPCC, 2021a, 2021b). For the ageing population, these disruptions can result in reduced access to affordable and nutritious food, particularly since older adults often rely on fixed incomes, making them especially susceptible to price fluctuations caused by food shortages (IPCC, 2021a, 2021b). Climate change can also alter the nutritional content of crops, reducing the availability of essential vitamins and minerals needed for maintaining health in older age (Myers et al., 2017).

Soil degradation and water scarcity further complicate food production challenges. Practices like intensive farming, deforestation, and overuse of mineral fertilizers have led to the depletion of arable land, while excessive groundwater extraction has brought about declining water tables (WEP, 2020a; FAO and UNCCD, 2020b). These issues disproportionately affect smallholder farmers in developing countries, many of whom are older adults. As their capacity to produce food diminishes, their ability to meet their own nutritional needs and contribute to local food systems also declines (HelpAge International, 2018a, 2018b). The ageing population faces unique challenges within the agricultural sector as well, particularly in rural areas where younger generations migrate to urban centers, leaving older adults to manage farms with limited physical capacity and access to modern technologies (HelpAge International, 2018a, 2018b). This labor shortage can lead to reduced agricultural productivity and a decline in the diversity of crops grown, which is critical for a balanced diet.

Furthermore, the global shift toward calorie-dense but nutrient-poor diets threatens nutrition security in older populations. Industrialized agriculture often prioritizes high-yield crops like wheat, rice, and maize over nutrient-rich foods such as fruits, vegetables, and legumes (Tilman and Clark, 2014). This trend contributes to the double burden of malnutrition, where older adults may experience both undernutrition and obesity-related health issues simultaneously. The lack of dietary diversity leads to micronutrient deficiencies, which are associated with increased risks of chronic diseases and weakened immune systems in older age (WHO, 2020e; FAO, 2020).

5.7.1 Cultural and ethnic considerations

As the global population ages, food and nutrition security becomes an increasingly significant concern, particularly when considering the cultural and ethnic factors that influence dietary patterns and health practices. Food security is a fundamental determinant of health, encompassing the availability, access, utilization, and stability of food. For ageing individuals, these factors can be influenced by their cultural and ethnic backgrounds, which shape dietary patterns, food preferences, and health practices (Bertheussen et al., 2019a, 2019b). The cultural and ethnic identities of ageing individuals can affect their food security in various ways, particularly regarding access to culturally appropriate foods, dietary patterns, and social practices that influence eating behaviors.

Many older adults maintain traditional diets shaped by their cultural and ethnic backgrounds. For instance, Hispanic, Asian, and African American communities in the United States have distinct dietary preferences based on historical, geographic, and cultural factors (Harris et al., 2015a, 2015b). These dietary patterns can impact food security in several ways: culturally specific foods may be harder to access, more expensive, or require preparation methods that not all communities can easily afford or replicate in a limited-resource environment (Guthrie et al., 2016). The Mediterranean diet, commonly followed in Southern European countries, emphasizes foods that may be difficult to obtain in areas with poor access to fresh produce, posing a barrier to maintaining nutritional security (Liu et al., 2020). Likewise, the traditional diets of African American seniors often include high-fat, high-sodium, and high-sugar foods, which can exacerbate health problems like hypertension, diabetes, and cardiovascular disease if not balanced by healthier food options (Guthrie et al., 2016).

Mainstream healthcare or food assistance programs may not always accommodate the food traditions, dietary preferences, and needs of diverse ethnic groups (Harris et al., 2015a, 2015b). Ignoring these cultural factors can hinder the effectiveness of interventions and reduce participation in nutrition programs.

Ethnicity also shapes attitudes toward health and nutrition. Cultural beliefs may encourage or discourage certain foods, and in some cultures, food is a central element of family gatherings, celebrations, and caregiving. These cultural practices can lead to overconsumption or inadequate nutrient intake, especially when health-promoting behaviors conflict with deeply ingrained social and emotional aspects of food (Chung et al., 2021).

For instance, food may be seen as an expression of love or care, leading to challenges in promoting healthy eating habits among older adults in some ethnic communities (Playdon et al., 2023).

Food insecurity among ageing individuals is influenced not only by cultural and ethnic dietary preferences but also by various social and economic barriers. Access to affordable, culturally appropriate foods is a major challenge faced by older adults from minority ethnic backgrounds. Discrimination and the lack of available culturally appropriate foods in certain neighborhoods exacerbates food insecurity, particularly in urban areas where ethnic minorities may live. These challenges complicate efforts to ensure adequate nutrition for ageing populations from diverse cultural backgrounds (González et al., 2019).

Additionally, older adults from minority ethnic groups may face language barriers, which can hinder their ability to access nutrition education, government assistance programs, or healthcare services related to food security. Studies have shown that older adults from immigrant populations, such as Latino or Asian American seniors, are less likely to participate in US food assistance programs due to language and cultural barriers, leading to underutilization of available resources (Vega et al., 2019). This creates a cycle of food insecurity that is difficult to break without tailored interventions that are culturally sensitive and take into account language differences.

The ageing process often leads to changes in physical and cognitive functions, which can further limit an individual's ability to purchase, prepare, or consume food. These challenges are compounded for seniors who rely on social networks for food preparation and consumption, especially in ethnic communities where communal meals are a cultural norm (Tsang et al., 2025). If these social structures weaken due to geographic separation, migration, or family changes, older adults may lose the support systems necessary to maintain adequate nutrition.

5.8 Technological and innovation gaps

The ageing global population presents unique challenges for food and nutrition security, particularly in the context of NCDs. While technological advancements and innovations have the potential to address some of these challenges, significant gaps remain in their development, accessibility, and implementation. Innovations such as fortified foods, easy-to-prepare and nutritioius meals, and digital tools could be pivotal in improving food and nutrition security for older adults and NCD patients. However, the lack of innovation in food production and delivery systems tailored to the needs of these populations, as well as barriers to accessing digital tools, hinders their potential benefits (FAO, 2011; Frishammar et al., 2023). Addressing these gaps is crucial to promoting healthy ageing and reducing the burden of NCDs.

- **Limited Access to Digital Health Tools:** Digital health tools, such as mobile apps, telehealth platforms, and wearable devices, can play a transformative role in promoting nutrition and managing NCDs among older adults. These technologies can help older adults access food delivery services, nutritional information, and medical advice remotely. However, many older adults face barriers to utilizing these tools, such as limited digital literacy, lack of internet connectivity, and affordability issues (Czaja et al., 2019). These barriers are particularly pronounced in low- and middle-income countries (LMICs), where infrastructure and resources are often inadequate.
- **Insufficient Development of Age-Friendly Technologies:** A significant issue is that many existing technologies are not designed with the specific needs of older adults in mind. For example, mobile apps and wearable devices may have complex interfaces that are difficult for older users to navigate. Additionally, few technologies address the unique nutritional requirements of older adults, such as the need for higher protein intake to prevent sarcopenia or tailored dietary recommendations for managing chronic conditions (Napierkowski et al., 2021). This gap limits the ability of older adults to access nutrition-specific technologies that could enhance their quality of life.
- **Gaps in Food Fortification and Functional Foods:** Food fortification and functional foods have the potential to address micronutrient deficiencies and support NCDs prevention among older adults. However, the development and availability of fortified foods tailored to the needs of older populations remain limited. Furthermore, there is a lack of awareness and acceptance of these products among older adults, particularly in LMICs where food fortification programs are often underfunded and poorly implemented (FAO, 2011). As a result, many older adults lack access to these critical nutritional interventions, contributing to ongoing food insecurity and poor health outcomes.
- **Limited Adoption of Precision Nutrition:** Precision nutrition, which involves tailoring dietary recommendations based on genetic, metabolic, and lifestyle factors, holds promise for improving health outcomes among older adults. Despite this potential, the adoption of precision nutrition

technologies is hindered by high costs, limited access to genetic testing, and a lack of evidence-based guidelines for older populations (Ordovas et al., 2018). These barriers prevent precision nutrition from reaching those who could benefit most from personalized dietary recommendations.

- **Inadequate Integration of Technology into Healthcare Systems:** While technological innovations have the potential to revolutionize nutrition and NCD management, their integration into healthcare systems remains inadequate. For example, electronic health records (EHRs) are rarely used to track and address the nutritional status of older adults, and telehealth platforms are underutilized for delivering nutrition counseling and monitoring chronic conditions (Bashshur et al., 2020a). The under integration of these technologies within healthcare systems limits their potential to support nutrition interventions and manage NCDs more effectively.

The failure to address these technological and innovation gaps has significant implications for public health and sustainable development. Older adults who lack access to digital health tools and age-friendly technologies are at greater risk of malnutrition and NCDs, which increases healthcare costs and reduces their quality of life.

5.9 Policy and governance gaps

Significant policy and governance gaps limit the effectiveness of interventions aimed at improving food and nutrition security for the ageing population. Many countries lack policies that address the specific food and nutrition needs of seniors and those with NCDs (Fanzo et al., 2020a, 2020b). Food distribution systems often fail to reach vulnerable groups, including the elderly and those with chronic illnesses, particularly in rural or marginalized communities (Pereira et al., 2022). Additionally, existing policies and governance structures often fail to adequately address the specific needs of older adults, resulting in significant gaps that hinder progress in improving food security:

- **Lack of Age-Specific Nutrition Guidelines:** Many countries lack dietary guidelines tailored to the nutritional needs of older adults. While general nutrition recommendations exist, they often do not account for age-related changes in metabolism, nutrient absorption, and chronic disease management (Napierkowski et al., 2021). The absence of age-specific guidelines leaves older populations underserved, at risk of malnutrition, and without adequate nutritional guidance.

- **Fragmented Health and Nutrition Policies:** Nutrition policies for older adults are often fragmented across health, social welfare, and agricultural sectors, leading to inefficiencies and gaps in service delivery. For instance, nutrition programs may focus predominantly on maternal and child health while neglecting older adults (Tohit et al., 2025). This lack of integration undermines efforts to address the complex interplay between nutrition, ageing, and NCDs. A more cohesive policy framework that addresses the needs of elderly people is crucial for improving food and nutrition security across populations.
- **Inadequate Social Protection Programs:** Social protection programs, such as food assistance and income support, are critical for ensuring food security among vulnerable older adults. However, many programs are underfunded, poorly targeted, or inaccessible to those in need. This is particularly true in LMICs, where high poverty rates among older adults exacerbate the challenges of accessing food and nutrition (FAO, 2012). The lack of targeted, effective social protection programs leaves many elderly individuals at risk of food insecurity and malnutrition.
- **Weak Multi-Sectoral Coordination:** Effective governance requires collaboration across multiple sectors, including health, agriculture, education, and social welfare. However, weak coordination mechanisms often result in siloed approaches that fail to address the root causes of food and nutrition insecurity (FAO, 2021a; Sackar et al., 2023). For example, agricultural policies may prioritize crop yields over nutritional quality, while health policies may focus on treatment rather than prevention. Strengthening multi-sectoral coordination is necessary to address the interconnected issues of nutrition, ageing, and NCDs.
- **Limited Engagement:** Older adults are often excluded from decision-making processes related to food and nutrition policies. This lack of engagement undermines the relevance and effectiveness of interventions aimed at addressing their needs (Kong et al., 2025). Community-based approaches that involve older adults in program design and implementation are essential for ensuring that interventions are culturally appropriate, tailored to their preferences and the local conditions, and more likely to succeed.
- **Insufficient Monitoring and Evaluation:** Many countries lack robust systems for monitoring and evaluating food and nutrition programs for older adults. Without reliable data, it is difficult to assess the impact of

interventions or identify areas for improvement. This gap is particularly concerning in LMICs, where resources are limited, and the burden of malnutrition and NCDs is high (Song and Song, 2014). Establishing better monitoring and evaluation systems will help optimize program effectiveness and guide policy decisions.

The failure to address these policy and governance gaps has significant consequences for public health and sustainable development. Malnutrition among older adults exacerbates the burden of NCDs, leading to increased healthcare costs, decreased quality of life, and greater demand for healthcare services. Furthermore, food insecurity contributes to social isolation and mental health issues, further compounding the challenges faced by ageing populations. These gaps also undermine global efforts to achieve the Sustainable Development Goals (SDGs), particularly Goal 2 (Zero Hunger) and Goal 3 (Good Health and Well-being).

5.10 Intersection of conflict situations, food security, ageing, and NCDs

The intersection of conflict situations, food and nutrition insecurity, and NCDs presents a complex and interconnected set of challenges, particularly for ageing populations. These issues are exacerbated in conflict zones, where the elderly face disproportionate vulnerabilities due to disrupted healthcare systems, food insecurity, and the rising prevalence of NCDs (Omoruyi et al., 2024).

The Rome Statute of the International Criminal Court (ICC) defines war crimes as violations of international humanitarian law such as targeting civilians, widespread destruction, and forced displacement. This specifically includes "intentionally using starvation of civilians as a method of warfare by depriving them of objects indispensable to their survival, including wilfully impeding relief supplies" (International Criminal Court ICC, 2021; see especially Article 8.2.b.xxv).

In conflict zones, these crimes severely disrupt access to essential services, including healthcare and food systems (HelpAge International, 2025). Elderly individuals often face discrimination, neglect, and physical harm during conflicts. For older adults, who are already vulnerable due to age-related health challenges, the consequences of conflicts can be devastating (HelpAge International, 2025). Forced displacement, in particular, leaves elderly individuals without shelter, food, or healthcare, worsening their physical and mental health outcomes (WHO, 2020h). Additionally, the breakdown of healthcare systems in war-torn regions disproportionately impacts older adults with NCDs, which require continuous care and management (WHO, 2020h).

Food and nutrition security are critical issues that intertwine with the challenges of conflict situations and ageing populations. Armed conflict often disrupts food systems, leading to food scarcity, rising prices, and widespread malnutrition. For older adults, malnutrition is particularly concerning as it can accelerate the onset of frailty, reduce immunity, and worsen the progression of chronic diseases. The elderly, who may already be suffering from weakened immune systems, are at higher risk of undernourishment during active conflict. Moreover, the lack of access to adequate nutrition can exacerbate the negative effects of NCDs, which are prevalent in older populations. For instance, food insecurity and poor nutrition can lead to uncontrolled blood sugar levels in individuals with diabetes, increase the risk of heart disease, and aggravate high blood pressure (WFP, 2019).

The management of NCDs becomes increasingly difficult during conflicts. Disruptions in healthcare, shortages of medications, and the breakdown of medical infrastructure can worsen the outcomes for older adults with chronic conditions (WHO, 2023). Older adults with NCDs face unique vulnerabilities. For example, the lack of access to essential medications, such as insulin for diabetes or antihypertensive drugs for hypertension, puts them at greater risk of severe complications. The breakdown of healthcare systems also prevents effective monitoring and management of these conditions, leading to higher mortality rates in conflict-affected areas.

The intersection of conflict situations, food insecurity, and NCDs creates a vicious cycle, particularly for older adults. Conflict-induced displacement disrupts food systems and healthcare services, leading to malnutrition and inadequate management of chronic diseases. The scarcity of nutritious food contributes to the development of diet-related NCDs, such as obesity and hypertension. Conversely, lack of access to medication and healthcare services in conflict zones prevents older adults from managing existing NCDs, which can then lead to more severe complications, including death (Greene-Cramer et al., 2020). Moreover, the psychological impact of living in conflict, including stress and trauma, exacerbates both physical and mental health challenges. Mental health issues such as depression and anxiety are common among displaced older adults and often intertwine with NCDs, further compounding their vulnerability.

Delivery of assistance during conflict should employ a "triple nexus approach," integrating humanitarian, development, and peace, and must take management of NCDs among elderly people into account. Addressing the negative synergies among conflict, food insecurity, and NCDs, particularly for ageing populations in conflict-affected areas, require coordinated efforts to prevent war crimes, resolve conflict justly, ensure food and nutrition security, and improve access to healthcare services (Infante, 2021).

5.10.1 Strategies for improving food and nutrition security among older adults

By developing age-sensitive nutrition policies, expanding social protection programs, and promoting community-based initiatives, governments can ensure that older adults remain healthy and food-secure, especially in the context of NCDs. Addressing these challenges is not only a moral imperative but also a necessary step toward building sustainable and inclusive food systems for all.

Moreover, guaranteeing food and nutrition security in the ageing population requires targeted interventions. Policymakers and healthcare providers have emphasized the importance of nutrition screening and education, which can help identify and mitigate risks associated with undernutrition in older adults (Visvanathan et al., 2015). Community-based initiatives, such as meal programs and grocery delivery services, have shown promise in improving food access for older individuals, particularly in vulnerable areas (Lloyd et al., 2017). Furthermore, integrating nutrition counseling into primary healthcare for older adults can improve dietary patterns and help manage or prevent NCDs (Hickson et al., 2016).

5.11 Global health initiatives

Population ageing presents significant challenges for global health systems, particularly in relation to food and nutrition security and the rising prevalence of NCDs among older adults. While global health initiatives such as the WHO Global Strategy on Diet, Physical Activity, and Health (WHO, 2021e), and the Integrated Care for Older People (ICOPE) program (WHO, 2020i) provide essential frameworks for addressing these challenges, further efforts are needed to ensure that older adults have access to the resources and support they require. Through coordinated action at local, national, and international levels, the health and well-being of the elderly can be improved, promoting healthier ageing and reducing the impact of NCDs.

The Global Burden of Disease study highlights the critical impact of poor-quality diets on global health, attributing one in five deaths to dietary factors (Zapata et al., 2023). Recognizing these challenges, several global health initiatives have been implemented to address food and nutrition insecurity among older adults.

The United Nations Decade of Action on Nutrition aims to eradicate hunger, end malnutrition, and reduce the burden of diet-related NCDs (Fedacko et al., 2022a, 2022b). One notable initiative is the WHO's Global Strategy on Diet, Physical Activity, and Health, which emphasizes the importance of balanced diets and physical activity in the prevention and management of NCDs among older adults (WHO, 2021e). Efforts include enhancing access to nutrient-dense foods, such as fruits, vegetables, and proteins while reducing the consumption of harmful substances like salt, sugar, and unhealthy fats.

Additionally, the UNICEF/WHO Framework for Action to End Malnutrition specifically addresses undernutrition and food insecurity by ensuring older adults, especially those in vulnerable communities, have access to nutritious food (UNICEF and World Health Organization WHO, 2020). This framework includes interventions such as increasing food access, improving nutritional education, and enhancing social protection systems to ensure older individuals receive adequate food. It also recognizes the importance of addressing financial constraints and physical limitations that may hinder access to nutritious meals.

To combat the increasing burden of NCDs, global health organizations have developed strategies to focus on prevention, early detection, and management tailored to the needs of older adults. For example, the WHO Global Action Plan for the Prevention and Control of NCDs 2013–2020 aims to reduce premature mortality from NCDs by targeting key risk factors, such as tobacco use, physical inactivity, and unhealthy diets (WHO, 2013). This plan encourages member countries to develop national policies to reduce NCD risk and promote health among older populations. Additionally, the Global Framework for the Prevention and Control of NCDs emphasizes a combination of primary prevention, such as promoting healthier lifestyles, and secondary prevention, including regular screenings for early detection (WHO, 2021e).

Furthermore, ICOPE provides a model for integrated care that includes the management of NCDs, nutritional support, and rehabilitation for older individuals (WHO, 2020i). This program focuses on managing common health conditions in older adults, such as frailty, hypertension, and cognitive decline, through a person-centered approach involving dietary modifications and medication management. It stresses the importance of integrating NCD prevention into public health strategies for the elderly, ensuring that policies address both the biological and socio-economic factors that contribute to NCDs.

The Global Action Plan for Healthy Lives and Well-being for All, endorsed by the United Nations in 2019, seeks to accelerate progress towards achieving the Sustainable Development Goals (SDGs), particularly SDG 3: Good Health and Well-Being (United Nations, 2019c). This initiative emphasizes the importance of addressing health inequalities, including those that disproportionately affect older populations, through collaboration among global health organizations, governments, and the private sector.

Despite these global health initiatives, several challenges remain in ensuring food and nutrition security and addressing NCDs among older populations. Barriers such as limited mobility, chronic health conditions, and financial constraints hinder older adults' ability to access nutritious food and appropriate healthcare services. Additionally, older adults in rural areas may encounter even greater difficulties due to limited infrastructure and healthcare facilities.

To address these challenges, it is crucial for global health initiatives to focus on increasing awareness and education regarding the unique nutritional needs of older adults. Governments and international organizations should collaborate to improve the health of older populations by ensuring well-equipped healthcare system, strengthening social safety nets, providing targeted subsidies for nutritious foods, and improving access to healthcare services, particularly in rural and underserved areas.

Moreover, community-based interventions that promote healthy ageing and support food security should be expanded. For instance, local governments can establish programs that provide older individuals with access to community gardens, meal delivery services, and health education. Fostering an environment where older adults are supported in maintaining a healthy diet and managing chronic conditions can improve the overall health and quality of life for older populations.

5.12 Policy and public health approaches

Addressing the challenges outlined in this chapter requires a life-course-oriented, multi-faceted policy approach that integrates healthcare, sustainable agriculture, community-based initiatives, and public policy. Governments and organizations must prioritize the creation of supportive environments that ensure older adults have consistent access to nutritious food, particularly in low-income or rural settings. Public health policies should promote age-friendly food systems, where initiatives such as food delivery services, nutrition assistance programs, and community gardens are integrated to support older adults' access to healthy meals (Hoare et al., 2020).

Additionally, strategies to improve the affordability of healthy foods, such as subsidies or social welfare programs, are critical in reducing food insecurity in ageing populations (Barrett et al., 2020). For example, in the United States, expanding access to programs like the Supplemental Nutrition Assistance Program (SNAP) for older adults can alleviate food insecurity and improve dietary quality (Bartfeld and Dunifon, 2006). These measures are essential for addressing disparities in food access and supporting the nutritional well-being of older populations.

The role of healthcare systems in addressing food and nutrition security cannot be overlooked. Incorporating nutritional assessments into routine health check-ups for older adults and individuals with NCDs, along with providing targeted interventions to address deficiencies, is essential for improving health outcomes (Tessema et al., 2023). Healthcare providers play a critical role in screening for malnutrition and providing dietary counseling to older patients, which can help prevent complications related to poor nutrition (Kaiser et al., 2010a, 2010b). Moreover, training healthcare professionals to recognize the signs of food insecurity and malnutrition can facilitate early intervention and support (Mullaney et al., 2021). Such efforts should also include integrating nutrition counseling into routine care for older adults to identify and address dietary deficiencies early. These approaches contribute to improving health outcomes and reducing the burden of NCDs.

Community-based programs have demonstrated effectiveness in addressing food insecurity and promoting healthy ageing. For instance, meal delivery services and senior nutrition programs not only improve dietary intake but also reduce social isolation among older adults (Thomas et al., 2016c, 2016d). Community initiatives, such as local food banks and programs promoting access to community gardens, can further support food security in ageing populations. Public health interventions also play a crucial role in promoting healthy diets and reducing the burden of NCDs among the elderly. Governments and healthcare providers should prioritize initiatives that address the social determinants of health, such as income support, transportation services, and social engagement programs.

Promoting early-life nutrition and ensuring food and nutrition security require a comprehensive, multi-sectoral approach. Public health interventions should focus on improving maternal and child nutrition through programs such as prenatal care, breastfeeding support, and school meal programs. These efforts lay the foundation for better health outcomes across the lifespan (Black et al., 2013). Policies addressing social determinants of health, such as poverty alleviation, education, and access to healthcare, are essential for reducing disparities in food and nutrition security. Additionally, integrating nutrition education into community-based programs can empower individuals to make informed dietary choices throughout their lives. By fostering a culture of informed nutrition practices and addressing systemic inequities, these interventions can significantly enhance food security and public health outcomes.

5.13 Addressing food and nutrition security to mitigate NCDs in ageing populations

A balanced approach to nutrition is essential for ageing populations to live healthier, more independent lives. Ensuring adequate intake of macronutrients and micronutrients is vital for maintaining health and preventing common age-related conditions. By focusing on protein-rich foods, healthy fats, complex carbohydrates, and key vitamins and minerals, older adults can support their overall well-being. Furthermore, maintaining proper hydration and adopting dietary strategies such as smaller, more frequent meals and nutrient-dense food choices can enhance health outcomes in ageing individuals.

Policy-level interventions are also critical for addressing food and nutrition security in ageing populations. Policies that support sustainable agriculture, food fortification, and social protection programs can enhance food availability and accessibility for older adults (Fanzo et al., 2018). Additionally, integrating nutrition services into healthcare systems can help identify and address nutritional deficiencies in older adults with NCDs (Wahlqvist et al., 2012b). Interventions aimed at preventing and managing NCDs should focus on promoting healthy dietary patterns, ensuring access to affordable and nutritious food, and addressing social determinants of health.

5.13.1 Strategies to address policy and governance gaps

Addressing policy and governance gaps in food and nutrition security for older adults is essential for promoting healthy ageing and reducing the burden of NCDs. By developing age-specific nutrition policies, strengthening multi-sectoral coordination, expanding social protection programs, promoting community engagement, and improving monitoring and evaluation, governments can create an enabling environment for food and nutrition security. Integrating food and nutrition security into public health strategies is essential to achieving global health goals and ensuring equity for older populations. Culturally tailored interventions, alongside efforts to address socioeconomic disparities, can significantly enhance food security and overall well-being among older adults.

- **Develop Age-Specific Nutrition Policies:** Governments should prioritize the development of dietary guidelines and nutrition programs tailored to the needs of older adults. These policies should address age-related changes in nutrient requirements and promote the prevention and management of NCDs.
- **Strengthen Multi-Sectoral Coordination:** Effective governance requires the establishment of effective multi-sectoral coordination mechanisms that bring together stakeholders from health, agriculture, social welfare, and other relevant sectors. Integrated approaches can ensure that food and nutrition policies are aligned with broader health and development goals.
- **Expand Social Protection Programs:** Governments should invest in social protection programs that provide food assistance, income support, and healthcare services to vulnerable older adults. These programs should be designed to reach those most in need, particularly in LMICs.
- **Promote Community Engagement:** Older adults themselves should be actively involved in the design, implementation, and evaluation of food and nutrition programs. Community-based approaches can help ensure that interventions are culturally appropriate and address the specific needs of older populations.
- **Improve Monitoring and Evaluation:** Robust systems for monitoring and evaluating food and nutrition programs are essential for ensuring accountability and driving continuous improvement. Governments should invest in data collection and analysis to track progress and identify gaps.

5.14 Strategies to bridge technological and innovation gaps

Technological and innovation gaps in food and nutrition security for older adults represent a missed opportunity to promote healthy ageing and reduce the burden of NCDs. By addressing these gaps through digital inclusion, age-friendly technologies, food fortification, precision nutrition, and healthcare integration, stakeholders can unlock the full potential of technology to improve health outcomes for older populations.

- **Promote Digital Inclusion for Older Adults:** Efforts to improve digital literacy and access to technology among older adults are essential for bridging the digital divide. Governments and organizations should invest in training programs, affordable internet access, and user-friendly devices designed specifically for older populations.
- **Develop Age-Friendly Technologies:** Technology developers should prioritize the creation of age-friendly tools that address the unique needs of older adults. This includes designing intuitive interfaces, incorporating voice-activated features, and providing personalized nutrition and NCD management solutions.
- **Expand Food Fortification Programs:** Governments and private sector stakeholders should invest in the development and promotion of fortified foods and functional foods tailored to the nutritional needs of older adults. Public awareness campaigns can help increase acceptance and utilization of these products.
- **Advance Precision Nutrition Research:** Research institutions and funding agencies should prioritize studies on precision nutrition for older adults. This includes developing cost-effective genetic testing methods, generating evidence-based guidelines, and exploring the role of gut microbiota, which refers to the diverse community of microorganisms—including bacteria, viruses, fungi, and other microbes—that live in the gastrointestinal tract, primarily in the intestines, in ageing and NCD prevention.
- **Integrate Technology into Healthcare Systems:** Healthcare systems should leverage technology to improve nutrition and NCD management for older adults. This includes incorporating nutrition tracking into EHRs, expanding telehealth services, and using data analytics to identify at-risk populations.

5.15 Addressing food security through culturally tailored approaches

Food and nutrition security is a critical issue among the ageing population, particularly in the context of cultural and ethnic considerations. Cultural food preferences, dietary habits, and health beliefs all play a significant role in shaping older adults' food security. As the ageing population continues to diversify, it is essential to recognize and address the unique barriers faced by ethnically diverse older adults in accessing nutritious and culturally appropriate foods. By adopting culturally sensitive approaches and tailoring nutrition assistance programs to meet the needs of these communities, we can enhance food security and improve the overall well-being of older adults.

Nutrition education programs that recognize the specific dietary needs and preferences of diverse ethnic groups are crucial. By integrating traditional food patterns with modern nutritional science, healthcare professionals can help older adults make informed choices that align with both their cultural values and health requirements. Partnerships with local ethnic markets and grocery stores can improve the availability of culturally relevant foods at affordable prices, helping to reduce food insecurity (Smith et al., 2016). Providing nutrition assistance in multiple languages, incorporating ethnic food preferences, and promoting cultural competency among program staff can help to increase participation in these essential programs.

In addition to policy changes, health practitioners must consider the role of culturally competent care when assessing the food security of older adults. For example, dietitians, physicians, and social workers should inquire about an individual's cultural food practices, preferences, as well as barriers to accessing food. Regular screenings for food insecurity, combined with culturally tailored solutions, can help ensure that ageing individuals receive the support they need to maintain good nutrition and health (Harris et al., 2015a).

5.16 Models for integrating healthcare and food systems

The integration of healthcare and food systems is crucial for addressing food and nutrition insecurity among the ageing population. Collaborative approaches involving medical-led nutrition programs, community-based initiatives, government interventions, and technological innovations all offer promising approaches to improving food security and health outcomes for older adults. Several models have shown promise in integrating these systems and improving food security for older adults.

- **Medical-Led Nutrition Programs:** One promising model is the incorporation of nutrition interventions into healthcare settings. For example, some healthcare providers have begun integrating dietitians into primary care teams, allowing for more comprehensive assessments of nutritional needs and the development of individualized care plans that include dietary guidance (Meyers et al., 2020). These interventions can ensure that seniors receive the proper nutrition to manage chronic conditions such as diabetes, heart disease, and osteoporosis. By including nutrition in healthcare protocols, healthcare providers can identify food insecurity and refer patients to appropriate community resources, such as food banks, meal delivery services, or government assistance programs.

- **Community-Based Food and Health Programs:** Another successful model is community-based initiatives that connect healthcare services with local food systems. Programs like the "Healthy Ageing Through Healthy Eating" initiative offer free or low-cost meals to seniors while providing nutrition education and health screenings (Smith et al., 2016). These programs often collaborate with local farms, food banks, and social service organizations to ensure that older adults have access to culturally appropriate and nutritious foods. These services not only address food security but also provide an opportunity for social interaction, reducing the risk of isolation and promoting mental and emotional well-being.
- **Government and Policy Interventions:** Government programs can play a critical role in facilitating the integration of healthcare and food systems for older adults. Expanding these programs to include more comprehensive healthcare screenings for food insecurity, offering culturally appropriate foods, and ensuring that healthcare providers are trained to assess and address food-related issues can strengthen the integration of food and healthcare systems.
- **Telemedicine and Online Food Delivery Services:** The rise of telemedicine and online food delivery services offers new opportunities for integrating healthcare and food systems. Seniors who are unable to travel to healthcare appointments or grocery stores can benefit from telehealth consultations with dietitians or healthcare providers who can offer dietary advice and refer them to local or online food delivery programs (Erickson et al., 2019a, 2019b). By combining remote healthcare services with access to nutritious food options delivered directly to the home, these models can overcome many of the logistical and transportation challenges that limit access to both healthcare and food for older adults.
- **Encouraging Research and Innovation:** Governments and research institutions should invest in studies to better understand the nutritional needs of older adults and develop innovative solutions to address food insecurity in ageing populations.

5.17 Addressing the environmental and agricultural challenges

The ageing population faces significant environmental and agricultural challenges that threaten their food and nutrition security. Climate change, soil degradation, and water scarcity are undermining agricultural productivity, while labor shortages and dietary shifts are exacerbating malnutrition among older adults. Addressing these issues requires a holistic approach that integrates sustainable agriculture, social protection, and community-based interventions to build resilient food systems. By prioritizing the needs of the ageing population, we can build a more resilient and inclusive food system that ensures food and nutrition security for all.

To address these challenges, a multi-faceted approach is required. Policymakers must prioritize sustainable agricultural practices that enhance resilience to climate change while ensuring the availability of nutritious food for all age groups. Agroecological methods, such as crop rotation, intercropping, and organic farming, can improve soil health and biodiversity while reducing reliance on chemical inputs (Altieri et al., 2015a, 2015b). Additionally, investments in water-saving technologies and infrastructure can help mitigate the impacts of water scarcity on food production. Also, empowering older farmers through training and access to resources can enhance their productivity and resilience to environmental and economic shocks.

5.18 Addressing the challenges of violent conflict

A number of steps are necessary to address the challenges of violent conflict, food insecurity, and NCDs in the ageing population. The humanitarian-development-peace triple nexus approach is essential to these efforts. First, international interventions must focus on holding perpetrators of conflicts accountable and ensuring the protection of civilians, including in particular elderly people. This can be achieved through strengthened international laws and greater enforcement of humanitarian laws designed to protect vulnerable populations during conflicts.

Second, improving food and nutrition security is critical, especially in conflict zones. Humanitarian agencies must ensure that food aid reaches vulnerable populations, including older adults, and is nutritionally adequate to meet the specific needs of those groups, including elderly people and children. Additionally, efforts must be made to rebuild local food systems and improve access to fresh and nutritious foods in post-conflict recovery efforts (WEP, 2024; FAO, 2016).

Lastly, addressing the rising burden of NCDs in the ageing population requires strengthening healthcare systems, particularly in conflict-prone regions. This includes ensuring the continued supply of medications, providing access to chronic disease management services, and offering mental health support to elderly individuals affected by conflict and trauma.

6. Conclusion

The intersection of food and nutrition security, ageing, and NCDs presents complex and interrelated challenges that demand urgent and comprehensive action. The ageing population faces unique vulnerabilities, including physiological changes, socioeconomic barriers, and the increasing burden of chronic diseases, all of which exacerbate the risk of malnutrition and food insecurity. Addressing these challenges requires a holistic, multi-sectoral approach that integrates policy reform, healthcare interventions, agriculture, social protection, community support, and sustainable food systems. Moreover, integrating food and nutrition security into public health frameworks and fostering collaborations across sectors can strengthen efforts to mitigate the challenges.

Ensuring access to safe, affordable, and nutritious food is essential for promoting healthy ageing and reducing the prevalence and severity of NCDs. Food and nutrition security for older adults is not only a moral imperative but also a necessary step toward achieving global health and development goals. Therefore, a comprehensive and coordinated effort is vital to build resilient food systems that ensure food and nutrition security for all, particularly in the context of ageing and the rising burden of NCDs. By adopting a life-course approach and prioritizing the nutritional needs of ageing populations, societies can promote healthy ageing, reduce the burden of NCDs, and improve the quality of life for older adults. Future research and policy efforts should continue to focus on innovative, inclusive, and culturally sensitive solutions to address the complex challenges at the intersection of ageing, nutrition, and chronic disease. Investments in the health and well-being of older populations will not only enhance their quality of life but also contribute to sustainable development and global health equity.

References

Afshin, A., Sur, P.J., Fay, K.A., Cornaby, L., Ferrara, G., Salama, J.S., et al., 2019. Health effects of dietary risks in 195 countries, 1990–2017: a systematic analysis for the Global Burden of Disease Study 2017. Lancet. 393 (10184), 1958–1972. https://doi.org/10.1016/S0140-6736(19)30041-8.

Ahmed, T., Haboubi, N., 2010a. Assessment and management of nutrition in older people and its importance to health. Clin. interv. Ageing. 5, 207–216. https://doi.org/10.2147/CIA.S9664.

Ahmed, T., Haboubi, N., 2010b. Nutritional assessment and support in older people. Clin. Med. 10 (1), 54–58. https://doi.org/10.7861/clinmedicine.10-1-54.

Altieri, M.A., Nicholls, C.I., Montalba, R., 2015a. Agroecological approaches to enhance resilience among small farmers. Agron. Sustain. Dev. 35 (3), 869–890. https://www.cabidigitallibrary.org/doi/full/10.5555/20173384847.

Altieri, M.A., Nicholls, C.I., Montalba, R., 2017. Technological approaches to sustainable agriculture at a crossroads: an agroecological perspective. Sustainability 9 (3), 349. https://doi.org/10.3390/su9030349.

Altieri, M.A., Nicholls, C.I., Henao, A., Lana, M.A., 2015b. Agroecology and the design of climate change-resilient farming systems. Agron. Sustain. Dev. 35 (3), 869–890. https://doi.org/10.1007/s13593-015-0285-2.

Barker, D.J.P., 2007. The origins of the developmental origins theory. J. Intern. Med. 261 (5), 412–419. https://doi.org/10.1111/j.1365-2796.2007.01809.x.

Barrett, G.S., Jones, M.C., Thomas, K., 2020. Food insecurity and its impact on the health of older adults. Int. J. Public Health 65 (6), 827–835. https://doi.org/10.1007/s00038-020-01456-w.

Bartfeld, J., Dunifon, R., 2006. State-level predictors of food insecurity among households with children. J. Policy Anal. Manag. 25 (4), 921–942. https://doi.org/10.1002/pam.20214.

Bashshur, R.L., Shannon, G.W., Krupinski, E.A., 2020a. The taxonomy of telemedicine. Telemed e-Health 26 (1), 1–10. https://doi.org/10.1089/tmj.2020.29040.rb.

Basiotis, P.P., Lino, M., 2020. The role of nutrition assistance programs in promoting food security. Food Security 12 (2), 487–495. https://doi.org/10.1007/s12571-020-01006-9.

Bauer, J., Biolo, G., Cederholm, T., Cesari, M., Cruz-Jentoft, A.J., Morley, J.E., et al., 2013b. Evidence-based recommendations for optimal dietary protein intake in older people: a position paper from the PROT-AGE Study Group. J. Am. Med. Dir. Assoc. 14 (8), 542–559. https://doi.org/10.1016/j.jamda.2013.05.021.

Bauer, J.M., Bosy-Westphal, A., Lühmann, D., 2013a. Sarcopenia and its clinical relevance. J. Nutrition, Health & Ageing 17 (9), 809–814. https://doi.org/10.1007/s12603-013-0023-4.

Baum, J.I., Kim, I.-Y., Wolfe, R.R., 2013a. Protein consumption and the elderly: what is the optimal level of intake? Nutrients 5 (8), 4067–4088. https://doi.org/10.3390/nu5084067.

Baum, P.A., Smith, T., Auld, M., 2013b. Impact of long-term medication use on nutritional status in older adults with cardiovascular disease. J. Clin. Nutr. 91 (5), 812–818. https://doi.org/10.1056/NEJMoa1211765.

Beaglehole, R., Bonita, R., Horton, R., Adams, C., Alleyne, G., Asaria, P., et al., 2011. Priority actions for the non-communicable disease crisis. Lancet. 377 (9775), 1438–1447. https://doi.org/10.1016/S0140-6736(11)60393-0.

Bertheussen, M., Røise, S.L., Røttingen, J.A., 2019b. Ethnic differences in dietary patterns among older adults. J. Ageing & Soc. Policy 31 (2), 123–141. https://doi.org/10.1080/08959420.2018.1531493.

Bertheussen, G.F., Romundstad, P.R., Landmark, T., Kaasa, S., Dale, O., Helbostad, J.L., 2019a. Associations between physical activity and physical and mental health—a HUNT 3 study. Med. & Sci. Sports & Exerc. 51 (4), 781–788. https://doi.org/10.1249/mss.0b013e318206c66e.

Black, R.E., Victora, C.G., Walker, S.P., Bhutta, Z.A., Christian, P., de Onis, M., et al., 2013. Maternal and child undernutrition and overweight in low-income and middle-income countries. Lancet. 382 (9890), 427–451. https://doi.org/10.1016/S0140-6736(13)60937-X.

Bloom, D.E., Canning, D., Sevilla, J., 2011b. The effect of human capital on economic growth: a review of the literature. World Bank Policy Res. Working Pap. 23 (5), 1556–1575.

Bloom, D.E., Cafiero, E.T., Jané-Llopis, E., Abrahams-Gessel, S., Bloom, L.R., Fathima, S., et al. (2011a). The global economic burden of noncommunicable diseases. World Economic Forum. https://www3.weforum.org/docs/WEF_Harvard_HE_GlobalEconomicBurdenNonCommunicableDiseases_2011.pdf.

Carrara, A., Schulz, P., 2018. The role of health literacy in predicting adherence to nutritional recommendations: a systematic review. Patient Educ Couns. 101. https://doi.org/10.1016/j.pec.2017.07.005.

Cederholm, T., Barazzoni, R., Austin, P., Ballmer, P., Biolo, G., Bischoff, S.C., et al., 2017a. ESPEN guidelines on definitions and terminology of clinical nutrition. Clin. Nutr. 36 (1), 49–64. https://doi.org/10.1016/j.clnu.2016.09.004.

Cederholm, T., Barazzoni, R., Austin, P., Ballmer, P., Biolo, G., Deutz, N., et al., 2017b. ESPEN guidelines on nutrition in geriatrics. Clin. Nutr. 36 (1), 1–10. https://doi.org/10.1016/j.clnu.2016.07.031.

Chung, M., Meyer, M., Zhang, Y., 2021. Understanding the role of food in health and social practices among older adults. J Ageing Health 33 (6), 953–974. https://doi.org/10.1177/0898264320917419.

Czaja, S.J., Boot, W.R., Charness, N., Rogers, W.A., 2019. Designing for Older Adults: Principles and Creative Human Factors Approaches, 3rd ed. CRC Press.

Dean, M., Raats, M.M., Grunert, K.G., Lumbers, M., 2014. Factors influencing eating a varied diet in old age. Public Health Nutr 12 (12), 2421–2427. https://doi.org/10.1017/S1368980009005318.

Dean, W.R., Sharkey, J.R., Johnson, C.M., 2015. Food insecurity is associated with social capital, perceived personal disparity, and partnership status among older and senior adults in a largely rural area of central Texas. J Nutr Gerontol. Geriatrics 34 (1), 1–63. https://doi.org/10.1080/21551197.2014.1002656.

Drewnowski, A., Schultz, J.M., 2001a. Impact of ageing on eating behaviors, food choices, nutrition, and health status. J Nutrition, Health & Ageing 5 (2), 75–79. https://europepmc.org/article/med/11426286#impact.

Drewnowski, A., Schultz, H., 2001b. The sensory and nutritional properties of low-fat, low-calorie foods. J Am Dietetic Assoc 101 (1), 59–64. https://doi.org/10.1016/S0002-8223(01)00014-2.

Drewnowski, A., Specter, S.E., 2004. Poverty and obesity: the role of energy density and energy costs. Am. J. Clin. Nutr. 79 (1), 6–16. https://doi.org/10.1093/ajcn/79.1.6.

Du Plooy, C., Schönfeldt, H.C., Hall, N., 2018a. The relationship between food environments and dietary intake in low- and middle-income countries: a systematic review. Public Health Nutr 21 (17), 3173–3185. https://doi.org/10.1017/S1368980018002071.

Du Plooy, Z., Schönfeldt, H., Hall, N., 2018b. The role of traditional foods in food-based dietary guidelines – A South African case study on maas (cultured milk). Food Chem 238. https://doi.org/10.1016/j.foodchem.2017.04.044.

Economic Research Service, U.S. Department of Agriculture. (2023). Nutrition security. https://www.ers.usda.gov.

Erickson, J., Sadeghirad, B., Lytvyn, L., 2019b. The effectiveness of telemedicine in the management of chronic diseases: a systematic review. J. Telemed Telecare 25 (3), 167–177. https://doi.org/10.1177/2054270416681747.

Erickson, J., Sadeghirad, B., Lytvyn, L., Slavin, J.L., Johnston, B.C., 2019a. The scientific basis of guideline recommendations on sugar intake: a systematic review. Ann. Intern. Med. 166 (4), 257–267. https://doi.org/10.7326/M16-2020.

Estruch, R., Ros, E., Salas-Salvadó, J., et al., 2018b. Primary prevention of cardiovascular disease with a Mediterranean diet. N. Engl. J. Med. 378 (25), 2441–2453. https://doi.org/10.1056/NEJMoa1200303.

Estruch, R., Ros, E., Salas-Salvadó, J., Covas, M.I., Corella, D., ... Arós, F., et al., 2018a. Primary prevention of cardiovascular disease with a Mediterranean diet supplemented with extra-virgin olive oil or nuts. N. Engl. J. Med. 378 (25), e34. https://doi.org/10.1056/NEJMoa1800389.

Fanzo, J., et al., 2020b. Food Systems, Nutrition, and Health: Pathways to Sustainability and Equity. World Health Organization.

Fanzo, J., Hawkes, C., Udomkesmalee, E., Afshin, A., Allemandi, L., Assery, O., et al., 2018. Global nutrition report: shining a light to spur action on nutrition. Development Initiatives 2018 International Food Policy Research Institute.

Fanzo, J., Hawkes, C., Udomkesmalee, E., Afshin, A., Allemandi, L., Assery, O., et al., 2020a. 2020 Global Nutrition Report: Action on Equity to End Malnutrition. Development Initiatives, Bristol, UK.

FAO and UNCCD. 2020b. Land Degradation Neutrality for Water Security and Combatting Drought. Bonn, Germany https://openknowledge.fao.org/items/6c67445a-574c-494c-b6b0-515f198a65dc.

FAO, IFAD, UNICEF, WFP and WHO. 2019. The State of Food Security and Nutrition in the World 2019. Safeguarding against economic slowdowns and downturns. Rome, FAO. Licence: CC BY-NC-SA 3.0 IGO.

FAO, IFAD, UNICEF, WFP and WHO. 2020. In Brief to The State of Food Security and Nutrition in the World 2020. Transforming food systems for affordable healthy diets. Rome, FAO. https://doi.org/10.4060/ca9699en.

FAO, IFAD, UNICEF, WFP and WHO. 2020a. The State of Food Security and Nutrition in the World 2020. Transforming food systems for affordable healthy diets. Rome, FAO. https://doi.org/10.4060/ca9692en.

FAO, IFAD, UNICEF, WFP and WHO. 2024. In Brief to The State of Food Security and Nutrition in the World 2024 – Financing to end hunger, food insecurity and malnutrition in all its forms. Rome, FAO. https://doi.org/10.4060/cd1276en.

Fedacko, J., Takahashi, T., Singh, R., Pella, D., Chibisov, S., Hristova, K., et al., 2022b. Western diets and risk of non-communicable diseases. Funct Foods Nutraceuticals Metab Non-Communicable Dis. https://doi.org/10.1016/B978-0-12-819815-5.00042-2.

Fedacko, J., Zapata, M., Malnutria, K., 2022a. The Global Burden of Malnutrition: Insights and Strategies. Global Health Initiatives Press.

Food and Agriculture Organization (FAO) 1996. World food summit: Rome declaration on world food security and world food summit plan ofaction. https://www.fao.org/4/w3613e/w3613e00.htm.

Food and Agriculture Organization (FAO), 2006. Food security: Policy brief. FAO. https://www.fao.org/food-security/en/.

Food and Agriculture Organization (FAO), 2011. Combating Micronutrient Defi ciencies: Food-based Approaches https://www.fao.org/4/am027e/am027e.pdf.

Food and Agriculture Organization (FAO), 2016. PEACE, CONFLICT AND FOOD SECURITY What do we know about the linkages? https://openknowledge.fao.org/server/api/core/bitstreams/3fc73834-753a-469b-8580-1edf00e875bd/content.

Food and Agriculture Organization (FAO), International Fund for Agricultural Development (IFAD), United Nations Children's Fund (UNICEF), World Food Programme (WFP), & World Health Organization (WHO), 2021a. The state of food security and nutrition in the world 2021: Transforming food systems for food security, improved nutrition, and affordable healthy diets for all. https://doi.org/10.4060/cb4474en.

Food and Agriculture Organization (FAO), International Fund for Agricultural Development (IFAD), United Nations Children's Fund (UNICEF), World Food Programme (WFP), & World Health Organization (WHO), 2025. The State of Food Security and Nutrition in the World 2025 – Addressing high food price inflation for food security and nutrition. Rome, FAO. https://doi.org/10.4060/cd6008en.

Food and Agriculture Organization (FAO), World Health Organization (WHO), & United Nations (UN), 2021b. Addressing food and nutrition security among ageing populations. Food Security and Nutrition Report, 19(4), 34–42.

Food and agriculture organization (FAO),2012. https://openknowledge.fao.org/server/api/core/bitstreams/9fadaf39-334a-42e5-abd4-1338737ba5b6/content.

Frishammar, J., Essén, A., Bergström, F., Ekman, T., 2023. Digital health platforms for the elderly? Key adoption and usage barriers and ways to address them. Technol. Forecast. Soc. Change 189, 122319.

Gianfredi, V., Nucci, D., Lattanzio, R., Piccinelli, S., Cicconi, G., Farfan, S., et al., 2024c. Food insecurity and body mass index among older people: a systematic review and meta-analysis. Arch. Gerontol. Geriatr. 128. https://doi.org/10.1016/j.archger.2024.105606.

Gianfredi, V., Provenzano, S., Leoni, O., 2024a. Nutrition and healthy ageing: a review of the role of food security in preventing non-communicable diseases. J. Ageing Health 36 (2), 145–160. https://doi.org/10.1177/08982643231234567.

Gianfredi, V., Provenzano, S., Santangelo, O.E., 2024b. Food insecurity and non-communicable diseases: a systematic review. Nutrients 16 (1), 123. https://doi.org/10.3390/nu16010123.

Gibson, R.S., 2006. Principles of Nutritional Assessment, 2nd ed. Oxford University Press.

Global Burden of Disease Collaborative Network, 2020. Global burden of NCDs: risk factors and prevention. Glob Health Metr Rev 1 (1), 10–19.

Global Burden of Disease Collaborative Network, 2020b. Global Burden of Disease Study 2019 (GBD 2019) results. Institute for Health Metrics and Evaluation. https://www.healthdata.org/research-analysis/gbd.

Gluckman, P.D., Hanson, M.A., Pinal, C., 2008b. The developmental origins of adult disease. Am J Clin Nutr 87 (4), 795–798. https://doi.org/10.1093/ajcn/87.4.795.

Gluckman, P.D., Hanson, M.A., Cooper, C., Thornburg, K.L., 2008a. Effect of in utero and early-life conditions on adult health and disease. N. Engl. J. Med. 359 (1), 61–73. https://doi.org/10.1056/NEJMra0708473.

González, H.M., Tarraf, W., Rodríguez, C.J., Gallo, L.C., Sacco, R.L., Talavera, G.A., et al., 2019. Cardiovascular health among diverse hispanics/latinos: hispanic community health study/study of latinos (HCHS/SOL) results. Circulation 139 (10), 1317–1334. https://doi.org/10.1016/j.ahj.2016.02.008.

Greene-Cramer, B., Summers, A., Lopes-Cardozo, B., Husain, F., Couture, A., Bilukha, O., 2020. Noncommunicable disease burden among conflict-affected adults in Ukraine: a cross-sectional study of prevalence, risk factors, and effect of conflict on severity of disease and access to care. PLoS One 15 (4), e0231899.

Gundersen, C., Ziliak, J.P., 2015. Food insecurity and health outcomes. Health Aff. 34 (11), 1830–1839. https://doi.org/10.1377/hlthaff.2015.0645.

Gustavsson, J., Cederberg, C., Sonesson, U., van Otterdijk, R., Meybeck, A., 2011. Global food losses and food waste: Extent, causes, and prevention. Food and Agriculture Organization of the United Nations. https://www.fao.org/3/i2697e/i2697e.pdf.

Guthrie, J.F., Lin, B.H., Frazao, E., 2016. Role of food prepared away from home in the American diet, 1977-78 versus 1994-96: changes and consequences. J. Nutr. Educ. Behav. 34 (3), 140–150. https://doi.org/10.1016/S1499-4046(06)60083-3.

Haddad, E.A., Bines, J.E., Corvalan, C., 2020. Nutritional needs of older adults and the challenges they face. Age Ageing 49 (3), 487–495. https://doi.org/10.1093/ageing/afz211.

Hanson, M.A., Gluckman, P.D., 2014. Early developmental conditioning of later health and disease: physiology or pathophysiology? Physiological Rev. 94 (4), 1027–1076. https://doi.org/10.1152/physrev.00029.2013.

Harris, J.L., Schwartz, M.B., Brownell, K.D., 2015a. Marketing foods to children and adolescents: licensed characters and other promotions on packaged foods in the supermarket. Public Health Nutr. 13 (3), 409–417. https://doi.org/10.1017/S1368980009991339.

Harris, D., Singh, A., Wang, R., 2015b. Culturally tailored nutrition interventions for minority older adults: a review of the literature. J. Nutr. Gerontol. Geriatr. 34 (2), 143–162. https://doi.org/10.1080/21551197.2015.1063102.

Harvard Health Publishing, 2012. Choosing good carbs with the glycemic index. Harvard Health. https://www.health.harvard.edu/staying-healthy/choosing-good-carbs-with-the-glycemic-index?.

Hawkley, L.C., Cacioppo, J.T., 2010a. Loneliness matters: a theoretical and empirical review of consequences and mechanisms. Ann. Behav. Med. 40 (2), 218–227. https://doi.org/10.1007/s12160-010-9210-8.

HelpAge International, 2018a. Ageing and the SDGs: Key messages to ensure age-inclusive policies. https://www.helpage.org/silo/files/ageing-and-the-sdgskey-messages.pdf.

HelpAge International, 2018b. The ageing of rural populations: evidence on older farmers in lowand middle-income countries. https://www.helpage.org/silo/files/the-ageing-of-rural-populations-evidence-on-older-farmers-in-low-and-middleincome-countries.pdf.

HelpAge International, 2018b. The ageing population in rural areas: Challenges for agricultural sustainability.

HelpAge International, 2023. World Food Day: The silent struggles of older people facing food insecurity. https://www.helpage.org/news/world-food-day-the-silent-struggles-of-older-people-facing-food-insecurity/.

HelpAge International, 2025. We're still here: Older people in Lebanon struggle to recover from conflict and crisis. https://www.helpage.org/news/were-still-here-older-people-in-lebanon-struggle-to-recover-from-conflict-and-crisis/.

Hickson, M., et al., 2016. Nutritional interventions for older adults with chronic disease: a review. Clin. Nutr. 35 (3), 437–448. https://doi.org/10.1016/j.clnu.2015.03.003.

Hinrichs, C.C., 2003. The practice and politics of food system localization. J. Rural Stud. 19 (1), 33–45. https://doi.org/10.1016/S0743-0167(02)00040-2.

Hoare, P., Tinson, A., Menzies, D., 2020. Strengthening food systems for healthy ageing: a policy perspective. Public Health Nutr. 23 (5), 876–884. https://doi.org/10.1017/S136898001900314X.

Infante, V, 2021. Transforming the systems that contribute to fragility and humanitarian crises: programming across the triple nexus. Oxfam Briefing Paper. Oxford: Oxfam. https://policy-practice.oxfam.org/resources/transforming-the-systems-that-contribute-to-fragility-and-humanitarian-crises-p-621203/.

Institute of Medicine, 2011. Dietary reference intakes for calcium and vitamin D. National Academies Press. https://doi.org/10.17226/13050.

Intergovernmental Panel on Climate Change (IPCC), 2019. Climate change and land: an IPCC special report on climate change, desertification, land degradation, sustainable land management, food security, and greenhouse gas fluxes in terrestrial ecosystems. https://www.ipcc.ch/srccl/.

Intergovernmental Panel on Climate Change (IPCC), 2021a. Climate change and food security: adaptation strategies for a changing world. https://www.ipcc.ch/srccl/chapter/chapter-5/.
Intergovernmental Panel on Climate Change (IPCC), 2021b. Climate change 2021: the physical science basis. Contribution of Working Group I to the Sixth Assessment Report of the Intergovernmental Panel on Climate Change. Cambridge University Press. https://www.ipcc.ch/report/ar6/wg1/downloads/report/IPCC_AR6_WGI_FullReport.pdf.
International Criminal Court (ICC), 2021. Rome statute of the international criminal court. The Hague: ICC. https://www.icc-cpi.int/sites/default/files/2024–05/Rome-Statute-eng.pdf.
International Osteoporosis Foundation (IOF), 2021. Vitamin D factsheet: public intake recommendations and health implications. https://www.osteoporosis.foundation/patients/prevention/vitamin-d.
Kahana, E., Bhatta, T., Lovegreen, L.D., Kahana, B., Midlarsky, E., 2012. Altruism, helping, and volunteering: pathways to well-being in late life. J. Ageing Health 25 (1), 159–187. https://doi.org/10.1177/0898264312469665.
Kaiser, M.J., Bauer, J.M., Ramsch, C., 2010a. Frequency of malnutrition in older adults: a multinational perspective. J. Am. Geriatr. Soc. 58 (9), 1734–1738. https://doi.org/10.1111/j.1532-5415.2010.03016.x.
Kaiser, M.J., Bauer, J.M., Ramsch, C., Uter, W., Guigoz, Y., Cederholm, T., et al., 2010b. Validation of the mini nutritional assessment short-form (MNA-SF): a practical tool for identification of nutritional status. J. Nutrition, Health & Ageing 14 (8), 782–788. https://doi.org/10.1007/s12603-010-0174-1.
Keller, H.H., 2012a. Malnutrition in older adults: prevalence, consequences, and causes. Can J Dietetic Pract Res 73 (1), e298–e304. https://doi.org/10.3148/73.1.2012.e298.
Keller, H.H., 2012b. Undernutrition in older adults. Am. J. Clin. Nutr. 96 (5), 1335S–1341S. https://doi.org/10.3945/ajcn.112.042734.
Keller, H.H., Duizer, L.M., Avenell, A., 2015b. Malnutrition in older adults: a critical review. J. Clin. Nutr. 14 (4), 185–197. https://doi.org/10.1016/j.clnu.2014.06.011.
Keller, H.H., Østbye, T., Goy, R., 2015a. Nutritional risk and quality of life in community-dwelling seniors: a cross-sectional study. J. Nutrition, Health & Ageing 19 (1), 5–12. https://doi.org/10.1007/s12603-014-0490-8.
Kim, K., Binns, M., 2019. Food security and health among older adults in rural areas: a review of challenges and interventions. J. Ageing Health 31 (10), 1776–1798. https://doi.org/10.1177/0898264318798234.
Kong, C., Reuter, A., Buffel, T., Xu, W., 2025. A scoping review on older adults' participation in policymaking. J. Aging & Soc. Policy. 1–31.
Labouvie-Vief, G., 2003a. Dynamic integration: affect, cognition, and the self in adulthood. Curr. Dir. Psychol. Sci. 12 (6), 201–206. https://doi.org/10.1046/j.0963-7214.2003.01262.x.
Labouvie-Vief, G., 2003b. Dynamic integration theory: a life-span perspective on cognitive and emotional development. In: Birren, J.E., Schaie, K.W. (Eds.), Handbook of the Psychology of Ageing, 6th ed. Elsevier Academic Press, pp. 494–510.
Larrea-Killinger, C., Muñoz, A., Mascaró, J., 2024. Food security and ageing: addressing the nutritional needs of older adults with chronic diseases. Public Health Nutr. 27 (1), 45–58. https://doi.org/10.1017/S1368980023001234.
Lee, J.S., Frongillo, E.A., Keating, J., 2021. Nutritional vulnerability among older adults: findings from a national survey. J. Nutr. Gerontol. Geriatr. 40 (2), 123–135. https://doi.org/10.1080/21551197.2021.1898623.
Lee, J.S., Frongillo, E.A., 2001a. Nutritional and health consequences are associated with food insecurity among U.S. elderly persons. J. Nutr. 131 (5), 1503–1509. https://doi.org/10.1093/jn/131.5.1503.

Lee, J.S., Frongillo, E.A., 2001b. Factors associated with food insecurity among U.S. elderly persons: implications for policy. J. Nutr. 131 (5), 1484–1488. https://doi.org/10.1093/geronb/56.2.S94.

Lee, D.E., Kim, K., 2022. Regional difference in the effect of food accessibility and affordability on vegetable and fruit acquisition and healthy eating behaviors for older adults. Int. J. Environ. Res. Public Health 19 (22), 14973.

Lindau, S.T., Schumm, L.P., Laumann, E.O., Levinson, W., O'Muircheartaigh, C.A., Waite, L.J., 2020a. A study of sexuality and health among older adults in the United States. N. Engl. J. Med. 357 (8), 762–774. https://doi.org/10.1056/NEJMoa067423.

Lindau, S.T., Schumm, L.P., Laumann, E.O., Levinson, W., 2020b. The role of health and social networks in ageing: a comprehensive overview. J Ageing & Soc Policy 32 (4), 322–341. https://doi.org/10.1080/08959420.2020.1806182.

Lips, P., 2006a. Vitamin D physiology. Prog. Biophys. Mol. Biol. 92 (1), 4–8. https://doi.org/10.1016/j.pbiomolbio.2006.02.016.

Lips, P., 2006b. Vitamin D deficiency and secondary hyperparathyroidism in the elderly: consequences for bone loss and fractures and therapeutic implications. Endocr. Rev. 27 (5), 564–594. https://doi.org/10.1210/er.2006-0006.

Liu, Y., Zhang, Y., Zhang, L., 2020. Social isolation and food insecurity among older adults: a systematic review. Public Health Nutr. 23 (14), 2536–2548. https://doi.org/10.1017/S1368980020001234.

Lloyd, J.L., Wellman, N.S., Anderson, A.K., 2017. Older adults and food insecurity: a review of challenges and interventions. J. Nutr. Gerontol. Geriatr. 36 (4), 171–193. https://doi.org/10.1080/21551197.2017.1398672.

Locher, J.L., et al., 2005b. The relationship between food security and nutritional status in older adults. J. Nutr. 135 (5), 1016–1022. https://doi.org/10.1093/jn/135.5.1016.

Locher, J.L., Ritchie, C.S., Roth, D.L., Baker, P.S., Bodner, E.V., Allman, R.M., 2005a. Social isolation, support, and capital and nutritional risk in an older sample: ethnic and gender differences. Soc. Sci. & Med. 60 (4), 747–761. https://doi.org/10.1016/j.socscimed.2004.06.023.

López-Otín, C., Blasco, M.A., Partridge, L., Serrano, M., Kroemer, G., 2013. The hallmarks of ageing. Cell 153 (6), 1194–1217. https://doi.org/10.1016/j.cell.2013.05.039.

Lüdecke, D., Kofahl, C., von dem Knesebeck, O., 2018. Socioeconomic status and food insecurity among older adults in Germany: a cross-sectional analysis. BMC Public Health 18 (1), 1234. https://doi.org/10.1186/s12889-018-6138-6.

Meyers, A.E., Swanson, M., Jones, L., 2020. Integrating nutrition and healthcare: the role of dietitians in promoting food security among older adults. J. Nutr. Educ. Behav. 52 (3), 193–201. https://doi.org/10.1016/j.jneb.2019.09.004.

Morley, J.E., 2012a. Undernutrition in older adults. Family Pract. 29 (1), i89–i93. https://doi.org/10.1093/fampra/cmr054.

Morley, J.E., 2012b. Nutrition and ageing. Curr. OpClNutr & Metab Care 15 (1), 1–5. https://doi.org/10.1097/MCO.0b013e32834dbdb9.

Mozaffarian, D., Fleischhacker, S., Andrés, J.R., 2021. Prioritizing nutrition security in the US. Jama 325 (16), 1605–1606. https://doi.org/10.1001/jama.2021.1915.

Mullaney, A., Smith, T., Martin, G., 2021. Healthcare approaches to address food insecurity in older populations. J. Ageing & Soc. Policy 33 (3), 345–360. https://doi.org/10.1080/08959420.2020.1778237.

Müller, O., Krawinkel, M., Jahn, A., 2022. Malnutrition and health in developing countries. CMAJ 173 (3), 279–286. https://doi.org/10.1503/cmaj.050342.

Mustafa Khalid, N., Haron, H., Shahar, S., Fenech, M., 2022. Current evidence on the association of micronutrient malnutrition with mild cognitive impairment, frailty, and cognitive frailty among older adults: a scoping review. Int. J. Environ. Res. Public Health 19 (23), 15722.

Myers, S.S., Gaffikin, L., Golden, C.D., Ostfeld, R.S., Redford, K.H., Ricketts, T.H., et al., 2017. Emerging threats to human health from global environmental change. Proc. Natl. Acad. Sci. 114 (13), 3463–3470. https://doi.org/10.1073/pnas.1604981114.

Napierkowski, D.B., Prado, K.B., 2021. Nutritional needs in the older adult, guidelines and prevention strategies to optimize health and avoid chronic disease. Geriatr. Gerontol. Aging 15, 1–4.

Nesbitt, P.D., Lee, J.S., Wolf, K.L., 2021. Nutritional challenges among older adults: a review of the literature. J. Ageing Res. 2021, 1–10. https://doi.org/10.1155/2021/6657415.

Nishida, C., Uauy, R., Kumanyika, S., 2004b. Nutrition and NCDs: pathways to prevention. WHO Bull 82 (9), 400–404.

Nishida, C., Uauy, R., Kumanyika, S., Shetty, P., 2004a. The joint WHO/FAO expert consultation on diet, nutrition, and the prevention of chronic diseases: process, product, and policy implications. Public Health Nutr. 7 (1A), 245–250. https://doi.org/10.1079/PHN2003592.

Odoms-Young, A., Brown, A., Agurs-Collins, T., Glanz, K., 2024. Food insecurity, neighborhood food environment, and health disparities: state of the science, research gaps and opportunities. Am. J. Clin. Nutr. 119. https://doi.org/10.1016/j.ajcnut.2023.12.019.

Omoruyi, A.J., Knipe, P., Durojaye, E., 2024. Preventing diet-related non-communicable diseases: a rights-based approach to improving health outcomes through food and nutrition security. South Afr. J. Hum. Rights. 40 (3-4), 245–267.

Ordovas, J.M., Ferguson, L.R., Tai, E.S., Mathers, J.C., 2018. Personalised nutrition and health. BMJ. https://doi.org/10.1136/bmj.k2173. 361, bmj.k2173.

Peng, F., Lin, X., Liu, S., Zou, H., Yi, C., Tian, Y., 2025. Epidemiological trends and disparities in iodine, vitamin A, and iron deficiencies among children aged 0–14 years globally, 1990–2021. Front Nutr. 12, 1622945.

Pereira, M.H., Pereira, M.L., Campos, G.C., Molina, M.C., 2019d. Food insecurity and nutritional status among older adults: a systematic review. Nutr. Rev. 80 (4), 631–644.

Pinquart, M., Sörensen, S., 2001. Influences on loneliness in older adults: a meta-analysis. Basic Appl. Soc. Psychol. 23 (4), 245–266. https://doi.org/10.1207/S15324834BASP2304_2.

Playdon, M., Rogers, T.N., Brooks, E., Petersen, E.M., Tavake-Pasi, F., Lopez, J.A., et al., 2023. Sociocultural influences on dietary behavior and meal timing among Native Hawaiian and Pacific Islander women at risk of endometrial cancer: a qualitative investigation. Cancer Causes & Control 34 (1), 23–ss.

Popkin, B.M., Adair, L.S., Ng, S.W., 2020b. The global nutrition transition: the pandemic of obesity in developing countries. Nutr. Rev. 61 (7), 83–94. https://doi.org/10.1111/j.1753-4887.2003.tb00094.x.

Popkin, B.M., Corvalan, C., Grummer-Strawn, L.M., 2020a. Dynamics of the double burden of malnutrition and the changing nutrition reality. Lancet. 395 (10217), 65–74. https://doi.org/10.1016/S0140-6736(19)32497-3.

Ren, H., Zhou, Y., Liu, J., 2025. Nutrition in early life and its impact through the life course. Nutrients 17 (4), 632.

Reynolds, A.N., Lang, J., Brand, A., Mann, J., 2025. Higher fiber higher carbohydrate diets better than lower carbohydrate lower fiber diets for diabetes management: rapid review with meta-analyses. Obes. Rev. 26 (1), e13837.

Ribeiro, S., Oliveira, A., Santos, C., 2022. Food insecurity and malnutrition among older adults: a systematic review. Nutrients 14 (3), 567. https://doi.org/10.3390/nu14030567.

Roseboom, T.J., van der Meulen, J.H., Osmond, C., Barker, D.J., Bleker, O.P., 2006b. Coronary heart disease after prenatal exposure to the Dutch famine, 1944–1945. Heart 92 (2), 189–190. https://doi.org/10.1136/hrt.2005.073339.

Roseboom, T.J., van der Meulen, J.H.P., Ravelli, A.C.J., Osmond, C., Barker, D.J.P., Bleker, O.P., 2006a. Effects of prenatal exposure to the Dutch famine on adult disease in later life: an overview. Mol Cell Endocrinol 185 (1-2), 93–98. https://doi.org/10.1016/S0303-7207(01)00721-3.

Sackar, S.A., Apprey, C., Aduku, L.N.E., Thow, A.M., Annan, R., 2023. Operationalising multi-sectoral food-and nutrition-related policies to curb the rise in obesity in Ghana. Public Health Nutr 26 (12), 3230–3238.

Saltzman, J.R., Russell, R.M., 1998. The aging gut: nutritional issues. Gastroenterol Clin North Am 27 (2), 309–324.

Santos, M., Araújo, L.A., Oliveira, L.P., 2020a. Omega-3 fatty acids in ageing: beneficial effects on brain and cardiovascular health. Curr OpClNutr & Metab Care 23 (1), 1–8. https://doi.org/10.1097/MCO.0000000000000623.

Santos, H.E., Ribeiro, S.M.L., Rogero, M.M., 2020b. Omega-3 fatty acids and inflammatory processes: from molecules to man. Biochemical Soc Trans 48 (5), 2075–2087. https://doi.org/10.1042/BST20200202.

Sayer, A.A., Cooper, C., 2005a. Early diet and growth: impact on ageing. Proc Nutr Soc 64 (3), 455–462. https://doi.org/10.1079/PNS2005447.

Sayer, A.A., Cooper, C., 2005b. Use it or lose it: the importance of physical activity in maintaining strength and function in old age. Proc Nutr Soc 64 (3), 476–483. https://doi.org/10.1079/PNS2005479.

Seligman, H.K., Laraia, B.A., Kushel, M.B., 2010. Food insecurity is associated with chronic disease among low-income NHANES participants. J Nutr 140 (2), 304–310. https://doi.org/10.3945/jn.109.112573.

Shankar, A., McMunn, A., Banks, J., Steptoe, A., 2020. Loneliness, social isolation, and behavioral and biological health indicators in older adults. Health Psychol 30 (4), 377–385. https://doi.org/10.1037/a0022826.

Sharkey, J.R., Dean, W.R., Nalty, C.C., 2012. Food access and dietary intake among rural older adults. J Nutr Gerontol. Geriatr. 31 (2), 97–119. https://doi.org/10.1080/21551197.2012.678214.

Sharkey, J.R., Johnson, C.M., Dean, W.R., 2010. Food access and perceptions of the community and household food environment as correlates of fruit and vegetable intake among rural seniors. BMC Geriatrics 10 (1), 32. https://doi.org/10.1186/1471-2318-10-32.

Shatenstein, B., Payette, H., Nadon, S., Gray-Donald, K., 2012. Diet quality and cognition among older adults from the NuAge study. Exp Gerontol. 47 (5), 353–360. https://doi.org/10.1016/j.exger.2012.02.002.

Slavin, J.L., 2013a. Dietary fiber and body weight. Nutrition 29 (3), 125–130. https://doi.org/10.1016/j.nut.2012.07.004.

Slavin, J.L., 2013b. Fiber and prebiotics: mechanisms and health benefits. Nutrients 5 (4), 1417–1435. https://doi.org/10.3390/nu5041417.

Smith, C., D'Angelo, C., Wang, Y., 2016. Community interventions to reduce food insecurity among ageing populations in urban areas. J. Public Health Policy 37 (2), 275–292. https://doi.org/10.1057/s41271-016-0002-7.

Smith, A.D., Refsum, H., Bottiglieri, T., 2012a. Homocysteine, B vitamins, and cognitive impairment. Annu. Rev. Nutr. 32, 211–239. https://doi.org/10.1146/annurev-nutr-071811-150741.

Smith, A.D., Refsum, H., Bottiglieri, T., 2012b. Homocysteine and dementia: an overview. Ageing Res. Rev. 11 (2), 132–144. https://doi.org/10.1016/j.arr.2011.06.004.

Song, S., Song, W.O., 2014. National nutrition surveys in Asian countries: surveillance and monitoring efforts to improve global health. Asia Pac. J. Clin. Nutr. 23 (4), 514–523.

Stelmach-Mardas, M., Kleiser, C., Uzhova, I., Peñalvo, J.L., La Torre, G., Palys, W., et al., 2021. Seasonality of food groups and total energy intake: a systematic review and meta-analysis. Nutr. Rev. 79 (6), 663–686. https://doi.org/10.1093/nutrit/nuaa107.

Swinburn, B.A., Kraak, V.I., Allender, S., Atkins, V.J., Baker, P.I., Bogard, J.R., et al., 2019. The global syndemic of obesity, undernutrition, and climate change: the Lancet Commission report. Lancet. 393 (10173), 791–846. https://doi.org/10.1016/S0140-6736(18)32822-8.

Tarasuk, V., Mitchell, A., Dachner, N., 2016. Household food insecurity in Canada, 2014. Research to Identify Policy Options to Reduce Food Insecurity (PROOF). https://proof.utoronto.ca.

Tessema, M., Demissie, T., Fikadu, M., 2023. Nutritional interventions in older adults: the role of healthcare providers in addressing food insecurity and NCDs. Nutr. & Dietetics 80 (2), 107–115. https://doi.org/10.1111/1747-0080.12824.

Thomas, K.S., Akobundu, U., Dosa, D., 2016d. More than a meal? A randomized control trial comparing the effects of home-delivered meals programs on participants' feelings of loneliness. J. Gerontol. Ser. B: Psychol. Sci. Soc. Sci. 71 (6), 1049–1058. https://doi.org/10.1093/geronb/gbv111.

Thomas, D.R., Bishop, N.L., Hatcher, L., 2016b. The importance of hydration for older adults. J. Am. Geriatrics Soc. 64 (10), 2037–2042. https://doi.org/10.1111/jgs.14492.

Thomas, D.R., Cote, T.R., Lawhorne, L., Levenson, S.A., Rubenstein, L.Z., Smith, D.A., et al., 2016a. Understanding clinical dehydration and its treatment. J. Am. Med. Dir. Assoc. 9 (5), 292–301. https://doi.org/10.1016/j.jamda.2008.03.006.

Thomas, D.R., Duthie, E.A., 2016c. Hydration and older adults. Nutr. Rev. 74 (6), 411–421. https://doi.org/10.1093/nutrit/nuw008.

Tilman, D., Clark, M., 2014. Global diets link environmental sustainability and human health. Nature 515 (7528), 518–522. https://doi.org/10.1038/nature13959.

Tohit, N.F.M., Ya, R.M., Haque, M., 2025. Seniors and sustenance: a scoping review on food security in the elderly. Adv. Hum. Biol. 15 (1), 47–65.

Tsang, J.P., Cheung, D.S., Liu, J.Y., 2025. Social factors, dietary intake and the nutritional status of community-dwelling Chinese older adults: a scoping review. Nutrients 17 (12), 2019.

UNICEF, & World Health Organization (WHO), 2020. Framework for action to end malnutrition. https://www.unicef.org/documents/framework-action-end-malnutrition.

United Nations Department of Economic and Social Affairs, Population Division (2023). World Population Ageing 2023: Challenges and opportunities of population ageing in the least developed countries. UN DESA/POP/2023/TR/NO.5. https://www.un.org/development/desa/pd/sites/www.un.org.development.desa.pd/files/undesa_pd_2024_wpa2023-report.pdf.

United Nations, (n.d.). Food. United Nations. https://www.un.org/en/global-issues/food.

United Nations, 2019a. World population prospects 2019: Highlights. United Nations Department of Economic and Social Affairs. https://population.un.org.

United Nations, 2019b. World population ageing 2019: Highlights. United Nations. https://www.un.org/development/desa/pd/content/World-Population-Ageing-2019.

United Nations, 2019c. Global action plan for healthy lives and well-being for all: aligning the work of the UN health agencies to support the SDGs. United Nations. https://www.un.org/sustainabledevelopment/health/.

United Nations, 2022. World Population Ageing 2022 Highlights. https://www.un.org/development/desa/pd/.

United Nations, Department of Economic and Social Affairs, Population Division, 2017. World Population Ageing 2017 Highlights. https://www.un.org/en/development/desa/population/publications/pdf/ageing/WPA2017_Highlights.pdf.

Vega, W.A., Rodriguez, M.A., Gruskin, E., 2019. Health disparities in the Latino population. Epidemiol Rev. 31 (1), 99–112. https://doi.org/10.1093/epirev/mxp008.

Ver Ploeg, M., Breneman, V., Farrigan, T., Hamrick, K., Hopkins, D., Kaufman, P., et al., 2009. Access to affordable and nutritious food: measuring and understanding food deserts and their consequences. USDA Economic Research Service. https://www.ers.usda.gov/publications/pub-details?pubid=42729.

Vermeulen, S.J., Campbell, B.M., Ingram, J.S.I., 2012. Climate change and food systems. Annu. Rev. Environ. Resour. 37, 195–222. https://doi.org/10.1146/annurev-environ-020411-130608.

Victora, C.G., Adair, L., Fall, C., Hallal, P.C., Martorell, R., Richter, L., et al., 2008. Maternal and child undernutrition: consequences for adult health and human capital. Lancet. 371 (9609), 340–357. https://doi.org/10.1016/S0140-6736(07)61692-4.

Visvanathan, R., et al., 2015. Nutritional interventions and screening in older adults: a review of guidelines. Clin. Nutr. 34 (1), 1–12. https://doi.org/10.1016/j.clnu.2014.03.001.

Volkert, D., Beck, A.M., Cederholm, T., Cruz-Jentoft, A., Goisser, S., Hooper, L., et al., 2019. ESPEN guideline on clinical nutrition and hydration in geriatrics. Clin. Nutr. 38 (1), 10–47. https://doi.org/10.1016/j.clnu.2018.05.024.

Volpi, E., Campbell, W.W., Josse, R., 2013b. Is the optimal level of protein intake for older adults greater than the recommended dietary allowance? J. Gerontol. Ser. A: Biol. Sci. Med. Sci. 68 (6), 677–681. https://doi.org/10.1093/gerona/gls228.

Volpi, E., Nazemi, R., Fujita, S., 2013a. Muscle tissue changes with ageing. Curr. OpClNutr & Metab. Care 16 (1), 21–26. https://doi.org/10.1097/MCO.0b013e32835a2032.

Wahlqvist, M.L., Kouris-Blazos, A., Wattanapenpaiboon, N., 2012b. The significance of eating patterns: an elderly Greek case study. Appetite 59 (2), 577–583. https://doi.org/10.1016/j.appet.2012.07.008.

Willett, W., Rockström, J., Loken, B., Springmann, M., Lang, T., Vermeulen, S., et al., 2019. Food in the anthropocene: the EAT–Lancet commission on healthy diets from sustainable food systems. Lancet 393 (10170), 447–492. https://doi.org/10.1016/S0140-6736(18)31788-4.

World Health Organization (WHO), 2015a. World report on ageing and health. https://www.who.int/publications/i/item/9789241565042.

World Health Organization (WHO), 2020a. Ageing and health. https://www.who.int/news-room/fact-sheets/detail/ageing-and-health.

World Health Organization (WHO), 2020b. Decade of Healthy Ageing 2020–2030. https://www.who.int/publications/b/56512.

World Health Organization (WHO), 2020d. Nutrition in older adults. https://www.who.int/nutrition/topics/ageing/en/.

World Health Organization (WHO), 2020h. Managing noncommunicable diseases in emergencies. https://www.emro.who.int/noncommunicable-diseases/publications/questions-and-answers-on-prevention-and-control-of-noncommunicable-diseases-in-emergencies.html.

World Health Organization (WHO), 2021b. Global report on health equity for persons with disabilities. https://www.who.int/publications/i/item/9789240063600.

World Health Organization (WHO), 2021d. Non-communicable diseases and ageing. https://www.who.int/westernpacific/about/governance/regional-director/ncds-and-ageing.

World Health Organization (WHO), 2021e. Global strategy on diet, physical activity, and health. https://www.who.int/dietphysicalactivity/en/.

World Health Organization (WHO), 2025c. Noncommunicable diseases: Key facts. https://www.who.int/news-room/fact-sheets/detail/noncommunicable-diseases.

World Health Organization (WHO), 2013. Global action plan for the prevention and control of noncommunicable diseases 2013–2020. https://www.who.int/nmh/publications/ncd-action-plan/en/.

World Health Organization (WHO), 2015b. Nutrition for older persons. https://www.who.int/nutrition/topics/ageing/en/.

World Health Organization (WHO), 2020c. Healthy ageing and functional ability. Retrieved from https://www.who.int/news-room/questions-and-answers/item/healthy-ageing-and-functional-ability.

World Health Organization (WHO), 2020e. Healthy diet for healthy ageing. Geneva: WHO. Retrieved from https://www.who.int.

World Health Organization (WHO), 2020i. Integrated care for older people (ICOPE) program. World Health Organization. https://www.who.int/ageing/health-systems/icope/en/.

World Health Organization (WHO), 2021a. World health statistics 2021: Monitoring health for the SDGs. https://www.who.int/data/gho/publications/world-health-statistics.

World Health Organization, 2023. Noncommunicable diseases in fragile and conflict-hit settings. WHO Eastern Mediterranean Regional Office. https://www.emro.who.int/media/news/noncommunicable-diseases-in-fragile-and-conflict-hit-settings.

Yilmaz, K., Wirth, R., Daubert, D., Pourhassan, M., 2024. Prevalence and determinants of micronutrient deficiencies in malnourished older hospitalized patients. J. Nutrition, Health Aging 28 (2), 100039.

Zapata, M., Cediel, G., Arrieta, E., Rovirosa, A., Carmuega, E., Monteiro, C., 2023. Ultra-processed foods consumption and diet quality among preschool children and women of reproductive age from Argentina. Public Health Nutr. 26. https://doi.org/10.1017/S1368980022002543.

Zhou, C., Miao, H., Zhao, Y., Wan, X., 2023. Food insecurity increases the risk of overweight and chronic diseases in adolescents: a systematic review and meta-analysis. Food Sci. Hum. Wellness. 12 (6), 1937–1947. https://doi.org/10.1016/j.fshw.2023.03.002.

Ziliak, J.P., Gundersen, C., 2020a. The health consequences of senior hunger in the United States: evidence from the 1999–2014 NHANES. J. Nutr. 150 (1), 135–141. https://doi.org/10.1093/jn/nxz230.

Ziliak, J.P., Gundersen, C., 2016. The state of senior hunger in America 2014: an annual report. National Foundation to End Senior Hunger. Retrieved from https://www.feedingamerica.org.

Ziliak, J.P., Gundersen, C., 2020b. The state of senior hunger in America 2018: an annual report. Feeding America. https://www.feedingamerica.org/research/senior-hunger-research/senior.

CHAPTER SIX

The role of pharmacists in addressing nutritional deficiencies in food-insecure countries

Zhinya Kawa Othman[a,*], Usman Abubakar Haruna[b], Shuaibu Saidu Musa[c,d], Olalekan John Okesanya[e,f], Mohamed Mustaf Ahmed[g], Jerico Bautista Ogaya[h,i], Adriana Viola Miranda[j], and Don Eliseo Lucero-Prisno III[k,l,m]

[a]Department of Pharmacy, Kurdistan Technical Institute, Sulaymaniyah, Kurdistan Region, Iraq
[b]Faculty of Pharmaceutical Sciences, Ahmadu Bello University, Zaria, Nigeria
[c]School of Global Health, Faculty of Medicine, Chulalongkorn University, Bangkok, Thailand
[d]Department of Nursing Science, Ahmadu Bello University, Zaria, Nigeria
[e]Department of Public Health and Maritime Transport, University of Thessaly, Volos, Greece
[f]Department of Medical Laboratory Science, Neuropsychiatric Hospital, Aro, Abeokuta, Ogun State, Nigeria
[g]Faculty of Medicine and Health Sciences, SIMAD University, Mogadishu, Somalia
[h]Department of Medical Technology, Institute of Health Sciences and Nursing, Far Eastern University, Manila, Philippines
[i]Center for University Research, University of Makati, Makati City, Philippines
[j]Global Health Focus Asia, Bandung, Indonesia
[k]Department of Global Health and Development, London School of Hygiene and Tropical Medicine, London, United Kingdom
[l]Research and Innovation Office, Southern Leyte State University, Leyte, Philippines
[m]Research and Development Office, Biliran Province State University, Biliran, Philippines
*Corresponding author. e-mail address: zhinya.kawa@kti.edu.iq

Contents

Advances in Food Security and Sustainability, Volume 11
ISSN 2452-2635, https://doi.org/10.1016/bs.af2s.2026.04.002

Abstract

Food insecurity is a major public health challenge that drives undernutrition and micronutrient deficiencies, particularly among children, pregnant women, and other vulnerable groups. In many food-insecure settings, pharmacists and community pharmacies are among the most accessible points of care and can support nutrition-related prevention and management through feasible and practice-based actions. This chapter synthesizes how pharmacists can contribute to reducing nutritional deficiencies by identifying risk profiles and warning signs, delivering brief and context-appropriate nutrition counselling, guiding safe and appropriate supplementation, and reviewing medicines for clinically relevant drug–nutrient interactions. Nutrient biomarkers are framed as decision-support tools that can strengthen targeting and referral when testing is available, while acknowledging that clinical risk assessment and symptom-based pathways often guide action when laboratory capacity is limited. The chapter also highlights pharmacists' roles in collaboration with other health professionals, engagement in fortification and public health initiatives, and participation in pragmatic monitoring and referral systems. Finally, recommendations are organized across short-, medium-, and long-term priorities to support implementation within constrained health systems and to strengthen pharmacists' contributions through training, supportive regulation, and integration into food security responses

1. Introduction

Malnutrition is a broad public health problem that includes undernutrition, micronutrient deficiencies (MNDs), and overnutrition (overweight and obesity). Undernutrition remains a major concern in food-insecure settings and is marked by inadequate intake of energy and essential nutrients (including proteins, essential amino acids, vitamins, and minerals), undermining growth, maintenance, and physiological function. In 2022, undernutrition affected approximately 390 million people worldwide (Malnutrition, 2024). Among children and adolescents (aged 5–19 years), 190 million experienced thinness. Additionally, 149 million children under the age of five suffered from stunting, with nearly half of under-five child deaths linked to undernutrition, primarily in low- and middle-income countries (LMICs) (Malnutrition, 2024).

Undernutrition encompasses both insufficient food quantity and poor-quality diets. The issue extends beyond food scarcity to include the consumption of energy-dense but nutrient-poor foods, which significantly affects developing countries and low-income populations (Matrins et al., 2011). It causes a broad spectrum of nutritional disorders, including deficiencies, excesses, and imbalances of nutrients, thereby increasing healthcare utilization, prolonging hospital stays, and reducing quality of life, while also heightening the risk of mortality. Individuals suffering from undernutrition experience delayed wound healing, muscle deterioration, and weakened immune function (van Wayenburg et al., 2005).

The association between insufficient nutritional consumption and disease prevalence establishes a detrimental loop that markedly exacerbates morbidity and mortality rates, particularly in developing nations. According to the UN Children's Fund's (UNICEF) conceptual framework on child malnutrition, undernutrition is driven by interconnected underlying determinants, notably household food insecurity, inadequate maternal and childcare and feeding practices, and unhealthy environments with limited access to quality healthcare services (Khan and Bhutta, 2010; UNICEF, 2021). The highest burden is observed in Asia, Africa, and Latin America (WHO, 2004).

Micronutrient deficiencies (MNDs) pose a critical public health and socioeconomic challenge globally, affecting both low-income and industrialized countries. Vulnerable populations, including women, children, middle-aged individuals, and the elderly, are disproportionately affected by these deficiencies. Over 2 billion individuals suffer from vitamin and mineral deficiencies, primarily iodine, iron, vitamin A, and zinc, as reported by the World Health Organization (WHO), and these lead to severe health consequences (Tulchinsky, 2010; WHO, 2007). Moreover, MNDs exacerbate the health issues of food-insecure individuals who already have a negative self-assessment of their health. Several factors lead to a suboptimal nutritional status, including insufficient food quality or quantity, elevated dietary requirements, increased metabolic losses, and compromised digestion and absorption. Prolonged intake of imbalanced or nutrient-deficient diets increases the likelihood of malnutrition (Precone et al., 2019).

A fundamental aspect of food insecurity is the lack of access to sufficient nutritious food, which poses a potential risk of malnutrition in both children and adults. When everyone, everywhere, has physical, social, and financial access to enough safe, nourishing food that satisfies their dietary needs and preferences for an active and healthy life, that situation is known as food security (FAO, 2018). According to the Food and Agriculture Organization of the United Nations (FAO), the core pillars of food security include food availability, access to food, utilization of food, and stability over time. These pillars are supported by enabling conditions such as sustainable food production systems (including soil management), economic access, and food safety and quality, which shape how food is used and its nutritional value (Moyo, 2007).

The global food security crisis remains a major public health concern. Recent editions of *The State of Food Security and Nutrition in the World* have shown that hunger has remained persistently high in recent years (UNICEF, 2024). As summarized in Table 1, updated harmonized estimates indicate that the number of undernourished people worldwide rose from 670.1 million in 2020–697.5 million in 2021, before declining modestly to 695.2 million in 2022, 688.4 million in 2023, and 673.2 million in 2024 (UNICEF, 2025). Despite this recent improvement, progress has been uneven, and hunger continues to rise in Africa (UNICEF, 2025). Overall, food insecurity remains a major global health and nutrition challenge, driven by poverty, armed conflict, climate change, demographic pressures, globalization-related disruptions, and financial constraints, with the highest prevalence seen in Africa and Asia (Iannotti et al., 2024).

The 2030 Agenda for Sustainable Development underscores the urgent need to address food insecurity, hunger, and malnutrition. This burden is particularly severe in Sub-Saharan Africa, where 63.2 % of the population experienced moderate or severe food insecurity in 2021, up from 61.8 % in 2020 (FAO, 2022). The scale of the crisis is further reflected in the *Global Report on Food Crises 2024*, which estimated that 281.6 million people across 59 food-crisis countries and territories faced high levels of acute food insecurity in 2023 (Global Network Against Food Crises, 2024). In the Middle East and North Africa region, 54.3 million people were undernourished in 2021, accounting for 12.2 % of the population, while climate

Table 1 Global undernourishment estimates, 2023–2024.

Year	**Estimated number of undernourished people (millions)**
2020	670.1
2021	697.5
2022	695.2
2023	688.4
2024	673.2

Source: *The State of Food Security and Nutrition in the World 2025* (UNICEF, 2025). Note*:* 2024 is a projected estimate.

change and economic downturns have further intensified food insecurity in the region (Middle East Council on Global Affairs, 2023.). These patterns underscore the need for practical, context-specific responses in the most affected settings.

In food-insecure countries, pharmacists are often among the most accessible health professionals and may be the first point of contact for communities facing constrained diets, limited clinical services, and irregular supply of medicines. This chapter, therefore, centers on pharmacists within the food insecurity context, emphasizing how pharmacy practice can support the prevention and management of nutritional deficiencies through risk identification, practical counselling, safe supplementation guidance, medication review for drug–nutrient interactions, and timely referral within local health system pathways. Pharmacists' contributions are particularly relevant where dietitian services are scarce, laboratory testing is limited, and the health impacts of food insecurity are compounded by infections, chronic diseases, and pregnancy-related nutritional demands (World Health Organization, 2024; Community Pharmacists, 2020).

Addressing these challenges involves teamwork amongst healthcare experts, including dietitians, pharmacists, and social workers, with enhanced communication and collaboration along with the vital roles of patients and families in these efforts (Alzahrani et al., 2024). Accordingly, this chapter aims to evaluate the role of pharmacists in mitigating nutritional deficiencies in food-insecure countries by linking key nutrition challenges to concrete, feasible pharmacy actions across community and clinical settings, while addressing the system-level barriers that shape what pharmacists can deliver in low-resource contexts. Moreover, it seeks to propose strategic recommendations to enhance pharmacists' capacity through policy reforms, specialized training, and integration into multidisciplinary nutrition initiatives.

2. Nutrient biomarkers

Nutritional biomarkers are measurable substances in blood, urine, and other biological samples that reflect dietary intake, nutrient status, and nutrition-related biological processes. In line with the chapter's focus on pharmacy practice in food-insecure settings, biomarkers are introduced here as decision-support information rather than as part of a specialized laboratory sciences discussion. Their purpose is to help pharmacists and pharmacy-led teams recognize likely deficiencies, understand when laboratory confirmation is useful or feasible, and use available results to guide practical steps, such as counselling, safe supplementation, monitoring for adverse effects, and referral. In settings where testing is limited, pharmacists may rely on risk profiles and clinical features to trigger referral pathways, whereas basic biomarkers, when available, can support targeted and cost-conscious decisions (Pande et al., 2018; Suchdev et al., 2016).

These biomarkers are influenced by both biological and laboratory variations. Key biological influences include infection and inflammation, underlying disease states, and individual variability, all of which may shift biomarker values independently of the diet. Laboratory influences include sample handling and the analytical procedures. Depending on the biomarker, the results may reflect recent intake, longer-term nutritional status, or functional effects of nutrient deficiency. Biomarkers may include nutrients themselves or their transport forms, as well as metabolites produced when the body processes food and nutrients. Precisely estimating nutrient intake using dietary methods alone is often difficult, particularly in settings without locally validated food composition databases, where analysts may rely on external references, such as the US Department of Agriculture nutrient database. For example, serum or plasma concentrations of key vitamins and minerals, taken from blood samples, can be used to evaluate nutritional exposure and status. However, biomarker values can be affected by specimen collection timing, storage conditions, seasonality, infection and inflammation, and laboratory procedures, which can introduce measurement errors. To minimize errors and improve comparability, studies should apply rigorous laboratory quality control measures, including blinded analyses, external pooled standards, and internal certified reference materials, to monitor assay accuracy and precision over time (Kiani et al., 2022).

2.1 Laboratory assessment of vitamin status

Vitamins are essential nutrients required in small amounts for normal body functions. Vitamin status can be assessed by measuring vitamin levels (or related markers) in blood and other samples, which helps identify deficiencies and guides supplementation or public health actions. In food-insecure settings, these measurements are most useful when they inform practical pharmacy actions, such as counselling on feasible dietary choices, recommending safe and appropriate supplementation where indicated, monitoring adverse effects, and referring patients for clinical assessment when red flags are present.

Vitamin A status is commonly assessed using blood markers that reflect its availability in the body. Because infections and inflammation can temporarily change these markers, the results are best interpreted alongside markers of inflammation to avoid over- or underestimating deficiency at the population level. For pharmacy practice, the key message is that low values may reflect true deficiency, but acute illness can also distort results; therefore, pharmacists should interpret findings alongside the clinical context and refer for further assessment when infection is suspected, or symptoms are significant (Tanumihardjo, 2011). Additional approaches (including specialized tests and breast milk measurements in lactating women) can provide a clearer picture for specific groups and settings (Pande et al., 2018).

Vitamin B12 deficiency can be missed if only one blood test is performed. Early deficiency may be better detected using "active" B12-related markers and functional indicators that increase when B12-dependent processes are impaired. From a pharmacy perspective, this matters because delayed recognition can allow neurological and hematological complications to progress, so pharmacists should consider referral when persistent symptoms coexist with dietary risk or limited animal-source food intake, even if a single routine test appears borderline (Harrington, 2019).

Vitamin D status is most assessed using blood markers that reflect both dietary intake and sunlight-related production. In practice, pharmacists can use vitamin D results to guide safe supplementation advice, reinforce adherence, and support referrals when severe deficiency is suspected or when symptoms and comorbidities indicate a higher risk. Some testing methods are more precise than others, and reliable assessment requires careful laboratory practice because different forms of vitamin D circulate in the body and can affect the interpretation of results (Carter and Card, 2019).

Vitamin K status is often evaluated through markers linked to blood clotting function, but these may only detect more severe deficiency. Direct measures of vitamin K in the blood can be more specific, although they require sensitive laboratory methods and careful sample handling because vitamin K levels are typically low. For pharmacy staff, it is essential to recognize people at higher risk, support the prevention and safe use of vitamin K–related products where appropriate, and prioritize referral when bleeding risk or anticoagulant use is present (Górska, 2019).

Folate status can be assessed using blood-based markers that reflect short-term intake and longer-term body stores, as well as functional indicators that increase when folate-dependent metabolism is impaired. The use of more than one marker helps distinguish folate deficiency from other causes of abnormal results. In food-insecure contexts, pharmacists can use folate-related results to reinforce prevention-focused counselling, support safe supplementation guidance, and trigger referrals for evaluation when the risk of anaemia is high or pregnancy-related needs are present (Sobczyn and Harrington, 2018).

2.2 Laboratory assessment of mineral status

In line with the chapter focus, mineral status is presented here as practical decision support for pharmacists, helping to identify risk, guide counselling on food-based strategies and safe supplementation, recognize when testing is demanded, and support timely referral through local pathways. Minerals are essential nutrients that support many core functions, including nerve and muscle activity, enzymatic function, and bone health. Mineral deficiencies remain common in many LMICs and can have major consequences for growth, pregnancy outcomes, infection risk, and overall health (Gupta and Gupta, 2014; Shankar, 2020).

Iron status is assessed using a combination of markers that reflect iron stores and the body's iron requirements. In routine practice, pharmacists can use these indicators to support targeted decisions, such as identifying individuals at a higher risk of iron deficiency, reinforcing adherence to prescribed iron therapy, monitoring tolerability concerns, and referring for clinical review when symptoms persist or when anaemia is suspected. Because inflammation can distort some markers, the use of multiple indicators improves accuracy, especially in settings where infection and chronic inflammation are common (Knovich et al., 2009).

Zinc status is often estimated using blood-based measures; however, results may undergo the influence of infection, inflammation, and the timing of sample collection. For pharmacists, the key value of zinc assessment is to support context-appropriate counselling and referral for high-risk groups, especially children, pregnant individuals, and people with recurrent infections, while recognizing that laboratory confirmation may be limited, and that clinical risk profiling may guide action in many low-resource settings. Reliable interpretation depends on appropriate sample handling and laboratory quality assurance when testing is performed (Laboratory Tests for the Assessment of Nutritional Status - 2nd edition, n.d).

Calcium status is commonly reported in clinical settings; however, its interpretation can be challenging because total calcium can be influenced by other blood components. From a pharmacy practice perspective, calcium assessment is most relevant for counselling on dietary sources, appropriate supplementation when indicated, identifying interactions with medicines that affect absorption, and supporting referrals for evaluation of symptoms or high-risk conditions affecting bone health. Clinical interpretation relies on context and, when needed, more specific measures (Principles of Nutritional Assessment (3rd ed.), 2025).

Iodine status is most often assessed using urine-based measures that reflect recent intakes of iodine. Because iodine is essential for thyroid hormone production, thyroid-related indicators can support assessments in specific populations, such as newborns, and guide prevention strategies where iodized salt coverage is incomplete. Thus, pharmacists can contribute by reinforcing the consistent use of iodized salt where available, providing patient education on iodine needs during pregnancy, and supporting referrals when thyroid dysfunction is suspected, particularly in areas with known deficiency risk (Höller et al., 2018).

3. Nutritional deficiencies and associated problems

In food-insecure settings, nutritional deficiencies are shaped by constrained diets, seasonal availability, displacement, infection, and limited access to preventive and clinical services. In practice, pharmacists can contribute by identifying high-risk groups, linking symptoms and risk profiles to likely deficiencies, providing feasible dietary and supplementation advice, checking for drug–nutrient interactions, and referring patients when severe disease or complications are suspected. Nutritional deficiencies include shortages of vitamins, minerals, and proteins. These deficiencies can impair growth, immunity, cognition, pregnancy outcomes, and long-term health, and they are often more severe in food-insecure settings where diets are limited in diversity and healthcare access is constrained (Singh and Bharti, 2021).

Vitamin A deficiency weakens immune protection and increases the risk of severe infections, especially in children. It can also cause eye problems that range from reduced night vision to preventable blindness, making it a major public health concern in vulnerable populations (Sommer and Vyas, 2012). Region-specific estimates show that vitamin A deficiency remains an important contributor to health loss in some settings, including parts of the MENA region (Safiri et al., 2024). However, excessive vitamin A intake (particularly from high-dose supplements) can be toxic and may lead to serious adverse outcomes (Olson et al., 2023). Clinical features can include night blindness and other eye manifestations, particularly in high-risk groups. Evidence also links deficiency with inflammatory processes that may worsen broader health outcomes (Wiseman et al., 2017). From a pharmacy perspective, children, pregnant women, and people with recurrent infections are priority groups for risk screening, counselling on locally available vitamin A sources, and timely referral when ocular symptoms or severe illness are present.

Vitamin B12 deficiency can cause anaemia and affect the nervous system, leading to symptoms that may occur even when anaemia is not obvious. The risk is higher when diets contain few animal-source foods, and deficiency during pregnancy and infancy can have serious developmental consequences (Green, 2017). Studies from different countries have shown that low B12 levels can be common and influenced by diet and broader health conditions (Elgormus et al., 2023; Melaku et al., 2024). Pharmacists can support prevention by identifying dietary risk, supporting safe supplementation when appropriate, and prioritizing referral for neurological symptoms, pregnancy, and infant feeding concerns.

Vitamin D deficiency is widespread and is associated with poor bone health and an increased risk of fractures. Severe deficiency can cause rickets in children and osteomalacia in adults. Some regions and clinical groups (including those with kidney or liver disease) are at particularly high risk, and observational evidence suggests broader associations with chronic diseases (Sahay and Sahay, 2012; Amrein et al., 2020; Cashman, 2020). In low-resource settings, pharmacists can focus on practical risk identification, counselling on feasible sun exposure and dietary sources when culturally appropriate, and referral for suspected rickets, fractures, or high-risk comorbidities.

Vitamin K deficiency is most clinically important because it increases bleeding risk. In newborns, deficiency can cause serious bleeding events if preventive measures are not implemented. Barriers in low-resource settings can limit access to prevention and timely care (Fiesack et al., 2021). Emerging evidence also explores how vitamin K deficiency may be related to broader health risks, but the prevention of bleeding remains the key priority (Kumar et al., 2019). Pharmacists can reinforce newborn prophylaxis messaging where relevant, recognize warning signs of bleeding, and support rapid referral pathways.

Folate deficiency is a major concern for women of reproductive age because it increases the risk of anaemia and neural tube defects (birth defects of the brain, spine, or spinal cord) in newborns. Population surveys show that deficiency varies widely by country, income level, and nutrition policies, and deficiency can also be linked to other adverse outcomes such as preterm birth (Rogers et al., 2018). Folate also supports neurological function, and its deficiency has been associated with visual and inflammatory eye conditions in some studies (Sijilmassi, 2019). Community and antenatal counselling are a feasible pharmacy contribution, particularly by promoting preconception and early pregnancy supplementation and addressing adherence barriers.

Iron deficiency is one of the most common nutrient deficiencies globally and a leading cause of anaemia. It affects energy, cognition, pregnancy outcomes, and economic productivity, and can occur with or without anaemia. The burden and risk vary across regions and groups, including adolescents, women, and individuals with chronic diseases (Safiri et al., 2025). Pharmacists can contribute by screening for likely iron deficiency, counselling on iron-rich foods and correct supplement use, and referring patients with severe anaemia symptoms or suspected ongoing blood loss.

Zinc deficiency can impair growth, immunity, and wound healing, and is common in low- and middle-income settings. Children and individuals with chronic illnesses may be particularly affected. Evidence also explores the links between zinc status and metabolic and cardiovascular outcomes, although the relationships can vary by context and population (Dembedza et al., 2024). Pharmacists can reinforce appropriate zinc use in specific clinical contexts and support counselling that avoids unnecessary supplementation while prioritizing children at risk of recurrent infections and poor growth.

Protein deficiency can weaken immunity, impair growth, and contribute to oedema (swelling due to fluid in body tissues) and poor recovery from illness, particularly in children and food-insecure populations. In clinical contexts, certain protein-related disorders can increase the risk of clotting complications, highlighting the need for appropriate diagnosis and management strategies (Gupta and Patibandla, 2023). In food-insecure contexts, pharmacists can recognize red-flag signs such as oedema and severe wasting, support referral, and provide feasible counselling on protein sources that align with local availability and household affordability.

4. The overall role of pharmacists in the management of nutritional deficiencies

Nutritional management has become increasingly important in public health, with growing awareness among individuals and efforts by governments to improve dietary choices through initiatives like fortified foods and school meal programs (FIP, 2025). Effective nutrition management involves multiple stakeholders, including governments, medical professionals, policymakers, and the public, working together to promote health. In many low- and middle-income settings, pharmacies and drug shops are often the first point of contact for care because they are geographically accessible, have long opening hours, and can be less costly than formal clinical visits. However, the quality of counselling, referral, and adherence to standards can be inconsistent due to workforce, incentive, and regulatory constraints (Miller and Goodman, 2016).

Primary healthcare providers (PHCPs), especially pharmacists, are essential in educating communities and promoting self-care for lifelong immune support. They offer accessible, evidence-based guidance on nutrition, physical activity, mental well-being, managing health supplements and nutraceuticals, and advising on drug-nutrient interactions. In food-insecure contexts, the most feasible pharmacy contributions typically sit at three levels: community-facing actions such as brief risk identification and counselling, clinical support actions such as medication review and safety monitoring, and system support actions such as participation in fortification monitoring and reporting systems when these exist. In addition, pharmacists can contribute to food security efforts by identifying nutrition-related needs during routine care, connecting patients to community food resources, and supporting prevention-oriented counselling as an approach that can be framed within the Social Determinants of Health (SDoH), where access to adequate food is a key driver of health outcomes (Chandra et al., 2025).

4.1 Nutritional counseling

Nutrition counselling is becoming more widely recognized for its role in preserving health and reducing disease risks, making it an essential component of modern healthcare. As highly accessible and trusted healthcare practitioners, pharmacists have a unique opportunity to incorporate nutrition counselling into their practice, offering patients a comprehensive approach to wellness. Poor dietary habits contribute significantly to chronic disorders such as diabetes, obesity, cardiovascular diseases, and certain malignancies, emphasizing the importance of preventative healthcare effort (WHO, 2025). To reflect realities in food-insecure settings, counselling should be framed as brief, practical, and culturally appropriate. A feasible approach is to prioritise short messages focused on affordable nutrient-dense choices available locally, safe supplementation when indicated, and clear triggers for referral such as pregnancy, severe symptoms, suspected anaemia, or recurrent infections. Pharmacists can assist patients to make informed decisions that benefit their overall health by providing personalized nutritional counsel, such as dietary recommendations, meal planning, and supplements advice (Prabahar and Alenazi, 2020).

For example, pregnant women's compliance with folate supplementation is critical for reducing neural tube defects, and pharmacists can influence adherence through medication counselling. A cross-sectional analytical study among 360 pregnant women in Tabuk, Saudi Arabia, found delayed initiation and low preconception use, with forgetfulness and insufficient knowledge as major reasons for noncompliance (Prabahar and Alenazi, 2020). In low-resource settings, adherence support should emphasise simple routines, reminder tools that fit local access to phones, and clear risk communication without overmedicalizing the message. The study found a significant association between pharmacist counseling and regular folate intake among pregnant women, reinforcing the positive impact of pharmacist-driven education and emphasizing the need for enhanced pharmacist involvement in prenatal care (Prabahar and Alenazi, 2020).

Many LMICs struggle to provide nutritionally adequate diets, leading to widespread micronutrient deficiencies. Food fortification is a proven and cost-effective intervention (Editors' note: see the chapter on Biofortification in this volume). Pharmacists can encourage consumption of fortified foods through their roles in patient education and distribution, and where relevant, they may support implementation through training and quality-focused practice. A systematic review of large-scale fortification programs reported reductions in vitamin A, iodine, and iron deficiencies, along with reductions in anaemia, goiter, and neural tube defects (Mkambula et al., 2020). Where pharmacists participate in fortification systems, their most realistic contributions are consumer education, detection of low uptake, and reporting of suspected quality problems through established monitoring channels rather than direct laboratory verification. Guidance on fortification programme monitoring highlights the importance of structured monitoring systems and quality assurance approaches for fortified food (WHO, 2021).

4.2 Medication therapy management

The interactions between food, nutrients, and medications are increasingly recognized in healthcare. Certain foods and nutrients can influence drug absorption, metabolism, and effectiveness. Drug–food interactions can occur with both prescription and over-the-counter medications, including antacids, vitamins, and iron supplements, and dietary supplements may interact with medicines, reinforcing the need for patients to inform clinicians and pharmacists (Drugs-nutrient interactions: a potential problem during adolescence - PubMed, 2000).

In food-insecure settings, medication therapy management should focus on a small set of high-yield safety checks. These include identifying medicines that worsen appetite or gastrointestinal tolerance, medicines that increase micronutrient depletion risk, and common interactions with iron, calcium, and vitamin K-related diets. For example, calcium and iron can reduce absorption of certain drugs, and diets high in vitamin K can reduce the anticoagulant effect of warfarin, so patients are advised to keep vitamin K intake consistent. Warfarin is a high-risk anticoagulant that can cause serious bleeding if not appropriately monitored, and it was historically introduced as an anticoagulant rodenticide before medical use (Pirmohamed, 2006).

Pharmacists play a critical role in identifying, monitoring, and educating patients about potential drug–food interactions to enhance efficacy and prevent adverse effects. A systematic review on clinical pharmacist services in Pakistan reported improvements in multiple outcomes and reductions in drug-related problems (Ahmed et al., 2021). However, evidence from low- and middle-income settings also shows that retail pharmacy practice can be constrained by limited time for counselling, inconsistent referral, and weak enforcement of standards, so implementation should be paired with realistic workflow design and supportive regulation (Miller and Goodman, 2016).

In accredited healthcare systems, pharmacy-related standards often require documentation and patient counselling processes, although implementation may vary across practice settings (Rich, 2004). In contexts with fragmented care and limited follow-up, pharmacists can still support safety by promoting clear dosing instructions, warning signs that require urgent review, and referral pathways that match local access patterns.

4.3 Advocacy and collaboration involvement

In an era of information overload, where contradictory views on nutritional supplements and therapies can create confusion, pharmacists can serve as reliable sources of evidence-based guidance. In food-insecure settings, this role is most credible when it is grounded in practical messages such as when supplementation is appropriate, how to avoid unsafe or non-verified products, and how to prioritise limited household resources toward the most cost-effective dietary improvements. The evidence base for pharmacist engagement in public health education and counselling supports this positioning (Slavcheva et al., 2025; Maher et al., 2014).

Pharmacists also play a role in advising policymakers on cost-effective treatment options and the public health impacts of diet and nutrition. To operationalise this in low-resource contexts, feasible contributions include participating in local technical working groups, supporting monitoring data for fortification or supplementation programmes, and sharing front-line observations about product availability, affordability, and community uptake with district health managers (Calder et al., 2023).

Moreover, pharmacists can collaborate with medical practitioners and nutritionists to develop nutritional interventions aligned with medical treatments, and to support management of undernutrition and micronutrient deficiencies (Singla et al., no.date). A multidisciplinary approach in osteoporosis care has shown improvements in vitamin D status, adherence, and outcomes through pharmacist–nutritionist collaboration integrating pharmacists and nutritionists into osteoporosis management teams to optimize care (Pharmacist-Nutritionist, 2020).

5. Recommendations

To strengthen pharmacists' contributions in food-insecure settings, recommendations should prioritize feasible actions within constrained systems while also outlining longer-term reforms. In the short term, pharmacists can integrate brief nutrition risk screening into routine encounters by asking simple questions about meal frequency, recent weight loss, pregnancy status, childhood growth concerns, and common deficiency symptoms, then using this information to provide targeted counselling and clear referral advice when danger signs are present. Where literacy is limited, counselling should rely on plain language, locally available foods, and visual or verbal instructions, and should be delivered in short, repeatable messages that fit high-volume pharmacy workflows. Pharmacists should also emphasize safe supplementation by discouraging unnecessary high-dose products, checking for contraindications, and advising patients on how to take supplements correctly to improve adherence and reduce adverse effects. These steps can be supported through evidence-based nutrition courses, online training, and scientific events that improve pharmacists' competence and confidence in counselling, supplementation guidance, and medicine–nutrition safety (Calder et al., 2023).

In the medium term, pharmacy practice can be strengthened through standardized counselling tools and referral pathways that link pharmacies with nearby primary care services, maternal and child health clinics, and community nutrition programs. Because weak regulation and informal medicine markets are common in many LMICs, pharmacists should advocate for basic standards for supplement quality, labelling, and dispensing, and participate in local reporting systems for adverse events related to supplements and nutraceuticals. Professional bodies and ministries can support this by embedding nutrition competencies into pharmacy curricula and continuing professional development, and by issuing practical scope-of-practice guidance tailored to community pharmacies in low-resource environments (Calder et al., 2023).

In the long term, pharmacists should be integrated into national and subnational nutrition governance and food security responses, including fortification programs and public health prevention strategies. Additionally, pharmacists can contribute to strengthening monitoring systems for fortification programs by supporting standardized quality control measures, routine inspections, and collaboration with regulatory bodies to improve tracking of fortified foods through supply chains. They can also help identify distribution bottlenecks that prevent fortified foods from reaching vulnerable populations and support culturally appropriate community education that improves acceptance and correct use of fortified products. Collaboration with global and local stakeholders, such as the WHO, UNICEF, and government health agencies, remains essential for scale-up and sustainability. Finally, embedding pharmacists in implementation research and routine data collection can help document what works in real-world food-insecure settings, enabling more credible, context-specific policy decisions and program improvements (Al Ghafli, 2024).

6. Conclusion

Nutritional deficiencies in food-insecure settings arise from constrained diets and are frequently compounded by infection, chronic disease, and pregnancy-related needs. Pharmacists can make a practical contribution by integrating brief nutrition risk identification into routine encounters, providing culturally appropriate counselling focused on feasible dietary choices, supporting safe supplementation, and strengthening medication safety through drug–nutrient interaction checks and adherence support. Where laboratory testing exists, biomarkers can be used as decision-support to prioritise referral and guide targeted actions; where testing is limited, risk profiling and clear referral triggers remain central. Strengthening pharmacists' impact requires realistic workflows, better linkage to primary care and maternal and child health services, and improved oversight of supplements and fortified products. In the longer term, integrating pharmacists into nutrition governance, fortification monitoring, and implementation research can help generate context-specific evidence and support more effective food security and public health strategies.

References

Ahmed, A., Saqlain, M., Tanveer, M., Blebil, A.Q., Dujaili, J.A., Hasan, S.S., 2021. The impact of clinical pharmacist services on patient health outcomes in Pakistan: a systematic review. BMC. Health Serv. Res. 21. https://doi.org/10.1186/s12913-021-06897-0.

Al Ghafli, M.A.M, 2024. Micronutrient fortification: a pharmacist's perspective on reducing global malnutrition. Power Syst. Technol. 48, 5121–5134. https://doi.org/10.52783/pst.1327.

Alzahrani, K.B.A., Alqattan, A.H., Alharbi, K.H., Alhassoon, A.H., Alrehaili, S.M., Alkuraiea, S.A., et al., 2024. The role of interdisciplinary healthcare teams in mitigating malnutrition among vulnerable populations: review of collaborative strategies among dietitians, pharmacists, dentists, and social workers. J. Ecohumanism 3, 12560–12568. https://doi.org/10.62754/joe.v3i8.5929.

Amrein, K., Scherkl, M., Hoffmann, M., Neuwersch-Sommeregger, S., Köstenberger, M., Tmava Berisha, A., et al., 2020. Vitamin D deficiency 2.0: an update on the current status worldwide. Eur. J. Clin. Nutr. 74, 1498–1513. https://doi.org/10.1038/s41430-020-0558-y.

Calder, P.C., Bach-Faig, A., Bevacqua, T., Caballero Lopez, C.G., Chen, Z.Y., Connolly, D., et al., 2023. Vital role for primary healthcare providers: urgent need to educate the community about daily nutritional self-care to support immune function and maintain health. BMJ. Nutr. Prev. Health 6, 392–401. https://doi.org/10.1136/bmjnph-2023-000755.

Carter, G., Card, D.J., 2019. Methods for assessment of vitamin D. Laboratory Assess Vitam Status 19, 49–77. https://doi.org/10.1016/B978-0-12-813050-6.00003-6.

Cashman, K.D., 2020. Vitamin D deficiency: defining, prevalence, causes, and strategies of addressing. Calcif. Tissue Int. 106, 14–29. https://doi.org/10.1007/s00223-019-00559-4.

Chandra, R.N., Baden, K.E.R., Chen, A.M.H., 2025. Moving upstream: the need for pharmacists to address food security. J. Am. Pharmacists Assoc. 65. https://doi.org/10.1016/j.japh.2025.102461.

Community pharmacists: on the frontline of health service against COVID19 in LMICs, 2020. https://www.aidshealth.org/wp-content/uploads/2023/06/Hedima-et-al-January-2021.pdf (Accessed 24 February 2026).

Dembedza, M.P., Chopera, P., Matsungo, T.M., 2024. Risk of zinc deficiency among children aged 0–59 months in sub-Saharan Africa: a narrative review. South Afr. J. Clin. Nutr. 37, 69–76. https://doi.org/10.1080/16070658.2023.2275932.

Drugs-nutrient interactions: a potential problem during adolescence - PubMed, 2000. https://pubmed.ncbi.nlm.nih.gov/10805043/ (Accessed 25 February 2026).

Elgormus, Y., Okuyan, O., Dumur, S., Sayili, U., Uzun, H., 2023. The epidemiology of deficiency of vitamin B12 in preschool children in Turkey. Medicina (Kaunas) 59. https://doi.org/10.3390/medicina59101809.

FAO, 2018. A review of studies examining the link between food insecurity and malnutrition. https://openknowledge.fao.org/server/api/core/bitstreams/8773b32e-e806–4a31-ae64-ebf6c2ed0fff/content (Accessed 24 February 2026).

FAO, 2022. The State of Food Security and Nutrition in the World 2022. https://openknowledge.fao.org/items/c0239a36–7f34–4170-87f7–2fcc179ef064 (Accessed 24 February 2026).

Fiesack, S., Smits, A., Rayyan, M., Allegaert, K., Alliet, P., Arts, W., et al., 2021. Belgian consensus recommendations to prevent vitamin K deficiency bleeding in the term and preterm infant. Nutrients 13. https://doi.org/10.3390/nu13114109.

FIP, 2025. Global Situation Report on Pharmacy 2025: Workforce, Practice, Policy. https://www.fip.org/files/content/publications/2025/FINAL_VF-CH3_Advancing_public_health_and_disease_prevention_through_pharmacy.pdf (Accessed 25 February 2026).

Global Network Against Food Crises, 2024. Global Report on Food Crises (GRFC) 2024. https://www.fightfoodcrises.net/report/global-report-food-crises-2024/ (Accessed 24 February 2026).

Górska, R.M., 2019. Methods for assessment of vitamin K. Laboratory Assess Vitam Status 107–147. https://doi.org/10.1016/B978-0-12-813050-6.00005-X.

Green, R., 2017. Vitamin B12 deficiency from the perspective of a practicing hematologist. Blood 129, 2603–2611. https://doi.org/10.1182/blood-2016-10-569186.

Gupta, A., Patibandla, S., 2026. Protein C Deficiency. [Updated 2023 Jul 4]. StatPearls [Internet]. StatPearls Publishing, Treasure Island Available from. https://www.ncbi.nlm.nih.gov/books/NBK542222/.

Gupta, U.C., Gupta, S.C., 2014. Sources and deficiency diseases of mineral nutrients in human health and nutrition: a review. Pedosphere 24, 13–38. https://doi.org/10.1016/S1002-0160(13)60077-6.

Harrington, D.J., 2019. Methods for assessment of vitamin B12. Laboratory Assess. Vitam. Status 265–299. https://doi.org/10.1016/B978-0-12-813050-6.00012-7.

Höller, U., Bakker, S.J.L., Düsterloh, A., Frei, B., Köhrle, J., Konz, T., et al., 2018. Micronutrient status assessment in humans: current methods of analysis and future trends. TrAC Trends Anal. Chem. 102, 110–122. https://doi.org/10.1016/j.trac.2018.02.001.

Iannotti, L., Kleban, E., Fracassi, P., Oenema, S., Lutter, C., 2024. Evidence for policies and practices to address global food insecurity. Annu. Rev. Public Health 45, 375–400. https://doi.org/10.1146/annurev-publhealth-060922-041451.

Khan, Y., Bhutta, Z.A., 2010. Nutritional deficiencies in the developing world: current status and opportunities for intervention. Pediatr. Clin. North Am. 57, 1409–1441. https://doi.org/10.1016/j.pcl.2010.09.016.

Kiani, A.K., Dhuli, K., Donato, K., Aquilanti, B., Velluti, V., Matera, G., et al., 2022. Main nutritional deficiencies. J. Prev. Med. Hyg. 63, E93–E101. https://doi.org/10.15167/2421-4248/jpmh2022.63.2S3.2752.

Knovich, M.A., Storey, J.A., Coffman, L.G., Torti, S.V., Torti, F.M., 2009. Ferritin for the clinician. Blood Rev. 23, 95–104. https://doi.org/10.1016/j.blre.2008.08.001.

Kumar, S., Choudhary, V., Suman, R.L., Meena, P., 2019. Prevalence and clinical manifestations of vitamin K deficiency in complicated severe acute malnutrition. Int. J. Contemp. Pediatrics 7, 57. https://doi.org/10.18203/2349-3291.ijcp20195726.

Laboratory tests for the assessment of nutritional status - 2nd edition, n.d. https://www.routledge.com/Laboratory-Tests-for-the-Assessment-of-Nutritional-Status/Sauberlich/p/book/9780849385063 (Accessed 24 February 2026).

Maher, J.H., Lowe, J., Hughes, R., 2014. Community pharmacy as a setting for public health nutrition action: Australian nutritionists' perspectives. Public Health Nutr. 18, 1864. https://doi.org/10.1017/S1368980014002201.

Matrins, V.J.B., Toledo Florêncio, T.M.M., Grillo, L.P., Franco, M. do C.P., Martins, P.A., Clemente, A.P.G., et al., 2011. Long-lasting effects of undernutrition. Int. J. Environ. Res. Public Health 8, 1817–1846. https://doi.org/10.3390/IJERPH8061817.

Melaku, E.E., Urgie, B.M., Tilahun, A.T., Assefa, H.K., Abebe, A.A., Tefera, A.S., 2024. Prevalence of vitamin B 12 deficiency and associated factors among primary school children: North East Ethiopia: multicenter cross-sectional study. J. Health Popul. Nutr. 43. https://doi.org/10.1186/s41043-024-00568-6.

Middle East Council on Global Affairs, 2023. Food insecurity in the Middle East and North Africa. https://mecouncil.org/publication/food-insecurity-in-the-middle-east-and-north-africa/ (Accessed 24 February 2026).

Miller, R., Goodman, C., 2016. Performance of retail pharmacies in low- and middle-income Asian settings: a systematic review. Health Policy Plan. 31, 940–953. https://doi.org/10.1093/heapol/czw007.

Mkambula, P., Mbuya, M.N.N., Rowe, L.A., Sablah, M., Friesen, V.M., Chadha, M., et al., 2020. The unfinished Agenda for food fortification in low- and middle-income countries: quantifying progress, gaps and potential opportunities. Nutrients 12. https://doi.org/10.3390/nu12020354.

Moyo, D., 2007. The future of food: elements of integrated food security strategy for south africa and food security status in Africa. Proc. ASIL. Annu. Meet. 101, 103–108. https://doi.org/10.1017/s0272503700025325.

Olson, J.M., Ameer, M.A., Goyal, A., 2023. Vitamin A toxicity. StatPearls.

Pande, S., Kratasyuk, V.A., Medvedeva, N.N., Kolenchukova, O.A., Salmina, A.B., 2018. Nutritional biomarkers: current view and future perspectives. Crit. Rev. Food Sci. Nutr. 58, 3055–3069. https://doi.org/10.1080/10408398.2017.1350136.

Pharmacist-nutritionist collaboration in optimizing vitamin D supplementation in osteoporotic patients, 2020. https://www.ijirmps.org/papers/2020/4/231764.pdf (Accessed 25 February 2026).

Pirmohamed, M., 2006. Warfarin: almost 60 years old and still causing problems. Br. J. Clin. Pharmacol. 62, 509. https://doi.org/10.1111/j.1365-2125.2006.02806.x.

Prabahar, K., Alenazi, T., 2020. Role of pharmacist's counseling on folate compliance. Saudi J. Health Sci. 9, 221. https://doi.org/10.4103/sjhs.sjhs_121_20.

Precone, V., Beccari, T., Stuppia, L., Baglivo, M., Paolacci, S., Manara, E., et al., 2019. Taste, olfactory and texture related genes and food choices: implications on health status. Eur. Rev. Med. Pharmacol. Sci. 23, 1305–1321. https://doi.org/10.26355/eurrev_201902_17026.

Principles of Nutritional Assessment (3rd ed.), 2025. https://nutritionalassessment.org/ (Accessed 24 February 2026).

Rich, D.S., 2004. New JCAHO medication management standards for 2004. Am. J. Health Syst. Pharm. 61, 1349–1358. https://doi.org/10.1093/ajhp/61.13.1349.

Rogers, L.M., Cordero, A.M., Pfeiffer, C.M., Hausman, D.B., Tsang, B.L., De-Regil, L.M., et al., 2018. Global folate status in women of reproductive age: a systematic review with emphasis on methodological issues. Ann. N.Y. Acad. Sci. 1431, 35–57. https://doi.org/10.1111/nyas.13963.

Safiri, S., Amiri, F., Karamzad, N., Sullman, M.J.M., Kolahi, A.A., Abdollahi, M., 2025. Burden and trends of dietary iron deficiency in the Middle East and North Africa region, 1990-2021. Front. Nutr. 11. https://doi.org/10.3389/fnut.2024.1517478.

Safiri, S., Mousavi, S.E., Nejadghaderi, S.A., Motlagh Asghari, K., Karamzad, N., Sullman, M.J.M., et al., 2024. Vitamin A deficiency in the MENA region: a 30-year analysis (1990–2019). Front. Nutr. 11, 1413617. https://doi.org/10.3389/fnut.2024.1413617.

Sahay, M., Sahay, R., 2012. Rickets–vitamin D deficiency and dependency. Indian J. Endocrinol. Metab. 16, 164. https://doi.org/10.4103/2230-8210.93732.

Shankar, A.H., 2020. Mineral deficiencies. Hunter's Tropical Med. Emerg. Infect. Dis. 10, 1048–1054. https://doi.org/10.1016/b978-0-323-55512-8.00145-9.

Sijilmassi, O., 2019. Folic acid deficiency and vision: a review. Graefes Arch. Clin. Exp. Ophthalmol. 257, 1573–1580. https://doi.org/10.1007/s00417-019-04304-3.

Singh, H., Bharti, J., 2021. Incredibly common nutrient deficiencies. EAS. J. Nutr. Food Sci. https://doi.org/10.36349/easjnfs.2021.v03i06.006.

Singla, N., Jindal, A., Mahapatra, D.K. (no.date). Role of pharmacist in nutrition management-the unexplored path. Indian J. Pharm. Pract. nd; 16. https://doi.org/10.5530/ijopp.16.2.14.

Slavcheva, K., Staynova, R., Neycheva, N., Kafalova, D., 2025. Community pharmacists' knowledge, attitudes, and readiness to provide counseling on food supplements—A scoping review. Nutrients 17. https://doi.org/10.3390/nu17233754.

Sobczyn, A., Harrington, D.J., 2018. Laboratory assessment of folate (vitamin B9) status. J. Clin. Pathol. 71, 949–956. https://doi.org/10.1136/jclinpath-2018-205048.

Sommer, A., Vyas, K.S., 2012. A global clinical view on vitamin A and carotenoids. Am. J. Clin. Nutr. 96, 1204S–1206S. https://doi.org/10.3945/ajcn.112.034868.

Suchdev, P.S., Namaste, S.M.L., Aaron, G.J., Raiten, D.J., Brown, K.H., Flores-Ayala, R., 2016. Overview of the biomarkers reflecting inflammation and nutritional determinants of anemia (BRINDA) project. Adv. Nutr. 7, 349. https://doi.org/10.3945/an.115.010215.

Tanumihardjo, S.A., 2011. Vitamin A: biomarkers of nutrition for development. Am. J. Clin. Nutr. 94. https://doi.org/10.3945/ajcn.110.005777.

Tulchinsky, T.H., 2010. Micronutrient deficiency conditions: global health issues. Public Health Rev. 32 (1), 243–255. https://doi.org/10.1007/BF03391600.

UNICEF, 2021. UNICEF conceptual framework on maternal and child nutrition. https://www.unicef.org/media/113291/file/UNICEFConceptualFramework.pdf (Accessed 24 February 2026).

UNICEF, 2024. The State of Food Security and Nutrition in the World 2024. UNICEF DATA. https://data.unicef.org/resources/sofi-2024/ (Accessed 24 February 2026).

UNICEF, 2025. The State of Food Security and Nutrition in the World 2025. UNICEF DATA. https://data.unicef.org/resources/sofi-2025/ (Accessed 24 February 2026).

van Wayenburg, C.A.M., van de Laar, F.A., van Weel, C., van Staveren, W.A., van Binsbergen, J.J., 2005. Nutritional deficiency in general practice: a systematic review. Eur. J. Clin. Nutr. 59 (1), S81–S88. https://doi.org/10.1038/sj.ejcn.1602178.

WHO, 2004. Comparative quantification of health risks: global and regional burden of disease attributable to selected major risk factors. Geneva: World Health Organization. https://www.who.int/publications/i/item/9241580313 (Accessed 24 February 2026).

WHO, 2007. Preventing and controlling micronutrient deficiencies in populations affected by an emergency. Geneva: World Health Organization. https://www.who.int/docs/default-source/nutritionlibrary/preventing-and-controlling-micronutrient-deficiencies-in-populations-affected-by-an-emergency.pdf (Accessed 24 February 2026).

WHO, 2021. Monitoring flour fortification to maximize health benefits: a manual for millers, regulators, and programme managers. https://www.who.int/news/item/13-09-2021-monitoring-flour-fortification-to-maximize-health-benefits-a-manual-for-millers-regulators-and-programme-managers (Accessed 25 February 2026).

WHO, 2024. Malnutrition. https://www.who.int/news-room/fact-sheets/detail/malnutrition (Accessed 24 February 2026).
WHO, 2025. Noncommunicable diseases. https://www.who.int/news-room/fact-sheets/detail/noncommunicable-diseases (Accessed 25 February 2026).
Wiseman, E.M., Bar-El Dadon, S., Reifen, R., 2017. The vicious cycle of vitamin a deficiency: a review. Crit. Rev. Food. Sci. Nutr. 57, 3703–3714. https://doi.org/10.1080/10408398.2016.1160362.
World Health Organization, 2024. Strategic framework for strengthening health laboratory services in the WHO Eastern Mediterranean Region, 2024-2029. https://applications.emro.who.int/docs/Health-laboratory-services-framework-eng.pdf (Accessed 24 February 2026).

CHAPTER SEVEN

Biofortification: Addressing the micronutrient deficiency of major food commodities of Africa

Stephen Chukwuemeka Igwe[a,*], Precious Kehinde Fadele[b], Kolawole Emmanuel Olayinka[c], Kolade Oluwakunmi Adekanye[d], Oluwadarasimi Babafemi Jaiyeoba[a], Mubarak Oladimeji Abdulsalam[a], Alexander Idu Entonu[a], and Don Eliseo Lucero-Prisno III[e,f,g]

[a]Department of Medicine and Surgery, University of Ilorin, Ilorin, Kwara State, Nigeria
[b]Department of Medicine and Surgery, Faculty of Clinical Sciences, College of Medicine, University of Nigeria, Enugu Campus, Enugu State, Nigeria
[c]Faculty of Engineering, University of Nigeria, Nsukka, Enugu State, Nigeria
[d]Department of Agricultural Economics, Landmark University, Omu Aran, Kwara State, Nigeria
[e]Department of Global Health and Development, Faculty of Public Health and Policy, London School of Hygiene and Tropical Medicine, London, United Kingdom
[f]Office for Research, Innovation and Extension Services, Southern Leyte State University, Sogod, Southern Leyte, Philippines
[g]Research and Innovation Office, Biliran Province State University, Naval, Biliran Province, Philippines
*Corresponding author. e-mail address: stephen.c.a.igwe@gmail.com

Contents

Advances in Food Security and Sustainability, Volume 11
ISSN 2452-2635, https://doi.org/10.1016/bs.af2s.2026.03.002

Abstract

Micronutrient deficiencies, or "hidden hunger," remain a major public health challenge in Africa, particularly affecting women and children. Biofortification offers a sustainable, cost-effective solution by increasing the nutrient content of staple crops through conventional breeding, agronomic practices, and biotechnology. Unlike traditional fortification, biofortification embeds essential vitamins and minerals directly into crops, ensuring long-term nutritional benefits for populations with limited dietary diversity. This chapter explores biofortification's role in addressing micronutrient deficiencies, focusing on key nutrients such as vitamin A, iron, and zinc. It examines successful case studies, including provitamin A-rich orange-fleshed sweet potatoes, iron-biofortified beans, and zinc-fortified rice, highlighting their impact on public health and food security. The chapter also discusses scientific advancements, consumer acceptance, and policy frameworks supporting biofortification. Despite its promise, biofortification faces challenges such as regulatory barriers, low farmer awareness, and socio-cultural resistance. Overcoming these obstacles requires coordinated efforts among governments, researchers, farmers, and policymakers to integrate biofortified crops into national nutrition policies. Scaling up biofortification efforts demands improved seed distribution, agricultural extension services, and targeted awareness campaigns to enhance adoption. By embedding vital nutrients into staple crops, biofortification offers a sustainable strategy to combat malnutrition in Africa. With continued research, policy support, and community engagement, it can significantly improve public health, strengthen food security, and contribute to global nutrition goals.

1. Introduction

1.1 The biofortification notion and its impact

1.1.1 Conceptualisation

Biofortification, a practice that increases the nutritional content of staple foods by using different farming practices, holds a critical position in combating malnutrition in Africa. Biofortification aims at raising the intake of important micronutrients such as vitamin A, iron, and zinc in the diet of the population by focusing on the increased content of these substances. Biofortification combats serious public health issues while providing the farmers and the consumer with serious economic benefits at the same time.

Biofortification constitutes an agricultural strategy for increasing the nutritional content of important staple foods by using different approaches that include conventional breeding practices, agronomy practices, and modern-day biotechnology tools. The initiative covers the widespread cause of unobserved deficiencies of micronutrients that affect the largest section of the global population, with particular intensity of severity in underdeveloped parts of Africa. The main purpose of biofortification is to increase the levels of vital vitamins and minerals or their precursors that are present within the parts of the staple foods that are eaten, and it improves the nutritional levels of the population that is consuming these staple foods. Unlike the traditional practices of staple food fortifications by the intentional introduction of the required nutrients into processed foods, the practice of biofortification creates the required nutrients within the staple foods by nature itself (Cheung et al., 2024; Garg et al., 2018; Saikanth et al., 2023; Siwela et al., 2020).

The significance of biofortification stems from its potential to combat malnutrition sustainably and cost-effectively.

Nuss et al. state that the main objective of biofortification transcends the mere increase of the intake of dietary micronutrients; it strives for the attainment of nutritional security for populations often characterised by restricted dietary diversity (Nuss et al., 2012). The challenge of nutritional deficiencies due to the unvaried intake of several micronutrients is more evident for rural populations of the sub-Saharan area of Africa due to the dependence upon staple foods (Hefferon, 2019; Severo et al., 2021). Hence, biofortification presents a scalable, cost-effective intervention that delivers essential micronutrients (iron, zinc, vitamin A, folate, etc.) through the very foods poor households already rely on daily, improving nutrition without requiring dietary changes or additional spending. This approach directly addresses the persistent global burden of micronutrient deficiencies, the primary form of malnutrition today, even among populations with adequate caloric intake from inexpensive staple grains and legumes. By overcoming the economic and logistical barriers that make nutrient-dense foods (animal-source foods, fruits, and vegetables) unaffordable and inaccessible for many, particularly where cold-chain infrastructure remains weak (FAO, 2015), these staple-focused strategies offer a practical and equitable pathway to better nutrition outcomes. Further, the practice of biofortification aligns with global health goals for the mitigation of malnutrition, intending to reduce the need for continuous interventions such as the delivery of supplements or external fortification interventions (Ojok et al., 2019; Koç & Karayiğit, 2021).

1.2 An examination of micronutrient deficiencies in Africa

The insufficiency of micronutrients represents a serious challenge to public health that affects nearly two billion people, with the impact thereof being most evident in Africa (Carvalho and Vasconcelos, 2013; Wang et al., 2016; Jha and Warkentin, 2020). Often referred to as hidden hunger, the situation largely entails deficiencies of crucial micronutrients such as vitamin A, zinc, and iron (Carvalho and Vasconcelos, 2013; Wang et al., 2016; Jha and Warkentin, 2020). Studies by health agencies project that such deficiencies are widespread and cause negative health consequences such as stunting of cognitive development, increased vulnerability to the contraction of infectious diseases, and weakened physical performance by the afflicted populations, especially children and women (Liberal et al., 2020; Kumar et al., 2019; Oh et al., 2021). The challenge is compounded by the fact that the main diet of such populations mainly entails foods that are naturally deficient in these critical nutrients.

Research indicates that the most dominant staple cereals in Africa are characterised by a remarkably low content of important micronutrients critical for providing proper nutrition for maximal health (Kashinath et al., 2024; Joy et al., 2015).

Moreover, the occurrence of acute and chronic illnesses caused by deficiencies of micronutrients is compounded by socio-economic determinants of poverty, hunger, and the widespread unavailability of diversified diets that are abundant in fruits and vegetables (Carvalho and Vasconcelos, 2013; De Steur et al., 2010). The consequences of these extend to impacting the health systems of the population and the viability of the country's economic systems (Jha and Warkentin, 2020; Koç and Karayiğit, 2021). Therefore, the solution to these challenges lies in enhancing the health of individuals, enhancing the resilience of the people at the population level, and enhancing the economic development of the country (Buturi et al., 2021; Garg et al., 2018).

1.3 Objectives and scope of the chapter

This chapter will thoroughly explore the critical importance of biofortification for combating the micronutrient deficiencies of staple foods in Africa. It will present a thorough synopsis and extensive analysis of the processes, success levels, and challenges of the practice. The particular goals are assessing the existing biofortification schemes for enhancing the nutritional quality of staple cereals, assessing the efficiency of different types of biofortification, and demarcating the socio-economic impacts of the interventions on the targeted populations.

Within the purview of this discussion, specific focus will be placed on the following topics: (1) a detailed discussion of both traditional and new biofortification methods, including genetic and agronomic considerations; (2) illustrative case studies of the effective application of biofortification methods in diverse African environments; (3) evaluation of the constraints and drivers affecting the uptake and effectiveness of biofortified food among African agricultural practitioners and consumers; and (4) an estimation of the potential for promoting biofortification programs as core components of broader strategies to drive food security and nutritional well-being. Through a focus on these areas, the chapter will identify the need for an overarching strategy to tackle hidden hunger through sustainable agriculture.

2. Background and historical development

2.1 The historical background of biofortification as a method

Biofortification has become the main strategy for overcoming persistent deficiencies of micronutrients, mainly in developing countries, whose diets often fall short of adequate vitamins and minerals. Micronutrient deficiencies—often termed "hidden hunger"—pose severe public health consequences, contributing to impaired cognitive development, weakened immunity, maternal mortality, increased susceptibility to infectious diseases, and reduced economic productivity. Deficiencies in iron, vitamin A, zinc, and iodine are among the leading causes of preventable illness and death globally, disproportionately affecting children and women of reproductive age in low-income settings

(Black et al., 2013; Global Nutrition Report, 2022). The roots of biofortification lie with ancient farming practices wherein farmers picked up the best traits of the crop; the intentional focus on the nutritional quality of the products came much later. The latter half of the 20th century saw the era of the Green Revolution, with the main thrust being the increase of productivity of agriculture through the introduction of new high-productive crop varieties at the cost of the nutritional quality of the products (Bouis, 2018).

The advancement of biofortification generated significant interest in the early 2000s, following concerns about hidden hunger, which impacts over two billion individuals worldwide.

Researchers have noted the increase in the levels of micronutrients within staple agricultural products as the most likely solution for successful nutrition improvement of malnutrition. The shift away from dependence upon the constant fortification and supplementation of processed foods represents the fundamental shift that agricultural science has taken toward the matching of practices of improvement of plants with the nutritional needs of humans (Wu et al., 2009; Bhattacharya et al., 2024). The World Health Organisation and several international agencies have started advocating for the cost-effective nature of the increased intake of vital nutrients using the practice of biofortification for areas with the constraint of restricted diet diversity and restricted access (Shahzad et al., 2014). The global average cost of a healthy diet rose to 4.46 PPP dollars per person per day in 2024, yet the number of people unable to afford such a diet decreased to 2.60 billion, driven largely by an uneven post-pandemic economic recovery. While affordability improved in Asia, Latin America, Europe, and Oceania, it worsened significantly in Africa, where the number of people unable to afford a healthy diet rose from 864 million to 1 billion, highlighting stark regional and income-level disparities (FAO, IFAD, UNICEF, WFP and WHO. 2025).

2.2 Timeline of major landmark achievements and developments

A notable improvement came with the launch of the HarvestPlus initiative in 2004 that focused on increasing staple cereals such as wheat, rice, and maize by increasing the content of these staple cereals with micronutrients such as iron, zinc, and provitamin A (Bouis, 2018). However, the conceptual and empirical groundwork for biofortification had been laid much earlier. Bouis began pioneering research on the integration of nutrition goals into agricultural breeding programs as early as 1993, advocating for the systematic enrichment of staple crops to address micronutrient malnutrition in low-income populations (Bouis and Welch, 2010). This early vision ultimately evolved into the coordinated HarvestPlus program, which operationalized large-scale biofortification through interdisciplinary collaborations between nutritionists, plant breeders, and policymakers, transforming the global approach to combating hidden hunger.

A notable success within the context of agricultural biotechnology is "Golden Rice," genetically engineered with increased levels of the vitamin A precursor β-carotene, effectively combating the nutritional deficiencies of the diet of countries that are highly dependent upon the cultivation of rice (Wu et al., 2024). The effort promoted the idea of genetic biofortification and prompted interest globally for the use of the biotechnological techniques for enhancing the nutritional quality of cereals.

Recent investigations expanded the number of crop types that are rich in important micronutrients zinc and iron present in different legumes and cereals. Iron- and zinc-fortified beans and maize are good examples that proved successful with early trials implemented in Africa and Latin America (Shahzad et al., 2014; Chaudhary et al., 2022).

The assessments were not just enhancing nutritional benefits but were taking into account the local functionality of consumption and the acceptability of diet practices; the biofortified products were accordingly consistent with the diet of the targeted population (Bhattacharya et al., 2024; Germ et al., 2019; Asif et al., 2024).

Moreover, advancements in agronomy practices, such as the improvement of the levels of nutrients in the soil by using organic amendments and optimising fertilisation processes, have contributed greatly to agronomic biofortification. The utilisation of biotechnological techniques using zinc solubilising bacteria (ZSB) (i.e. the use of bacteria to enhance plant accessibility to zinc by converting its insoluble forms into usable ones) has proved the ability of zinc solubilisation for most crops and contributed highly toward increasing the fertility of the soil and increasing the productivity of agriculture (Upadhayay et al., 2022a; Upadhayay et al., 2022b). The successful implementation of these practices presents the ability for the nutritional content of the food to be enhanced sustainably.

2.3 Key actors and affected stakeholders

The biofortification industry is marked by a wide array of stakeholders, ranging from international agencies, government ministries, and scientists to non-governmental institutions and local community organisations. HarvestPlus itself has played a pivotal role in leading the world in biofortified crops through research support, funding, and facilitating partnerships between sectors (Bouis, 2018). Inter-university and research centre collaborations like the International Rice Research Institute (IRRI) and the International Food Policy Research Institute (IFPRI) have also contributed meaningfully towards the enrichment of theoretical thought processes along with field-level practices of biofortification procedures (Hu et al., 2023). National NGOs and government agencies also need to contribute by facilitating community-level uptake of biofortified foods. For example, the distribution of biofortified sweet potatoes in certain areas of Africa has necessitated mass community outreach to get released varieties that are well adapted to local feeding patterns and farming systems (Wu et al., 2009; Laîné et al., 2019). The Food and Agriculture Organization of the United Nations (FAO) and World Health Organization (WHO, 2014) have been responsible for supporting and directing the formulation of policy to incorporate the implementation of biofortification intervention in national nutrition and food schemes, having valued its significant contribution towards the management of malnutrition (Bouis, 2018; Shahzad et al., 2014). Recent trends indicate that private sector actors have begun to initiate interest in biofortification following the realisation of the potential market advantages associated with crops that possess improved nutritional content.

Agrochemical and seed companies are now investing in research aimed at the development and commercialisation of biofortified crops to develop a sustainable supply chain that can provide nutrient-enriched foods for human consumption (Olum et al., 2018; Laîné et al., 2019). Overall, the various strategies adopted in biofortification—ranging from research activities, technological development, community mobilisation, and policy-making—reflect the collective nature of this program and the potential impact that it can have in preventing micronutrient deficiency across the world.

The findings and conclusions that are reported here are targeted at scholars, policymakers, and practitioners of agriculture, nutrition, and public health. The purpose here is to allow for informed discourse about the most appropriate strategies for the solution of the serious problem of Africa's micronutrient malnutrition.

3. The scientific basis of biofortification

3.1 Genetic and agronomic techniques

Genetic engineering and agronomic methods are two critical scientific techniques used in the biofortification of plant crops.

Genetic engineering can be used to increase the bioavailability of nutrients in plants (Lima et al., 2017). For instance, a publication by Arizona State University reported how recombinant DNA technique was used to produce provitamin A-enriched rice by two scientists, Potrykus, a plant biotechnologist, and Beyer, a biochemist, who started the Golden Rice project in 1993 (Baranski, 2013). In order to enhance human health, Ingo Potrykus and Peter Beyer created Golden Rice in the 1990s by modifying regular rice. The genome of golden rice has a designed multi-gene metabolic pathway. When metabolized by humans, beta-carotene, a chemical produced by this route, forms vitamin A (Tang et al., 2009). Beta-carotene is responsible for the golden hue of golden rice. Beta-carotene is not expressed in the endosperm, the starchy and largest portion of the rice seed, which is often off-white in normal rice (*Oryza sativa*). However, Tang et al. (2009) demonstrated that beta-carotene was effectively converted to vitamin A in humans.

Three distinct genes—two from bacteria and one from daffodils—are present in Golden Rice embryos.

Potrykus' lab tried several approaches to the transformation because no one had ever been able to properly express three genes in a food crop. Using bacterial transfer or particle bombardment, the genes were initially inserted into the rice embryo. By inserting its DNA into the targeted rice plant embryos, Potrykus' lab employed an Agrobacterium-mediated transformation. All three genes were present in this DNA: lycopene beta-cyclase (lcy, from daffodils), phytoene synthase (psy, from daffodils), and phytoene desaturase (crtI, from bacteria). To trace the implanted DNA, scientists also added marker genes and additional DNA fragments that the genes need to operate in the cell. The embryos were then grown, chosen, and examined for beta-carotene by the scientists. When the rice plants reached maturity, their starch generated and stored beta-carotene (Baranski, 2013). Since the first rice trials, scientists have modified other crops, such as maize and potatoes, to manufacture beta-carotene through distinct biochemical processes.

Additionally, Obukosia (2014) reported the biofortification of staple food crops to limit micronutrient deficiency in developing countries through biotechnology techniques to improve four staple crops (rice, sorghum, cassava, and bananas). This effort has been particularly for consumers in arid and semi-arid regions of Africa with limited access to fruits and vegetables during the dry seasons. It is also relevant for rural farmers and consumers who do not purchase commercially fortified foods. However, it is important to note that much of the progress in biofortification to date has been achieved through conventional breeding approaches rather than genetic modification. Conventional breeding remains the primary method due to its cost-effectiveness, public acceptance, and regulatory simplicity compared to transgenic approaches. Through selective cross-breeding, researchers have successfully enhanced the levels of iron, zinc, and provitamin A in key staple crops such as beans, maize, sweet potatoes, and pearl millet (Bouis and Saltzman, 2017; Garg et al., 2018). These conventionally bred varieties have already reached millions of households across sub-Saharan Africa and South Asia, demonstrating the scalability and sustainability of this approach. Biofortification via genetic transformation is a useful addition to ongoing biofortification efforts through conventional breeding and fortification (Obukosia, 2014). Conventional breeding improves crop nutritional attributes while maintaining other agronomic aspects (Ofori et al., 2022). According to Jha and Warkentin (2020), plant breeding is a more sustainable and cost effective biofortification strategy compared to other approaches, with financial investments occurring only during research and development stages. In addition, unlike agronomic biofortification, plant breeding has minimal to no environmental impact (Marques et al., 2021).

Furthermore, using the agronomic method, nutrient absorption and storage in crop edible sections are increased by administering micronutrient-rich fertilisers to the soil or plant leaves. For instance, applying fertilisers enhanced with iron, zinc, and selenium has been demonstrated to successfully raise the micronutrient content in crops like maize and wheat (De Valença et al., 2017). This technique is sometimes referred to as "agronomic biofortification." In addition, the necessity of integrated soil fertility management is shown by the fact that micronutrient fertilisation works best when combined with NPK, organic fertilisers, and enhanced crop types. While genetic biofortification could ultimately be more economical, agronomic biofortification offers a quick and efficient way to increase the concentrations of micronutrients in edible agricultural products (De Valença et al., 2017). Agronomic biofortification has been effective in Finland, a selenium-poor region where since 1985, selenium has been added to the fertilizers which has directly translated into an increased average dietary intake from 0.04 mg Se/day/10 MJ in 1985 to a present plateau of 0.08 mg Se/day/10 MJ, which is significantly more than the current dietary guidelines. The national addition of fertilisers with sodium selenate has demonstrated to be effective and safe in raising the Se intake of the entire population (Alfthan et al., 2014).

Transgenic breeding is also a means of biofortication that involves the addition of specific micronutrients not naturally synthesised in a crop via genetic engineering to the crop. It is formulated by adding completely novel genes, overexpressing the genes currently existing, downregulating the expression of specific genes, or stopping the production pathway genes of inhibitors (Malik and Maqbool, 2020). A successful example is the production of golden rice fortified with provitamin-A as no rice variety has vitamin A precursors (Ye et al., 2000). The exploration of transgenic approaches is essential due to weaknesses in the following: the lack of elite genotype for desired trait within the specie (e.g provitamin A in rice), or the inability to target nutritional traits to specific organs, the length of time needed to introduce single or multiple traits (pyramiding traits) into locally adapted elite varieties, incongruous varieties for conventional breeding (due to a lack of sexuality; e.g., banana), and insufficient knowledge of genetic and environmental interactions (Dhaliwal et al., 2022).

3.2 Case studies of biofortified crops: Pro-vitamin A sweet potatoes, iron-fortified beans, zinc rice

In various studies, biofortified crops such as vitamin A-enriched sweet potatoes, iron-fortified beans, and zinc-fortified rice have shown promise in combating "hidden hunger"—a term that describes micronutrient deficiencies that often go unnoticed yet significantly affect health outcomes. The introduction of biofortified sweet potatoes, particularly orange-fleshed varieties, has addressed vitamin A deficiencies in several African countries, resulting in increased serum retinol levels among consumers (Siwela et al., 2020). This is critical in targeting populations, especially children and pregnant women, who are at higher risk of vitamin A deficiency and its associated health issues (Senguttuvel et al., 2023).

Iron biofortification efforts have prominently aimed at legumes like beans, which are staple foods in many African diets. Programs focusing on iron-fortified beans have shown significant increases in iron content, thereby helping to alleviate iron deficiency anaemia among vulnerable populations (Zubair et al., 2023). Moreover, zinc biofortification, particularly in rice, has garnered attention as a sustainable means to improve health outcomes. Rice is a primary staple for many in Africa, and the low zinc content in standard rice varieties has contributed to widespread zinc deficiency. By increasing zinc concentration in rice through biofortification, health metrics such as immune function and cognitive development can be significantly improved (Zubair et al., 2023; Majumder et al., 2019; Liu et al., 2017).

3.3 Nutritional efficacy and absorption

The nutritional efficacy of biofortified crops goes beyond mere increases in micronutrient concentrations; it is equally essential to consider the absorption and bioavailability of these nutrients within the human body. Research has indicated that biofortified crops can lead to higher absorption rates of essential micronutrients due to their improved availability and lower levels of antinutritional factors that inhibit absorption (Nahar et al., 2020; Sharma and Singh, 2022). For instance, studies have reported that zinc biofortification in rice can increase zinc bioavailability, which is crucial given the high prevalence of zinc deficiency in regions where rice constitutes a major part of the diet (Nahar et al., 2020).

To further substantiate this, it is necessary to analyse the mechanistic pathways through which biofortified crops interact with human physiology. The interaction of phytates and zinc is a vital aspect to consider—lower phytate-to-zinc ratios in biofortified rice enhance the absorption of zinc (Qin et al., 2012; White and Broadley, 2011). A controlled study illustrated that biofortification efforts that increased the levels of zinc in rice consequently improved the efficiency of its absorption, ultimately benefiting individuals with limited dietary diversity and access to nutrient-rich foods (Zubair et al., 2023; Sharma and Singh, 2022).

Comprehensive zinc biofortification methods have been developed, including the employment of agronomic approaches that combine proper fertilisation techniques with breeding. Such methods have effectively raised the zinc concentrations in rice, thus meeting public health targets and addressing deficiencies in populations relying on these staples (Swamy et al., 2018; Woods et al., 2020). Furthermore, longitudinal studies have indicated that regular consumption of biofortified foods correlates with improved micronutrient levels in populations, leading to a measurable reduction in deficiency-related morbidities (Sazawal et al., 2018).

The advancements in breeding techniques, which allow for the increased genetic expression of micronutrient transporters, have shown potential for enhancing both iron and zinc delivery in staple crops, making them a pivotal part of global food-security strategies (Ashokkumar et al., 2020; Kawakami and Bhullar, 2018; Boonyaves et al., 2017). The research conducted within this framework supports the argument that biofortification is not only feasible but also presents an effective long-term strategy for addressing micronutrient deficiencies among Africa's most vulnerable populations.

4. Implementation strategies for biofortification of staple foods in Africa

4.1 Crop development and breeding programs

4.1.1 Crop development

Despite the existence of at least 50,000 plant species fit for human consumption, fewer than 300 reach the market. A 2016 report highlights the need for greater dietary diversity to combat malnutrition and the rising threat of food insecurity. Expanding the consumption of underutilised crops, often termed "orphan" or "minor" crops, could be key to future food security (Massawe et al., 2016). These crops often have high nutrient profiles and resilience to harsh climatic conditions, making them valuable in addressing micronutrient deficiencies in Africa.

Before the 1960s, agricultural production in Europe was characterised by diverse farming techniques. However, since the 1970s, monoculture farming—fuelled by agrochemicals, mechanisation, and high-yield plant varieties—has dominated, leading to reduced crop diversity, increased chemical usage, and negative environmental impacts (Zabala et al., 2023). Similar trends have been observed in Africa, where high-yielding monocultures have often replaced diverse cropping systems, reducing the availability of nutrient-dense traditional crops.

In Africa, biofortification efforts require a multidisciplinary approach linking agricultural research with human health and nutrition sectors. This necessitates collaboration between National Agricultural Research Institutes (NARIs) and international research bodies to develop biofortified crops such as sweet potatoes, cassava, pearl millet, beans, and maize. These biofortified crops undergo multi-locational trials to assess adaptability, stability, and agronomic performance before formal release. Additionally, participatory breeding programs that involve local farmers in the selection of biofortified varieties ensure that these crops align with cultural preferences, thereby increasing adoption rates (Nestel et al., 2006; African Union, 2020).

4.1.2 Breeding strategies

Biofortification through plant breeding is a sustainable strategy to combat micronutrient deficiencies, particularly among vulnerable populations such as women and children. Research has demonstrated the genetic potential of staple crops, including rice, wheat, maize, beans, and cassava, to be enriched with micronutrients like iron, zinc, and provitamin A without compromising yield (Welch and Graham, 2004).

Breeding strategies include increasing bioavailable micronutrients while reducing antinutrients. Agronomic biofortification involves applying fertilizers to enrich soil and plant tissues with essential minerals, though its cost can be prohibitive. Genetic biofortification, on the other hand, involves conventional and transgenic breeding techniques to enhance micronutrient accumulation, offering a cost-effective and sustainable approach (Ludwig and Slamet-Loedin, 2019; Tumbare and Maphosa, 2023). Advances in genome editing tools such as CRISPR-Cas9 have further enhanced the ability to develop biofortified crops with precision, ensuring that desirable traits such as higher nutrient density and pest resistance are effectively integrated into staple food crops.

4.2 Field trials and pilot projects

HarvestPlus, part of the Consultative Group on International Agricultural Research (CGIAR) Program on Agriculture for Nutrition and Health (A4NH), collaborates with over 200 organisations to develop and promote biofortified crops. These efforts have resulted in millions of farmers growing vitamin- and mineral-rich crops across Africa, Asia, and Latin America (Andersson, 2017).

For example, the International Potato Centre (CIP) and its partners developed the first biofortified crop, the orange-fleshed sweet potato (OSP), which contains high levels of provitamin A and has been successfully adopted in Uganda and Kenya. The introduction of OSP in these countries has helped reduce vitamin A deficiency, particularly among children under five and pregnant women (Mwaniki, 2007). The breeding of OSP was complemented by extensive awareness campaigns and farmer training programs to ensure widespread adoption and sustained cultivation.

In Zambia, the challenge of integrating provitamin-rich maize with local high-yielding varieties was overcome through rigorous breeding programs. The first three provitamin A maize varieties were released in 2012, with seed companies commercialising them. Lessons from this experience emphasised the importance of ensuring seed producibility and availability. Large-scale farmer education initiatives were launched to promote the nutritional benefits of provitamin A maize, ensuring greater market demand and continued cultivation (Tanumihardjo et al., 2017).

Nigeria's biofortification program, initiated in 2011, led to the official release of provitamin A cassava and maize varieties, as well as iron pearl millet and zinc rice. Collaboration between the National Agricultural Research and Extension Systems (NARES) and CGIAR facilitated the scaling of biofortified crops into national agricultural policies and private sector portfolios. Notably, Nigeria's approach emphasised integrating biofortification with existing agricultural extension services, enabling smallholder farmers to easily access information and planting materials. By 2021, biofortified crops had gained significant traction, with increasing numbers of households incorporating them into their diets (Tanumihardjo et al., 2017).

4.3 Scaling up production and distribution

Scaling up biofortification entails expanding interventions to achieve widespread impact. Strategies include increasing demand for biofortified crops, advocating for supportive policies, and accelerating seed commercialisation (Mulongo et al., 2021). The success of biofortified crops depends on both the supply of high-quality seeds and the creation of market demand through consumer education and policy support.

Advocacy has played a crucial role in scaling biofortification efforts. The 3.5-year RAC project (2011–2015) in Tanzania, Mozambique, Nigeria, Ghana, and Burkina Faso promoted investments in orange-fleshed sweet potatoes, fostering policy change, institutional capacity, and proven technologies. The project demonstrated that sustained advocacy efforts could influence national policies, leading to increased government investment in biofortified crop development and dissemination (Mulongo et al., 2021).

Distribution mechanisms also influence the success of biofortification programs. In Nigeria, provitamin A cassava stems were initially distributed through social and farmer-to-farmer networks. Over time, a commercial stem distribution model emerged, ensuring market sustainability. By 2018, a self-sustaining provitamin A cassava market had been established, aided by quality control initiatives supported by the Gates Foundation and the International Institute of Tropical Agriculture (IITA). The transition from a subsidised to a commercial seed distribution system highlights the importance of private sector engagement in sustaining biofortification efforts (Foley et al., 2021).

In addition to cassava, efforts to scale up the production of iron-rich beans in Rwanda and zinc rice in Bangladesh have demonstrated the feasibility of integrating biofortified crops into existing agricultural value chains. Lessons from these projects indicate that government buy-in, financial incentives for seed producers, and targeted consumer awareness campaigns are crucial in driving adoption. Future efforts should focus on expanding partnerships with food processors to incorporate biofortified crops into widely consumed products, thereby increasing consumer exposure and acceptance.

Efforts must continue to refine scaling models to enhance biofortified crop adoption, ensuring long-term sustainability and widespread nutritional impact. A key area of focus should be strengthening supply chain logistics, ensuring that biofortified seeds and planting materials reach remote farming communities. Moreover, innovative financing mechanisms, such as micro-loans for biofortified seed production and distribution, could further accelerate the reach of these interventions.

As biofortification moves from pilot projects to large-scale implementation, collaboration between research institutions, policymakers, private sector stakeholders, and farmers will be essential in ensuring its success. Addressing micronutrient deficiencies through biofortified crops represents a significant step towards improving food security and public health outcomes across Africa.

5. Adoption and acceptance of biofortified crops in Africa

Malnutrition constitutes a pressing public health issue in Africa, as millions of individuals are deficient in essential micronutrients such as vitamin A, iron, and zinc (Sun, 2024). Dietary supplementation and food fortification strategies have yielded significant advancements; however, they are limited by issues of accessibility, affordability, and sustainability (Sandhu, et al., 2023). The integration of selective breeding and modern biotechnology in genetic biofortification (GB) constitutes a promising strategy to augment the nutritional content of staple crops (Sen et al., 2024). Biofortification facilitates sustainable and cost-effective prevention of hidden hunger by augmenting nutrient levels in staple foods such as rice, cassava, sweet potatoes, and maize (Mmbando and Missanga, 2024).

The effective execution of biofortification requires farmers' endorsement and consumers' acceptance of its advantages. Farmers must comprehend that biofortification technology enhances crop yields and that their crops exhibit resilience and significant market value. Consumers must embrace biofortified crops in their diets, notwithstanding any challenges related to taste, appearance, or cultural attitudes. The widespread adoption and utilisation of biofortified crops necessitate coordinated efforts among governments, research institutions, market participants, and agricultural extension programs to ensure these crops are accessible and economically viable (Bouis and Saltzman, 2017).

This section evaluates the current status of biofortified crop implementation on African farms by analysing farmer behaviour, consumer responses to nutritionally enhanced foods, cultural influences, and governmental support initiatives. In this section, the chapter also presents the major barriers to general acceptance as well as strategic suggestions to boost adoption levels. Through real-world examples, adoption trend research, and consumer preference measurements, this chapter evaluates both possible advantages and obstacles concerning biofortified crop implementation in African agriculture.

5.1 Farmer adoption: Rates and challenges

5.1.1 Understanding farmer adoption

Biofortification shows great promise to transform African food security through its ability to overcome obstacles that prevent public acceptance and adoption of fortified food products (Mmbando and Missanga, 2024) and (Adeyeye and Idowu-Adebayo, 2019). The knowledge of biofortification success and failure drivers will be essential for expanding its reach to target communities effectively.

The scientific community widely acknowledges biofortification as a proven method to enhance nutritional outcomes across Africa (Msungu et al., 2022). Scientific research demonstrates that sustained biofortified staple crop consumption leads to observable health improvements for humans (Bouis and Saltzman, 2017). The adoption speed of these crops by farmers depends on multiple elements that encompass biofortified seed availability along with farmer understanding of the advantages and cost of inputs and yield potential and farmer readiness for novel agricultural methods (Siwela et al., 2020).

Undoubtedly, successful adoption of biofortified crop varieties hinges on their productivity being comparable to or exceeding that of existing local varieties. Research indicates that a biofortified variety must demonstrate equivalent or superior agronomic performance in similar growing conditions for farmers to consider adoption (Islam et al., 2024). For instance, the development of the ACI dhan1 rice variety in Bangladesh focuses not only on enhancing iron and zinc content but also on maintaining high yield potential, thereby addressing both nutritional and productivity needs (Islam et al., 2024). Similarly, varieties like NamChe 4 and NamChe 5 have shown significant agricultural performance while addressing nutritional deficiencies, demonstrating that successful biofortification should not disrupt existing farming practices (Kankwatsa et al., 2019).

The imperative for competitive yield is underscored by farmer reluctance to adopt underperforming varieties (Jalal et al., 2023). Furthermore, the biofortification process should improve the nutritional content of staple crops already cultivated by communities, thereby facilitating the alleviation of micronutrient deficiencies without requiring a complete overhaul of agricultural practices (Hassan et al., 2017). This principle highlights the adaptability of biofortified crops across diverse agronomic environments, which is essential for fostering farmer acceptance (Huang et al., 2024).

The adoption process increases when farmers recognise both economic benefits and improved agricultural outcomes (Kuehne et al., 2017). For instance, the widespread adoption of provitamin A-rich orange-fleshed sweet potatoes (OFSP) in certain regions occurred because these plants demonstrate drought tolerance and attract high market prices (Low et al., 2020). The data regarding biofortified crop variety adoption rates appears in Table 1 across African countries.

Additionally, HarvestPlus reports that as of 2024, approximately million households in Africa are growing biofortified crops (HarvestPlus, 2025), demonstrating their increasing acceptance and potential for further expansion.

5.1.2 Challenges hindering adoption

5.1.2.1 Lack of awareness and education

The agricultural community lacks information about biofortification together with its advantages (Sandhu, et al., 2023). The adoption process requires awareness as its starting point, but many farmers remain uninformed about biofortification. Research indicates that the adoption rates of OFSP in Ghana and Nigeria could reach 61 % and 42 % through enhanced dissemination and exposure programs (Adekambi et al., 2020).

5.1.2.2 Market and economic uncertainty

The potential market risks associated with biofortified crops cause farmers to avoid planting them (Adekambi et al., 2020). Biofortified varieties lose market appeal if they do not maintain stable demand or receive premium prices since farmers choose conventional crops with more reliable returns.

5.1.2.3 Environmental and climatic challenges

Climate change poses significant challenges to the adoption of biofortified crops in Africa, as agricultural yields are projected to decline, negatively impacting nutrient quality. Research indicates potential yield losses ranging from 1.2 % to

Table 1 Adoption rates of biofortified crops in Africa.

S/N	Country	Staple crop	Variety developed	Adoption rate	References
1.	Mozambique, Uganda	Sweet potatoes	Orange-fleshed sweet potatoes (OFSP)	61–68 %	Eozenou et al. (2018)
2.	Ghana, Nigeria	Sweet potatoes	Orange-fleshed sweet potatoes (OFSP)	51–33 %	Adekambi et al. (2020)
3.	Rwanda	Beans	Iron-biofortified beans	58–68 %	Vaiknoras et al. (2019)
4.	Uganda, Kenya, Tanzania	Maize	Quality Protein Maize (QPM)	73 %, 25 %, and none, respectively.	De Groote et al. (2016)

Source: Mmbando and Missanga (2024).

4.5 % across various East African countries due to climate-related factors such as soil erosion, which limits agricultural productivity Okorie et al. (2023). These environmental stressors can also compromise the bioavailability of critical nutrients in crops, highlighting the need for adaptive breeding strategies to sustain and enhance nutritional quality amid these changes (Petry et al., 2015).

Moreover, maintenance breeding is crucial for all crops, including biofortified varieties, because traits can diminish over time, particularly when seeds are saved for subsequent planting (Beebe, 2020; Boy, 2017). While iron-rich beans have shown promise in improving nutrition in regions like Rwanda, their adaptability may be limited, and these challenges are common among all bean crops due to regional agro-ecological conditions (Ejigu et al., 2024; Peláez et al., 2022). This underscores the importance of localized breeding programs focused on resilience and sustainability, which are essential for the widespread adoption of biofortified crops in the face of climate change (Lockyer et al., 2018).

5.2 Consumer acceptance and market dynamics

5.2.1 Factors influencing consumer acceptance

Consumer acceptance of biofortified crops is influenced by various factors, including taste, appearance, cooking properties, affordability, and cultural preferences (Talsma et al., 2017). If a biofortified crop differs significantly from traditional varieties, it may face resistance (Birol et al., 2015). For instance, some consumers initially rejected orange-fleshed sweet potatoes (OFSP) due to their unfamiliar orange color, associating it with carrots rather than sweet potatoes (Low et al., 2015). In many African regions where white rice and maize is preferred over their yellow counterparts, which are biofortified with provitamin A, as the latter is sometimes perceived as animal feed. On the other hand, iron-rich beans, which closely resemble conventional varieties, have been more widely accepted due to minimal differences in taste and appearance (Philipo, 2021).

5.2.2 Market dynamics of biofortified crops

Price sensitivity plays a major role in low-income areas (Nassar et al., 2021). This is because consumers need to choose affordable food options instead of nutritious alternatives because of their restricted financial means. The cost of biofortified foods stands as a vital aspect when communities decide to adopt these products. The implementation of effective branding and public awareness campaigns serves to boost biofortified food demand by raising consumer understanding about their nutritional value and market advantages.

The consumption of biofortified foods will rise and reach mainstream status when biofortified ingredients become part of commercial food products (Mitra-Ganguli et al., 2022). The process includes adding pro-vitamin A-rich maize to flour and iron-rich beans to canned goods and other nutrient-enhanced crops to processed foods. Biofortified foods will become household essentials, especially for regions where micronutrient deficiencies are widespread through this approach.

The availability of biofortified crops in retail stores stands as an essential factor for making these crops accessible to consumers. Biofortified foods need to be accessible through markets and supermarkets along with local food stores to enable consumers to easily buy them for dietary inclusion. The continuous availability of biofortified products depends on joint efforts between food manufacturers, distributors, and retailers. Biofortified foods establish sustainability as a solution for micronutrient deficiency management and public health improvement through proper consideration of these aspects.

5.2.3 Strategies to improve consumer acceptance

These are some strategies that can be put in place to improve consumers' acceptance of biofortified crops:

- Consumer Education and Taste Trials: Food tastings and demonstrations aid in acquainting consumers with biofortified crops (Singh et al., 2023).
- Marketing and Rebranding: Promoting yellow maize as a high-end health food, for example, can help change how people view biofortified crops (Pillay et al., 2011).
- Policy Support: Biofortified crops can be incorporated by governments into food assistance programs and school nutrition programs (Bouis and Saltzman, 2017).
- Cooperation with food processors: Promoting the use of biofortified ingredients by food manufacturers can lead to a wider market uptake (Osendarp et al., 2018).

5.3 Socio-cultural factors influencing adoption

5.3.1 Cultural beliefs and dietary preferences

The acceptance of biofortified crops in Africa heavily depends on socio-cultural elements (Bell et al., 2024). Traditional African cuisines need to accept these crops for successful adoption to occur. Where sweet potatoes represent a dietary staple, orange-fleshed sweet potatoes (OFSP) gain better acceptance among local communities. The adoption of these crops becomes more successful when local culinary customs integrate them through strategies that blend the crops with familiar eating habits.

The public needs clear information about biofortification and natural breeding techniques to gain consumer trust and overcome misconceptions. Community outreach programs and educational efforts need to deliver factual information about biofortified crops while demonstrating their practical health advantages. Better health outcomes observed by communities typically lead them to recognize biofortified crops as essential dietary improvements.

5.3.2 Role of women in adoption

The adoption of biofortified crops depends heavily on women because they make most nutrition-related decisions within households. (Samuel et al., 2024). Programs that provide women with training about biofortified crop cultivation and explain their nutritional value have led to higher adoption rates. Women face multiple barriers because they have restricted access to resources, and traditional cultural standards restrict their participation in farm decisions (Quisumbing and Pandolfelli, 2010).

5.3.3 Cultural beliefs and myths about biofortified crops

The false ideas that biofortified crops are genetically engineered creatures or dangerous to health limit their acceptability in some areas (Mmbando and Missanga, 2024). Biofortified crops must be promoted through proactive communication tactics, which include educational seminars alongside radio broadcasts and community gatherings to address concerns, directly explain misunderstandings, and show the safety and benefits through evidence-based information.

The involvement of community leaders together with religious figures and traditional authorities serves as a crucial method to build trust, which validates biofortification projects while giving these respected figures the role of advocates to explain the safety and advantages of biofortified crops to their communities. When local farmers participate in pilot initiatives and demonstrate their accomplishments, it eliminates skepticism among their communities and stimulates others to adopt biofortified crops because direct observation of health gains and agricultural growth encourages widespread acceptance.

The inclusion of farmers and consumers in local decision-making processes leads to reduced cultural resistance since they feel accountable for biofortification projects, which results in continued support for these efforts. Collaborative strategy formulation, such as picking crops that match local food preferences or addressing specific nutritional needs, ensures biofortification projects align with cultural values and receive general acceptability. Biofortified crops will achieve successful integration into African agriculture and diets when stakeholders actively address traditional beliefs and myths.

5.4 Case study and regional examples

5.4.1 Uganda: Orange-fleshed sweet potato (OFSP)

The Ugandan government implemented Orange Fleshed Sweet Potato (OFSP) as a solution to fight vitamin A deficiency. (Low et al., 2017). The adoption of OFSP by farmers and consumers increased through educational initiatives and distribution programs because the root vegetables matched traditional dishes like regular sweet potatoes. (Jenkins et al., 2015).

5.4.2 Nigeria: Biofortified cassava

The International Institute of Tropical Agriculture (IITA) led Nigerian acceptance of cassava fortified with provitamin A through their biofortification initiatives (Chidiebere-Mark and GodsPower, 2020). Widespread consumer adoption of biofortified crops occurred after community demonstrations showed that the products maintained a familiar taste and texture.

5.4.3 Zambia: Provitamin-A maize (PVAM)

The government of Zambia integrated biofortified maize into their national agricultural initiatives (Lividini and Fiedler, 2015). Government backing combined with farmer training programs and financial support from the government helped farmers adopt biofortified crops, which led to better food security.

6. Health impact

Throughout Africa, several promising clinical trials and research initiatives are underway to develop and implement sustainable breakthroughs in food biofortification. These efforts aim to impact the health of the continent's youngest populations, who face the highest risk of malnutrition.

Notable clinical trials are currently demonstrating the effectiveness of fortifying staple foods such as cereals, legumes, oilseeds, vegetables, and fruits. The selection of these staples is based on the most prevalent diets in different sub-regions of Africa. Numerous national and multinational organisations recognise food fortification as a vital step toward enhancing food security, highlighting its significance in improving food quality as we advance into the 21st century and beyond. These initiatives are crucial for ensuring a healthier future for everyone.

6.1 Clinical trials

In West Africa, a study was undertaken in Nigeria by Afolami et al. (2020) to determine the efficacy of biofortified cassava in improving the vitamin A status of Nigerian preschool children. The randomised controlled trial had 159 children, of which the yellow cassava group was 80 and the white cassava was 79. After controlling for baseline retinol concentrations, inflammation, and asymptomatic malaria status, the study concluded that the daily intake of β-carotene from biofortified cassava modestly enhanced serum retinol and haemoglobin levels in Nigerian preschool children.

Another study carried out in Nigeria in a bid to increase the consumption of orange-fleshed sweet potatoes among children inferred that when sweet potatoes are introduced with behavioural cues like songs or associations with positive role models, children eat more on average, thus reducing the chances of vitamin A deficiency and its associated risks (Just et al., 2021).

In East Africa, several programs and studies have been undertaken on the effect of biofortified staple foods across the region. Wenger et al. (2019) carried out a randomised controlled trial to determine if changes in iron status are related to changes in brain activity and behaviour in Rwandan female university students, and the result showed that iron level affects both brain activity and behavioural performance as determined by electroencephalogram. In a sample of female university students, eating iron-biofortified beans for 18 weeks improved both parameters compared to consuming a reference bean. Additionally, the findings provide credence to the proposition that the relationship between alterations in iron indicators and cognitive changes is mediated by modifications in brain activity brought on by eating biofortified beans.

Children and women in farmer group member households significantly increased their dietary consumption of β-Carotene-rich Orange Sweet Potato (OSP) as a result of a large-scale intervention to introduce OSP to rural Ugandan communities. This large increase in vitamin A intake amongst the study population resulted in a 9.5 % decrease in the prevalence of low serum retinol (Hotz et al., 2012).

Talsma et al. (2015) carried out a randomised controlled trial in Kenya on biofortified yellow cassava and the vitamin A status among 342 children with a primary outcome of measuring the serum retinol concentration and pre-specified secondary outcomes such as haemoglobin concentration and serum concentrations of β-carotene, retinol-binding protein, and prealbumin. Their findings revealed supplementing with yellow cassava and beta-carotene raised serum retinol levels, which in turn increased serum β-carotene levels by 5-times, although retinol-binding protein, prealbumin serum concentrations, and haemoglobin concentration were unaffected (Talsma et al. 2015).

A study was carried out by Haas et al. (2016) among Rwandan women to increase their iron levels through feeding with iron-fortified beans. During the 128 research days, there was a substantial 4.2-g/L rise in haemoglobin for every 1 g of iron eaten from beans (Haas et al. 2016).

A randomised controlled trial was done to assess if Zambian women and their nursing babies could enhance milk retinol concentration and infant Total Body Stores (TBS) by consuming biofortified or industrially fortified maize. The retinol content of breast milk improved for Zambian mothers who consumed Provitamin A maize twice daily for three months, and the prevalence of low retinol concentration in breast milk decreased by more than 50 %. Conclusively, consuming both fortified and biofortified maize increased the content of retinol in milk. This did not result in higher newborn TBS, most likely because baseline TBS was sufficient (Palmer et al., 2020).

Conversely, in South Africa, biofortification has helped to increase the availability of provitamin A-rich foods. However, despite the successful biofortification of sweet potatoes, the prevalence of vitamin A deficiency (VAD) in the country remains unacceptable. This could be attributed to the low acceptance of orange-fleshed sweet potatoes (OFSP) among adults, who make decisions at the household level (Siwela et al., 2020).

6.2 Impact on vitamin a deficiency, anaemia, and overall health

A study undertaken in Zambia by Gannon et al. (2014) to assess the efficacy of biofortified orange maize on preschool children found that the children's vitamin A status was much enhanced when ordinary maize was swapped out for Provitamin A maize. Increased vitamin A levels have been linked with improvements in vision in low-light conditions, cell growth and development, maintaining a healthy immune system, encouraging healthy skin and mucous membranes, and functioning as an antioxidant to guard against cell damage. Such benefits are significant in the growing and aged population who are more at risk of infections such as diarrhoea, which is the second leading cause of death of young children in low- and middle-income countries (Jones and de Brauw, 2015).

Furthermore, a randomised controlled trial conducted in Rwanda by Murray-Kolb et al. (2017) sought to evaluate the effectiveness of iron-biofortified beans compared to control beans in enhancing cognitive function among women of reproductive age (WRA) aged 18–27 with low iron levels. The findings indicated that iron status significantly influences cognitive abilities. Notably, the cognitive performance of young adult women—particularly their efficiency in searching for and swiftly retrieving information during memory tasks—showed improvement after consuming iron-biofortified beans for eighteen weeks. This increase in cognitive function enhances their chances of reaching their potential at work, in school, and at home.

Luna et al. assessed 125 iron-depleted female Rwandan university students to correlate whether physical performance can be enhanced by increased iron status via iron biofortification. According to the study, eating iron-biofortified beans significantly improves iron status, and these gains have an impact on work efficiency and physical performance during moderately intense exercise (Luna et al., 2020). This indicates that enhanced iron status can positively impact workplace productivity as well as economic output.

7. Economic and environmental impacts

7.1 The economic benefits for farmers and communities

The economic benefits of biofortification are multi-fold, with the quality and yield of the crops being the first benefit. Farmers growing such biofortified crops are likely to reap higher yields since these crops are genetically bred for tolerance of different abiotic and biotic stresses and are likely to be more resistant to failures of the crops. Studies suggest that staple foods that are biofortified are responsible for increased productivity of agriculture due to the increased nutritional quality and tolerance to the different conditions of the environment (Talsma et al., 2017; Rai et al., 2024). Iron-biofortified beans and zinc-biofortified corn are good examples of such staple foods that are not just of superior quality but are more marketable due to this quality improvement, eventually leading to increased revenues for farmers (Gautam et al., 2021; Glahn and Noh, 2021).

Moreover, the marketing of biofortified crops can increase consumer demand for nutrition-dense foods, allowing farmers to charge premium rates for products. With the marketing of biofortified foods as "healthier" products, it is feasible that the demand for such products will increase more substantially, especially for areas that are facing malnutrition. The increased demand can generate huge economic returns for farmers, leading to the development of a robust agriculture market that encourages farmers with small holdings to undertake investments in biofortification (Sheoran et al., 2022; Birol and Bouis, 2023). The use of biofortified crops emphasizes the ability of new farming practices to balance economic viability with nutritional improvement, providing concrete economic benefits for farmers (Talsma et al., 2017).

In addition, biofortification can stimulate local economies by generating new sources of employment within the farming industry. With the increasing demand for biofortified staples, follows the heightened need for workers within the spectrum of farming operations from cultivation to harvesting (Saikanth et al., 2023). The increase can stimulate economic activity within rural areas that are lacking with regard to occupational possibilities. Programs such as HarvestPlus have shown that the input of resources into biofortification can yield a more extensive economic impact that transcends just the farming output, leading to economic empowerment and resilience of the farming population (Van Der Straeten et al., 2020).

Beyond the direct benefits to farmers, biofortification can enhance community nutritional security at large. When farmers grow nutrient-rich crops, they contribute not only to their families' health but also to the nutritional standards of their local communities. This ripple effect can lead to reduced healthcare costs associated with micronutrient deficiency diseases—costs which often burden rural economies. Studies reveal that each invested dollar in biofortified crops yields substantial health benefits that can diminish healthcare expenditures in communities affected by malnutrition (Sheoran et al., 2022). Such economic shifts underscore biofortification as an investment in both health and economic welfare.

The enhancement of nutritional standards within populations due to the adaptation of biofortified crops can be a great contribution toward better performance at schools. Properly nourished individuals are well-equipped with the cognitive and physical skills necessary for children's learning ability. The well-fed workforce that results from such enhanced nutritional levels can then impact the local economies positively by forming a cycle of progress based on increased health, education, and productivity of the younger population. Thus, the interlinkages of agriculture, nutrition, and education are complexly integrated into pillars for successful community development (Saikanth et al., 2023).

Further support for biofortification comes from its sustainability as an agricultural practice. Unlike short-term interventions such as food aid, biofortification offers sustained benefits over time as nutrient-dense traits are inherited by subsequent generations of crops. This characteristic ensures that investments in biofortified crops continually yield returns for farmers (Rai et al., 2024; Sida-Arreola et al., 2015). However, it is important to note that the sustainability of these benefits may vary depending on the crop type and breeding system. In hybrid varieties, subsequent generations can experience a loss of hybrid vigor—resulting in reduced yield, nutrient density, or resilience—if seeds are saved and replanted rather than regenerated through controlled breeding (Bouis and Saltzman, 2017; Garg et al., 2018). By integrating biofortification into farming systems, farmers can ensure long-term food security, enhanced human capital through better health, and economic productivity—all critical elements for poverty alleviation in rural African communities.

Furthermore, one should not downplay the importance of involving local knowledge and practices in the growing of biofortified crops. The sharing of knowledge between farmers regarding the cultural adaptation and farming viability of different biofortified crops plays an important role in maximizing the effect of these innovations (Birol et al., 2015). With farmers engaging with the research and extension agencies to adapt the practices of biofortification within their specific environments, they build the ability that resonates within the surrounding populations. The people-centered approach helps build resilience, economic stability, and innovation (Kumar et al., 2023).

In the pursuit of eco-sustainability, conservation agriculture practices are complemented by the principles of biofortification, emphasizing the health of the soil and conservation of water resources. With the cultivation of more nutritious, stress-resilient crops promoted by it, it becomes feasible to encourage sustainable agriculture practices that conserve natural resources while addressing the critical need for increased nutrition for subsistence farming-dependent populations (Rehman et al., 2020; Sharma et al., 2020). With such an integrated approach, agriculture becomes economical while it addresses the most significant nutritional deficiencies. Biofortification efforts have revealed hundreds of success tales in Africa due to the introduction of diversified schemes that encourage farming households to adopt more robust crop varieties. Studies suggest that countries such as South Africa and Uganda are leading the revolution of the era of biofortification by displaying great success in the improvement of provitamin A-enhanced maize and beans with increased iron content (Siwela et al., 2020; Zuma et al., 2018). All these success tales are model cases for the adaptation of such technologies with the ability for economic impact in different areas of life.

7.2 Environmental sustainability of biofortified crops

Emphasising environmental sustainability, biofortified crops provide a dual benefit: enhancing nutritional content while promoting agricultural practices that conserve resources and foster biodiversity. Through genetic improvement methods, such as selective breeding, biofortified crops can be developed with traits that enhance resilience to climatic stresses—such as drought and pests—thereby reducing the need for synthetic inputs and irrigation (Bouis and Welch, 2010; Mandal et al., 2024). Research indicates that crops like biofortified peas and maize contribute to improved soil health by increasing the soil's organic matter and reducing dependence on synthetic fertilisers through traits such as nitrogen fixation (Garg et al., 2018; Kihara et al., 2020).

Moreover, when biofortified crops are integrated into existing agricultural systems, they can help enhance overall system productivity without exacerbating environmental degradation (Sheoran et al., 2022). Studies suggest that cultivating biofortified maize (which is enriched with provitamin A) not only addresses vitamin A deficiency but also offers competitive yields, thus promoting both food security and nutritional enhancement within sustainable frameworks (Birol and Bouis, 2023; Zuma et al., 2018). Furthermore, these crops can play a role in diversifying farmers' livelihoods, as they command higher market prices due to their enhanced nutritional profiles, providing economic incentives for sustainable practices (Konlan et al., 2022; Lividini and Fiedler, 2015).

7.3 Comparison with other nutritional interventions

In comparison to traditional nutritional interventions, such as supplementation and industrial fortification, biofortification provides a more sustainable, field-based solution that can be integrated into the diets of vulnerable populations without requiring ongoing external resources (Garg et al., 2018; Rai et al., 2024). While supplementation programs, including micronutrient pills, can be effective in the short term, they often lack sustainability and require constant infrastructure and funding (Bandumula et al., 2016). Similarly, food fortification practices face challenges such as affordability, accessibility, and acceptance within local communities, which can hinder widespread implementation (Mutuku et al., 2020; Chawafambira et al., 2021). However, fortification of salt with iodine remains the primary and most successful global strategy for preventing iodine deficiency disorders, endorsed by the World Health Organization and UNICEF as a safe, cost-effective, and scalable intervention. Supplementation also serves as an important complementary measure in specific populations, and in the case of iodine, these two approaches remain the cornerstone of prevention and control efforts (WHO, 2015).

Biofortification, on the other hand, ensures nutrients are embedded within staple foods, creating a self-sustaining cycle of nutrition enhancement directly at the agricultural source (Finkelstein et al., 2017; Saikanth et al., 2023). Researchers have shown that crops such as iron-biofortified beans are particularly favourable in regions where dietary diversity is low, as they can meaningfully contribute to improving iron status among at-risk populations (Kihara et al., 2020; Finkelstein et al., 2019). By allowing farmers to cultivate crops that naturally yield higher micronutrient concentrations through already existing farming practices without requiring significant new infrastructures, biofortification changes long-term agricultural practices in ways that traditional methods have yet to achieve (Sanahuja et al., 2013; Bajaj, 2023).

The economic implications of biofortification further reinforce its efficiency as an intervention. For instance, the costs associated with developing biofortified crop varieties are largely one-time investments compared to the continual costs linked to ongoing supplementation and fortification programs, which cannot often reach the most marginalised communities effectively (Mutuku et al., 2020; Mogendi et al., 2016). As countries like Zambia continue to explore agricultural strategies, the return on investment from biofortification appears to be significantly higher than the alternatives, leading to improved nutritional outcomes and thus economic benefits for communities reliant on these staple foods (Lividini and Fiedler, 2015; Bouis and Saltzman, 2017).

7.3.1 Interlinked approaches to nutrition security

The relationship between agricultural practices and nutritional health necessitates a multifaceted approach to addressing micronutrient deficiencies. Historical data indicates that regions adopting biofortification strategies in tandem with other nutritional interventions have experienced pronounced improvements in public health metrics (Reddy et al., 2024; Bajaj, 2023). For instance, integrating biofortified crops with educational campaigns on dietary diversity maximises their potential impact, leveraging the existing cultural practices surrounding staple foods (Jada and Berg, 2024; Birol and Bouis, 2023).

Innovations such as mobile apps and community participatory programs further reinforce these interventions, enhancing awareness and acceptance of biofortified foods (Bachewe et al., 2023; Bouis and Welch, 2010). The emergence of social norms around the benefits of biofortification within various communities can drive demand for these crops, transforming local economies and food systems (Baral and Birol, 2020; Jada and Berg, 2024). The introduction of biofortified grains, root vegetables, and legumes into diets enriches existing food matrices, ensuring that populations consume necessary nutrients efficiently (Saikanth et al., 2023; Garg et al., 2018).

As public health frameworks in African nations increasingly incorporate biofortification into their strategies, the potential for agricultural policies to align with health objectives becomes apparent (Zuma et al., 2018; Bouis et al., 2014). Evaluating the success of biofortification hinges not only on agricultural yield statistics but also on comprehensive assessments of nutritional indicators within populations over time (Sheoran et al., 2022; Finkelstein et al., 2019).

8. Challenges and barriers

In our world today, micronutrient deficiencies afflict over two billion individuals, or one in every three persons, globally (Weffort and Lamounier, 2024). These deficiencies occur when the intake and absorption of vitamins and minerals become very low to sustain healthy growth and development. Historically, in the past five decades, agricultural research in many developing countries has increased the production and availability of staple crops that are high in calories, whereas the production of micronutrient-rich non-staples, such as vegetables, pulses, and animal products, has not increased in the same proportion. As a result of this, non-staple food prices have been noted to increase at a steady and substantial rate, making it difficult for the poor to afford a balanced diet (Bouis et al., 2011a). It is on this note that the concept of a complete biofortification of food has become a growing interest. While the practice of biofortification is gaining scientific attention, the limitations, however, have also not been unnoticed.

8.1 Technical and logistical challenges

One of the problems with the technical side of biofortification in Africa is that new crops can't always be used because of differences in the soil and climate, even in close regions. This is the case of a biofortified crop finding it difficult to thrive in every region, probably due to drought, soil type, and other agroecological conditions. A typical example is the innovation around the orange-fleshed sweet potatoes (OFSP) both in Kenya and Uganda, an attempt towards resolution of vitamin A deficiency in the diet. The success of this research had its limitations in that due to factors like the variability of soil types and looming drought conditions in some other parts of these regions, the crop yield turned out to be very poor, thereby limiting its adoption by farmers (Low et al., 2017). Similarly, in Uganda, successful adoption of iron-biofortified beans was inhibited in this region because some of the varieties were faced with low disease resistance (Petry et al., 2020). It is also a fact that the art of biofortification thrives on conventional breeding or genetic modification to enhance nutrient content, so there is usually the challenge of compromise in terms of yield and drought tolerance, as in the case of Zambia and their zinc-biofortified maize varieties. Hence, while zinc-biofortified maize shows promise in improving dietary zinc intake, its adoption is hindered by the need for extensive field testing to ensure stability across different environments (Goredema-Matongera et al., 2021). Other instances are nutrient degradation at storage and processing, seen in Nigeria's provitamin A-fortified cassava (Ilona et al., 2017), as well as the Golden Rice variety in global Africa as opposed to Asia in terms of performance. Generally, these scenarios point to the absence of advanced biotechnology infrastructure in numerous African nations as a major restriction to the success of biofortification; the development and testing of genetically modified biofortified crops remains contingent (Bouis and Saltzman, 2017).

Logistics present another significant obstacle in terms of biofortification progress for Africa. The successful implementation of biofortification requires efficient seed distribution systems and widespread farmer adoption. In many African countries, fragmented agricultural supply chains and limited access to quality seeds hinder the dissemination of biofortified crops. For example, despite the success of provitamin A-biofortified maize in Zimbabwe, many small-holder farmers lack access to certified seeds due to high costs and inadequate distribution networks (Mulongo et al., 2021; Mudyazvivi et al., 2020). Additionally, farmers' unwillingness to adopt new crops, which is typically due to a lack of understanding, risk perception (Muvhuringi and Chigede, 2021), or a preference for formerly established kinds, undermines scaling attempts. Consumer acceptance is also important since it has been demonstrated in many situations to be vital to the success of biofortified crops. In some situations, biofortified variants may differ in colour, flavour, or texture from regular types, discouraging consumption. For example, orange-fleshed sweet potatoes in Ghana were initially met with opposition from consumers who were more used to white or yellow types. Similarly, iron-biofortified beans in Rwanda were met with mistrust due to their darker hue, which was seen as less appealing (Petry et al., 2020). Thus, these cases call for attention, addressing these cultural preferences through extensive awareness campaigns and community engagement, and no doubt are predicted to be resource-intensive.

8.1.1 Policy and regulatory barriers

Biofortification in Africa is fraught with limiting biotechnology policies. Although these regulations are being put in place for safety reasons, there can be obvious implications if not properly managed. A lot of biofortified crops, such as provitamin A-enriched maize and cassava, have been developed through genetic modification (GM) or conventional breeding. However, due to safety concerns and public resistance (Komen and Mignouna, 2022), GM crops are usually faced with stringent regulations in several African countries. In countries like Kenya, Zambia, and Nigeria, due to public health concern, there were imposed bans or severe restrictions on genetically modified organisms (GMOs), limiting the introduction of biofortified crops that rely on genetic engineering (Jenkins et al., 2021; Adenle et al., 2018). Delays in bringing novel crop varieties to market are also an issue, especially when those varieties are created through genetic engineering and need complex regulatory processes (Mmbando, 2023; Gbashi et al., 2021). Furthermore, trade barriers and strict import/export rules impede the cross-border movement of biofortified seeds and products. (Mmbando, 2023; Gbadegesin et al., 2022). Similarly, fortified crops may be less accessible to farmers and consumers due to variances in legislation and certification requirements across borders.

In other cases, a major hindrance is a lack of comprehensive policies, with engaging institutions serving as coordinating agencies. Many African countries lack adequate policies to fully promote biofortification. While several countries have included biofortification in their nutrition plans, there are frequently inadequate or absent institutions to carry out these policies across multiple sectors. This lack of coordination across agencies and sectors such as agriculture and health leads to fragmented efforts, slowing the implementation of biofortification programs (Talsma et al., 2017).

Regulatory hurdles, particularly cumbersome seed certification processes and the high costs of establishing intellectual property rights, make it difficult to scale up biofortified crops in Africa (Gouse et al. 2016). Although an important component for assuring the quality and availability of biofortified crops, many African countries face severe issues with their seed certification systems, which are frequently delayed, inefficient, and underfunded. For example, in Nigeria, the certification of provitamin A-biofortified cassava has been delayed due to bureaucratic inefficiencies and a lack of scientific expertise among regulatory organisations. According to Ilona et al. (2017), these delays have slowed the distribution of biofortified cassava varieties, restricting farmer uptake. Similarly, in Ethiopia, the certification process for iron-biofortified teff has proven difficult. The lack of clear criteria and regulated procedures for certifying biofortified seeds has caused uncertainty for seed producers, inhibiting private-sector investment. As a result, the availability of certified biofortified seeds remains limited, which slows their uptake and impact.

While biofortification has considerable promise for combating malnutrition in Africa, policy and regulatory constraints continue to impede its growth. These obstacles are exacerbated by restrictive biotechnology policies, lengthy licensing processes, inadequate institutional support, public perception issues, trade limitations, and many more.

8.2 Strategies to overcome these barriers

Agriculture is critical to Africa's survival and is regarded as its lifeline. It is the backbone of many economies, giving a living to a large portion of the population (Rafael, 2023). However, it has technical and logistical hurdles, as well as legislation and other types of regulation, that are sufficient to halt its progress. While the biofortification of crops remains a game-changing solution in the history of Africa's agriculture, efforts are being projected as strategies to ensure that these barriers are remediated.

Inconsistent trade restrictions and certification requirements for biofortified crops throughout Africa hinder seed availability and crop scaling potential. A typical example is the difficulty in adopting provitamin A-fortified maize in East Africa due to differing import regulations, as well as its eventual high cost to poor farmers. (Mudyazvivi et al., 2020). To address such situations, regional economic communities such as the East African Community (EAC) and the Southern African Development Community (SADC) should work to harmonise trade policies and seed certification standards, as this would aid farmers' accessibility and address cost implications when adopting fortified crops.

Similarly, certain biofortified crops go through a substandard certification process, which frequently causes delays during the distribution phase. In Nigeria, the certification process for provitamin A-fortified cassava was delayed due to a lack of process management expertise. It will be in the best interests of the government to build seed regulatory training and processes for workers, as well as to acquire digital tools that aid in the monitoring process (Ilona et al., 2017). Another strategy for removing these barriers is full government support through an enabling policy environment that includes the entire value chain in the development and deployment of multi-nutrient cultivars, as well as strengthening seed systems to ensure easy access to subsidised biofortified seeds and inputs. However, in many Sub-Saharan African countries, cultivars made using genetic engineering are outlawed (Lalani et al., 2019). Therefore, policymakers could assist alleviate hunger in high-risk societies by participating in awareness and educational campaigns that call for policy reform to permit genome-edited or transgenic multi-nutrient maize.

Another issue is farmers' resistance to using biofortified crops. Usually, the farmers' readiness to adopt biofortified crops is linked to longstanding beliefs as well as practices and patterns surrounding their vocation (Bouis and Saltzman, 2017). There is the notion that agriculture has traditional implications for farmers, and this is a major reason why they prefer their familiar varieties over the newly developed ones. Sometimes, this notion could also be woven around the perception that biofortified crops possess low-yield traits (BÄnziger and Long, 2000). The orange-fleshed sweet potatoes (OFSP) in Ghana present one such scenario; the hostility meted out at this development was that both farmers and consumers were unfamiliar with the crop's nutritional benefits (Low et al., 2017). To overcome this, governments and NGOs should invest in farmer training programs and public awareness campaigns. These initiatives can educate farmers and consumers about the benefits of biofortified crops, addressing cultural preferences and increasing adoption rates.

9. Future directions and policy recommendations

9.1 Innovations in biofortification research

Future developments in biofortified crop varieties have great potential to address the malnutrition issues facing Africa. More robust and nutrient-dense crop varieties have been developed as a result of recent improvements in breeding techniques that have revolutionised biofortification efforts. Breeders may find and choose traits linked to increased nutrient content, disease resistance, and yield by using methods like marker-assisted selection (MAS) and other genomic selection (Prasanna et al., 2020; Goredema-Matongera et al., 2021). Furthermore, Clustered Regularly Interspaced Short Palindromic

Repeats (CRISPR-Cas9) and other gene editing technologies provide accurate and effective ways to incorporate advantageous traits into crop genomes, accelerating the breeding process (Medina-Lozano and Díaz, 2022).

Using these modern breeding techniques, it is feasible to create biofortified crops with precise increases in critical nutrients such as vitamin A precursors, iron, zinc, and folate. These crops can then be used to help combat malnutrition in Africa and other parts of the world. Furthermore, generating crops with higher nutritional content, disease resistance, and yield potential would necessitate ongoing investment in breeding techniques like MAS and gene editing. To give comprehensive solutions to diverse micronutrient deficits, researchers are focusing on producing biofortified cultivars with a variety of nutritional characteristics (Siwela et al. 2020; Goredema-Matongera et al. 2021). Similarly, the incorporation of climate-smart agricultural techniques and the advancement of biofortified crops that are heat and drought-tolerant will strengthen food security amidst shifting climate patterns (Goredema-Matongera et al., 2021). Africa could benefit significantly from ongoing biofortification innovation and research in terms of increased nutrition, health outcomes, and agricultural sustainability.

There are clear examples of where these improvements have been exhibited in the development of new seed crops. One of the most effective efforts has been the production of OFSP cultivars, which are high in beta-carotene. These types have been widely adopted, with Kenya and Uganda leading the way (Low et al. 2017). Similarly, the MAS-based innovation technique led to the invention of iron-fortified beans, which are now widely produced and consumed in countries such as the Republic of Congo and Rwanda (Haas et al. 2016). According to Ilona et al. (2017), biofortified conventional breeding has also enabled the spread of provitamin A-fortified cassava in Nigeria and Ghana. Hence, collaboration with research institutions and organisations is critical to the success of the African biofortification program (Goredema-Matongera et al., 2021; Beattie and Sallu, 2021). Academic institutions, international organisations, and agricultural research centres must also collaborate to develop new biofortified crop types to further this advancement (Dowd-Uribe et al., 2024).

9.2 Policy frameworks to support biofortification

While the impact of micronutrient insufficiency varies by location (Stevens et al., 2022), crop biofortification using approaches such as conventional plant breeding, transgenic techniques, or agronomic practices is a food-based solution that may help ease this burden. However, biofortification initiatives often encompass a wide range of activities across crop-specific agricultural value chains; for example, the agricultural research for varietal development, varietal release, agricultural input supply, farming, aggregation, processing, and retailing/public distribution, is dependent on a strong policy framework that is specifically tailored to context (Bouis and Saltzman, 2017). This section seeks to highlight some of the frameworks that have been implemented in some African countries to drive biofortification into full accomplishment.

In terms of incorporating biofortification into national agricultural and nutrition policies, Nigeria has taken a bold step to include certain developed crops like provitamin A-maize and iron-fortified beans into the national food security policy and the agricultural promotion policy of the years 2016–2020. These policy ideas were to ensure that the production and the large intake of these newly developed crops by consumers were prioritised for nutrient deficiency remediation. Similarly, the seventh national development plan of 2017 into 2021, projected by the Zambian government, ensured a partnership with HarvestPlus to promote the upscaling of their provitamin A-fortified maize crops (Talsma et al., 2017).

There is also the African Union's Comprehensive African Agriculture Development Programme, which has seen the continental integration of biofortified crops in agricultural development plans. As such, countries were motivated to show performance for nutrition-centred agriculture. It was also in line with this that the Malabo Declaration of 2014 was established to aid the commitment of governments to ten per cent at least of their yearly budgets for agriculture and its advancements (African Union, 2014). According to Bouis et al., the monitoring and evaluation framework is also essential for tracking the progress of biofortification programs across Africa (Bouis et al., 2019). Countries such as Zambia and Nigeria utilise their national food and nutrition commissions and national biofortification programs, respectively, to assess the adoption of biofortified crops through indicator mechanisms for dietary and micronutrient intake.

While the African Seed Biotechnology Program (ASBP) seeks to unify the regulation and certification processes for biofortified seeds within the African Union (Mabaya et al., 2019), the African Seed Access Index (TASAI), on the other hand, as a framework, analyses the performance of seed sectors in Africa. It also seeks to ensure that there is availability of improved agricultural crop varieties, including biofortified ones, through the participation of governments in the form of investments in seed multiplication programs and partnerships with local seed companies for upscaling benefits. The sole aim of this is to ensure that new seeds are high in quality, affordable, and available to small farmholders (Mabaya et al., 2019).

These frameworks attempt to address agricultural research and development, certification and regulatory approval, farmer adoption, consumer acceptance, and market integration throughout Africa.

9.3 Policies to support adoption and food systems transformation

Policies play a crucial role in boosting the acceptance and long-term sustainability of biofortified foods (Bouis and Saltzman, 2017). A significant instance is in India, where legislation prohibits the release of new pearl millet varieties that do not meet the iron content of biofortified pearl millets, in order to maintain consistent nutrient levels post-breeding. (Birol et al., 2015). The integration of regulatory oversight with public health messaging is essential to foster trust and interest in biofortified foods among farmers and consumers.

From an agricultural standpoint, it is crucial to disseminate agronomic messages via agricultural promotion that emphasizes yield advantages; all biofortified varieties must achieve at least equivalent yields (Garg et al., 2018). The rapid adoption of zinc wheat in HarvestPlus Pakistan's work demonstrates that farmers have embraced this new technology due to its tolerance and superior performance (Lowe et al., 2022).

In the commercial value chain, procurers and food processors must ensure that acquired grain meets specified nutrient standards. Hence the reason why the British Standards Institute (BSI) has established international standards that allow buyers to integrate nutrient grades and specifications into their procurement documentation (Bouis et al., 2024).

Furthermore, customer labeling requirements are crucial at the household level, enabling families to verify nutrient claims. India and Nigeria have established labeling systems to enhance transparency and customer options (United States Department of Agriculture Foreign Agricultural Service, 2026; Iheme et al., 2025).

The integration of several policy instruments, including regulatory thresholds and labeling, can facilitate adoption and transformation across the food system.

9.4 Recommendations for stakeholders

In Africa, the biofortification of staple foods dates back several decades. Concerning the progress accomplished thus far, it is clear that more cautious attention will be required. It is also important to mention that the duty to drive the entire advancement of biofortification in Africa necessitates as many efforts as the list of actions. Every participant in the operational chain, including government and non-government agencies, corporate partners, academic institutions and research centres, farmers, and eventually consumers, should promote full engagement.

The government stands at the forefront of all decision-making processes. In the context of biofortification, national governments serve as the policy-making and enforcement agents. It is proposed that national governments prioritise biofortification practices by financing research institutions and providing infrastructure for the storage and preservation of biofortified commodities. This effort can also be demonstrated in the subsidy or free allocation of seed varieties to farmers at the first usage or attempt, or the recompense of any loss whatsoever at harvest. Ultimately, biofortification should be included in agricultural plans and nutritional recommendations, with regular monitoring.

In terms of academic and research institutions, focusing on the creation of staple crop varieties that farmers and consumers are familiar with will produce better results. It is also recommended that research institutes collaborate with existing partners—such as both government and non-government agencies—to strengthen cooperation and overcome the technical expertise gap that has been identified in some circumstances. It is also critical that research institutes provide training and scientific seminars for farmers and extension workers on seed planting and the need to adopt new seed varieties.

The role of non-government organisations cannot be overstated; these organisations should ensure the distribution of seeds and monitor the levels of adoption of these distributed seeds by location. It is also crucial that the NGOs at intervals embark on awareness campaigns, surveys, and data collection on the perception of farmers and consumers of newly developed seed varieties, as this will inform better approaches and also foster cooperation among stakeholders.

Lastly, the farmers and the consumers determine the success of biofortified varieties. Farmers should be encouraged to attend seminars and workshops that educate on best agronomic and post-harvest practices, learn about the adoption of new varieties, and understand their benefits in terms of yield and nutritional content. Ultimately, they should be very ready to provide feedback to researchers and NGOs. For their part, consumers should show interest in the consumption of these varieties and involvement in awareness and educational campaigns through feedback from personal experience.

10. Conclusion

Biofortification plays a crucial role in addressing micronutrient deficiencies across Africa by enhancing the nutritional value of staple foods through conventional breeding, agronomic techniques, and modern biotechnology. The development of crops such as Provitamin A-enriched sweet potatoes, iron-fortified beans, and zinc-fortified rice demonstrates its potential to improve public health and reduce the burden of hidden hunger. However, the widespread adoption of these crops remains constrained by factors such as low awareness, limited access to biofortified seeds, high costs, and environmental challenges.

More awareness should be created for biofortified seeds as they are cost-effective in prices as regards non-biofortified seeds. They can be saved and replanted, thereby having to save time. The goal is to produce nutrient-rich crops without added expense, benefiting farmers and consumers. Biofortified varieties match or exceed existing crop yields and performance, ensuring seamless adoption without requiring changes to traditional farming practices.

To maximize its impact, governments, research institutions, and private stakeholders must collaborate to promote and scale up biofortification initiatives. This requires sustained investment in research, awareness campaigns, farmer training, and policy support to enhance access, affordability, and consumer acceptance. With the right strategies and partnerships, biofortification can become a cornerstone of sustainable nutrition in Africa helping to combat micronutrient deficiencies, improve health outcomes, and build a more resilient and equitable food system for future generations, thereby improving public health and food security through targeted interventions and combined efforts that will be needed to address unique challenges, separate from general agricultural challenges which the help to maximize nutritional impact.

References

Adekambi, S.A., Okello, J.J., Abidin, P.E., Carey, E., 2020. Effect of exposure to biofortified crops on smallholder farm household adoption decisions: the case of orange-fleshed sweet potato in Ghana and Nigeria. Sci. Afr. 8, e00362.

Adenle, A.A., Aworh, O.C., Akromah, R., Parayil, G., 2018. Developing GM super cassava for improved health and food security: future challenges in Africa. Agric. Food Secur. 7 (1), 1–15. https://doi.org/10.1186/s40066-018-0165-5.

Adeyeye, S.A.O., Idowu-Adebayo, F., 2019. Genetically modified and biofortified crops and food security in developing countries: a review. Nutr. Food Sci. 49 (5), 978–986.

Afolami, I., Mwangi, M.N., Samuel, F., Boy, E., Ilona, P., Talsma, E.F., et al., 2020. Daily consumption of pro-vitamin A biofortified (yellow) cassava improves serum retinol concentrations in preschool children in Nigeria: a randomised controlled trial. Am. J. Clin. Nutr. 113 (1), 221–231. https://doi.org/10.1093/ajcn/nqaa290.

African Union, 2014. Malabo declaration on accelerated agricultural growth and transformation for shared prosperity and improved livelihoods. Retrieved from https://au.int.

African Union. (2020). Upscaling biofortification in Africa: A roadmap. https://au.int/sites/default/files/documents/41149-doc-Roadmap_-Upscaling_Biofortification_in_Africa-Final-_Eng.pdf.

Alfthan, G., Eurola, M., Ekholm, P., Venäläinen, E., Root, T., Korkalainen, K., ... Aro, A., 2014. Effects of nationwide addition of selenium to fertilizers on foods, and animal and human health in Finland: from deficiency to optimal selenium status of the population. J. Trace Elem. Med. Biol. 31, 142–147. https://doi.org/10.1016/j.jtemb.2014.04.009.

Andersson, M., 2017. Progress update: crop development of biofortified staple food crops under HarvestPlus. Afr. J. Food Agric. Nutr. Dev. 17 (02), 11905–11935. https://doi.org/10.18697/ajfand.78.harvestplus05.

Ashokkumar, K., Govindaraj, M., Karthikeyan, A., Shobhana, V., Warkentin, T., 2020. Genomics-integrated breeding for carotenoids and folates in staple cereal grains to reduce malnutrition. Front. Genet. 11. https://doi.org/10.3389/fgene.2020.00414.

Asif, M., Asad, M.S., Safdar, M.E., Khan, I., Akhtar, N., Hassan, M.A., et al., 2024. Agronomic bio-fortification of zinc improves the yield and quality of fodder oat. J. Ecol. Eng. 25 (6).

Bachewe, F., Genye, T., Girma, M., Samuel, A., Warner, J., Zyl, C., 2023. Biofortification in Ethiopia: opportunities and challenges. Food Nutr. Bull. 44 (3), 151–161. https://doi.org/10.1177/03795721231188913.

Bajaj, R., 2023. Food systems that prioritise nutrition and crops enhanced with biofortification. tjjpt 44 (5), 785–792. https://doi.org/10.52783/tjjpt.v44.i5.2671.

Bandumula, N., Babu, V., Neeraja, C., Waris, A., Pandurangan, M., Rao, D., 2016. Linking agriculture and nutrition: an ex-ante analysis of zinc biofortification of rice in India. Agric. Econ. Res. Rev. 29 (conf), 171. https://doi.org/10.5958/0974-0279.2016.00044.6.

BÄnziger, M., Long, J., 2000. The potential for increasing the iron and zinc density of maize through plant breeding. Food Nutr. Bull 21, 397–400 ([CrossRef]).

Baral, A., Birol, E., 2020. Catalysing the Scale-Up of Crop Biofortification. International Food Policy Research Institute, Washington, DC. https://doi.org/10.2499/p15738coll2.133959.

Baranski, M., 2013. Golden Rice. Embryo Project EncyclopediaArizona State University. https://hdl.handle.net/10776/6275.

Beattie, S., Sallu, S.M., 2021. How does nutrition feature in climate-smart agricultural policy in southern Africa? A systematic policy review. Sustainability 13, 2785.

Beebe, S., 2020. Biofortification of common bean for higher iron concentration. Front. Sustain. Food Syst. 4. https://doi.org/10.3389/fsufs.2020.573449.

Bell, V., Rodrigues, A.R., Ferrão, J., Varzakas, T., Fernandes, T.H., 2024. The policy of compulsory large-scale food fortification in Sub-Saharan Africa. Foods 13 (15), 2438.

Bhattacharya, S., Sathiyabalan, A., Amir, M., 2024. Advancing nutritional quality through genomic approaches for biofortification in cereal crops: a review. Plant Cell Biotechnol. Mol. Biol. 25 (5-6), 110–123. https://doi.org/10.56557/pcbmb/2024/v25i5-68713.

Birol, E., Asare-Marfo, D., Karandikar, B., Roy, D., Diressie, M.T., 2015. Investigating demand for biofortified seeds in developing countries: high-iron pearl millet in India. J. Agribusiness Dev. Emerg. Econ. 5 (1), 24–43.

Birol, E., Bouis, H., 2023. Role of socio-economic research in developing, delivering, and scaling new crop varieties: the case of staple crop biofortification. Front. Plant Sci. 14. https://doi.org/10.3389/fpls.2023.1099496.

Birol, E., Meenakshi, J., Oparinde, A., Pérez, S., Tomlins, K., 2015. Developing country consumers' acceptance of biofortified foods: a synthesis. Food Secur. 7 (3), 555–568. https://doi.org/10.1007/s12571-015-0464-7.

Black, R.E., Victora, C.G., Walker, S.P., Bhutta, Z.A., Christian, P., de Onis, M., et al., 2013. Maternal and child undernutrition and overweight in low-income and middle-income countries. Lancet (London, England) 382 (9890), 427–451. https://doi.org/10.1016/S0140-6736(13)60937-X.

Boonyaves, K., Wu, T., Gruissem, W., Bhullar, N., 2017. Enhanced grain iron levels in rice express an iron-regulated metal transporter, nicotianamine synthase, and ferritin gene cassette. Front. Plant Sci. 8. https://doi.org/10.3389/fpls.2017.00130.

Bouis, H., 2018. Reducing mineral and vitamin deficiencies through biofortification: progress under harvestplus. World Rev Nutr Diet 112–122. https://doi.org/10.1159/000484342.

Bouis, H.E., Eozenou, P., Rahman, A., 2011a. Food prices, household income, and resource allocation: socioeconomic perspectives on their effects on dietary quality and nutritional status. Food Nutr. Bull 32 (1), S14–S23.

Bouis, H., Foley, J., Lividini, K., Jumrani, J., Reinke, R., Van Der Straeten, D., et al., 2024. Biofortification: future challenges for a newly emerging technology to improve nutrition security sustainably. Curr. Dev. Nutr. 8 (12), 104478. https://doi.org/10.1016/j.cdnut.2024.104478.

Bouis, H., Hotz, C., McClafferty, B., Meenakshi, J., Pfeiffer, W., 2014. Biofortification: a new tool to reduce micronutrient malnutrition. Food Nutr Bull 202–215. https://doi.org/10.1079/9781780642994.0202.

Bouis, H.E., Saltzman, A., Birol, E., 2019. Biofortification: A global strategy to reduce micronutrient malnutrition. European Journal of Clinical Nutrition 73 (2), 209–211. https://doi.org/10.1038/s41430-018-0371-3.

Bouis, H., Saltzman, A., 2017. Improving nutrition through biofortification: a review of evidence from HarvestPlus, 2003 through 2016. Global Food Security 12, 49–58. https://doi.org/10.1016/j.gfs.2017.01.009.

Bouis, H., Welch, R., 2010. Biofortification—a sustainable agricultural strategy for reducing micronutrient malnutrition in the global south. Crop Science 50 (S1). https://doi.org/10.2135/cropsci2009.09.0531.

Boy, E., 2017. Efficacy of iron-biofortified crops. Afr. J. Food Agric. Nutr. Dev. 17 (02), 11879–11892. https://doi.org/10.18697/ajfand.78.harvestplus03.

Buturi, C., Mauro, R., Fogliano, V., Leonardi, C., Giuffrida, F., 2021. Mineral biofortification of vegetables as a tool to improve the human diet. Foods 10 (2), 223. https://doi.org/10.3390/foods10020223.

Carvalho, S., Vasconcelos, M., 2013. Producing more with less: strategies and novel technologies for plant-based food biofortification. Food Res. Int. 54 (1), 961–971. https://doi.org/10.1016/j.foodres.2012.12.021.

Chaudhary, M., Katel, S., Parajuli, M., Pandey, S., Mandal, H., 2022. Biofortification of major cereal crops: a review. Plant Physiol. Soil Chem. 2 (1), 01–07. https://doi.org/10.26480/ppsc.01.2022.01.07.

Chawafambira, A., Nyoni, Q., Mkungunugwa, T., 2021. The potential of utilising provitamin-a-biofortified maize in producing mutwiwa, a Zimbabwean traditional fermented food. Food Sci. Nutr. 9 (3), 1521–1529. https://doi.org/10.1002/fsn3.2125.

Cheung, Y., Zheng, B., Rehman, Y., Zheng, Z., Rangan, A., 2024. Iron content of wheat and rice in Australia: a scoping review. Foods 13 (4), 547. https://doi.org/10.3390/foods13040547.

Chidiebere-Mark, N.M., GodsPower, C.A., 2020. Farmers' preferences and perceptions influencing decision to grow bio-fortified provitamin a cassava. J. Agric. Food Sci. 18 (1), 130–142.

De Groote, H., Gunaratna, N.S., BiofortifiedKebebe, E.G., Mmbando, F., Friesen, D., 2016. The effectiveness of extension strategies for increasing the adoption of biofortified crops: the case of quality protein maize in East Africa. Food Secur. 8 (6), 1101–1121. https://doi.org/10.1007/s12571-016-0621-7.

De Steur, H., Gellynck, X., Storozhenko, S., Liqun, G., Lambert, W., Van Der Straeten, D., et al., 2010. Health impact in China of folate-biofortified rice. Nature Biotechnol. 28 (6), 554–556. https://doi.org/10.1038/nbt0610-554.

De ValenÇa, A.W., Bake, A., Brouwer, I.D., Giller, K.E., 2017. Agronomic biofortification of crops to fight hidden hunger in sub-Saharan Africa. Global Food Secur. 12, 8–14. https://doi.org/10.1016/j.gfs.2016.12.001.

Dhaliwal, S.S., Sharma, V., Shukla, A.K., Verma, V., Kaur, M., Shivay, Y.S., et al., 2022. Biofortification—A frontier novel approach to enrich micronutrients in field crops to encounter the nutritional security. Molecules 27 (4), 1340. https://doi.org/10.3390/molecules27041340.

Dowd-Uribe, B., Rock, J.S., Spreadbury, T., Chiril, P., Uminsky, D., 2024. Bridging the gap? Public–private partnerships and genetically modified crop development for smallholder farmers in Africa. Plants People Planet 6, 437–451.

Ejigu, G., Glahn, R., Mukankusi, C., Fenta, B., Wiesinger, J., 2024. Genetic and the environmental influences on the concentrations iron and zinc in small seeded common bean (Phaseolus vulgaris l.) varieties and advanced lines from ethiopia. Legume Sci. 6 (2). https://doi.org/10.1002/leg3.221.

Eozenou, P., Gilligan, D.O., Hotz, C., Kumar, N., Meenakshi, J.V., 2018. Biofortification, crop adoption and health information: impact pathways in mozambique and Uganda. Am. J. Agric. Econ. 100 (3), 906–930. https://doi.org/10.1093/ajae/aay005.

FAO, IFAD, UNICEF, WFP and WHO, 2025. The State of Food Security and Nutrition in the World 2025 – Addressing High Food Price Inflation for Food Security and Nutrition. FAO, Rome. https://doi.org/10.4060/cd6008en.

FAO, IFAD, WFP, 2015. The State of Food Insecurity in the World 2015. FAO, Rome.

Finkelstein, J., Fothergill, A., Hackl, L., Haas, J., Mehta, S., 2019. Iron biofortification interventions to improve iron status and functional outcomes. Proc. Nutr. Soc. 78 (02), 197–207. https://doi.org/10.1017/s0029665118002847.

Finkelstein, J., Haas, J., Mehta, S., 2017. Iron-biofortified staple food crops for improving iron status: a review of the current evidence. Curr. Opin. Biotechnol. 44, 138–145. https://doi.org/10.1016/j.copbio.2017.01.003.

Foley, J.K., Michaux, K.D., Mudyahoto, B., Kyazike, L., Cherian, B., Kalejaiye, O., et al., 2021. Scaling up delivery of biofortified staple food crops globally: paths to nourishing millions. Food Nut Bull 42 (1), 116–132. https://doi.org/10.1177/0379572120982501.

Gannon, B., Kaliwile, C., Arscott, S.A., Schmaelzle, S., Chileshe, J., Kalungwana, N., et al., 2014. Biofortified orange maize is as efficacious as a vitamin a supplement in zambian children even in the presence of high liver reserves of vitamin A: a community-based, randomized placebo-controlled trial. Am. J. Clin. Nutr. 100 (6), 1541–1550. https://doi.org/10.3945/ajcn.114.087379.

Garg, M., Sharma, N., Sharma, S., Kapoor, P., Kumar, A., Chunduri, V., et al., 2018. Biofortified crops generated by breeding, agronomy, and transgenic approaches are improving lives of millions of people around the world. Front. Nutr. 5, 12. https://doi.org/10.3389/fnut.2018.00012.

Gautam, D., Shrestha, B., Subedi, B., 2021. Bio-fortified maize: cornerstone in plant breeding to combat hidden hunger in developing countries. Arch. Agric. Environ. Sci. 6 (1), 100–107. https://doi.org/10.26832/24566632.2021.0601013.

Gbadegesin, L.A., Ayeni, E.A., Tettey, C.K., Uyanga, V.A., Aluko, O.O., Ahiakpa, J.K., et al., 2022. GMOs in Africa: status, adoption and public acceptance. Food Control 141, 109193.

Gbashi, S., Adebo, O., Adebiyi, J.A., Targuma, S., Tebele, S., Areo, O.M., et al., 2021. Food safety, food security and genetically modified organisms in Africa: a current perspective. Biotechnol. Genet. Eng. Rev. 37 (1), 30–63.

Germ, M., Stibilj, V., Šircelj, H., Jerše, A., Kroflič, A., Golob, A., et al., 2019. Biofortification of common buckwheat microgreens and seeds with different forms of selenium and iodine. J. Sci. Food Agric. 99 (9), 4353–4362.

Glahn, R., Noh, H., 2021. Redefining bean iron biofortification: a review of the evidence for moving to a high iron bioavailability approach. Front. Sustain. Food Syst. 5. https://doi.org/10.3389/fsufs.2021.682130.

Global Nutrition Report, 2022. Stronger Commitments for Greater ActionDevelopment Initiatives, Bristol, UK. https://globalnutritionreport.org/reports/2022-global-nutrition-report/.

Goredema-Matongera, N., Ndhlela, T., Magorokosho, C., Kamutando, C.N., van Biljon, A., Labuschagne, M., 2021. Multinutrient biofortification of maize (Zea mays L.) in Africa: current status, opportunities and limitations. Nutrients 13, 1039.

Gouse, M., Pray, C., Kirsten, J., Schimmelpfennig, D., 2016. Three seasons of subsistence insect-resistant maize in South Africa: Have smallholders benefited? AgBioForum 19 (2), 113–122. https://agbioforum.org.

Haas, J.D., Luna, S.V., Lung'aho, M.G., Wenger, M.J., Murray-Kolb, L.E., Beebe, S., et al., 2016. Consuming iron biofortified beans increases iron status in Rwandan Women after 128 days in a randomized controlled feeding trial. J. Nutr. 146 (8), 1586–1592. https://doi.org/10.3945/jn.115.224741.

HarvestPlus, 2025. HarvestPlus 2024 annual report: Scaling impact, enriching lives. International Food Policy Research Institute. https://www.harvestplus.org/wp-content/uploads/2025/07/2024-HarvestPlus-Annual-Report.pdf.

Hassan, M., Khan, İ., Chattha, M., Mahmood, A., Chattha, M., Nawaz, M., et al., 2017. Biofortification of wheat cultivars to combat zinc deficiency. Front. Plant Sci. 8. https://doi.org/10.3389/fpls.2017.00281.

Hefferon, K., 2019. Biotechnological approaches for generating zinc-enriched crops to combat malnutrition. Nutrients 11 (2), 253. https://doi.org/10.3390/nu11020253.

Hotz, C., Loechl, C., Lubowa, A., Tumwine, J.K., Ndeezi, G., Masawi, A.N., et al., 2012. Introduction of β-Carotene–rich orange sweet potato in rural Uganda resulted in increased vitamin A intakes among children and women and improved vitamin a status among children3. J. Nutr. 142 (10), 1871–1880. https://doi.org/10.3945/jn.111.151829.

Hu, J., Tao, R., Cao, C., Xie, J., Gao, Y., Hu, H., et al., 2023. Effect of leaf surface regulation of zinc fertilizer on absorption of cadmium, plumbum and zinc in rice (Oryza sativa L.). Sustainability 15 (3), 1877.

Huang, R., Bañuelos, G., Zhao, J., Wang, Z., Farooq, M., Yuling, Y., et al., 2024. Comprehensive evaluation of factors influencing selenium fertilization biofortification. J. Sci. Food Agric. 104 (10), 6100–6107. https://doi.org/10.1002/jsfa.13442.

Iheme, G.O., Egechizuorom, I.M., Edafioghor, L.O., Okonkwo, E.M., Olah, L.E., Onyeonu, B.A., et al., 2025. Compliance of pre-packaged food products with Nigerian food labelling guidelines: the NigeFE study. Front. Nutr. 12, 1644344. https://doi.org/10.3389/fnut.2025.1644344.

Ilona, P., Bouis, H.E., Palenberg, M., Moursi, M., Oparinde, A., 2017. Vitamin A cassava in Nigeria: crop development and delivery. Afr. J. Food, Agric., Nutr. Dev. 17 (2), 12000–12025. https://doi.org/10.18697/ajfand.78.HarvestPlus05.

Islam, M., Hasan, M., Akter, T., Islam, S., 2024. Development of iron and zinc biofortified rice variety for nutrition improvement in bangladesh. Food Sci. Eng. 322–330. https://doi.org/10.37256/fse.5220245087.

Jada, K., Berg, M., 2024. Social norms, nutrition messaging, and demand for biofortified staple crops: evidence from a discrete choice experiment in ethiopia. Agric. Econ. 55 (5), 695–713. https://doi.org/10.1111/agec.12845.

Jalal, A., Oliveira, C., Bastos, A., Fernandes, G., Lima, B., Furlani, E., et al., 2023. Nanozinc and plant growth-promoting bacteria improve biochemical and metabolic attributes of maize in tropical cerrado. Front. Plant Sci. 13. https://doi.org/10.3389/fpls.2022.1046642.

Jenkins, M., Byker Shanks, C., Houghtaling, B., 2015. Orange-fleshed sweet potato: successes and remaining challenges of the introduction of a nutritionally superior staple crop in Mozambique. Food Nutr. Bull. 36 (3), 327–353.

Jenkins, M., Shapira, T., Adekola, K.A., 2021. Barriers to the adoption of biofortified crops in sub-Saharan Africa: policy, regulatory, and socio-economic perspectives. Agric. Food Secur. 10 (1), 23–45.

Jha, A., Warkentin, T., 2020. Biofortification of pulse crops: status and future perspectives. Plants 9 (1), 73. https://doi.org/10.3390/plants9010073.

Jones, K.M., de Brauw, A., 2015. Using agriculture to improve child health: promoting orange sweet potatoes reduces diarrhea. World Dev. 74, 15–24. https://doi.org/10.1016/j.worlddev.2015.04.007.

Joy, E., Stein, A., Young, S., Ander, E., Watts, M., Broadley, M., 2015. Zinc-enriched fertilisers as a potential public health intervention in africa. Plant Soil 389 (1-2), 1–24. https://doi.org/10.1007/s11104-015-2430-8.

Just, D.R., Okello, J.J., Gabrielyan, G., Adekambi, S., Kwikiriza, N., Abidin, P.E., et al., 2021. A behavioral intervention increases consumption of a new biofortified food by school children: evidence from a field experiment in Nigeria. Eur. J. Dev. Res. 34 (1), 124–146. https://doi.org/10.1057/s41287-021-00363-7.

Kankwatsa, P., Muzira, R., Mutenyo, H., Lamo, J., 2019. Improved upland rice: adaptability, agronomic and farmer acceptability assessment under semi-arid conditions of south western uganda. Oalib 06 (12), 1–5. https://doi.org/10.4236/oalib.1105660.

Kashinath, J., Teja, C., Bhor, T., 2024. Genetic variability, heritability, and correlation analysis of rice (Oryza sativa) for iron and zinc content: implications for biofortification and yield improvement. Int. J. Adv. Biochem. Res. 8 (5), 39–44. https://doi.org/10.33545/26174693.2024.v8.i5a.1054.

Kawakami, Y., Bhullar, N., 2018. Molecular processes in iron and zinc homeostasis and their modulation for biofortification in rice. J. Integr. Plant Biol. 60 (12), 1181–1198. https://doi.org/10.1111/jipb.12751.

Kihara, J., Bolo, P., Kinyua, M., Rurinda, J., Piikki, K., 2020. Micronutrient deficiencies in african soils and the human nutritional nexus: opportunities with staple crops. Environ. Geochem. Health 42 (9), 3015–3033. https://doi.org/10.1007/s10653-019-00499-w.

Koç, E., Karayiğit, B., 2021. Assessment of biofortification approaches used to improve micronutrient-dense plants that are a sustainable solution to combat hidden hunger. J. Soil Sci. Plant Nutr. 22 (1), 475–500. https://doi.org/10.1007/s42729-021-00663-1.

Komen, J., Mignouna, D., 2022. Regulatory challenges of biofortified crops in Africa: a policy perspective. Front. Nutr. 9, 112–130.

Konlan, A., Assumang, I., Abe-Inge, V., 2022. Perspective chapter: crop biofortification - a key determinant towards fighting micronutrient malnutrition in northern ghana 2022 https://doi.org/10.5772/intechopen.104460.

Kuehne, G., Llewellyn, R., Pannell, D.J., Wilkinson, R., Dolling, P., Ouzman, J., et al., 2017. Predicting farmer uptake of new agricultural practices: a tool for research, extension and policy. Agric. Syst. 156, 115–125.

Kumar, S., Palve, A., Joshi, C., Srivastava, R., 2019. Crop biofortification for iron (fe), zinc (zn) and vitamin a with transgenic approaches. Heliyon 5 (6), e01914. https://doi.org/10.1016/j.heliyon.2019.e01914.

Kumar, M.V., Ramya, V., Maheshwaramma, S., Ganapathy, K.N., Govindaraj, M., Kavitha, K., et al., 2023. Exploiting Indian landraces to develop biofortified grain sorghum with high protein and minerals. Front. Nutr. 10, 1228422.

Laîné, P., Haddad, C., Arkoun, M., Yvin, J., Étienne, P., 2019. Silicon promotes agronomic performance in Brassica napus cultivated under field conditions with two nitrogen fertilizer inputs. Plants 8 (5), 137. https://doi.org/10.3390/plants8050137.

Lalani, B., Bechoff, A., Bennett, B., 2019. Which choice of delivery model(s) works best to deliver fortified foods? Nutrients 11 (7), 1594. https://doi.org/10.3390/nu11071594.
Liberal, Â., Pinela, J., Quintana, A., Ferreira, I., Barros, L., 2020. Fighting iron-deficiency anemia: innovations in food fortificants and biofortification strategies. Foods 9 (12), 1871. https://doi.org/10.3390/foods9121871.
Lima, R.M., Santos, C.S., Vasconcelos, M.W., 2017. The use of genetic engineering to improve the nutritional profile of traditional plant foods. 233–258. https://doi.org/10.1002/9781119079972.ch6.
Liu, D., Liu, Y., Zhang, W., Chen, X., Zou, C., 2017. Agronomic approach of zinc biofortification can increase zinc bioavailability in wheat flour and thereby reduce zinc deficiency in humans. Nutrients 9 (5), 465. https://doi.org/10.3390/nu9050465.
Lividini, K., Fiedler, J., 2015. Assessing the promise of biofortification: a case study of high provitamin a maize in zambia. Food Policy 54, 65–77. https://doi.org/10.1016/j.foodpol.2015.04.007.
Lockyer, S., White, A., Buttriss, J., 2018. Biofortified crops for tackling micronutrient deficiencies – what impact are these having in developing countries and could they be of relevance within europe? Nutr. Bull. 43 (4), 319–357. https://doi.org/10.1111/nbu.12347.
Low, J., Ball, A., Van Jaarsveld, P.J., Namutebi, A., Faber, Grant, F.K., 2015. Assessing nutritional value and changing behaviours regarding orange-fleshed sweetpotato use in sub-Saharan Africa. Potato and Sweetpotato in Africa: Transforming the Value Chains for Food and Nutrition Security. CABI International and the International Potato Center, pp. 551–579.
Low, J.W., Mwanga, R.O., Andrade, M., Carey, E., Ball, A.M., 2017. Tackling vitamin, A deficiency with biofortified sweet potato in sub-Saharan Africa. Glob. Food Secur. 14, 23–30. https://doi.org/10.1016/j.gfs.2017.01.004.
Low, J.W., Ortiz, R., Vandamme, E., Andrade, M., Biazin, B., Grüneberg, W.J., 2020. Nutrient-dense orange-fleshed sweetpotato: advances in drought-tolerance breeding and understanding of management practices for sustainable next-generation cropping systems in sub-Saharan Africa. Front. Sustain. Food Syst. 4, 50.
Lowe, N.M., Zaman, M., Khan, M.J., Brazier, A.K.M., Shahzad, B., Ullah, U., et al., 2022. Biofortified wheat increases dietary zinc intake: a randomised controlled efficacy study of zincol-2016 in rural Pakistan. Front. Nutr. 8, 809783. https://doi.org/10.3389/fnut.2021.809783.
Ludwig, Y., Slamet-Loedin, I.H., 2019. Genetic biofortification to enrich rice and wheat grain iron: From genes to product. Frontiers in Plant Science 10 (833). https://doi.org/10.3389/fpls.2019.00833.
Luna, S.V., Pompano, L.M., Lung'aho, M., Gahutu, J.B., Haas, J.D., 2020. Increased iron status during a feeding trial of iron-biofortified beans increases physical work efficiency in rwandan women. J. Nutr. 150 (5), 1093–1099. https://doi.org/10.1093/jn/nxaa016.
Mabaya, E., Wainaina, P., Waithaka, M., 2019. Harmonizing seed regulations in Africa: a comparative analysis of regional seed policies. Food Secur. 11 (6), 1297–1310. https://doi.org/10.1007/s12571-019-00964-5.
Majumder, S., Datta, K., Datta, S., 2019. Rice biofortification: high iron, zinc, and vitamin-a to fight against "hidden hunger". Agronomy 9 (12), 803. https://doi.org/10.3390/agronomy9120803.
Malik, K.A., Maqbool, A., 2020. Transgenic crops for biofortification. Front. Sustain. Food Syst. 4. https://doi.org/10.3389/fsufs.2020.57140.
Mandal, O., Krishnamoorthi, A., Singh, A., Kaur, R., Kalaiselvi, P., Tiwari, U., et al., 2024. Biofortification: enhancing nutritional content in crops through biotechnology and fighting climate change. J. Adv. Biol. Biotechnol. 27 (2), 186–210.
Marques, E., Darby, H.M., Kraft, J., 2021. Benefits and limitations of non-transgenic micronutrient biofortification approaches. Agronomy 11 (3), 464. https://doi.org/10.3390/agronomy11030464.

Massawe, F., Mayes, S., Cheng, A., 2016. Crop diversity: An unexploited treasure trove for food security. Trends in Plant Science 21 (5), 365–368. https://doi.org/10.1016/j.tplants.2016.02.006.

Medina-Lozano, I., Díaz, A., 2022. Applications of genomic tools in plant breeding: crop biofortification. Int J Mol Sci 23, 3086.

Mitra-Ganguli, T., Pfeiffer, W.H., Walton, J., 2022. The global regulatory framework for the commercialization of nutrient enriched biofortified foods. Ann. NY Acad Sci. 1517 (1), 154–166.

Mmbando, G.S., 2023. The legal aspect of the current use of genetically modified organisms in Kenya, Tanzania, and Uganda. GM Crops Food 14, 1–12.

Mmbando, G.S., Missanga, J., 2024. The current status of genetic biofortification in alleviating malnutrition in Africa. J. Genet. Eng. Biotechnol. 22 (4), 100445. https://doi.org/10.1016/j.jgeb.2024.100445.

Mogendi, J., Steur, H., Gellynck, X., Makokha, A., 2016. Modelling protection behaviour towards micronutrient deficiencies: case of iodine biofortified vegetable legumes as health intervention for school-going children. Nutr. Res. Pract. 10 (1), 56. https://doi.org/10.4162/nrp.2016.10.1.56.

Msungu, S.D., Mushongi, A.A., Venkataramana, P.B., Mbega, E.R., 2022. A review on the trends of maize biofortification in alleviating hidden hunger in sub-Sahara Africa. Sci. Hortic. 299, 111029.

Mudyazvivi, E., Makuvaro, V., Chagonda, I., 2020. Challenges and opportunities for biofortification of maize in Zimbabwe: a review. J. Agric. Food Res. 2, 100052. https://doi.org/10.1016/j.jafr.2020.100052.

Mulongo, G., Munyua, H., Mbabu, A., Maru, J., 2021. What is required to scale-up and sustain biofortification? Achievements, challenges and lessons from scaling-up Orange-Fleshed Sweetpotato in Sub-Sahara Africa. J. Agric. Food Res. 4, 100102. https://doi.org/10.1016/j.jafr.2021.100102.

Murray-Kolb, L.E., Wenger, M.J., Scott, S.P., Rhoten, S.E., Lung'aho, M.G., Haas, J.D., 2017. Consumption of iron-biofortified beans positively affects cognitive performance in 18- to 27-Year-Old Rwandan female college students in an 18-Week randomized controlled efficacy trial. J. Nutr. 147 (11), 2109–2117. https://doi.org/10.3945/jn.117.255356.

Mutuku, J., Mwaniki, M., Onjong, H., Michira, J., 2020. The biofortification continuum: implications for food and nutrition security in developing countries. Afr. J. Food Agric. Nutr. Dev. 20 (01), 15317–15330. https://doi.org/10.18697/ajfand.89.18445.

Muvhuringi, P.B., Chigede, N., 2021. Trends in production and consumption of selected biofortified crops by rural communities in Zimbabwe. Cog. Food Agric 7, 1894760.

Mwaniki, A., 2007. Biofortification as a vitamin A deficiency intervention in Kenya (Case Study NoCornell University, Program on Food Policy for Developing Countries. https://ecommons.cornell.edu/bitstream/1813/55679/1/3-7.pdf.

Nahar, K., Jahiruddin, M., Islam, M., Nayem, Z., 2020. Improvement of nutrient concentration in rice grain by zinc biofortification. Asian J. Adv Agric. Res. 41–47. https://doi.org/10.9734/ajaar/2020/v14i230129.

Nassar, Y., Gad, G., Kortam, W., 2021. The effect of demographic variables on price sensitivity of customers A field study. Arch. Bus. Res 9, 101–142.

Nestel, P., Bouis, H.E., Meenakshi, J., Pfeiffer, W., 2006. Biofortification of staple food crops. J. Nutr. 136 (4), 1064–1067. https://doi.org/10.1093/jn/136.4.1064.

Nuss, E.T., Arscott, S.A., Bresnahan, K., Pixley, K.V., Rocheford, T., Hotz, C., et al., 2012. Comparative intake of white-versus orange-colored maize by Zambian children in the context of promotion of biofortified maize. Food Nutr. Bull. 33 (1), 63–71.

Obukosia, S., 2014. Biotechnology of nutritionally enhanced food crops targeting malnutrition in rural agricultural populations: the case study of Africa biofortified sorghum. Biotechnology in Africa. Science Policy Reports. Springer, Cham, pp. 157–177. https://doi.org/10.1007/978-3-319-04001-1_9.

Ofori, K.F., Antoniello, S., English, M.M., Aryee, A.N., 2022. Improving nutrition through biofortification–A systematic review. Frontiers in Nutrition 9, 1043655. https://doi.org/10.3389/fnut.2022.1043655.

Oh, S., Cave, G., Lu, C., 2021. Vitamin B12 (cobalamin) and micronutrient fortification in food crops using nanoparticle technology. Front. Plant Sci. 12. https://doi.org/10.3389/fpls.2021.668819.

Ojok, J., Omara, P., Opolot, E., Odongo, W., Olum, S., Gijs, D.L., et al., 2019. Iodine agronomic biofortification of cabbage (Brassica oleracea var. capitata) and cowpea (Vigna unguiculata L.) is effective under farmer field conditions. Agronomy 9 (12), 797.

Okorie, I., Afuecheta, E., Nadarajah, S., 2023. Time series and power law analysis of crop yield in some east african countries. Plos One 18 (6), e0287011. https://doi.org/10.1371/journal.pone.0287011.

Olum, S., Gellynck, X., Okello, C., Webale, D., Odongo, W., Ongeng, D., et al., 2018. Stakeholders' perceptions of agronomic iodine biofortification: a SWOT-AHP analysis in Northern Uganda. Nutrients 10 (4), 407.

Osendarp, S.J., Martinez, H., Garrett, G.S., Neufeld, L.M., De-Regil, L.M., Vossenaar, M., et al., 2018. Large-scale food fortification and biofortification in low-and middle-income countries: a review of programs, trends, challenges, and evidence gaps. Food Nutr. Bull. 39 (2), 315–331.

Palmer, A.C., Jobarteh, M.L., Chipili, M., Greene, M.D., Oxley, A., Lietz, G., et al., 2020. Biofortified and fortified maize consumption reduces prevalence of low milk retinol, but does not increase vitamin A stores of breastfeeding Zambian infants with adequate reserves: a randomized controlled trial. Am. J. Clin.l Nutr. 113 (5), 1209–1220. https://doi.org/10.1093/ajcn/nqaa429.

Peláez, D., Aguilar, P., Mercado, M., López-HernÁndez, F., Guzman, M., Burbano-Erazo, E., et al., 2022. Genotype selection, and seed uniformity and multiplication to ensure common bean (Phaseolus vulgaris l.) var. liborino. Agronomy 12 (10), 2285. https://doi.org/10.3390/agronomy12102285.

Petry, N., Boy, E., Wirth, J., Hurrell, R., 2015. Review: the potential of the common bean (Phaseolus vulgaris) as a vehicle for iron biofortification. Nutrients 7 (2), 1144–1173. https://doi.org/10.3390/nu7021144.

Petry, N., Boy, E., Wirth, J.P., Hurrell, R.F., 2020. Review: the potential of the common bean (Phaseolus vulgaris) as a vehicle for iron biofortification. Nutrients 12 (5), 1234. https://doi.org/10.3390/nu12051234.

Philipo, M., 2021. Iron and zinc genetic biofortification of yellow common bean (Phaseolus vulgaris l.) genotypes in Tanzania (Doctoral dissertation, NM-AIST).

Pillay, K., Derera, J., Siwela, M., Veldman, F.J., 2011. Consumer acceptance of yellow, provitamin A-biofortified maize in KwaZulu-Natal: original research. S. Afr. J. Clin. Nutr. 24 (4), 186–191.

Prasanna, B.M., Palacios-Rojas, N., Hossain, F., Muthusamy, V., Menkir, A., Dhliwayo, T., et al., 2020. Molecular breeding for nutritionally enriched maize: status and prospects. Front. Genet. 10, 1392.

Qin, Y., Melse-Boonstra, A., Yuan, B., Pan, X., Dai, Y., Zhou, M., et al., 2012. Zinc biofortification of rice in China: a simulation of zinc intake with different dietary patterns. Nutrients 4 (6), 517–528. https://doi.org/10.3390/nu4060517.

Quisumbing, A.R., Pandolfelli, L., 2010. Promising approaches to address the needs of poor female farmers: resources, constraints, and interventions. World Dev. 38 (4), 581–592.

Rafael, B.M., 2023. The importance of agricultural development projects: a focus on sustenance and employment creation in Kenya, Malawi, Namibia, Rwanda, and Uganda. J Agric Chem Environ 12 (2), 152–170. https://doi.org/10.4236/jacen.2023.122013.

Rai, D., Chaudhary, C., Khatak, A., Banyal, S., 2024. A sustainable approach to combat micronutrient deficiencies and ensure global food security through biofortification. Eur. J. Nutr. Food Saf. 16 (4), 15–30. https://doi.org/10.9734/ejnfs/2024/v16i41404.

Reddy, K., Alekhya, G., Gundreddy, R., Kalwala, S., Mogallapu, U., 2024. Agronomic biofortification of millets: new way to alleviate malnutrition. https://doi.org/10.5772/intechopen.110805.

Rehman, A., Farooq, M., Ullah, A., Nadeem, F., Im, S.Y., Park, S.K., et al., 2020. Agronomic biofortification of zinc in Pakistan: status, benefits, and constraints. Front.Sustain. Food Syst. 4, 591722. https://doi.org/10.3389/fsufs.2020.591722.

Saikanth, D., Kesarwani, A., Nain, M., Pandey, S.K., Singh, B.V., Maurya, C., 2023. Biofortification and its impact on global health. J. Exp. Agric. Int 45 (10), 106–115.

Samuel, L., de Barcellos, M.D., Watabaji, M.D., De Steur, H., 2024. Factors affecting farmers' acceptance and adoption of biofortified crops: a systematic review. Outlook Agric. 53 (1), 15–29.

Sanahuja, G., Farré, G., Berman, J., Zorrilla-López, U., Twyman, R.M., Capell, T., et al., 2013. A question of balance: achieving appropriate nutrient levels in biofortified staple crops. Nutr. Res. Rev. 26 (2), 235–245. https://doi.org/10.1017/s0954422413000176.

Sandhu, R., Chaudhary, N., Shams, R., Singh, K., Pandey, V.K., 2023. A critical review on integrating biofortification in crops for sustainable agricultural development and nutritional security. J. Agric. Food Res. 14, 100830.

Sazawal, S., Dhingra, U., Dhingra, P., Dutta, A., Deb, S., Kumar, J., et al., 2018. Efficacy of high zinc biofortified wheat in improvement of micronutrient status, and prevention of morbidity among preschool children and women-a double masked, randomized, controlled trial. Nutr. J. 17 (1), 86. https://doi.org/10.1186/s12937-018-0391-5.

Sen, H., Kumar, A., Janeja, H.S., 2024. Biofortification of major crops through conventional and modern biotechnological approaches to fight hidden hunger: an overview. J. Adv. Biol. Biotechnol. 27 (7), 96–113.

Senguttuvel, P., Padmavathi, G., Jasmine, C., Sanjeeva, R.D., Neeraja, C.N., Jaldhani, V., et al., 2023. Rice biofortification: breeding and genomic approaches for genetic enhancement of grain zinc and iron contents. Front Plant Sci. https://doi.org/10.3389/fpls.2023.1138408.

Severo, J., Santos, F., Samborski, T., Rodrigues, R., Mello, A., 2021. Biofortified sweet potatoes as a tool to combat vitamin a deficiency: effect of food processing in carotenoid content. Rev. Chil. Nutr 48 (3), 414–424. https://doi.org/10.4067/s0717-75182021000300414.

Shahzad, Z., Rouached, H., Rakha, A., 2014. Combating mineral malnutrition through iron and zinc biofortification of cereals. Compr. Rev. Food Sci. Food Saf. 13 (3), 329–346. https://doi.org/10.1111/1541-4337.12063.

Sharma, D., Ghimire, P., Bhattarai, S., Adhikari, U., 2020. Biofortification of wheat: genetic and agronomic approaches and strategies to combat iron and zinc deficiency. Sustain. Food Agric. 1 (1), 48–54. https://doi.org/10.26480/sfna.01.2020.48.54.

Sharma, S., Singh, D., 2022. Zinc for enhancing crop growth of rice-a brief review. J. Nat. Res. Conserv. Manag. 3 (2), 125–129. https://doi.org/10.51396/anrcm.3.2.2022.125-129.

Sheoran, S., Kumar, S., Ramtekey, V., Kar, P., Meena, R., Jangir, C., 2022. Current status and potential of biofortification to enhance crop nutritional quality: an overview. Sustainability 14 (6), 3301. https://doi.org/10.3390/su14063301.

Sida-Arreola, J., Sánchez, E., Ávila-Quezada, G., Zamudio-Flores, P., Acosta-Muñíz, C., 2015. Can iron biofortification improve antioxidant response, yield and nutritional quality in green beans? Agric. Sci. 06 (11), 1324–1332. https://doi.org/10.4236/as.2015.611127.

Singh, A., Palanisamy, V., Raman, R.V., Velayudhan, P.K., Singh, A.K., 2023. Exploring the prevalence of undernutrition and consumers' knowledge, preferences, and willingness to pay for biofortified food. Achieving Zero Hunger in India: Challenges and Policies. Springer Nature Singapore, Singapore, pp. 171–209.

Siwela, M., Pillay, K., Govender, L., Lottering, S., Mudau, F.N., Modi, A.T., et al., 2020. Biofortified crops for combating hidden hunger in South Africa: availability, acceptability, micronutrient retention and bioavailability. Foods 9 (6), 815. https://doi.org/10.3390/foods9060815.

Stevens, G.A., Beal, T., Mbuya, M.N.N., Luo, H., Neufeld, L.M., Addo, O.Y., et al., 2022. Micronutrient deficiencies among preschool-aged children and women of reproductive age worldwide: a pooled analysis of individual level data from population-representative surveys. Lancet Glob. Health 10 (11), e1590–e1599.

Sun, Y., 2024. Malnutrition problems caused by vitamin deficiency in African countries. Biological Sciences and Environmental Health. CRC Press, pp. 314–321.

Swamy, B.M., Descalsota, G.I.L., Nha, C.T., Amparado, A., Inabangan-Asilo, M.A., Manito, C., et al., 2018. Identification of genomic regions associated with agronomic and biofortification traits in DH populations of rice. PloS ONE 13 (8), e0201756. https://doi.org/10.1371/journal.pone.0201756.

Talsma, E.F., Brouwer, I.D., Verhoef, H., Mbera, G.N., Mwangi, A.M., Demir, A.Y., et al., 2015. Biofortified yellow cassava and vitamin A status of Kenyan children: a randomized controlled trial. Am. J. Clin. Nutr. 103 (1), 258–267. https://doi.org/10.3945/ajcn.114.100164.

Talsma, E.F., Melse-Boonstra, A., Brouwer, I.D., 2017. Acceptance and adoption of biofortified crops in low- and middle-income countries: a systematic review. Nutrition reviews 75 (10), 798–829. https://doi.org/10.1093/nutrit/nux037.

Tang, G., Qin, J., Dolnikowski, G.G., Russell, R.M., Grusak, M.A., 2009. Golden rice is an effective source of vitamin A. Am. J. Clin. Nutr. 89 (6), 1776–1783. https://doi.org/10.3945/ajcn.2008.27119.

Tanumihardjo, S.A., Ball, A., Kaliwile, C., Pixley, K.V., 2017. The research and implementation continuum of biofortified sweet potato and maize in Africa. Ann. NY Acad. Sci. 1390 (1), 88–103. https://doi.org/10.1111/nyas.13315.

Tumbare, T., Maphosa, M., 2023. Genetic bio-fortification of cereals from a plant breeding perspective. J. Plant Breeding Crop Sci. 15 (1), 42–56. https://doi.org/10.5897/jpbcs2022.0997.

United States Department of Agriculture Foreign Agricultural Service, 2026. India publishes policy amendments for food labeling regulations. (January 15). https://apps.fas.usda.gov/newgainapi/api/Report/DownloadReportByFileName?fileName=India%20Publishes%20Policy%20Amendments%20for%20Food%20Labeling%20Regulations_New%20Delhi_India_IN2026-0001.pdf.

Upadhayay, V., Singh, A., Khan, A., 2022a. Cross talk between zinc-solubilizing bacteria and plants: a short tale of bacterial-assisted zinc biofortification. Front. Soil Sci. 1. https://doi.org/10.3389/fsoil.2021.788170.

Upadhayay, V., Singh, A., Khan, A., Singh, J., Pareek, N., Raghav, A., 2022b. Fe-sem/edx based zinc mobilization analysis of burkholderia cepacia and pantoea rodasii and their functional annotation in crop productivity, soil quality, and zinc biofortification of paddy. Front. Microbiol. 13. https://doi.org/10.3389/fmicb.2022.852192.

Vaiknoras, K., Larochelle, C., Birol, E., Asare-Marfo, D., Herrington, C., 2019. Promoting rapid and sustained adoption of biofortified crops: what we learned from iron-biofortified bean delivery approaches in Rwanda. Food Policy 83, 271–284. https://doi.org/10.1016/j.foodpol.2018.11.003.

Van Der Straeten, D., Bhullar, N.K., De Steur, H., Gruissem, W., MacKenzie, D., Pfeiffer, W., et al., 2020. Multiplying the efficiency and impact of biofortification through metabolic engineering. Nat. Commun. 11 (1), 5203. https://doi.org/10.1038/s41467-020-19020-4.

Wang, Y.H., Zou, C.Q., Mirza, Z., Li, H., Zhang, Z.Z., Li, D.P., et al., 2016. Cost of agronomic biofortification of wheat with zinc in China. Agron. Sustain. Dev. 36 (3), 44. https://doi.org/10.1007/s13593-016-0382-x.

Weffort, V.R.S., Lamounier, J.A., 2024. Hidden hunger - a narrative review. J. de Pediatria 100 (1), S10–S17. https://doi.org/10.1016/j.jped.2023.08.009.

Welch, R.M., Graham, R.D., 2004. Breeding for micronutrients in staple food crops from a human nutrition perspective. J.Exp. Bot. 55 (396), 353–364. https://doi.org/10.1093/jxb/erh064.

Wenger, M.J., Rhoten, S.E., Murray-Kolb, L.E., Scott, S.P., Boy, E., Gahutu, J., et al., 2019. Changes in iron status are related to changes in brain activity and behaviour in rwandan female university students: results from a randomised controlled efficacy trial involving iron-biofortified beans. J. Nutr. 149 (4), 687–697. https://doi.org/10.1093/jn/nxy265.

White, P., Broadley, M., 2011. Physiological limits to zinc biofortification of edible crops. Front. Plant Sci. 2. https://doi.org/10.3389/fpls.2011.00080.

Woods, B., Gallego-Castillo, S., Talsma, E., Alvarez, D., 2020. The acceptance of zinc biofortified rice in Latin America: a consumer sensory study and grain quality characterisation. Plos ONE 15 (11), e0242202. https://doi.org/10.1371/journal.pone.0242202.

World Health Organization, 2014. Guideline: fortification of food-grade salt with iodine for the prevention and control of iodine deficiency disorders. https://iris.who.int/bitstream/handle/10665/136908/9789241507929_eng.pdf?sequence=1.

Wu, W., Qi, D., Chen, Y., Wang, J., Wang, Q., Yang, Y., et al., 2024. Enhancement of nutrient, trace element, and organic selenium contents of ratooning rice grains and straw through foliar application of selenite. Foods 13 (22), 3637. https://doi.org/10.3390/foods13223637.

Wu, J., Salisbury, C., Graham, R., Lyons, G., Fenech, M., 2009. Increased consumption of wheat biofortified with selenium does not modify biomarkers of cancer risk, oxidative stress, or immune function in healthy Australian males. Environ. Mol. Mutagenesis 50 (6), 489–501. https://doi.org/10.1002/em.20490.

Ye, X., Al-Babili, S., KlöTi, A., Zhang, J., Lucca, P., Beyer, P., et al., 2000. Engineering the provitamin A (β-Carotene) biosynthetic pathway into (carotenoid-free) rice endosperm. Science 287 (5451), 303–305. https://doi.org/10.1126/science.287.5451.303.

Zabala, J.A., Martínez-García, V., Martínez-Paz, J.M., López-Becerra, E.I., Nasso, M., Díaz-Pereira, E., et al., 2023. Crop diversification practices in Europe: an economic cross-case study comparison. Sustain. Sci. 18 (6), 2691–2706. https://doi.org/10.1007/s11625-023-01413-1.

Zubair, A., Zaib, S., Ghani, N., Mehboob, S., 2023. Zinc-fortified rice: approach to improve dietary zinc consumption and effect on human health. Acs Agric. Sci. Technol. 3 (11), 926–938. https://doi.org/10.1021/acsagscitech.3c00140.

Zuma, M., Kolanisi, U., Modi, A., 2018. The potential of integrating provitamin-a-biofortified maize in smallholder farming systems to reduce malnourishment in South Africa. Int. J. Environ. Res. Public Health 15 (4), 805. https://doi.org/10.3390/ijerph15040805.

Further reading

Dias, D., Costa, N., Nutti, M., Tako, E., Martino, H., 2017. Advantages and limitations of in vitro and in vivo methods of iron and zinc bioavailability evaluation in the assessment of biofortification program effectiveness. Crit. Rev. Food Sci. Nutr. 58 (13), 2136–2146. https://doi.org/10.1080/10408398.2017.1306484.

Finkelstein, J., Mehta, S., Udipi, S., et al., 2015. A randomized trial of iron-biofortified pearl millet in school children in India. J. Nutr. https://doi.org/10.3945/jn.114.208009.

Ijaz, S., Iqbal, J., Abbasi, B.A., Tufail, A., Ullah, Z., Sharifi-Rad, J., et al., 2023. Biofortification: lessons from the past and strategies for future food security. Legumes Biofortification. Springer International Publishing, Cham, pp. 521–545.

Kumar, S., Hash, C., Thirunavukkarasu, N., Singh, G., Rajaram, V., Rathore, A., et al., 2016. Mapping quantitative trait loci controlling high iron and zinc content in self and open pollinated grains of pearl millet [pennisetum glaucum (l.) r. br.]. Front. Plant Sci. 7. https://doi.org/10.3389/fpls.2016.01636.

Rock, J.S., Schnurr, M.A., Kingiri, A., Glover, D., Stone, G.D., Ely, A., et al., 2023. Beyond the genome: genetically modified crops in Africa and the implications for genome editing. Development and Change 54 (1), 117–142.

Welch, R.M., 2002. Breeding strategies for biofortified staple plant foods to reduce micronutrient malnutrition globally. J. Nutr. 132 (3), 495S–499S. https://doi.org/10.1093/jn/132.3.495s.

PART 2

Resilient Food Systems

CHAPTER EIGHT

Revisiting Green Revolution (1965–2000)

Wendell L. Ultra[a,b,*] and **Cherry I. Ultra**[c]
[a]College of Agriculture and Technology, Northwest Samar State University, Calbayog City, Samar, Philippines
[b]College of Agriculture and Food Science, Visayas State University, Visca, Baybay City, Leyte, Philippines
[c]Office of the President, University of Eastern Philippines, Catarman, Northern Samar, Philippines
*Corresponding author. e-mail address: wendell.ultra@nwssu.edu.ph

Contents

Abstract

For decades, the Green Revolution (GR) has brought about significant changes worldwide in agriculture, the environment, and society. Spanning roughly from the mid-1960s to the end of the 20th century, it represents one of the most profound agricultural transformations in global history. This chapter revisits the GR by reviewing its history, its milestones, regional trajectories, and long-term socio-economic and environmental implications. Drawing on evidence from the Asian, African, European, North American, and Latin American regions, the chapter examines how the introduction of high-yielding crop varieties, purchased synthetic inputs, mechanization, expanded irrigation systems, and gender disparities transformed food production and rural livelihoods. Attention is given to the uneven regional outcomes of the GR, highlighting success in enhancing food security and agricultural productivity in Asia, in contrast to the more limited impacts observed in Africa due to weak infrastructure, institutional constraints, and agroecological diversity. The chapter also revisits the environmental legacy, including soil degradation, declining soil organic matter, soil

Advances in Food Security and Sustainability, Volume 11
ISSN 2452-2635, https://doi.org/10.1016/bs.af2s.2026.06.002

nutrient imbalances, and biodiversity loss, which have raised concerns regarding the sustainability of input-sensitive farming systems. Furthermore, the role of policy frameworks is examined to understand how institutional interventions shaped farmer behavior and production outcomes. By synthesizing historical insights with contemporary sustainability challenges, this chapter reframes the GR as a multifaceted process rather than a singular technological success. The analysis underscores the need to move beyond yield-centric paradigms toward more inclusive, resource-efficient, and environmentally resilient agricultural strategies for future food systems.

1. Introduction

The Green Revolution (GR) is the most globally impactful modern agricultural transformation achieved from the 1960s to the early 2000s, enhancing global food production (Evans and Lawson, 2020). It was a countermeasure to address the socioeconomic problems, particularly in global food insecurity and an exponential increase in population. Conventional agricultural practices emerged during this period (Evans and Lawson, 2020; Muchhadiya et al., 2024). Revisiting the GR is important because of its legacy in agricultural development, food policies and security, and environmental outcomes. However, it is interpreted through a simplified narrative of success. This has diminished attention to the important social effects of the GR and stifled debate about competing approaches that emphasize the complexity and context-specificity of agricultural development (van Etten, 2022). The American agronomist Norman Borlaug, one of the pioneers of the GR, is often hailed as the "Father of the Green Revolution" and is known for successfully developing disease-resistant, geographically adaptable, and high-yielding wheat varieties, preventing global hunger and starvation (Annesley, 2022). Technological advancements in agriculture have rapidly spread globally.

The GR contributed to food security and alleviating hunger, leading to substantial economic development in some parts of the world (Pingali, 2023). However, while the GR brought successes to agricultural development and other sectors, it also faced criticisms because of its detrimental environmental impacts caused by the excessive use of synthetic fertilizers and pesticides, attenuation of water resources, and loss of biodiversity (Ebel et al., 2021). Furthermore, the benefits of the revolution were unevenly distributed, and social inequalities within rural communities were observed, as wealthier farmers could easily access new technologies and inputs (Harwood, 2020). In this chapter, we will revisit the multifaceted legacy of the GR, which continues to inform discussions on food security, sustainable agriculture, and socio-economic and rural development. The revisiting of was conducted through a socioeconomic and technical perspective and an interdisciplinary synthesis of historical studies, contemporary literature, government reports, and published journal articles. Rather than treating it as a single technological event, the analysis examined it as a long-term development process shaped by economy, institutions, politics, and regional contexts.

2. Green Revolution in regional contexts

The GR significantly impacted global agriculture by introducing conventional techniques such as synthetic inputs, high-yielding crop varieties (HYVs) produced by plant breeding, and modern irrigation techniques (Hamdan et al., 2022). However, its impact was not felt globally; Asia and Latin America experienced an immediate increase in production of irrigated rice and wheat. The remaining developing countries, particularly those in Africa, hardly benefited from the revolution and struggled to adopt the model of high-input technologies, especially irrigation systems (Wise, 2020).

Factors contributing to the fate of the GR in a regional context include crop adaptation, temperature, soil composition, prevalent diseases, solar radiation, precipitation, wind, competition, physiological tolerance, and climate change, all of which influence crop growth and development (Liliane and Charles, 2020). In the socioeconomic and political context, the success of the GR is also dependent on the national government's capacity and the quality of its agricultural research systems, including local and university research programs, highlighting the implementation and establishment of adaptive breeding and HYVs (Gollin et al., 2021).

The GR produced multifaceted impacts that reshaped agricultural landscapes, rural economies, and ecological systems across regions. On one hand, it delivered historic gains in cereal productivity, transforming regions such as South Asia from food-deficient zones into largely self-sufficient producers. Most importantly, the GR meant low and stable prices for staple grains, benefiting consumers and the workforce available for broader economic development on and off the farm. Lower food prices combined with rising incomes changed salary spending ratios, allowing for higher spending capacity particularly for the poor and facilitating rapid economic growth (Connor et al., 2023). At the same time, the GR's technological package of HYVs, synthetic inputs, and water-intensive practices introduced significant regional disparities in who benefited, as seen in Fig. 1.

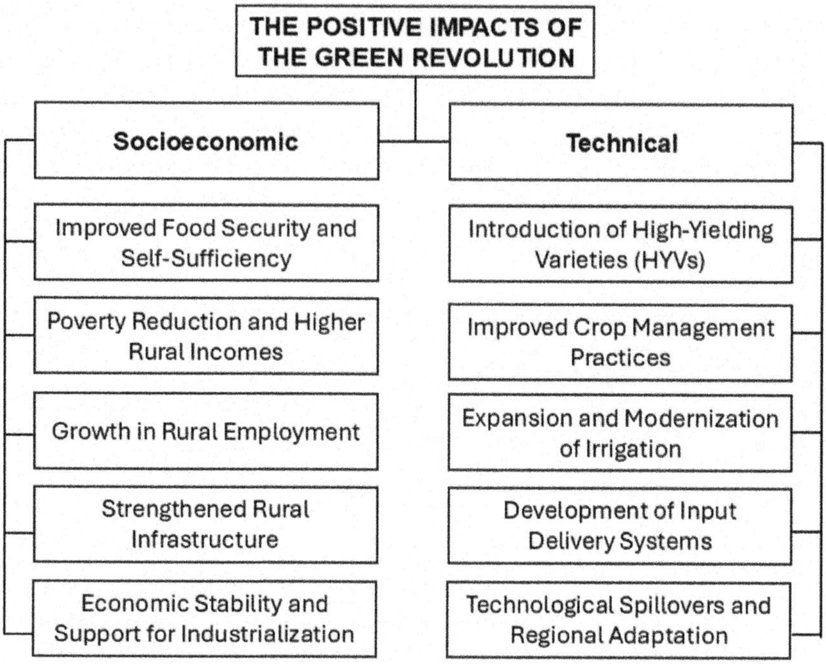

Fig. 1 The socioeconomic and technical impacts of the Green Revolution *(original interpretation by the authors).*

2.1 Green Revolution in Asia

2.1.1 Successes of GR in Asia

2.1.1.1 Irrigation

Rice and wheat were the key crops during the GR in Asia (Jha et al., 2010). Because of the GR, various irrigation technologies and techniques were introduced, therefore achieving food self-sufficiency during the 1970s–80s (Mukherji et al., 2009). In the case of South and Southeast Asian nations, the value of water has increased since the globalization era, beginning in 1990. The dominant irrigation systems have been pumps and wells (Barker, 2004). In India, the micro-irrigation technique was practiced, a common farming method that delivers water to plants using drip irrigation and sprinkler systems. This approach has seen extensive use because it can effectively provide water to plants (Deshpande et al., 2024). Pakistan has one of the most irrigated agricultural areas (Qureshi et al., 2023) and an extensive irrigated agricultural system (Ghumman et al., 2011). Irrigation sources for agriculture in South Asia include groundwater and meltwater (Lutz et al., 2022). Sources for irrigation in South Korea, a major rice-growing country, come from abundant rainfall, up to 1.6 times higher than the global average (Hazell, 2010). However, due to the large population living in a small country, water resources remain scarce; nevertheless, the water intake and processing facilities are well-equipped and maintained. Furthermore, dams, rivers, and groundwater comprise South Korea's main water supply (Choi, 2020). The GR in Asia transformed irrigation through shifts in technology, institutions, and hydrology, promoting agricultural growth while introducing challenges like water scarcity and groundwater depletion, impacting governance, as explained in Fig. 2.

2.1.1.2 Synthetic fertilizers

Fertilizer use was also extensive during this era, with an average of 41.33 kg/ha in 1970 and 146.79 kg/ha in 1995, meaning that synthetic fertilizer utilization in Asia increased 255 % in a 25-year span. China and South Korea utilized the most synthetic fertilizers during this era. According to the National Bureau of Statistics of the People's Republic of China (2020), China's per capita arable land is 0.089 ha, which is below the global average. Yield gaps were observed due to low organic matter content in the soil and poor soil conditions (Li et al., 2019). Furthermore, organic fertilizer was widely used; however, due to nutrient limitations, scalability constraints, and slow nutrient release, organic fertilizer alone was insufficient to address agricultural productivity challenges. Consequently, the widespread adoption of high-synthetic inputs and intensive farming systems in China was regarded as essential to achieve the substantial increases in cereal production necessary for food security (Liu et al., 2022; Hou et al., 2015). Use of synthetic fertilizer played an important role in

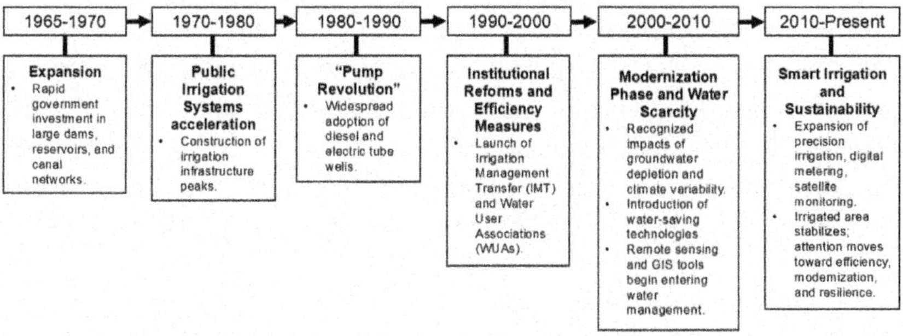

Fig. 2 The timeline and phases of irrigation in the Green Revolution in Asia *(original interpretation by the authors).*

increasing the production of grains, namely rice, wheat, and corn, hence agricultural development (Jin et al., 2006). The extensive use of synthetic fertilizers in South Korea stemmed from the government's vision of rural modernization, which resulted in increased crop production (Lee, 2023). Additionally, in order to stabilize farmgate prices, governments also strengthened farm finance programs and subsidized essential inputs, including synthetic fertilizers. Numerous governments made use of their actions to prevent small farmers from falling behind (Hazell, 2009).

2.1.1.3 Cultivated modern varieties and percentage of harvest

During the years of 1965–1975, there were steady and significant increases in wheat production in South Asia, followed by corn and then rice (Evenson and Gollin, 2003). Wheat successfully thrived in the temperate regions of South Asia, with Northwestern India (Punjab, Haryana, and Uttar Pradesh) and the Punjab region of Pakistan exhibiting ideal growing conditions for HYVs of wheat. Rice requires heavy monsoon rains, which occur less frequently in the region (Singh, 2000).

From the mid-1970s to the early 2000s, rice production expanded rapidly across Southeast Asia, driven primarily by strong government support, irrigation investments, and the widespread adoption of HYVs developed through international research programs. As a result, rice became the dominant staple crop in the region and a central pillar of food security strategies (Pingali et al., 1997). However, this expansion did not displace corn production. On the contrary, corn output grew significantly in East and Southeast Asia, supported by rising demand for livestock feed associated with increasing meat consumption and dietary transformation (Pingali, 2012). In countries such as the Philippines and China, government initiatives including hybrid corn promotion programs and the expansion of poultry and pork industries further accelerated corn production (Gonzales, 1993; Huang et al., 1999). Moreover, the adaptability of corn to diverse agroecological conditions, particularly in upland and rainfed areas unsuitable for paddy rice, facilitated its expansion into marginal lands (Pingali and Rosegrant, 1994). Thus, rice and corn growth during this period reflected complementary, rather than competing, dynamics within evolving Asian agricultural systems.

2.1.2 Challenges of GR in Asia

2.1.2.1 Extreme population pressure and land fragmentation

The Asian GR occurred in the context of some of the highest rural population densities worldwide. This demographic pressure necessitated the rapid adoption of high-input technology packages to achieve a tripling of rice output and secure aggregate economic growth. The rapid population growth intensified pressure on already limited arable Asian land. Over time, inheritance systems led to highly fragmented and small landholdings, particularly in South and East Asia. While HYVs raised productivity, the small scale of farms constrained mechanization efficiency, investment capacity, and income growth per household. The productivity gains felt did not always convert into substantial improvements in the livelihood of millions of smallholder farmers (Pingali, 2012).

2.1.2.2 Intensification, environmental degradation, and resource depletion

The dominance of irrigated rice–wheat systems in Asia, especially in the Indo-Gangetic Plains and parts of China, created ecological stresses unique to these monoculture rotations (Gupta et al., 2003). The continuous double-cropping reduced soil biodiversity, contributed to zinc and iron deficiencies, and increased pest resistance (Conway, 1998; Nazari et al., 2024). Unlike more diversified systems elsewhere, these intensive cereal rotations locked the farmers into narrow production cycles that were difficult to diversify.

Furthermore, the monoculture system destroyed the barriers of pest movement (Zou et al., 2024; Shah et al., 2021; Mortensen and Smith, 2020). The intensive input requirements of these systems, particularly for water and synthetic inputs, led to severe environmental degradation, including groundwater depletion, soil salinization, and reduced agrobiodiversity (Jhariya et al., 2021; Hashemi et al., 2022). The intensive use of synthetic inputs led to soil nutrient imbalances, contamination of water bodies, and declining soil organic matter. In major production zones such as the Indo-Gangetic Plains. Excessive nitrogen application reduced nutrient-use efficiency and contributed to groundwater pollution. Irrigation expansion, though critical for crop yield, resulted in groundwater source depletion, salinization, and waterlogging in several regions. These environmental stresses have raised concerns about the long-term sustainability of high-input cereal systems. The current production system clearly shows diminishing marginal returns, necessitating a shift toward more balanced and ecologically sustainable approaches (Byerlee, 1992; Neelam et al., 2022).

2.1.2.3 Groundwater overexploitation

Groundwater overexploitation became one of the most significant environmental consequences of the GR in Asia (Pingali and Rosegrant, 1994;

Hazell, 2009; Shah, 2007). The rapid adoption of HYVs of rice and wheat required timely and reliable irrigation, especially during critical growth stages (Mallareddy et al., 2023; Zhu et al., 2024; Debangshi and Ghosh, 2022). While canal irrigation played an important role, the widespread introduction of private tube wells from the 1970s onward dramatically expanded access to groundwater, particularly in India, Pakistan, Bangladesh, and northern China (Shah, 2007; Shah et al., 2003; Qureshi, 2020). Furthermore, government subsidies for rural electrification and diesel pumps reduced the extraction cost of groundwater (Smith and Urpelainen, 2016; Scott and Sharma, 2009). The minimum support prices and procurement policies for rice and wheat encouraged year-round cereal cultivation, increasing irrigation demand (Pingali and Shah, 2001). Progressively, groundwater extraction started to surpass natural replenishment rates in key production areas. In the Indo-Gangetic and the North China Plains, declining water tables became common (Dangar and Mishra, 2024; Mall et al., 2006; Closas and Molle, 2016). Excessive extraction of groundwater contributed to saltwater intrusion, leading to soil salinization, reduced river base flows, and water quality deterioration (Lu et al., 2025).

2.1.2.4 State-Led subsidy dependence

State-led input subsidies were central to the Asian GR; with government provision of inputs, irrigation equipment, electricity for groundwater pumping, and credit. These interventions benefitted farmers through reduced production risks, lowered input costs, and accelerated dispersal of modern agricultural practices (Hazell, 2009; Byerlee, 1992). However, subsidy systems became firmly established. Farmers based their cropping decisions around the availability of subsidized inputs and price guarantees, particularly for rice and wheat (Bellmann, 2019; Yang et al., 2023). This reduced the incentives for diversification into other crops such as pulses, oilseeds, or less water-intensive crops. Moreover, despite the role of the subsidies in significantly enhanced rice productivity, agricultural public investment, power consumption, and synthetic inputs consume substantial portions of government agriculture budgets. Thus, while state-led subsidies were instrumental in initiating Asia's GR, they also established path dependence that complicates efforts to achieve fiscal efficiency, sustainability, and crop diversification at the present time (Lele and Goswami, 2020; Wu et al., 2025).

2.2 Green Revolution in Africa

2.2.1 Successes of GR in Africa

2.2.1.1 Expansion of improved seed varieties (Corn, Rice, Wheat)

One of the most notable successes associated with Africa's Green Revolution efforts has been the expansion of improved seed varieties of corn, rice, and wheat (Eicher, 1995; Wise, 2021). Building on international research collaborations and national breeding programs, new high-yielding and stress-tolerant varieties were developed in response to the diverse agro-ecological conditions of Africa (DeVries, 2011; Effa et al., 2025). In corn, open-pollinated varieties and improved hybrids raised yield potential in Eastern and Southern Africa, particularly where complementary inputs such as fertilizers were accessible (Sibanda et al., 2016). In rice, the creation and dissemination of New Rice for Africa (NERICA) varieties combined the excellent yield of Asian rice with the robustness of African rice types, enhancing efficiency in upland and low-resource settings (Alou, 2020; Diagne et al., 2011). Moreover, wheat breeding programs similarly introduced disease-resistant and heat-tolerant varieties well-adapted to highland and irrigated systems in countries such as Sudan and Ethiopia (Tadesse et al., 2017, 2022; Tadesse et al., 2019). Even though adoption rates have varied across regions due to infrastructure, market access, and policy constraints, the expansion of improved cultivars has increased productivity and bolstered resilience to biotic and abiotic stresses, and contributed to improved food availability in several African nations (Atera et al., 2011; Diagne et al., 2011; Somado and Guei, 2008).

2.2.1.2 Strategies for a sustainable GR in Africa

A number of strategies have been proposed and implemented to overcome the challenges in Africa including the adoption of improved seed varieties, integrated soil fertility management (ISFM), investment in irrigation, and enhanced farmer training and extension services. More resilient crop varieties were necessary and developed for climatic adaptation (Lobell et al., 2008). The International Maize and Wheat Improvement Center (CIMMYT) and the International Institute of Tropical Agriculture (IITA) developed cereal and cassava varieties that are drought-resilient and adaptable to poor soil fertility (Pingali, 2012). Despite showing promise, the Drought-Tolerant Maize for Africa (DTMA) project did not lead to widespread adoption because of the high cost of seeds (Cairns et al., 2013). IITA and the International Center for Tropical Agriculture (CIAT) introduced resilient high-yielding cassava varieties

(FAO, 2013). Successful adoption of hybrid corn in Kenya led to significant yield increases (Hassan et al., 2001), while government support helped expand corn production in Zimbabwe until political and economic instability undermined this initiative (Smale and Jayne, 2003).

ISFM techniques were employed in some parts of Africa. This includes the use of synthetic fertilizer along with organic amendments such as manure, compost, and crop residues to enhance soil microbial activity, water retention, and organic matter (Sanginga and Woomer, 2009). Furthermore, intercropping and crop rotation have helped with nitrogen fixation, reducing dependence on synthetic fertilizers (Bationo et al., 2006). Researchers promoted legume intercropping combined with farmyard manure and synthetic fertilizers for corn, which resulted in significant yield increases in western Kenya (Vanlauwe et al., 2015). Government-supported subsidy schemes in Malawi reduced the cost of fertilizers. These synthetic fertilizers are then combined with organic fertilizers, leading to improved corn yields (Dorward and Chirwa, 2012). In order to maximize productivity and cut expenses, farmers in Burkina Faso and Niger used fertilizer microdosing, which involves applying small quantities of synthetic fertilizers (nitrogen, phosphorus, and potassium) straight to plant roots (Tabo et al., 2007). A project in Zambia integrated ISFM and agroforestry (planting crops together with trees) using a nitrogen-fixing legume species, *Faidherbia albida,* mixed into cropping systems, thereby reducing reliance on synthetic fertilizers (Garrity et al., 2010). The incorporation of both organic and synthetic inputs showed more promising results than those with sole applications; for example, in Rwanda, smallholder farmers using both synthetic fertilizers and compost achieved 30%–50% higher corn yields (Haggblade et al., 2017).

2.2.1.3 Emergence of regional agricultural policy frameworks (CAADP)

A notable institutional development accompanying Africa's Green Revolution efforts has been the creation of regional agricultural policy frameworks designed to coordinate investment, enhance accountability, and foster long-term sectoral transformation (Otsuka and Kijima, 2010; Wise, 2020; Dare Kolawole, 2012). Leading among these is the Comprehensive Africa Agriculture Development Programme (CAADP), inaugurated in 2003 under the African Union as part of the New Partnership for Africa's Development (NEPAD) (African Union, 2003; Waal, 2002). CAADP created a continent-wide pledge for governments to allocate a minimum of 10% of their national budgets to agriculture and to aim for yearly agricultural growth of 6 % (Brüntrup, 2011; Kimenyi et al., 2013). Evidence-based planning, peer review, and regional coordination frameworks were introduced beyond the financial and growth targets. In this way, CAADP encouraged alignment between national agricultural investment plans and broader development goals, while fostering partnerships among various agencies, including governments, research institutions, and the private sector (Afun-Ogidan, 2012). CAADP has been instrumental in shaping a more coordinated and strategic approach to agricultural modernization across Africa, strengthening goals related to productivity, food security, and market integration (Badiane et al., 2020).

2.2.2 Challenges to the GR in Africa

2.2.2.1 Political, infrastructure, and technological challenges

The GR in Africa emerged later than in other regions due to a combination of factors, including political, socioeconomic, infrastructure, and technological challenges as shown in Fig. 3 (see also Toenniessen et al., 2008).

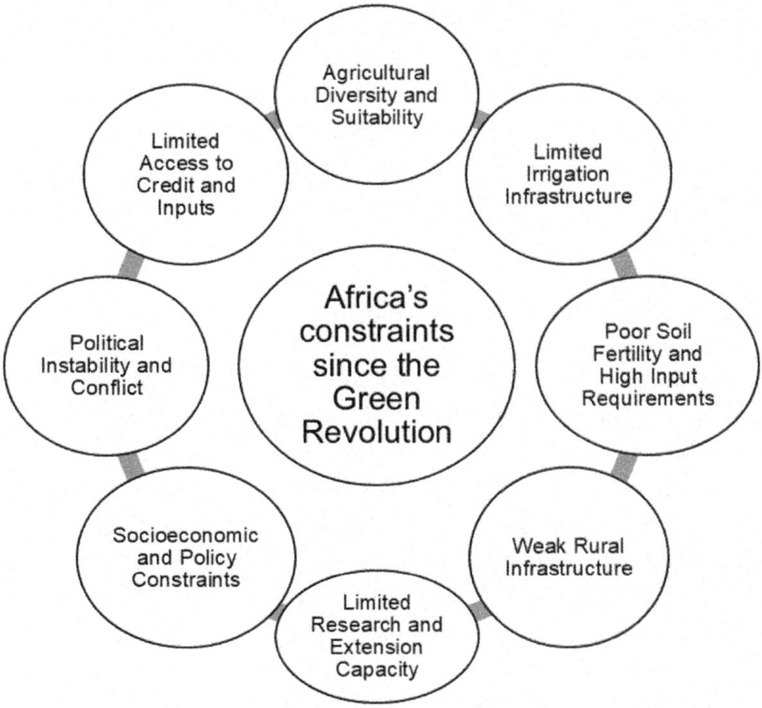

Fig. 3 The multifaceted constraints in the African Green Revolution *(original interpretation by the authors).*

Political challenges include inconsistencies in policy implementation and governance issues, ownership and land tenure issues, and lack of government support and subsidies. While Asia and Latin America experienced significant yield increases from the 1960s onwards, Africa stagnated in agriculture (Evenson and Gollin, 2003).

Africa also faces significant infrastructure challenges. According to the World Bank (2008), the region suffered from poor transportation, limiting the farmers' access to markets for inputs such as fertilizers, pesticides, improved seeds, and planting materials, and to markets for their produce (see also Jayne et al., 2014). Crucially, Africa also had limited irrigation systems (Morris et al., 2007), and insufficient distribution and storage facilities meant that high post-harvest losses were a huge additional problem for farmers (FAO, 2013).

In terms of technological challenges, the low adoption rate of high-yielding crop varieties among many African farmers stemmed from limited accessibility to improved seeds and also the limited adaptability of the improved varieties to local conditions and environments (Smale et al., 2012). Furthermore, knowledge gaps were also observed, and the GR was not implemented properly due to the lack of necessary education and training for many smallholder farmers in Africa (Conway and Toenniessen, 2003).

In the 2000s, Africa's leaders sought to address these constraints. They initiated the Alliance for a GR in Africa (AGRA) to tackle poverty and disparities in society, and to promote agricultural transformation (Toenniessen et al., 2008).

2.2.2.2 Agricultural development challenges in Africa

Africa faced challenges in implementing a GR, including climate variability, soil degradation, limited access to inputs, and infrastructure deficits. According to Cooper et al. (2008), African agriculture still heavily relies on rainfall, while Asia has large-scale irrigation playing a key role in the success of the GR techniques. Crop yields were affected because of unpredictable growing seasons and irregular rainfall patterns. Specifically, the Sahel region in Africa faced the challenges brought by rainfall variability, including erratic precipitation patterns and frequent droughts, thus contributing to food insecurity and hunger (Nicholson, 2001). All these factors made the adoption of GR technologies extremely difficult (Hulme, 2001; Rockström et al., 2009). Heat stress and increasing temperature were also evident in sub-Saharan Africa, reducing productivity for temperature-sensitive crops such as corn and wheat (Lobell et al., 2008). Furthermore, high temperature causes higher evaporation rates, making it difficult for the soil to retain moisture during long periods of drought (Schlenker and Lobell, 2010). Although there is a diffusion of modern varieties in Africa, it was significantly lower than in Asia and Latin America (Evenson and Gollin, 2003).

Irrigation was of key importance during the GR, as evident in its implementation and success in Asia. However, over-dependence on rainfall in Africa resulted in soil degradation (Lal, 2004; Rockström et al., 2009). Traditional shifting farming practices have been abandoned in favor of monocropping and continuous cropping, leading to soil nutrient depletion, due to the nutrient-demanding physiology of high-yield crop varieties (Bationo et al., 2006). This increased the pressure on the land to provide more nutrients, further promoting the depletion of soil nutrients. The degraded soil produced lower yields due to decreases in essential macronutrients in soil such as nitrogen, phosphorus, and potassium (Sanchez et al., 1997; Sanchez, 2002).

Several countries engaged in soil conservation practices during the GR. However, African farmers lacked soil management training compared to those in Asia, where soil conservation programs and terracing were promoted and encouraged (Tittonell and Giller, 2013).

All these factors led to lower corn and wheat yields in sub-Saharan Africa (FAO, 2015). Thus, some African farmers relied on synthetic fertilizers to compensate for the soil fertility losses. However, widespread adoption of technology was prevented due to high fertilizer costs and limited availability (Jayne et al., 2014). Soil degradation in sub-Saharan Africa contributed to desertification, forcing farmers to vacate unproductive lands (Mortimore, 1998); 83.6% of countries on the African continent are at risk of desertification (Stallwood, 2024). The African continent loses 3 million hectares of its forestlands annually, based on the African Forest Landscape Restoration Initiative (AFR100).

2.3 Green Revolution in Europe

2.3.1 Successes of GR in Europe

2.3.1.1 Mechanization and industrialization of agriculture

The industrialization of agriculture during the GR in Europe was characterized by rapid mechanization, which played a crucial role in contributing to agricultural productivity by reducing dependence on manual labor and streamlining activities from sowing to processing. Various implements were developed and adopted, such as tractors and combine harvesters, which made European farmers more efficient (Mazoyer and Roudart, 2006). Modernization in farming practices was a result of a shift in government policies and incentives, thus reducing labor demands and leading to a serious migration of rural settlement workers to urban regions for jobs in industrial sectors (Swinnen et al., 2009). Countries like the UK, Germany, and France saw significant reductions in their agricultural workforce due to the replacement of farm labor with machinery that could perform tasks more efficiently.

Aside from mechanization, the revolution also led to the increased adoption and utilization of synthetic inputs such as fertilizers and pesticides, and improved irrigation methods similar to those of Asia. Ammonium nitrate was used for consistent crop production and soil fertility enhancement (Tilman et al., 2002).

The transformation of European agriculture ensured food security in post-World War II Europe and contributed to the region's economic growth Despite the success in its implementations, these changes resulted in significant environmental problems, including water pollution, biodiversity loss, and soil degradation (Matson et al., 1997; Tripathi et al., 2020).

In Central Europe, the GR coincided with industrial reorganization and rapid mechanization, particularly during the socialist period and the subsequent transition to market economies (Bański and Mazur, 2021; Harcsa et al., 2002). In countries such as Hungary, Poland, and former Czechoslovakia, agricultural modernization was closely tied to state-led industrial strategies that emphasized large-scale machinery use and centralized policies and supply systems (Schlett, 2015). State farms, especially in East Germany under the German Democratic Republic, operated on a substantial scale, enabling the adoption of irrigation, harvesting, and mechanized plowing technologies (Koester, 1999). Increased labor productivity and wide-area production of cereals were the outcome of this industrial model, positioning agriculture closely with industrial principles. However, this led to the rural labor displacement, causing rural community restructures and demographic shifts.

After the 1989–1991 political transformations, several Central European countries looked to the complex task of privatization of lands, modernizing aging machinery, and restructuring former collective farms while integrating into the agricultural framework of the European Union (Swinnen and Mathijs, 2018). Thus, agricultural industrialization was not merely a technological shift but a deeply institutional transformation shaped by post-socialist transition dynamics and political economy. Central and Eastern Europe shared a common socialist agricultural framework but diverged in outcomes due to differences in institutions, farm structure, and reform paths (Lerman, 2001; Csáki and Lerman, 1994; Bański and Mazur, 2021).

2.3.1.2 Adoption of high-yield crop varieties

The GR saw the introduction of HYVs, which brought about significant changes in European agriculture, boosting productivity and shaping farming practices. During the 1960s and 1970s, countries in Europe, including France and Germany, adopted HYVs of corn and wheat, including semi-dwarf varieties. This resulted in greater yields and alterations in crop management practices (Harwood, 2012). Nonetheless, the swift uptake of these varieties resulted in diminished agricultural biodiversity, increasing crops' vulnerability to diseases and pests. This transition to monoculture farming required increased use of agrochemicals like fertilizers and pesticides in order to sustain the health of crops and their yields.

Due to the environmental and economic challenges posed by intensive chemical use, European policymakers began advocating for sustainable agricultural practices. To reduce adverse effects and simultaneously uphold productivity, approaches like organic farming, crop rotation, and integrated pest management (IPM) were implemented (Lechenet et al., 2017; Boix-Fayos and De Vente, 2023). The goal of these changes in policy was to strike a balance between the advantages of HYVs and the necessity of eco-friendly farming practices and environmental preservation. In summary, the introduction of HYVs during the GR was crucial in molding European agriculture, offering opportunities and posing challenges that still affect contemporary farming practices.

2.3.1.3 Role of the Common Agricultural Policy (CAP)

One of the policies reinforced in 1962 to industrialize Western European agriculture is the Common Agricultural Policy (CAP), targeting the modernization of agricultural methods, the stabilization of food prices, and a rise in self-sufficiency. Farmers were able to invest in modern agricultural technologies and infrastructure due to financial subsidies provided by CAP (Fusco, 2021; Scown and Nicholas, 2020). This led to a shift in European nations from subsistence farming on a small scale to agribusiness models that were large-scale and industrialized. For instance, in France and Germany, farms expanded in size due to land consolidation initiatives aimed at promoting efficiency and competitiveness in agricultural production.

Since its inception in 1962, the CAP has served as a foundational element of the European Union's (EU) agricultural policy, with objectives including farmer support, food supply stability, and sustainable rural development promotion. The CAP was crucial in changing European agriculture during the GR, which lasted from the 1960s to the 1980s and was marked by major boosts in agricultural productivity due to the use of HYVs and contemporary farming methods. A key aim of the CAP was to raise farming productivity in order to guarantee a stable supply of reasonably priced food. The policy aimed to provide farmers with income support, allowing them to invest in GR technologies and practices in order to achieve this. This backing made it easier to adopt HYVs, synthetic inputs, and advanced irrigation methods on a large scale, resulting in significant increases in crop yields throughout Europe as summarized in Fig. 4. Moreover, the CAP introduced market measures aimed at stabilizing prices and guaranteeing farmers fair compensation, thereby promoting the uptake of innovative farming practices (European Commission, 2025).

Nevertheless, the agriculture intensification prompted by the CAP also resulted in environmental challenges. The augmented application of synthetic inputs played a role in soil degradation, water pollution, and the decline of biodiversity. In light of these issues, the CAP was reformed several times with the aim of fostering environmental sustainability.

In the past few years, the CAP has kept developing in order to deal with current challenges. With a focus on sustainability and in alignment with the European Green Deal, the CAP 2023–2027 framework introduces enhanced conditionality for farmers, mandating compliance with stricter

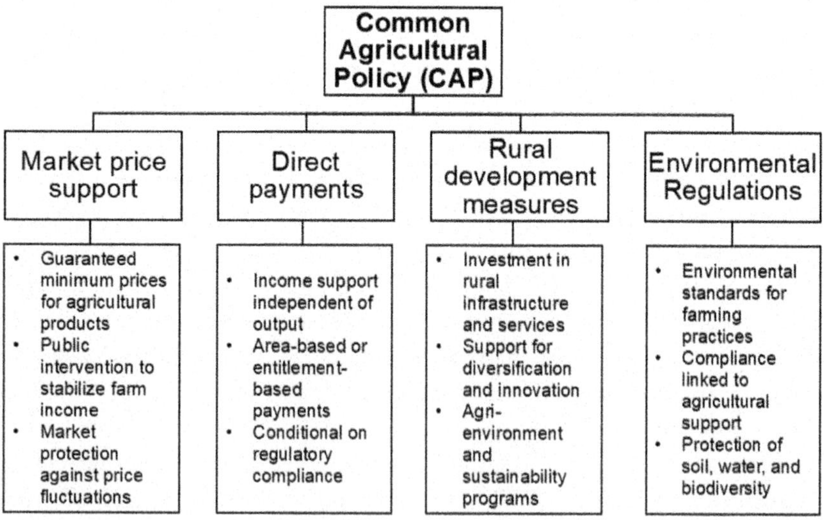

Fig. 4 The role of the Common Agricultural Policy in European Green Revolution *(original interpretation by the authors).*

environmental and climate standards to qualify for subsidies. Additionally, it supports eco-schemes that provide rewards for farming practices that are advantageous to the environment, like preserving permanent grasslands and promoting biodiversity. Even with these attempts, criticisms continue regarding the CAP's effectiveness in meeting environmental objectives, particularly its capacity to reconcile agricultural productivity with ecological sustainability (Zieliński et al., 2025; Sotte and Vergamini, 2025).

To summarize, the CAP offered vital support during the GR that allowed European farmers to embrace HYVs and modernize their agricultural practices, resulting in enhanced productivity and food security. This intensification, however, also resulted in environmental challenges, leading to ongoing reforms focused on incorporating sustainability into European agriculture. The evolution of the CAP illustrates an ongoing attempt to reconcile the goals of agricultural development with those of environmental preservation (Daugbjerg and Swinbank, 2006; Pe'er et al., 2019).

2.3.1.4 Expansion of irrigation and land reforms

The GR in Europe included the enlargement of irrigation systems and the establishment of land reforms. The aim of these changes, which were largely backed by the CAP, was to modernize European agriculture and boost food production. The GR's effects were most pronounced in areas with substantial agricultural sectors, like southern Europe, where water scarcity presented a major obstacle to crop production. To tackle these challenges, it was crucial to expand irrigation infrastructure—such as dams, reservoirs, and canal systems – to guarantee the year-round cultivation of water-intensive crops like cereals, vegetables, and fruits, even in arid areas (Angelakıs et al., 2020).

During this period, irrigation practices in Mediterranean Europe were quite sophisticated. The GR led to significant investments in modern water management systems in areas such as Spain and Italy, which had depended on traditional irrigation methods that were often ineffective. The implementation of sophisticated irrigation networks, including drip and sprinkler systems, facilitated more effective water use, minimized waste, and allowed farmers to optimize their yield potential. In areas such as Almería, where dependable water supplies backed intensive greenhouse agriculture, this advancement in technology was particularly important (Caparrós-Martínez et al., 2020). Consequently, the agricultural productivity in these regions increased significantly, ensuring a reliable supply of crops for domestic use and export markets. The development of irrigation helped lessen the impact of water scarcity, but it also caused concerns about the sustainability of water resources. In certain instances, the over-extraction of groundwater and the excessive use of irrigation systems resulted in the salinization and depletion of nearby aquifers (Angelakis et al., 2020).

Moreover, land consolidation resulted in the displacement of smaller family farms, impacting rural communities and traditional agricultural practices. In certain areas, smallholder farmers found it difficult to compete against larger, mechanized operations. This resulted in a decrease in the number of farms and an increase in urbanization. As a response to these challenges, European agricultural policies (including the CAP) started to incorporate more sustainable practices in the late 20th and early 21st centuries.

Simultaneously with irrigation expansion, many European countries implemented land reforms aimed at improving the efficiency and productivity of farms. A crucial aspect of these reforms was land consolidation, which involved merging small, fragmented holdings into larger, more manageable farms. This enabled the use of mechanized farming tools, which enhanced efficiency and production. As an example, in Italy and France, initiatives aimed at land consolidation contributed to lowering the expenses related to small-scale farming and facilitated the widespread use of contemporary machinery such as tractors and combine harvesters (European Parliament, 2016; Baldock et al., 2001). Due to the consolidation, economies of scale emerged that allowed larger farms to take advantage of improved access to resources like credit, technology, and expertise. This made them more competitive in the global agricultural market.

2.3.1.5 Increased food production and self-sufficiency

From 1960 to 2000, European agriculture experienced a major transformation, characterized by heightened food production and advances toward self-sufficiency. This era, frequently linked to the GR, brought forth HYVs, improved irrigation methods, and a greater reliance on synthetic fertilizers and pesticides—all of which played a role in boosting agricultural productivity. A key advancement of this time was the creation of the CAP in 1962. The objectives of the CAP included enhancing agricultural productivity, providing farmers with a fair standard of living, market stabilization, guaranteeing food availability, and ensuring that consumers had access to food at acceptable prices. The CAP encouraged farmers to boost production through a system of subsidies and price supports, resulting in significant surpluses of various agricultural commodities (Rizov et al., 2013;

Giuliani and Baron, 2025). Yield improvement was greatly facilitated by the adoption of modern farming practices, such as mechanization and synthetic inputs. In Spain, the state encouraged the farming of high-yield crops like barley and wheat to reduce dependence on imports, which also resulted in wheat surpluses (Martínez-Moreno et al., 2024). According to González de Molina et al. (2017), Spain produced around 16.5 million tons of cereals in the 1990s, compared to 7.6 million tons in the 1960s. According to FAO (2025) data, Spain's output rose from 7.5 million tons in 1961–18.8 million tons in 1990, a 151 % increase.

Figures suggest that France's cereal production increased by 163 % between 1961 and 1990, from 20.9 million tons to 55 million tons. Germany increased its production by 119 %, from 17.2 million tons to 37.6 million tons. In comparison to Western Europe, the USSR's average yields per hectare were lower, notwithstanding its enormous output in absolute terms (1961 – 118.9 million tons, 1990 – 209 million tons, a 76 % increase). Among the four countries mentioned, France recorded the largest absolute increase in cereal production, reflecting rapid yield gains and intensification under postwar agricultural modernization policies.

Similar to this, Germany's agricultural industry saw significant structural change after World War II, marked by mechanization, farm consolidation, and productivity growth, all of which helped to sustainably raise grain output. Postwar efforts concentrated on attaining self-sufficiency in food production. The number of farms in West Germany declined from 1.6 million in 1950–630,000 in 1990, indicating a trend towards consolidation and modernization. During the socialist regime in East Germany, collectivization resulted in the formation of large agricultural production units. These units were subsequently restructured after German reunification. Due to these changes, efficiency and productivity in German agriculture increased (Hill et al., 2004).

2.3.2 Challenges of GR in Europe

2.3.2.1 Persistent overproduction and market distortions

Yield-enhancing inputs, mechanization, and improved crop varieties significantly increased sugar, grain, dairy, and livestock production. However, the guaranteed pricing, intervention buying, and export subsidies provided by frameworks such as the CAP shielded producers from typical market signals (Tracy, 1989). As a result, production frequently surpassed domestic demand, resulting in significant surpluses that necessitated costly storage, disposal, or subsidized exports (Gardner, 2006). These distorted price formations impaired international trade competitiveness and generated criticism and negotiations in global trade, notably within the World Trade Organization (WTO) framework. The fiscal cost of managing surpluses put increasing strain on European budgets, forcing a series of policy reforms aimed at lowering intervention prices, establishing production quotas, and separating subsidies from output (Daugbjerg and Swinbank, 2009). Thus, throughout Europe, the GR's technological successes generated paradoxical abundance, resulting in economic inefficiencies and governmental distortions in the place of concerns about food scarcity.

2.3.2.2 Environmental externalities from high-input intensive farming

The GR in Europe was characterized by capital-intensive, high-input farming systems that significantly increased agricultural productivity but also brought substantial environmental externalities. Widespread use of synthetic inputs contributed to soil degradation, nutrient imbalances, and contamination of surface and groundwater bodies through nitrate leaching and agrochemical runoff (Uphoff, 2013). Even with these improvements, the escalation of modern agricultural practices prompted concerns about the environment. Due to the intensive application of synthetic fertilizers and pesticides, the soil became degraded, water sources became polluted, and biodiversity was diminished. In recent years, attempts to revise the CAP sought to tackle these problems by integrating environmental factors into agricultural policies (Langlais, 2023). Furthermore, the intensive livestock production facilitated the strains on ecosystems, leading to increased ammonia emissions and manure excesses in animal farming-concentrated regions (Menzi et al., 2010). Land consolidation, mechanization, and monocropping simplification, accelerated biodiversity loss (Kremen and Merenlender, 2018).

These trends were closely linked to the CAP production incentives, which historically rewarded expansion more than environmental management. Intensive and high-input production secured food supply, but it simultaneously imposed ecological costs that eventually reshaped agricultural policy toward environmental accountability and sustainability. The accumulation of environmental impacts over the years prompted public concerns from the 1980s onward, eventually leading to regulatory responses such as agri-environmental measures and the Nitrates Directive (van Grinsven et al., 2012; Buller et al., 2017). It was in the 2000s that CAP had reforms made to improve market competitiveness, food safety and quality, environmental sustainability, and rural area development (Bernini and Galli, 2024).

2.3.2.3 Fiscal pressures from long-term agricultural subsidy systems

Due to the long duration of the agricultural subsidy systems of the GR in Europe, several fiscal pressures emerged which became a structural consequence. The design of the European Economic Community (currently known as the European Union) policy regimes was to set guaranteed minimum prices, purchasing interventions, export refunds, and subsidies for inputs in order to ensure food security and stabilize farmer incomes. While there was significant expansion in innovation and HYVs, these support policies produced fiscal obligations and increasing strain on public budgets (Ackrill, 2000; Bureau and Swinnen, 2018). The agricultural expenditure accounted for a substantial share of the European budget in the 1980s-1990s, which raises issues over distributed equity and long-term sustainability. The MacSharry reforms of 1992 were catalyzed in order to mitigate the fiscal pressures, reduce market distortions, reorient support toward rural development, control expenditure growth, and promote environmental goals (Ackrill, 1992; Levy and Stancich, 1997). In summary, while subsidy systems initially underpinned agricultural modernization and rural stability, their aggregate fiscal burden prompted structural adjustments to Europe's agricultural policy framework.

2.4 Green Revolution in North America

2.4.1 Successes of GR in North America

2.4.1.1 Widespread adoption of HYVs

The introduction and spread of HYVs were key to the GR's effects in North America. The focus of researchers was on developing crop varieties that demonstrated greater yields, resilience to disease, and adaptability across various climate conditions. As a result of these endeavors, semi-dwarf wheat varieties were developed that were shorter and less likely to lodge, thus aiding in the increase of grain production. Likewise, progress in rice breeding led to the development of varieties with quicker maturation times and greater yields, thus bolstering food security (Jobson et al., 2019; Jiao et al., 2024; Ferrero-Serrano et al., 2019; Pingali, 2012). Aside from improvements in wheat varieties, hybrid corn also expanded in North America as early as the 1930s-1940s, expanding worldwide during the 1960s, enhancing yields and providing raw materials for ethanol industries and livestock (Pingali, 2012). The adaptation of these HYVs was accompanied by considerable changes in agricultural practices. To make the most of the new crop varieties, farmers became more reliant on synthetic fertilizers and pesticides. Moreover, the development of irrigation systems and mechanization made it possible to cultivate larger areas of land and enhanced the efficiency of planting and harvesting (Hamdan et al., 2022; Chi et al., 2021).

HYVs had implications that went beyond the mere increase of yields, which include changing patterns of land use. Due to the increased productivity per unit area, there was less necessity for agricultural expansion into natural ecosystems, which helped in conserving biodiversity and curbing deforestation (Stevenson et al., 2013). However, Peterson (2025) and Chen et al. (2019) note that the agriculture industry underwent structural change despite yield intensification. Rising productivity was accompanied by significant land consolidation and farm expansion in countries like the United States and Canada, where larger enterprises absorbed land and smaller farms left production. Therefore, intensification hastened the vertical restructuring of farmland ownership and scale while also moderating horizontal expansion into forests and grasslands. The GR's focus on HYVs had economic ramifications, resulting in higher farming incomes and aiding rural development. Nevertheless, the advantages were not spread out evenly. Farmers with large-scale operations and access to resources and capital were in a better position to implement the new technologies, whereas small-scale farmers frequently encountered difficulties in obtaining the required inputs and infrastructure (Delvaux et al., 2020).

2.4.1.2 Mechanization in North America

Farmers in North America were able to cultivate larger areas more efficiently due to the adoption of tractors, combine harvesters, and other specialized equipment. The transition from manual work to mechanized operations not only accelerated agricultural processes but also helped to alleviate labor shortages, enabling timely planting and harvesting (United States Department of Agriculture Economic Research Service, 2025). The expansion of monoculture practices, especially for staple crops such as wheat, corn, and rice, was facilitated by mechanization, resulting in substantial yield improvements (Hamilton et al., 2022). Mechanization had a deep impact on the economy. Farmers who operated on a large scale and had access to capital were found in a better position to invest in costly machinery, which allowed them to realize economies of scale and enhance their profitability.

2.4.2 *Challenges of GR in North America*

2.4.2.1 The Downsides of North American Mechanization

While there are significant economic advantages produced by mechanization, the small-scale farmers frequently encountered financial obstacles that hindered their ability to adopt mechanization, thereby worsening economic inequalities in the agriculture industry. This gap underscored the necessity for policies aimed at ensuring fair access to agricultural innovations and assistance for smallholder farmers (Duffy, 2009; Zhang, 2024; Dhillon and Moncur, 2023). As mechanization intensified, environmental considerations also emerged. The greater reliance on heavy machinery led to soil compaction and erosion, sparking conversations about sustainable farming methods (Keller et al., 2019).

North American dependence on machinery and monocultures sparked worries regarding environmental sustainability and soil health (Kaur et al., 2024). To reduce the ecological effects of mechanized farming, conservation tillage and precision agriculture methods were investigated (Idowu and Angadi, 2013; Jug et al., 2025; Getahun et al., 2024). Furthermore, mechanization also encouraged intensive irrigation use, which accelerated the depletion of groundwater resources, particularly in regions relying on the Ogallala Aquifer and the states it spans (Chen et al., 2023). Thus, the intensive irrigation reliance led to a notable supply rift, where the rates of groundwater extraction exceeded natural recharge rates, resulting in substantial long-term economic and ecological challenges (Hornbeck and Keskin, 2014).

2.4.2.2 Commodity overproduction and market volatility

In North American farms, structural overproduction was evident in wheat, corn, soybeans, and dairy. Output increases led to substantial surpluses, driving down prices. This pressed farm incomes despite increasing productivity (Winders, 2024; Katsikis, 2023). The incongruity became clear: individual farms adopted the latest technologies to improve yields and stay competitive, but with the downside of reduced prices (Bonnen and Schweikhardt, 1998).

In order to offset declining prices, producers increased their production area or scaled up their output, further aggravating the oversupply (Alston et al., 2009; Trostle, 2011). Furthermore, mechanization amplified this loop by enabling rapid scaling, reducing marginal production costs. The 1980s farm crisis demonstrated the vulnerability of highly capitalized and indebted agricultural systems to falling prices and increasing interest rates

(Barnett, 2000). Many who had expanded during periods of high prices faced bankruptcy when export demand weakened and credit conditions were tightened and reinforced (Scott, 2000).

Policies sought to insulate farmers from these impacts. Farm bills in the United States helped in managing instabilities by implementing extensive government support systems such as price floors, crop insurance, supply management programs, and deficiency payments (Johnson and Monke, 2008). Quota-based systems in Canada were implemented and intended for the dairy and poultry sector. Although these responses in the two countries alleviated extreme losses. they exacerbated fiscal exposure and production incentives, making attempts to address structural imbalances more difficult (Rajsic, 2025; Chow and Sheldon, 2023). Therefore, the GR in North America meant technology-driven abundance but did not eliminate economic insecurity and hurdles. Instead, high productivity entrenched overproduction cycles, policy dependence, and volatility, at high cost to the national treasuries.

2.5 Green Revolution in Latin America

2.5.1 Successes of GR in Latin America

2.5.1.1 HYVs, expansion of inputs, institutional support, and exportation

According to (Hurt, 2020; Lorek, 2022), the GR signified a transformative phase in Latin American agriculture, mainly through the introduction of HYVs of staple crops like wheat, rice, and maize. This transformation in agriculture began in the middle of the 20th century, and Mexico was crucial to this process as an early adopter and innovator. The semi-dwarf wheat varieties developed by Norman Borlaug in Mexico were revolutionary and established the foundation for the GR globally, which resulted in major yield increases.

Among the institutions that facilitated the dissemination of HYVs were the International Maize and Wheat Improvement Center (CIMMYT) in El Batán, Mexico, and the International Rice Research Institute (IRRI), in Los Baños, the Philippines. Both are among the public international agricultural research centers supported by the Consultative Group on International Agricultural Research (CGIAR). These institutions were instrumental in promoting new crop varieties and breeding suited to diverse climatic regimes, thus enhancing agricultural productivity and food security in Latin America (Lorek, 2023). They worked closely with national public agricultural research centers, such as Mexico's INIFAP (National Agricultural, Livestock, and Forestry Research Institute), Brazil's

EMBRAPA (Brazilian Agricultural Research Corporation), and Colombia's ICA (Colombian Agricultural Institute). The use of HYVs was accompanied by the implementation of modern farming techniques in these three countries and elsewhere in the region. The shift to HYVs and synthetic inputs not only boosted crop yields but also led to a more intensive and mechanized agricultural practices (Picado, 2022).

Another important facet of the GR in Latin America was the transition to export-oriented agriculture. Mechanization and HYVs improved output capacity, allowing nations to concentrate on export-oriented cash crops like sugarcane and its byproducts (including ethanol) in Brazil (Martines-Filho et al., 2006; Rogers, 2015; Crago et al., 2010); coffee and cut flowers in Colombia, which also became a major producer of coca and opium and exporter of illicit drugs, such as cocaine and heroin (Hair, 2019; Mendez, 1991; Vargas, 2002; O'Connor, 2009; Mejia and Restrepo, 2013); and soybeans in Argentina and Brazil (Klein and Luna, 2021). Infrastructure advancements and legislative frameworks that promoted large-scale commercial farming, frequently at the expense of smallholders, bolstered this export-oriented strategy. Global market needs and the Cold War's geopolitical backdrop also had an impact on the emphasis on exports, as they pushed Latin American nations to sell their agricultural products in order to join the global economy (Williamson, 2024; Picado, 2022).

Although the GR had huge benefits for Latin America, there were winners and losers. Large landowners and commercial businesses were better positioned to take advantage of the new technologies and market opportunities. The marginalization or displacement of many small-scale farmers exacerbated social tensions and rural inequality. Furthermore, criticisms of the GR's long-term viability grew as the environmental effects of intensive farming methods, such as soil erosion, water pollution, and biodiversity loss—became more obvious (Reveles, 2012; Pingali, 2012).

2.5.1.2 The American Agronomist Norman Borlaug

Due in large part to his pioneering work in Mexico in the 1940s, American Agronomist Norman Borlaug was instrumental in launching the GR in Latin America (Ullah et al., 2024). George Harrar, Borlaug's boss, opposed this expansion. Aside from the additional expenses of doubling the work, Borlaug's idea violated an agronomic concept that has subsequently been shown to be false. It was thought that seeds need a rest period following harvest in order to conserve energy for germination before planting. Borlaug quit after Harrar disapproved his plan. A project visitor named

Elvin Stakman helped to defuse the crisis by persuading Harrar to permit the double wheat season and Borlaug rescinded his resignation. He developed "shuttle breeding," leading to the creation of semi-dwarf wheat varieties with higher yields and less tendency to lodge, which is a common problem in wheat with fragile stems (Miller, 2012). By the 1960s, Mexico had become self-sufficient and eventually an exporter of wheat due to the success of Borlaug's breeding program. His efforts also led to the establishment of CIMMYT, which continues to serve as a center for corn and wheat research and global distribution of GR innovations (Braun, 2011; CIMMYT, 2011, CIMMYT, 2016). In 1970, Borlaug received the Nobel Peace Prize for his efforts to combat world hunger (Nobel Prize Outreach, 2026). In 1986, he established the World Food Prize as the "Nobel Prize for Food and Agriculture" (World Food Prize Foundation, 2026).

2.5.1.3 Global market integration and GMO introduction

In addition to increasing agricultural output, the GR in Latin America set the stage for more market integration. Countries like Brazil, Argentina, and Mexico transitioned from subsistence farming to export-oriented agribusiness between the 1960s and the early 2000s as they progressively matched their agricultural policies and products with the demands of global trade. The advent of technologies such as automation, synthetic inputs, and high-yielding crop types, made this shift easier (Evenson and Gollin, 2003). With the introduction of genetically modified organisms (GMOs) in the 1990s and early 2000s, the modernization expansion accelerated, especially in nations like Argentina and Brazil that emerged as world leaders in the production of transgenic corn and soybeans (Qaim and Traxler, 2005). North American agriculture acted as a primary technological driver for Latin America, although this is not exclusive to Latin America, but is also the case in parts of Asia and Africa as well. In the USA, patents and sales agreements with farmers safeguard the sale and utilization of RoundupReady® (RR) technology (which produces herbicide-tolerant varieties), whereas Argentina does not employ either type of intellectual property protection. Unlike in the United States, RR soybeans in Argentina can be easily sourced from suppliers besides Monsanto, and farmers in the country incur a modest price increase. Argentine farmers have the legal right to utilize seeds saved from their own harvests. The distribution of counterfeit seeds, including sales in Brazil and Paraguay, is prevalent (Qaim and Traxler, 2005; Traxler, 2006). All these developments allowed Latin American agriculture to compete in increasingly liberalized international markets.

Critics pointed to a number of issues. They noted that the adoption of GMOs frequently favored large-scale commercial farms with access to cash and international markets, excluding smallholders who lacked the financial means and infrastructure necessary to take advantage of such technology (Altieri and Rosset, 2002). Moreover, global firms that developed GMO seeds, such as Monsanto, sought to patent them, causing legal battles and concerns about farmers' right to save and reuse the seeds (Kloppenburg, 2005). Concerns about the environment and human health also emerged, such as worries about the loss of biodiversity and the contamination of local species (Otero and Lapegna, 2016).

2.5.2 Challenges of GR in Latin America

2.5.2.1 The socioeconomic and ecological downsides of HYVs

While successes due to HYVs were profound, problems such as social and environmental challenges (e.g., soil degradation and water contamination) were encountered due to synthetic input usage. Furthermore, the region engages in practices similar to those in North America, such as monoculture, thereby reducing agricultural biodiversity. Smallholder farmers often faced hurdles such as limited access to technical support and credit (Hurt, 2020). Although HYVs are designed for increased productivity, they are embedded in broader technological packages in order to become more efficient (Das et al., 2023; De Janvry et al., 2005). Their performance thrives and they are highly responsive to these inputs as introduced by Norman Borlaug - shifting from low to high input farming systems, affecting farmer's financial resources, creating a technology dependency (Sarmah et al., 2025). In commodity and marketing contexts, this dependency exposed the farmers to in financial risks and contributed to rising indebtedness (Green, 2003; Cypher, 2010; Altieri and Toledo, 2011). Furthermore, various studies support the idea that genetic homogeneity, monocropping, and high-input production encouraged by HYVs narrowed and contracted the genetic base available for future breeding, raising concerns about sustainable adaptive capacity in changing climates and increases in crop vulnerability (Snowdon et al., 2021; Singer et al., 2021; Mercer and Perales, 2010; Cooper et al., 2001; Evans and Evans; 1996). While there are successes in increasing caloric output in Latin America, their adoption reformed farming practices toward greater input intensity and ecological simplification (Pingali, 2012). As in other regions, the distribution and benefits of HYVs was uneven. Farmers with access to credit, land security, irrigation, and extension services were better positioned to achieve productivity gains, while the smallholders often faced constraints in crop and technology adoption, making HYVs socioeconomically selective (Picado, 2022). Thus, the GR made agriculture more deeply embedded with a capital-intensive and input dependent systems that compromise long-term sustainability.

2.5.2.2 Adapting HYVs to diverse agroecologies

Norman Borlaug's challenge during his tenure at CIMMYT was to tailor the corn and wheat varieties in various agroecological zones, such as highlands, irrigated valleys, and semi-arid regions, particularly in Mexico (Borlaug, 1986; Rajaram et al., 2002; Reynolds and Borlaug, 2006a; Reynolds and Borlaug, 2006b). One of the techniques employed and initiated by Borlaug and his colleagues in the challenging agroecological zones is shuttle breeding, which produced photoperiod-insensitive and disease-resistant varieties, requiring frequent experimentation (Borlaug, 1964; Phillips, 2013; Ortiz et al., 2007). The introduction of hybrid corn varieties required harmonizing high yield potential with farmers' preferences for traits such as drought tolerance, post-harvest quality, taste, and suitability for traditional foods. Furthermore, genetic erosion poses a threat in the conservation of landrace varieties (i.e., locally adapted varieties), which is important to climate change adaptation (Guzzon et al., 2021; Edmeades et al., 2017). Cultivation constraints were also experienced in the highland Andean zones due to cold temperatures, which limit imported hybrid performance (Sevilla, 2006; Garcia et al., 2025). Therefore, ecological calibration as well as genetic improvement presented a difficulty. In low-input, rainfed, or stress-prone areas, varieties bred for greatest nitrogen- and irrigation-responsiveness frequently performed poorly. This resulted in eographical disparity in productivity gains: smallholder-dominated, ecologically marginal areas experienced modest improvements, whereas irrigated zones maximized their output. This demonstrates the importance of analyzing gene-environment interaction and its implications to diverse agroecologies (Zaidi et al., 2026).

3. Achieving caloric sufficiency in the GR and its dietary implications

The GR's primary goal was simple but historically pressing: to prevent widespread starvation by expanding the supply of energy from staple crops. The rapid population expansion in Asia and Latin America in the middle of the 20th century heightened concerns that agricultural output would not be able to meet demand. The ability of a country to consistently produce enough staple grains to meet caloric sufficiency emerged as a key policy goal. Sukhatme (1961) estimated that cereal production would need to more than double and animal product output would need to increase six-fold by the year 2000 to ensure an adequate level of nutrition for the population. However, the direction of GR centered on calories from cereals, and did not focus on animal-sourced protein (Eliazer Nelson et al., 2019; Sahu, 2020; Pingali and Abraham, 2022; Rashid, 2019; Roy et al., 2017). Furthermore, while it is not directly causative, the GR's cereal-centric paradigm reinforced agricultural structures that later aligned with grain-based dietary models such as the USDA Food Pyramid of 1992 (Nestle, 1998). Several countries adopted the food pyramid, including Chile, Panama, Mexico, and the Philippines (Painter et al., 2002; Philippines, no date).

Over time, several countries revised their dietary guidelines. In the United States, for example, the Department of Agriculture promoted the MyPlate guidelines from 2011 to 2026. Cereal-grains remain dominant, but these recommendations put greater emphasis on the incorporation of whole-grains (Santillán-Urquiza et al., 2017; Marcus, 2013). Globally, the emphasis has focused on achieving caloric sufficiency through staple cereal crops, but rather than leading to overall nutritional adequacy, this approach has instead, contributed to the "double burden of malnutrition", turning nutrient-dense and protective foods into nutrient-poor foods, thereby increasing obesity prevalence (Milani et al., 2022). Sekhri and Shastry (2024) also discovered that men born following the introduction of HYVs had higher rates of diabetes in places where new staple varieties were extensively adopted. One of the products of the GR is inexpensive and calorie-dense ultra-processed foods (UPFs) (Williams et al., 2025); while these have helped to avoid the dreadful Malthusian projections of population growth outstripping food production, the UPFs have received a lot of criticisms because of their various detrimental effects on human health (Hall, 2023; Marino et al., 2021).

In 2026, the emphasis on restricting UPFs is a welcome addition. Nonetheless, an array of the newest US Dietary Guidelines' features give rise to serious issues (Revenga Frauca and Soriano del Castillo, 2026). Most notably, since protein intake already surpasses needs in economically developed nations, the prescription to increase intake of protein sourced from animals seems unreasonable. Additionally, it ignores long-standing research showing that eating plenty of red and processed meat raises the risk of certain types of cancers and cardiometabolic diseases. On the contrary, the plant-based dietary patterns, particularly the Mediterranean Diet, which has been demonstrated to have a lower incidence of cardiovascular disease and type 2 diabetes, are strongly supported by the scientific literature promoting health (Giosuè et al., 2026).

In summary, while caloric-sufficiency was achieved by the GR through staple cereals, problems were identified. Through time, several countries have kept changing their dietary guidelines as new health problems arise, but some of the advice remains problematic.

4. A gendered perspective of the GR: Impacts, equity, and progress

The evolution of the GR transcends mere gains in yield and macroeconomic shifts; it fundamentally reshaped gender dynamics and labor divisions within the agrarian social construct. The GR's labor and structural consequences were not gender neutral, even though yield increases, calorie sufficiency, and macroeconomic growth are commonly used to evaluate it. In agriculture, technological advancements changed labor demand, resource availability, and marginalization patterns in ways that frequently disproportionately impacted women. According to Sobha (2007), women were mainly peasants in the third world countries, having a close relationship with the forest and the trees. While men's perspective and prospect about the forest is about commercialization, women view it as a source of necessities for the household. In India, most workers involved in agricultural labor are male; only one out of four is a woman. Despite their limited access to production inputs, services, and information, and very limited control over their income, women played a significant role in agricultural development (De, 2015). The findings of D'Agostino (2017) suggest that microdata and census outcomes indicate that women have shifted their workload from wage work to unpaid work on their own farms and in their homes. Additionally, they have increased their time spent on home production. Furthermore, female farmers were given little attention, and the agricultural research and extension systems are more inclined to males. Men dominated the external economic domain, while women dominated the household domain, particularly in Western models of the division of labor (Reijntjes et al., 1992).

The gender biases during the GR are closely associated with the HYVs. Men participate in local social and economic institutions more frequently than women do, and this is positively correlated with new HYVs, particularly in wheat (Jaleta et al., 2023). Complementary inputs, irrigation, credit, and extension services were needed for HYVs. However: Men controlled the majority of land titles; formal land ownership was frequently needed as collateral for credit systems; male "household heads" were the focus of extension programs; and women were frequently shut out of decision-making processes and input access in situations where they maintained plots within household systems or engaged in informal land cultivation (Tsige et al., 2020; Kang et al., 2020; Ankrah et al., 2020). Benefits were therefore not equally dispersed across gender lines, even in cases where productivity improved. Additionally, besides social susceptibilities, women also faced biological susceptibilities (Cycon, 2013). Because of traditional gender roles, household tasks, including cleaning and spraying equipment and their spouses' pesticide-soaked clothing, storing pesticides, or discarding containers, exposed women to more pesticides. In Vietnam, girls were more likely to get exposed to pesticides than boys when cleaning spraying tanks. Furthermore, women are more susceptible than men in terms of health and reproductive issues due to women having a higher level of hormonally sensitive tissues (Heinrich Böll Foundation, 2025). According to Cycon (2013), preterm delivery, spontaneous abortion, and breast milk contamination are the most frequent reproductive effects of exposure to DDT.[2] Its use in the GR was amplified even though it existed in the 19th century.

In recognition of women's role in agriculture, international institutions such as FAO (Raney et al., 2011), IFPRI (Jackson, 2005), and CGIAR (van der Burg, 2019) from the 1990s onward began documenting women's substantive role in the agricultural development. Rather than being viewed as a demographic footnote, gender started to be acknowledged as a key analytical aspect in research and policymaking. The disparities mentioned above led to gender-sensitive programs by governments and development agencies, encompassing extension services, training, and targeted loans for

[2] DDT (dichlorodiphenyltrichloroethane) is an organochloride pesticide used extensively in the past for insect control, but was banned in the U.S. in 1972 due to its high persistence and environmental damage, although still used to control malaria-carrying mosquitoes (DDT, (1979) and its derivatives (United States Environmental Protection Agency, 2025; International Programme on Chemical Safety, 1992)).

women farmers. Thus, it is only imperative to incorporate the role of women's empowerment for productivity, gender equitability, and agricultural development (Alkire et al., 2013; Botreau and Cohen, 2020; Quisumbing et al., 2022; Anik and Rahman, 2021).

5. GR influences and impacts on soil health and insect biodiversity

While there have been significant impacts of the GR in the past decades, there is a serious environmental tradeoff. The environmental and economic growth effects must be perceived in a circular system as explained in Fig. 5. The ability of soil to continue to support plants, animals, and people as a viable living ecosystem is known as soil health. While some activities resulted in agricultural prosperity during the GR, one of the problems faced, consequently, is soil degradation. It links agricultural and soil research to policy, stakeholder demands, and sustainable supply-chain management. While crop production was the primary focus of soil evaluations in the past, soil health now encompasses the function of soil in water quality, climate change, and human health. Despite an increasing understanding of the significance of soil biodiversity, chemical markers continue to dominate the quantification of soil health due to a lack of efficient techniques and a lack of functional knowledge (Lehmann et al., 2020).

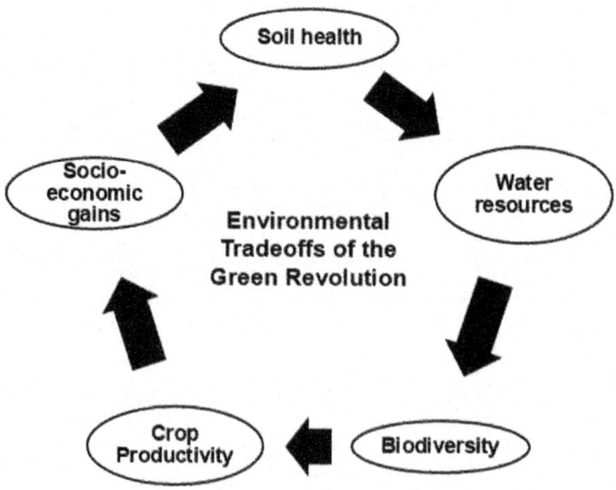

Fig. 5 Environmental tradeoffs of the GR *(original interpretation by the authors).*

The extensive use of synthetic fertilizers during the GR (1960–2000) negatively impacted soil health. Overuse of phosphate, potassium, and nitrogen fertilizers upset soil microbial populations, which decreased biodiversity and changed nutrient cycling mechanisms. Additionally, prolonged use of these fertilizers led to soil acidification, a decline in organic matter, and weakened soil structure, which over time made the soils less fertile and more prone to erosion (Ogidi and Akpan, 2023; Pahalvi et al., 2021).

Nitrogen fertilizers, especially, contribute to soil acidification. In many areas, soils became acidified due to the prolonged and excessive use of synthetic nitrogen fertilizers, especially urea and ammonium sulfate (Guo et al., 2010). Under acidic conditions, the availability of vital nutrients like phosphorus decreases, while the solubility of harmful substances such as aluminum and manganese increases—this adversely impacts plant root growth and microbial communities (Lindsay and Walthall, 2020; Miguel et al., 2017; Rahman et al., 2018).

The practice of continuous monoculture cropping associated with the GR, along with minimal return of organic matter to soils worsened nutrient imbalances. In alkaline soils, the excessive application of phosphate fertilizers can lead to the precipitation of insoluble phosphorus compounds, which reduces nutrient availability even when application rates are high (Zhao et al., 2009). These chemical disruptions not only harmed plant health and productivity but also reduced the soils' buffering capacity, rendering them more susceptible to further degradation.

Furthermore, synthetic pesticides and herbicides, also used in conjunction with HYVs, often persist in the environment and gradually build up in soils, causing long-term harm to soil health. Pesticide and herbicide residues can interfere with normal nutrient cycles, reduce soil fertility, and restrict soil microbial activity (Rose et al., 2016; Mandal et al., 2020; Tripathi et al., 2020; Nieder et al., 2018; Ogidi and Akpan, 2023).

Synthetic inputs also had a detrimental effect on the soil organic matter (SOM). SOM has been depleted as a result of a greater reliance on synthetic fertilizers, monocropping, and fewer organic inputs. SOM is essential for soil structure, water retention, and nutrient cycling, and soil degradation may result from its loss. The GR's intensive farming approaches have gradually decreased SOM levels, given reduced crop rotation and removal of organic wastes from the soil (Manlay et al., 2007; Cotrufo et al., 2022; John and Babu, 2021; Feller et al., 2012).

Synthetic inputs also played a role in soil erosion and physical degradation. One of the main characteristics of the GR, monocropping, diminished the variety of plant root systems that naturally aid in binding soil particles, weakening the soil structure. As a result, it became more susceptible to erosion from wind and water. Furthermore, the growth of irrigation, frequently without adequate water management, resulted in waterlogging and salinization, which further deteriorated the soil condition. Over time, erosion-induced topsoil loss can drastically lower soil fertility and limit agricultural productivity. Due to the loss of essential soil nutrients and organic matter brought on by GR-intensive farming methods, the soil is now less able to support crops in the future. Additionally, physical deterioration was made worse by soil compaction from intensive mechanical use, which decreased root penetration and water infiltration (Trébuil, 1995; Singh, 2000; Rahman, 2015; Meena et al., 2013).

According Massah and Azadegan (2016), compaction of the soil negatively impacts almost every physical, chemical, and biological soil characteristic.[3] Soil compaction reduces yields by 40 %, accessible water by 34 %, and permeability by 81.4 %.

The excess use of fertilizers beyond what is advised results in the development, buildup, and concentration of fertilizer mineral salts, which, over time, cause soil degradation and a compaction layer (Pahalvi et al., 2021). While increasing bulk density and soil penetration resistance, high compaction reduces porosity and aeration. Additionally, lowering water and nutrient intake will restrict root development and plant growth, lowering yields. Soil biodiversity is also affected by frequent use of synthetic inputs, which upsets soil microbial activity and populations (Bünemann et al., 2006; Gupta et al., 2022; Murphy et al., 2007) and reduced organic residue inputs and root diversity further lessen the complexity of the ecosystem and the range of resources that soil organisms can use (Thiele-Bruhn et al., 2012; Bünemann et al., 2006; Sylvain and Wall, 2011).

Conventional agriculture contributed to decreased soil organic carbon and nitrogen through time, and the only way to mitigate this problem is through alternative practices (Tilman, 1998). A lot of factors are affected by synthetic inputs; however, infrastructure, such as irrigation channels, also plays a role in reducing soil quality. During the GR, irrigation expanded quickly, frequently without proper drainage systems. Raised water levels caused salts to be mobilized in the soil profile. This salt accumulation eventually had a detrimental impact on plant growth and soil structure,

[3] In noncompacted and severely compacted soils, respectively, the bulk density increased from 1.34 to 1.80 Mg.m−3 and the penetrating resistance increased from 0.89 to 3.54 MPa.

which resulted in land degradation and a decline in agricultural productivity (Singh, 2000; Den Herder et al., 2010; Ameen and Raza, 2017).

The significance of pH monitoring and remediation was frequently overlooked in long-term soil management within the framework of GR practices. This neglect has led to a decrease in soil fertility and a threat to long-term agricultural sustainability. Acidification of the soil is often connected to heavy metal toxicity introduced by agrochemicals (Alengebawy et al., 2021; Zou et al., 2018; Uchimiya et al., 2020). Soil fertility decreases and crops may become toxic as a result of heavy metals, which have an adverse impact on soil microbial activity (Abdu et al., 2017; Naz et al., 2022), enzyme function (Aponte et al., 2020; Yeboah, 2021), and nutrient cycling (Li et al., 2024; Yousuf et al., 2022). Prolonged exposure to these metals via food chains can lead to significant health risks, highlighting the necessity for sustainable nutrient management and routine monitoring of soil quality (Sarker et al., 2022).

Aside from the soil degradation, pollinators and beneficial insects were also affected by various GR technologies. The intensification of agriculture and dependence on HYVs diminished habitat diversity and raised chemical exposure, leading to declines in insect populations and local extinctions of many species (Abudulai et al., 2022; Osumanu and Kosoe, 2023; Gepts, 2023). A significant consequence was the reduction of beneficial insect populations, including pollinators and natural predators, because of indiscriminate pesticide use (Sharma et al., 2018). Research has recorded reductions in both the abundance and diversity of pollinators in areas subjected to intensive agriculture, especially those that depend on synthetic agrochemicals in large amounts; the mechanism of the decline includes compromising the physiological and foraging insect behaviors (Sponsler et al., 2022; Ara and Haque, 2021). Furthermore, the development of resistance in pest species due to repeated pesticide exposure has led to increased pesticide use and worsened the negative effects on non-target organisms (Hawkins et al., 2019; Akhter et al., 2024).

6. Sustainability: Green Revolution or organic agriculture?

The legacy of the GR is greatly increased productivity and efficiency, especially in enhancing crop yield. Despite the positive impact on humanity, the problems caused will surely affect generations to come. What are the remedies and mitigation strategies for soil degradation and insect species decline? According to Schrama et al. (2018), conventional agricultural practices tend to decrease crop yields as compared to organic practices. The World Bank reported that health outcomes have not improved, despite the success of the GR (John and Babu, 2021).

Following the GR, a number of significant adverse effects emerged. Before implementation, there was little consideration of benefits and harms. Some of the GR interventions diverge from the natural laws of balance and functioning, resulting in practices that cannot be sustained. In order to prevent negative consequences of certain developments, it is essential to review these interventions (John and Babu, 2021). Recent studies are examining innovative ways to mitigate these problems, such as sustainable intensification of agriculture through intercropping (Martin-Guay et al., 2018).

In the current situation, organic agriculture cannot easily feed the world and cannot replace conventional methods introduced by the GR (Badgley and Perfecto, 2007; Connor, 2008). Nevertheless, there should be modifications to production methods, which could include hybrid approaches that integrate conventional and organic agriculture. An example of a hybrid technique is Integrated Soil Fertility Management (ISFM). ISFM has been practiced as a means of maximizing agronomic efficiency and enhancing crop productivity; in addition, this can contribute to sustainable intensification and address natural resource degradation and rural poverty (Vanlauwe et al., 2015). Farmers who are planning to go total organic will need a general transition period of 36 months from conventional farming (Reganold and Wachter, 2016; Menalled et al., 2009; Lotter, 2003; Zinati, 2002). Aside from ISFM, practices such as Integrated Pest Management are also encouraged; this practice incorporates various strategies to control pests and diseases while relying on synthetic pesticides as a last resort (Tiwari, 2024). Organic farming can and must contribute significantly to addressing future challenges in food production. By minimizing external inputs and using sustainability knowledge, we can reduce environmental contamination and increase food production for more people without harming our environment. Organic agriculture encompasses not just farming as a production practice but also processing, trading, and consumption. Organic farming needs to continually develop in order to address new challenges. To bolster organic food and farming as a solution to future challenges, it is essential to acquire science-based knowledge through committed research (Rahmann et al., 2017).

7. Concluding remarks

Our effort to revisit the Green Revolution (1965–2000) was intended to provide a nuanced understanding of one of the most transformative periods in agricultural history. This accomplishment, which undeniably prevented widespread famine and led to tremendous increases in food production in various regions of the developing world, incurred considerable ecological and social costs. Soil degradation, water scarcity, loss of biodiversity, and increasing inequalities among farmers and the nonfarm rural populace were consequences of excessive dependence on synthetic inputs, monoculture farming practices, intensive irrigation, and use of technologies requiring access to capital and substantially less labor.

Recommendations include the hybrid approaches, such as ISFM. Where organic agriculture alone is insufficient to meet caloric demands, ISFM offers a balanced approach. By combining organic amendments with judicious use of synthetic inputs, farmers can improve nutrient-use efficiency and rebuild soil organic matter without sacrificing yields. Furthermore, to address chemical toxicity and biodiversity loss, Integrated Pest Management should be established as the operational standard. This framework prioritizes crop rotation, biological controls, and mechanical interventions, relegating synthetic pesticides to a "last resort" measure used only when damage exceeds the economic threshold and injury levels. Lastly, gender-responsive research and policy: environmental degradation is inextricably linked to social inequalities. Future agricultural research, extension services, and policymaking must always incorporate gender analysis into their foundational design. Ensuring that interventions are gender-sensitive optimizes the adoption of sustainable practices and leverages the labor and knowledge of all stakeholders in the transition to regenerative systems.

This chapter has underscored the technological progress and yield increases, as well as the unforeseen repercussions that have arisen over time. These encompass long-term deterioration of soil health, buildup of heavy metals and agrochemical residues, and disturbances to traditional agricultural systems and rural economies. When contemplating the legacy of the GR, it becomes evident that future agricultural development needs to move toward approaches that are more sustainable, inclusive, and resilient. The lessons learned from this era highlight the necessity of incorporating ecological principles, and fair resource access into the design of the next phase of agricultural innovation. It is essential to understand the lessons of the GR, not just for the sake of historical analysis, but also as a vital step toward shaping future policies that seek to nourish the global population while upholding environmental integrity and social justice.

Declaration of AI and AI-assisted technologies in the writing process

During the preparation of this work the authors used OpenAI's ChapGPT (GPT-5.3 and GPT-5.3 Instant) in order to reformat the reference lists from APA to Harvard system, and some guides in making the Figures. Grammarly (v.1.2.237.1845) was also used as a writing assistant, particularly in minor grammar correction. After using this tool/service, the author(s) reviewed and edited the content as needed and take(s) full responsibility for the content of the publication.

References

Abdu, N., Abdullahi, A.A., Abdulkadir, A., 2017. Heavy metals and soil microbes. Environ. Chem. Lett. 15 (1), 65–84. https://doi.org/10.1007/s10311-016-0587-x.

Abudulai, M., Nboyine, J.A., Quandahor, P., Seidu, A., Traore, F., Abudulai, M., 2022. Agricultural intensification causes decline in insect biodiversity. Glob. Decline Insects 1, 43–64.

Ackrill, R., 1992. The EC Budget and Agricultural Policy Reforms, with Special Reference to Cereals (Doctoral dissertation). University of Nottingham.

Ackrill, R., 2000. Common agricultural policy. A&C Black 9.

African Union, 2003. Comprehensive Africa Agriculture Development Programme. NEPAD, Midrand, South Africa.

Afun-Ogidan, D., 2012. CAADP in a nutshell: brief overview of the comprehensive Africa Agriculture Development Programme's rationale, principles and objectives. Great Insights, Volume 1, Issue 7, September 2012. Available at: https://ecdpm.org/work/caadp-and-food-security-volume-1-issue-7-september-2012/caadp-in-a-nutshell-brief-overview-of-the-comprehensive-africa-agriculture-development-programmes-rationale-principles-and-objec [Accessed 01 March 2026].

Akhter, S., Naik, V.K., Naladi, B.J., Rathore, A., Yadav, P., Lal, D., 2024. The ecological impact of pesticides on non-target organisms in agricultural ecosystems. Advances in. Bioresearch 15 (4), 322–334. https://doi.org/10.15515/abr.0976-4585.15.4.322334.

Alengebawy, A., Abdelkhalek, S.T., Qureshi, S.R., Wang, M.Q., 2021. Heavy metals and pesticides toxicity in agricultural soil and plants: ecological risks and human health implications. Toxics 9 (3), 42.

Alkire, S., Meinzen-Dick, R., Peterman, A., Quisumbing, A., Seymour, G., Vaz, A., 2013. The women's empowerment in agriculture index. World Dev. 52, 71–91. https://doi.org/10.1016/j.worlddev.2013.06.007.

Alou, I.N., 2020. Strategies to Bridge Yield Gaps in Equatorial African Upland Rice (Oryza sativa L.) Production Systems (Doctoral dissertation) University of Pretoria. https://repository.up.ac.za/items/fc433e0c-0e1d-40c9-8a96-67b0f2c50953.

Alston, J.M., Pardey, P.G., James, J.S., Andersen, M.A., 2009. The economics of agricultural R&D. Annu. Rev. Resour. Econ. 1 (1), 537–566. https://doi.org/10.1146/annurev.resource.050708.144137.

Altieri, M.A., Rosset, P., 2002. Ten reasons why biotechnology will not ensure food security, protect the environment and reduce poverty in the developing world. Ethical issues Biotechnol. 175–182.

Altieri, M.A., Toledo, V.M., 2011. The agroecological revolution in Latin America: rescuing nature, ensuring food sovereignty and empowering peasants. J. Peasant. Stud. 38 (3), 587–612. https://doi.org/10.1080/03066150.2011.582947.

Ameen, A., Raza, S., 2017. Green Revolution: a review. Int. J. Adv. Sci. Res. 3 (12), 129–137. https://doi.org/10.7439/ijasr.v3i12.4410.

Angelakıs, A.N., Zaccaria, D., Krasilnikoff, J., Salgot, M., Bazza, M., Roccaro, P., et al., 2020. Irrigation of world agricultural lands: Evolution through the millennia. Water 12 (5), 1285. https://doi.org/10.3390/w12051285.

Anik, A.R., Rahman, S., 2021. Women's empowerment in agriculture: level, inequality, progress, and impact on productivity and efficiency. J. Dev. Stud. 57 (6), 930–948. https://doi.org/10.1080/00220388.2020.1817393.

Ankrah, D.A., Freeman, C.Y., Afful, A., 2020. Gendered access to productive resources–evidence from small holder farmers in Awutu Senya West District of Ghana. Sci. Afr. 10, e00604. https://doi.org/10.1016/j.sciaf.2020.e00604.

Annesley, T.M., 2022. Everyday people. Clin. Chem. 68 (6), 864–865. https://doi.org/10.1093/clinchem/hvac052.

Aponte, H., Meli, P., Butler, B., Paolini, J., Matus, F., Merino, C., et al., 2020. Meta-analysis of heavy metal effects on soil enzyme activities. Sci. Total. Environ. 737, 139744. https://doi.org/10.1016/j.scitotenv.2020.139744.

Ara, Z.G., Haque, A.R., 2021. A comprehensive review on synthetic insecticides: toxicity to pollinators, associated risk to food security, and management approaches. J. Biosyst. Eng. 46 (3), 254–272. https://doi.org/10.1007/s42853-021-00104-y.

Atera, E.A., Onyango, J.C., Azuma, T., Asanuma, S., Itoh, K., 2011. Field evaluation of selected NERICA rice cultivars in Western Kenya. Afr. J. Agric. Res. 6 (1), 60–66.

Badgley, C., Perfecto, I., 2007. Can organic agriculture feed the world? Renew. Agriculture Food Syst. 22 (2), 80–86. https://doi.org/10.1017/S1742170507001986.

Badiane, O., Collins, J., Ulimwengu, J.M., 2020. The past, present and future of agriculture policy in Africa. https://doi.org/10.2499/9780896293946_02.

Baldock, D., Dwyer, J., Lowe, P., Petersen, J.E. and Ward, N., 2001. The nature of rural development: towards a sustainable integrated rural policy in Europe. *Synthesis Report*.

Bański, J., Mazur, M., 2021. Transformation of Agricultural Sector in the Central and Eastern Europe after 1989. Springer. https://doi.org/10.1007/978-3-030-73766-5.

Barker, R., 2004. Evolution of irrigation in South and Southeast Asia (Vol. 5). Iwmi.

Barnett, B.J., 2000. The US farm financial crisis of the 1980s. Agric. history 74 (2), 366–380.

Bationo, A., Waswa, B., Kihara, J., Kimetu, J., 2006. Advances in integrated soil fertility management in sub-Saharan Africa: challenges and opportunities. Nutrient Cycl. Agroecosystems 76 (2-3), 2–8.

Bellmann, C., 2019. Subsidies and Sustainable Agriculture: Mapping the Policy Landscape. Chatham House, London, UK.

Bernini, C., Galli, F., 2024. Economic and environmental efficiency, subsidies and spatio-temporal effects in agriculture. Ecol. Econ. 218, 108120. https://doi.org/10.1016/j.ecolecon.2024.108120.

Boix-Fayos, C., De Vente, J., 2023. Challenges and potential pathways towards sustainable agriculture within the European Green Deal. Agric. Syst. 207, 103634. https://doi.org/10.1016/j.agsy.2023.103634.

Bonnen, J.T., Schweikhardt, D.B., 1998. The future of US agricultural policy: reflections on the disappearance of the "farm problem". Appl. Econ. Perspect. Policy 20 (1), 2–36.

Borlaug, N.E., 1986. The Green Revolution: The Role of CIMMYT and What Lies Ahead. D.F.: International Maize and Wheat Improvement Center (CIMMYT), Mexico.

Borlaug, N.E., 1964. Basic concepts which influence the choice of methods for use in breeding for disease resistance in cross-pollinated and self-pollinated crop plants. In NATO/NSF Advance. Symposium held the Pennsylvania State University, august (Vol. 30, pp. 327–348).

Botreau, H., Cohen, M.J., 2020. Gender equality and food insecurity: a dozen years after the food price crisis, rural women still bear the brunt of poverty and hunger. In: Cohen, M.J. (Ed.), Advances in Food Security and Sustainability 5. Academic Press, Cambridge, MA, pp. 53–117.

Braun, H.J., 2011. Norman Borlaug's legacy and the urgent need for continuing innovative wheat technology. CIMMYT. https://repository.cimmyt.org/entities/publication/c5dcb7c2-ed90-4e1c-aabe-5e385c28593f.

Brüntrup, M., 2011. The Comprehensive Africa Agriculture Development Programme (CAADP)–An Assessment of a Pan-African Attempt to Revitalise Agriculture–. Q. J. Int. Agriculture 50 (1), 79–106. https://doi.org/10.22004/ag.econ.155492.

Buller, H., Wilson, G., Holl, A. (Eds.), 2017. Agri-Environmental Policy in the European Union. Routledge.

Bünemann, E.K., Schwenke, G.D., Van Zwieten, L., 2006. Impact of agricultural inputs on soil organisms—a review. Soil. Res. 44 (4), 379–406.

Bureau, J.C., Swinnen, J., 2018. EU policies and global food security. Glob. food security 16, 106–115. https://doi.org/10.1016/j.gfs.2017.12.001.

Byerlee, D., 1992. Technical change, productivity, and sustainability in irrigated cropping systems of South Asia: Emerging issues in the post-green revolution Era. J. Int. Dev. 4 (5), 477–496. https://doi.org/10.1002/jid.3380040502.

Cairns, J.E., Hellin, J., Sonder, K., Araus, J.L., MacRobert, J.F., Thierfelder, C., et al., 2013. Adapting maize production to climate change in sub-Saharan Africa. Food Security 5 (3), 345–360.

Caparrós-Martínez, J.L., Rueda-Lópe, N., Milán-García, J., de Pablo Valenciano, J., 2020. Public policies for sustainability and water security: the case of Almeria (Spain). Glob. Ecol. Conserv. 23, e01037. https://doi.org/10.1016/j.gecco.2020.e01037.

Chen, F., Zhao, H., Roberts, D., Van de Voorde, T., Batelaan, O., Fan, T., et al., 2023. Mapping center pivot irrigation systems in global arid regions using instance segmentation and analyzing their spatial relationship with freshwater resources. Remote. Sens. Environ. 297, 113760. https://doi.org/10.1016/j.rse.2023.113760.

Chen, H., Weersink, A., Beaulieu, M., Lee, Y.N., Nagelschmitz, K., 2019. A historical review of changes in farm size in Canada. Working Paper WP(19–03), Institute for the Advanced Study of Food and Agricultural Policy, University of Guelph. Available at: https://ageconsearch.umn.edu/record/283563/files/A%20Historical%20Review_2019_03.pdf [Accessed 01 March 2026].

Chi, Y., Zhou, W., Wang, Z., Hu, Y., Han, X., 2021. The influence paths of agricultural mechanization on green agricultural development. Sustainability 13 (23), 12984.

Choi, K., 2020. New Solutions for Water Resources Management in South Korea. Netherlands Enterprise Agency [Accessed 01 March 2026]. https://www.rvo.nl/sites/default/files/2022/03/Water-Report-New-solutions-for-water-resources-management-in-South-Korea%20DEF.pdf.

Chow, D.C., Sheldon, I.M., 2023. The United States' challenge (s) to Canada's dairy import tariff rate quotas: what is all the fuss about? Ohio St. Bus. LJ 18, 353 [Accessed 01 March 2026].

CIMMYT, 2011. Norman Borlaug's Legacy and the Urgent Need for Continuing Innovative Wheat Technology. https://repository.cimmyt.org/server/api/core/bitstreams/8ef7d09e-9847–4ee3–921e-2bf2167907bf/content [Accessed 01 March 2026].

CIMMYT, 2016. CIMMYT and SAGARPA Honor Legacy of Norman Borlaug. https://www.cimmyt.org/news/cimmyt-and-sagarpa-honor-legacy-of-norman-borlaug/ [Accessed 01 March 2026].

Closas, A. and Molle, F., 2016. Groundwater Governance in Asia and the Pacific. IWMI Project Report No. 4. Colombo, Sri Lanka: International Water Management Institute (IWMI).

Connor, D.J., 2008. Organic agriculture cannot feed the world. Field Crop. Res. 106 (2), 187–190. https://doi.org/10.1016/j.fcr.2007.11.010.

Connor, M., Gummert, M., Singleton, G.R., 2023. Closing Rice Yield Gaps in Asia: Innovations, Scaling, and Policies for Environmentally Sustainable Lowland Rice Production. Springer Nature, pp. 259.

Conway, G., 1998. The Doubly Green Revolution: Food for all in the Twenty-First Century. Cornell University Press.

Conway, G., Toenniessen, G., 2003. Science for African food security. Science 299 (5610), 1187–1188.

Cooper, P.J., Dimes, J., Rao, K.P.C., Shapiro, B., Shiferaw, B., Twomlow, S., 2008. Coping better with current climatic variability in the rain-fed farming systems of sub-Saharan Africa: An essential first step in adapting to future climate change? Agric. Ecosyst. Environ. 126 (1-2), 24–35.

Cooper, H.D., Spillane, C., Hodgkin, T., Cooper, H., 2001. Broadening the genetic base of crops: an overview. Broadening Genet. base Crop. Prod. 1–23.

Cotrufo, M.F., Haddix, M.L., Kroeger, M.E., Stewart, C.E., 2022. The role of plant input physical-chemical properties, and microbial and soil chemical diversity on the formation of particulate and mineral-associated organic matter. Soil. Biol. Biochem. 168, 108648.

Crago, C.L., Khanna, M., Barton, J., Giuliani, E., Amaral, W., 2010. Competitiveness of Brazilian sugarcane ethanol compared to US corn ethanol. Energy Policy 38 (11), 7404–7415. https://doi.org/10.1016/j.enpol.2010.08.016.

Csáki, C., Lerman, Z., 1994. Land reform and farm sector restructuring in the former socialist countries in Europe. Eur. Rev. Agric. Econ. 21 (3-4), 553–576. https://doi.org/10.1093/erae/21.3-4.553.

Cycon, S.M., 2013. Fertile lands and bodies: connecting the Green Revolution, pesticides, and women's reproductive health. https://scholarship.claremont.edu/pitzer_theses/38. [Accessed 01 March 2026].

Cypher, J.M., 2010. South America's commodities boom: developmental opportunity or path dependent reversion? Can. J. Dev. Studies/Revue canadienne d'études du. développement 30 (3-4), 635–662. https://doi.org/10.1080/02255189.2010.9669319.

D'Agostino, A., 2017. Technical change and gender wage inequality: long-run effects of India's Green Revolution. Available at SSRN 3400889. http://dx.doi.org/10.2139/ssrn.3400889.

Dangar, S., Mishra, V., 2024. Groundwater sustainability in India through nonrice-dominated cropping pattern. PNAS nexus 3 (9), pgae378. https://doi.org/10.1093/pnasnexus/pgae378.

Dare Kolawole, O., 2012. Agrarian reforms and the African green revolution. World J. Sci. Technol. Sustain. Dev. 9 (4), 301–316. https://doi.org/10.1108/20425941211271504.

Das, A., Saha, S., Layek, J., Babu, S., Saxena, R., Ramkrushna, G.I., 2023. Agricultural technologies. In: Ghosh, P.K., Das, A., Saxena, R., Banerjee, K., Kar, G., Vijay, D. (Eds.), Trajectory of 75 Years of Indian Agriculture after Independence. Springer Nature Singapore, Singapore, pp. 57–78.

Daugbjerg, C., Swinbank, A., 2006. The 2003 CAP reform: Accommodating WTO pressures. Comp. Eur. Polit. 4 (1), 47–64.

Daugbjerg, C., Swinbank, A., 2009. Ideas, Institutions, and Trade: The WTO and the Curious Role of EU Farm Policy in Trade Liberalization. Oxford University Press.

DDT and its derivatives (EHC 9, 1979) (no date). https://www.inchem.org/documents/ehc/ehc/ehc009.htm. [Accessed 01 March 2026].

De, D., 2015. The impact of Green Revolution on adivasi women of Eastern India. Asian Man. (The)-An Int. J. 9 (1), 92–101. https://doi.org/10.5958/0975-6884.2015.00015.8.

De Janvry, A., Graff, G., Sadoulet, E., Zilberman, D., 2005. Technological change in agriculture and poverty reduction: the potential role of biotechnology. In: Cooper, J., Lipper, L.M., Zilberman, D. (Eds.), Agricultural Biodiversity and Biotechnology in Economic Development. Springer US, Boston, MA, pp. 361–386.

Debangshi, U., Ghosh, P., 2022. Rice wheat cropping systems-constraints and strategies: a review. Plant. Arch. (09725210) 22 (1). https://doi.org/10.51470/PLANTARCHIVES.2022.v22.no1.004.

Delvaux, P.A.G., Riesgo, L., Paloma, S.G.Y., 2020. Are small farms more performant than larger ones in developing countries? Sci. Adv. 6 (41).

Den Herder, G., Van Isterdael, G., Beeckman, T., De Smet, I., 2010. The roots of a new Green Revolution. Trends plant. Sci. 15 (11), 600–607.

Deshpande, V.P., Ahmad, I.S.H.T.I.Y.A.Q., Singh, C.K., 2024. Sustainable groundwater management through micro irrigation: a critical review on challenges and solutions. J. Landsc. Ecol. 17 (1), 16–34.

DeVries, J., 2011, May. The exciting role of public agricultural research in Africa's Green Revolution. In Consortium of Higher Education and Research for Agriculture Conference (P. 56).

Dhillon, R., Moncur, Q., 2023. Small-scale farming: a review of challenges and potential opportunities offered by technological advancements. Sustainability 15 (21), 15478.

Diagne, A., Midingoyi, S.K.G., Wopereis, M., Akintayo, I., 2011. Increasing rice productivity and strengthening food security through new rice for Africa (NERICA). Yes Afr. can: Success. stories a Dyn. Cont. 253, 253.

Dorward, A., Chirwa, E., 2012. The Malawi agricultural input subsidy programme: 2005/06–2008/09. Sustain. Intensif. 232–247.

Duffy, M., 2009. Economies of size in production agriculture. J. Hunger. & Environ. Nutr. 4 (3-4), 375–392.

Ebel, R., Menalled, F., Ahmed, S., Gingrich, S., Baldinelli, G.M., Félix, G.F., 2021. How biodiversity loss affects society. In: James, H.S. (Ed.), Handbook on the Human Impact of Agriculture. Edward Elgar Publishing, pp. 352–376.

Edmeades, G.O., Trevisan, W., Prasanna, B.M., Campos, H., 2017. Tropical maize (Zea mays L.). In: Campos, H., Caligari, P.D.S. (Eds.), Genetic Improvement of Tropical Crops. Springer International Publishing, Cham, pp. 57–109. https://doi.org/10.1007/978-3-319-59819-2_3.

Effa, K., Fana, D.M., Nigussie, M., Geleti, D., Abebe, N., Dechassa, N., et al., 2025. The irrigated wheat initiative of Ethiopia: a new paradigm emulating Asia's Green Revolution in Africa. Environ. Dev. Sustain. 27 (1), 2161–2186. https://doi.org/10.1007/s10668-023-03961-z.

Eicher, C.K., 1995. Zimbabwe's maize-based Green Revolution: preconditions for replication. World Dev. 23 (5), 805–818. https://doi.org/10.1016/0305-750X(95)93983-R.

Eliazer Nelson, A.R.L., Ravichandran, K., Antony, U., 2019. The impact of the Green Revolution on indigenous crops of India. J. Ethnic Foods 6 (1), 8. https://doi.org/10.1186/s42779-019-0011-9.

European Commission, 2025. Common Agricultural Policy (CAP) at a glance. Available at: https://agriculture.ec.europa.eu/common-agricultural-policy/cap-overview/cap-glance_en [Accessed 01 March 2026].

European Parliament 2016. https://www.europarl.europa.eu/RegData/etudes/STUD/2016/573428/IPOL_STU(2016)573428_EN.pdf.

Evans, L.T., Evans, L.T., 1996. Crop Evolution, Adaptation and Yield. Cambridge University Press.
Evans, J.R., Lawson, T., 2020. From green to gold: agricultural revolution for food security. J. Exp. Botany 71 (7), 2211–2215.
Evenson, R.E., Gollin, D., 2003. Assessing the impact of the Green Revolution, 1960–2000. Science 300 (5620), 758–762.
FAO, 2025. FAOSTAT Database: Cereal Production [online]. Food and Agriculture Organization of the United Nations [Accessed 01 March 2026]. https://www.fao.org/faostat/en/#data/QC/visualize.
FAO, 2013a. Save and Grow: Cassava - A guide to sustainable production intensification. Food and Agriculture Organization [Accessed 01 March 2026].
FAO, 2013b. The State of Food and Agriculture 2013: Food systems for better nutrition. Food and Agriculture Organization [Accessed 01 March 2026].
FAO, 2015. The state of food insecurity in the world 2015: Meeting the 2015 international hunger targets. Food and Agriculture Organization [Accessed 01 March 2026].
Feller, C., Blanchart, E., Bernoux, M., Lal, R., Manlay, R., 2012. Soil fertility concepts over the past two centuries: the importance attributed to soil organic matter in developed and developing countries. Arch. Agron. Soil. Sci. 58 (sup1), S3–S21.
Ferrero-Serrano, Á., Cantos, C., Assmann, S.M., 2019. The role of dwarfing traits in historical and modern agriculture with a focus on rice. Cold Spring Harb. Perspect. Biol. 11 (11), a034645. https://doi.org/10.1101/cshperspect.a034645.
Fusco, G., 2021. Twenty years of common agricultural policy in Europe: a bibliometric analysis. Sustainability 13 (19), 10650. https://doi.org/10.3390/su131910650.
Garcia, G., Montero, F., Torres, M.E., Alvarez, S., Vasquez, W., Villantoy, A., et al., 2025. Adaptability, yield stability, and agronomic performance of improved purple corn (Zea mays L.) hybrids across diverse agro-ecological zones in Peru. Int. J. Plant. Biol. 17 (1), 3. https://doi.org/10.3390/ijpb17010003.
Gardner, B., 2006. European Agriculture: Policies, Production and Trade. Routledge.
Garrity, D.P., Akinnifesi, F.K., Ajayi, O.C., Weldesemayat, S.G., Mowo, J.G., Kalinganire, A., et al., 2010. Evergreen agriculture: a robust approach to sustainable food security in Africa. Food Security 2 (3), 197–214. https://doi.org/10.1007/s12751-010-0070-7.
Gepts, P., 2023. Biocultural diversity and crop improvement. Emerg. Top. Life Sci. 7 (2), 151–196.
Getahun, S., Kefale, H., Gelaye, Y., 2024. Application of precision agriculture technologies for sustainable crop production and environmental sustainability: a systematic review. Sci. World J. 2024 (1), 2126734. https://doi.org/10.1155/2024/2126734.
Ghumman, A.R., Khan, R.A., Hashmi, H.N., Kamal, M.A., Khan, I.T., 2011. Performance assessment of canal irrigation in Pakistan. Afr. J. Agric. Res. 6 (12), 2692–2698.
Giosuè, A., Vitale, M., Riccardi, G., 2026. A critical evaluation of the 2025–2030 dietary guidelines for Americans. Nutrition, Metab. Cardiovascular Dis. 104614. https://doi.org/10.1016/j.numecd.2026.104614.
Giuliani, A., Baron, H., 2025. The CAP (common agricultural policy): a short history of crises and major transformations of European agriculture. Forum Soc. Econ. 54 (1), 68–94. https://doi.org/10.1080/07360932.2023.2259618. (Routledge).
Gollin, D., Hansen, C.W., Wingender, A.M., 2021. Two blades of grass: the impact of the Green Revolution. J. Polit. Econ. 129 (8), 2344–2384.
Gonzales, L.A., 1993. Economic incentives and comparative advantage in Indonesian food crop production. Intl. Food Policy Res. Inst. 93. https://hdl.handle.net/10568/156836.
González de Molina, M., Soto Fernández, D., Infante-Amate, J., Aguilera, E., Vila Traver, J., Guzmán, G.I., 2017. Decoupling food from land: the evolution of Spanish agriculture from 1960 to 2010. Sustainability 9 (12), 2348. https://doi.org/10.3390/su9122348.

Green, D., 2003. Silent Revolution: The Rise and Crisis of Market Economics in Latin America. NYU Press.

Guo, J.H., Liu, X.J., Zhang, Y., Shen, J.L., Han, W.X., Zhang, W.F., et al., 2010. Significant acidification in major Chinese croplands. Science 327 (5968), 1008–1010.

Gupta, R.K., Naresh, R.K., Hobbs, P.R., Jiaguo, Z., Ladha, J.K., 2003. Sustainability of post-Green Revolution agriculture: the rice–wheat cropping systems of the Indo-Gangetic plains and China. Improving Product. Sustainability Rice-Wheat Systems: Issues Impacts 65, 1–25. https://doi.org/10.2134/asaspecpub65.c1.

Gupta, A., Singh, U.B., Sahu, P.K., Paul, S., Kumar, A., Malviya, D., et al., 2022. Linking soil microbial diversity to modern agriculture practices: a review. Int. J. Environ. Res. Public. Health 19 (5), 3141.

Guzzon, F., Arandia Rios, L.W., Caviedes Cepeda, G.M., Céspedes Polo, M., Chavez Cabrera, A., Muriel Figueroa, J., et al., 2021. Conservation and use of Latin American maize diversity: Pillar of nutrition security and cultural heritage of humanity. Agronomy 11 (1), 172. https://doi.org/10.3390/agronomy11010172.

Haggblade, S., Smale, M., Kergna, A., Theriault, V., Assima, A., 2017. Causes and consequences of increasing herbicide use in Mali. Eur. J. Dev. Res. 29 (3), 648–674.

Hair, C., 2019. Roses along the equator: Situating Ecuador and Colombia within the global cut-flower market. South. Q. 57 (1), 50–67.

Hall, K.D., 2023. From dearth to excess: the rise of obesity in an ultra-processed food system. Philos. Trans. R. Soc. B: Biol. Sci. 378 (1885). https://doi.org/10.1098/rstb.2022.0214.

Hamdan, M.F., Mohd Noor, S.N., Abd-Aziz, N., Pua, T.L., Tan, B.C., 2022. Green revolution to gene revolution: technological advances in agriculture to feed the world. Plants 11 (10), 1297.

Hamilton, S.F., Richards, T.J., Shafran, A.P., Vasilaky, K.N., 2022. Farm labor productivity and the impact of mechanization. Am. J. Agric. Econ. 104 (4), 1435–1459.

Harcsa, I., Kovách, I., Szelényi, I., 2002. The Hungarian agricultural 'miracle'and the limits of socialist reforms. Privatizing the Land. Routledge, pp. 37–58.

Harwood, J., 2012. Europe's Green Revolution and Its Successors: The Rise and Fall of Peasant-Friendly Plant Breeding. Routledge.

Harwood, J., 2020. Could the adverse consequences of the green revolution have been foreseen? How experts responded to unwelcome evidence. Agroecol. Sustain. Food Syst. 44. pp. 509–535.

Hashemi, S.Z., Darzi-Naftchali, A., Karandish, F., Ritzema, H., Solaimani, K., 2022. Assessing agro-environmental sustainability of intensive agricultural systems. Sci. Total. Environ. 831, 154994. https://doi.org/10.1016/j.scitotenv.2022.154994.

Hassan, R.M., Mekuria, M., Mwangi, W.M., 2001. Maize breeding research in Eastern and Southern Africa: Current status and impact of past investments made by the public and private Sectors, 1966–97. CIMMYT.

Hawkins, N.J., Bass, C., Dixon, A., Neve, P., 2019. The evolutionary origins of pesticide resistance. Biol. Rev. 94 (1), 135–155.

Hazell, P.B., 2009. The Asian Green Revolution. International Food Policy Research Institute, pp. 911.

Hazell, P.B.R., 2010. Asia's Green Revolution: Past achievements and future challenges. In: Dawe, D. (Ed.), Rice in the Global Economy: Strategic Research and Policy Issues for Food Security. International Rice Research Institute (IRRI), Los Baños, Philippines, pp. 37–54.

Heinrich Böll Foundation, 2025. Pesticide Atlas: Gender and pesticide exposure. Available at: https://eu.boell.org/en/PesticideAtlas-gender#:~:text=Due%20to%20traditional%20gender%20roles,mammary%20carcinogens%20and%20tumour%20promoters [Accessed 01 March 2026].

Hill, B., Greer, A., Heckelei, T., Witze, P., 2004. Agriculture, forestry and fisheries. Handbook of Public Policy in Europe: Britain, France and Germany. Palgrave Macmillan UK, London, pp. 175–193. https://doi.org/10.1057/9780230522756_17.

Hornbeck, R., Keskin, P., 2014. The historically evolving impact of the ogallala aquifer: agricultural adaptation to groundwater and drought. Am. Econ. J.: Appl. Econ. 6 (1), 190–219.

Hou, C.H., Xu, X.C., Wang, H.B., 2015. Establishment of green fertilizer industrial system and its scientific problems. Chin. Sci. Bull. 60 (36), 3535–3542. https://doi.org/10.1360/N972015-00770.

Huang, J., Rozelle, S., Rosegrant, M.W., 1999. China's food economy to the twenty-first century: supply, demand, and trade. Economic Dev. Cultural Change 47 (4), 737–766.

Hulme, M., 2001. Climatic perspectives on Sahelian desiccation: 1973–1998. Glob. Environ. Change 11 (1), 19–29. https://doi.org/10.1016/S0959-3780(00)00042-X.

Hurt, R.D., 2020. The Green Revolution in the Global South: Science, Politics, and Unintended Consequences. University Alabama Press.

Idowu, J., Angadi, S., 2013. Understanding and Managing Soil Compaction in Agricultural Fields. NM State University, Cooperative Extension Service, pp. 9.

International Programme on Chemical Safety (IPCS), 1992. Environmental Health Criteria 9: Pesticide residues in food – 1986 evaluations. Available at: https://www.inchem.org/documents/ehc/ehc/ehc009.htm [Accessed 01 March 2026].

Jackson, C., 2005. Strengthening food policy through gender and intrahousehold analysis: Impact assessment of IFPRI multicountry research. https://hdl.handle.net/10568/160720 [Accessed 01 March 2026].

Jaleta, M., Euler, M., Gartaula, H., Krishna, V., 2023. Gender differences in smallholders' socioeconomic networks and acquisition of seed of improved wheat varieties in Ethiopia. Front. Sustain. Food Syst. 6, 1080401. https://doi.org/10.3389/fsufs.2022.1080401.

Jayne, T.S., Chamberlin, J., Headey, D.D., 2014. Land pressures, the evolution of farming systems, and development strategies in Africa: A synthesis. Food policy 48, 1–17.

Jha, S., Roland-Holst, D., Sriboonchitta, S., Behnke, D., 2010. Regional trade opportunities for Asian agriculture. In: Beladi, H., Choi, K. (Eds.), New Developments in Computable General Equilibrium Analysis for Trade Policy. England: Emerald Publishing Limited, pp. 273–302. https://doi.org/10.1108/S1574-8715(2010)0000007013.

Jhariya, M.K., Banerjee, A., Meena, R.S., Kumar, S., Raj, A., 2021. Sustainable intensification for agroecosystem services and management: an overview. In: Jhariya, M.K., Banerjee, A., Meena, R.S., Kumar, S., Raj, A. (Eds.), Sustainable Intensification for Agroecosystem Services and Management, pp. 1–35. https://doi.org/10.1007/978-981-16-3207-5_1.

Jiao, S., Li, Q., Zhang, F., Tao, Y., Yu, Y., Yao, F., et al., 2024. Artificial selection of the Green Revolution gene Semidwarf 1 is implicated in upland rice breeding. J. Integr. Agriculture 23 (3), 769–780.

Jin, J.Y., Li, Jia-kang, Li, S., 2006. Chemical fertilizer and food security. J. Plant. Nutr. Fertilizers 12 (5), 601–609.

Jobson, E.M., Johnston, R.E., Oiestad, A.J., Martin, J.M., Giroux, M.J., 2019. The impact of the wheat RHT-B1B Semi-Dwarfing allele on photosynthesis and seed development under field conditions. Front. Plant. Sci. 10, 51.

John, D.A., Babu, G.R., 2021. Lessons from the aftermaths of green revolution on food system and health. Front. Sustain. food Syst. 5, 644559. https://doi.org/10.3389/fsufs.2021.644559.

Johnson, R., Monke, J., 2008. What is the farm bill?. Congressional Research Service, the Library of Congress. https://nationalaglawcenter.org/wp-content/uploads/assets/crs/remove.pdf. [Accessed 01 March 2026].

Jug, D., Jug, I., Brozović, B., Šeremešić, S., Dolijanović, Ž., Zsembeli, J., et al., 2025. Conservation soil tillage: bridging science and farmer expectations—An overview from southern to northern Europe. Agriculture 15 (3), 260.

Kang, M., Schwab, B., Yu, J., 2020. Gender differences in the relationship between land ownership and managerial rights: implications for intrahousehold farm labor allocation. World Dev. 125, 104669. https://doi.org/10.1016/j.worlddev.2019.104669.

Katsikis, N., 2023. The horizontal factory. The operationalisation of the US corn and soy belt. Ext. Urbanisation: Tracing Planet. Struggles 121–158.

Kaur, S., Bedi, M., Singh, S., Kour, N., Bhatti, S.S., Bhatia, A., et al., 2024. Monoculture of crops: a challenge in attaining food security. In: Sharma, A., Kumar, M., Sharma, P. (Eds.), Advances in Food Security and Sustainability 9. Elsevier, pp. 197–213.

Keller, T., Sandin, M., Colombi, T., Horn, R., Or, D., 2019. Historical increase in agricultural machinery weights enhanced soil stress levels and adversely affected soil functioning. Soil. Tillage Res. 194, 104293.

Kimenyi, M.S., Routman, B., Westbury, A., Omiti, J., Akande, T., 2013. CAADP at 10: Progress Towards Agricultural Prosperity. Africa Growth Initiative at Brookings.

Klein, H.S., Luna, F.V., 2021. The growth of the soybean frontier in South America: the case of Brazil and Argentina. Rev. de. Historia Economica-Journal Iber. Lat. Am. Economic History 39 (3), 427–468. https://doi.org/10.1017/S0212610920000269.

Kloppenburg, J.R., 2005. First the Seed: The Political Economy of Plant Biotechnology. Univ of Wisconsin Press.

Koester, U., 1999, June. The evolving farm structure in East Germany. In Second World Bank EU Accession Workshop in the Rural Sector. Warsaw/Poland.

Kremen, C., Merenlender, A.M., 2018. Landscapes that work for biodiversity and people. Science 362 (6412), eaau6020. https://doi.org/10.1126/science.aau6020.

Lal, R., 2004. Soil carbon sequestration impacts on global climate change and food security. Science 304 (5677), 1623–1627. https://doi.org/10.1126/science.1097396.

Langlais, A., 2023. The new Common Agricultural Policy: reflecting an agro-ecological transition. The legal perspective. Rev. Agric. Food Environ. Stud. 104 (1), 51–66.

Lechenet, M., Deytieux, V., Antichi, D., Aubertot, J.N., Barberi, P., Bertrand, M., et al., 2017. Diversity of methodologies to experiment integrated pest management in arable cropping systems: analysis and reflections based on a European network. Eur. J. Agron. 83, 86–99. https://doi.org/10.1016/j.eja.2016.09.012.

Lee, J., 2023. Preparatory labor for chemical fertilizer: rural modernity and the practices of South Korean farmers in the 1960s. History Sci. 61 (4), 588–607.

Lehmann, J., Bossio, D.A., Kögel-Knabner, I., Rillig, M.C., 2020. The concept and future prospects of soil health. Nat. Rev. Earth & Environ. 1 (10), 544–553.

Lele, U., Goswami, S., 2020. Agricultural policy reforms: roles of markets and states in China and India. Glob. food security 26, 100371. https://doi.org/10.1016/j.gfs.2020.100371.

Lerman, Z., 2001. Agriculture in transition economies: from common heritage to divergence. Agric. Econ. 26 (2), 95–114. https://doi.org/10.1016/S0169-5150(00)00107-9.

Levy, R.P., Stancich, L., 1997. The exception that proves the rule? The case of the MacSharry reforms for the CAP. Proceedings of the European Union Studies Association (EUSA), Seattle, WA, USA, 29.

Li, S., Lei, Y., Zhang, Y., Liu, J., Shi, X., Jia, H., et al., 2019. Rational trade-offs between yield increase and fertilizer inputs are essential for sustainable intensification: a case study in wheat–maize cropping systems in China. Sci. Total. Environ. 679, 328–336.

Li, T., Wang, S., Zhao, L., Yuan, X., Gao, Y., Fu, D., et al., 2024. Improvement of soil nutrient cycling by dominant plants in natural restoration of heavy metal polluted areas. Environ. Res. 263, 120030.

Liliane, T.N., Charles, M.S., 2020. Factors affecting yield of crops. Agronomy-Climate Change & Food Security 9, 9–24.

Lindsay, W.L., Walthall, P.M., 2020. The solubility of aluminum in soils. In: Sposito, G. (Ed.), The Environmental Chemistry of Aluminum. CRC Press, pp. 333–361.

Liu, M., Linna, C., Ma, S., Ma, Q., Song, W., Shen, M., et al., 2022. Biochar combined with organic and inorganic fertilizers promoted the rapeseed nutrient uptake and improved the purple soil quality. Front. Nutr. 9, 997151. https://doi.org/10.3389/fnut.2022.997151.

Lobell, D.B., Burke, M.B., Tebaldi, C., Mastrandrea, M.D., Falcon, W.P., Naylor, R.L., 2008. Prioritizing climate change adaptation needs for food security in 2030. Science 319 (5863), 607–610.

Lorek, T.W., 2022. The Green Revolution in Latin America. In: Research Encyclopedia of Latin American History. Oxford University Press, In: Oxford.

Lorek, T.W., 2023. Making the Green Revolution: Agriculture and Conflict in Colombia. University of North Carolina Press.

Lotter, D.W., 2003. Organic agriculture. J. Sustain. Agric. 21 (4), 59–128. https://doi.org/10.1300/J064v21n04_06.

Lu, T., Luo, P., Wang, J., Lu, Y., Huo, A., Liu, L., 2025. Soil salinity accumulation and groundwater degradation due to overexploitation over recent 40-year period in Yaoba Oasis, China. Soil. Tillage Res. 248, 106398. https://doi.org/10.1016/j.still.2024.106398.

Lutz, A.F., Immerzeel, W.W., Siderius, C., Wijngaard, R.R., Nepal, S., Shrestha, A.B., et al., 2022. South Asian agriculture increasingly dependent on meltwater and groundwater. Nat. Clim. Change 12 (6), 566–573.

Mall, R.K., Gupta, A., Singh, R., Singh, R.S., Rathore, L.S., 2006. Water resources and climate change: an Indian perspective. Curr. Sci. 1610–1626.

Mallareddy, M., Thirumalaikumar, R., Balasubramanian, P., Naseeruddin, R., Nithya, N., Mariadoss, A., et al., 2023. Maximizing water use efficiency in rice farming: a comprehensive review of innovative irrigation management technologies. Water 15 (10), 1802. https://doi.org/10.3390/w15101802.

Mandal, A., Sarkar, B., Mandal, S., Vithanage, M., Patra, A.K., Manna, M.C., 2020. Impact of agrochemicals on soil health. In: Prasad, M.N.V. (Ed.), Agrochemicals Detection, Treatment and Remediation. Butterworth-Heinemann, pp. 161–187.

Manlay, R.J., Feller, C., Swift, M.J., 2007. Historical evolution of soil organic matter concepts and their relationships with the fertility and sustainability of cropping systems. Agric. Ecosyst. Environ. 119 (3-4), 217–233.

Marcus, J.B., 2013. Culinary arts basics: healthy cooking fundamentals: the culinary competencies of healthy food selection, preparation and food service. Culin. Nutr. 99–147. https://doi.org/10.1016/B978-0-443-16004-2.00003-3.

Marino, M., Puppo, F., Del Bo', C., Vinelli, V., Riso, P., Porrini, M., et al., 2021. A systematic review of worldwide consumption of ultra-processed foods: findings and criticisms. Nutrients 13 (8), 2778. https://doi.org/10.3390/nu13082778.

Martines-Filho, J., Burnquist, H.L., Vian, C.E., 2006. Bioenergy and the rise of sugarcane-based ethanol in Brazil. Choices 21 (2), 91–96.

Martínez-Moreno, F., Solís, I., Igartua, E., 2024. Barley history and breeding in Spain. Agriculture 14 (10), 1674. https://doi.org/10.3390/agriculture14101674.

Martin-Guay, M.O., Paquette, A., Dupras, J., Rivest, D., 2018. The new green revolution: sustainable intensification of agriculture by intercropping. Sci. Total. Environ. 615, 767–772.

Massah, J., Azadegan, B., 2016. Effect of chemical fertilizers on soil compaction and degradation. Agric. Mechanization Asia, Afr. Lat. Am. 47 (1), 44–50.

Matson, P.A., Parton, W.J., Power, A.G., Swift, M.J., 1997. Agricultural intensification and ecosystem properties. Science 504–509. https://doi.org/10.1126/science.277.5325.504.

Mazoyer, M., Roudart, L., 2006. A History of World Agriculture: From the Neolithic Age to the Current Crisis. NYU Press.

Meena, R.N., Yadav, L., Ghilotia, Y.K., Meena, R.K., 2013. Food security and agricultural sustainability–an impact of Green Revolution. Environ. Ecol. 31 (2C), 1190–1197.

Mejia, D., Restrepo, P., 2013. Bushes and bullets: Illegal cocaine markets and violence in Colombia. Documento CEDE (2013-53).

Menalled, F., Jones, C., Buschena, D., Miller, P., 2009. From Conventional to Organic Cropping: What to Expect During the Transition Years. Montana State University Extension Service MontGuide MT200901AG, Bozeman, MT.

Mendez, J.A., 1991. The Development of the Colombian Cut Flower Industry. Country Economics Department, World Bank, Washington, DC, USA.

Menzi, H., Oenema, O., Burton, C., Shipin, O., Gerber, P., Robinson, T., et al., 2010. Impacts of intensive livestock production and manure management on the environment. Livest. a Changing Landsc 1, 139–163.

Mercer, K.L., Perales, H.R., 2010. Evolutionary response of landraces to climate change in centers of crop diversity. Evolut. Appl. 3 (5-6), 480–493. https://doi.org/10.1111/j.1752-4571.2010.00137.x.

Miguel, M.G., Barreto, R.P., Pereira, S.Y., 2017. Study of a tropical soil in order to use it to retain aluminum, iron, manganese and fluoride from acid mine drainage. J. Environ. Manag. 204, 563–570.

Milani, P., Torres-Aguilar, P., Hamaker, B., Manary, M., Abushamma, S., Laar, A., et al., 2022. The whole grain manifesto: from green revolution to grain evolution. Glob. Food Security 34, 100649. https://doi.org/10.1016/j.gfs.2022.100649.

Miller, H.I., 2012. Norman Borlaug: The Genius Behind The Green Revolution. Forbes URL http://www.forbes./com/sites/henrymiller/2012/01/18/norman-borlaug-the-geniusbehind-the-green-revolution. [Accessed 01 March 2026].

Morris, M., Kelly, V.A., Kopicki, R.J., Byerlee, D., 2007. Fertilizer Use in African Agriculture: Lessons Learned and Good Practice Guidelines. World Bank Publications. https://doi.org/10.1596/978-0-8213-6880-0.

Mortensen, D.A., Smith, R.G., 2020. Confronting barriers to cropping system diversification. Front. Sustain. Food Syst. 4, 564197. https://doi.org/10.3389/fsufs.2020.564197.

Mortimore, M., 1998. Roots in the African Dust: Sustaining the Sub-Saharan Drylands. Cambridge University Press.

Muchhadiya, R.M., Gohil, B., Yadahalli, V., Hm, A.U.R., Siddiqua, A., Khayum, A., et al., 2024. Feeding the world: agronomic innovations to meet the challenges of a growing population. Int. J. Res. Agron 7 (7), 790–802.

Mukherji, A., Facon, T., Burke, J.J., De Fraiture, C., Faures, J., Fuleki, B., et al., 2009. Revitalizing Asia's Irrigation: To Sustainably Meet Tomorrow's Food Needs. IWMI Books Reports.

Murphy, D.V., Stockdale, E.A., Brookes, P.C., Goulding, K.W., 2007. Impact of microorganisms on chemical transformations in soil. In: Abbott, L.K., Murphy, D. (Eds.), Soil Biological Fertility: A Key to Sustainable Land Use in Agriculture. Springer Netherlands, Dordrecht, pp. 37–59.

National Bureau of Statistics of the People's Republic of China, 2020. National data. http://data.stats.gov.cn/index.htm [Accessed 07 Oct 2025].

Naz, M., Dai, Z., Hussain, S., Tariq, M., Danish, S., Khan, I.U., et al., 2022. The soil pH and heavy metals revealed their impact on soil microbial community. J. Environ. Manag. 321, 115770.

Nazari, A.H., Abdullah, Z.K., Malakzai, M.Y., 2024. Exploring the multi-dimensional impact of insects on the Green Revolution. J. Nat. Sci. Rev. 2, 221–238. https://doi.org/10.62810/jnsr.v2iSpecial.Issue.126.

Neelam, P., Dorin, B., Nagaich, R., 2022. A new paradigm for Indian agriculture. Agroindustry Agroecology. https://hal.science/hal-04351735v1 [Accessed 01 Mar 2026].

Nestle, M., 1998. defense of the USDA food guide pyramid. Nutr. Today 33, 189–197.

Nicholson, S.E., 2001. Climatic and environmental change in Africa during the last two centuries. Clim. Res. 17, 123–144.

Nieder, R., Benbi, D.K., Reichl, F.X., 2018. Health risks associated with pesticides in soils. Soil Components and Human Health. Springer Netherlands, Dordrecht, pp. 503–573. https://doi.org/10.1007/978-94-024-1222-2_10.

Nobel Prize Outreach, 2026. Nobel Peace Prize 2025: Maria Corina Machado. NobelPrize.org [online]. Available at: https://www.nobelpeaceprize.org/laureates/.

O'Connor, D., 2009. The political economy of Colombia's cocaine industry. Pap. Político 14 (1), 81–106. https://hdl.handle.net/20.500.14330/CLA01000351870 [Accessed 01 Mar 2026].

Ogidi, O.I., Akpan, U.M., 2023. Impacts of chemical use in agricultural practices: perspectives of soil microorganisms and vegetation. In: Ogwu, M.C., Izah, S.C. (Eds.), One Health Implications of Agrochemicals and Their Sustainable Alternatives. Springer Nature Singapore, Singapore, pp. 765–792.

Ortiz, R., Mowbray, D., Dowswell, C., Rajaram, S., 2007. Dedication: Norman E. Borlaug the humanitarian plant scientist who changed the world. Plant. Breed. Rev. 1–37.

Osumanu, I.K., Kosoe, E.A., 2023. Global biodiversity decline and loss from agricultural intensification through agrochemical application. One Health Implications of Agrochemicals and Their Sustainable Alternatives. Springer Nature Singapore, Singapore, pp. 77–103.

Otero, G., Lapegna, P., 2016. Transgenic crops in Latin America: expropriation, negative value and the state. J. Agrarian Change 16 (4), 665–674.

Otsuka, K., Kijima, Y., 2010. Technology policies for a green revolution and agricultural transformation in Africa. J. Afr. Economies 19 (_2), ii60–ii76. https://doi.org/10.1093/jae/ejp025.

Pahalvi, H.N., Rafiya, L., Rashid, S., Nisar, B., Kamili, A.N., 2021. Chemical fertilizers and their impact on soil health. In: Dar, G.H., Bhat, R.A., Mehmood, M.A., Hakeem, K.R. (Eds.), Microbiota and Biofertilizers, Vol 2: Ecofriendly Tools for Reclamation of Degraded Soil Environs. Springer International Publishing, Cham, pp. 1–20.

Painter, J., Rah, J.H., Lee, Y.K., 2002. Comparison of international food guide pictorial representations. J. Am. Dietetic Assoc. 102 (4), 483–489. https://doi.org/10.1016/S0002-8223(02)90113-6.

Pe'er, G., Bonn, A., Bruelheide, H., Dieker, P., Eisenhauer, N., Feindt, P.H., Hagedorn, G., Hansjürgens, B., Herzon, I., Lomba, Â., Marquard, E., Moreira, F., Nitsch, H., Oppermann, R., Perino, G., Röder, N., Schmidt, J., Siebert, R., Settele, J., Sutcliffe, L.M.E., Tzanopoulos, J., Vadineanu, A., von Wehrden, H., Lakner, S., 2019. Action needed for the EU Common Agricultural Policy to address sustainability challenges. People and. Nature 2 (2), 305–316.

Peterson, E.W.F., 2025. Structure U.S. Agriculture. Cornhusker Econ., Dep. Agric. Economics. University of Nebraska–Lincoln [Accessed 7 October 2025]. https://agecon.unl.edu/structure-us-agriculture/.

Philippines (no date). https://www.fao.org/nutrition/education/food-dietary-guidelines/regions/countries/philippines/en/#:~:text=Official%20name,support%20messages%20for%20the%20pyramid [Accessed 01 March 2026].

Phillips, R.L., 2013. Norman Ernest Borlaug, 25 March 1914–12 September 2009. Biographical Memoirs of Fellows of the Royal Society, 59, pp.59–72.

Picado, W., 2022. Evaluating the Green Revolution dominant narrative for Latin America: technology, geopolitics, and institutions. In: Barahona, A. (Ed.), Handbook of the Historiography of Latin American Studies on the Life Sciences and Medicine. Springer International Publishing, Cham, pp. 355–373. https://doi.org/10.1007/978-3-030-74723-7_14.

Pingali, P.L., 2012. Green Revolution: Impacts, limits, and the path ahead. Proc. Natl. Acad. Sci. 109 (31), 12302–12308.

Pingali, P., 2023. Are the lessons from the Green Revolution relevant for agricultural growth and food security in the twenty-first century? In: Estudillo, J.P., Kijima, Y., Sonobe, T. (Eds.), Agricultural Development in Asia and Africa: Essays in Honor of Keijiro Otsuka. Springer Nature Singapore, Singapore, pp. 21–32.

Pingali, P., Abraham, M., 2022. Food systems transformation in Asia–A brief economic history. Agric. Econ. 53 (6), 895–910. https://doi.org/10.1111/agec.12734.

Pingali, P.L., Hossain, M., Gerpacio, R.V., 1997. Asian Rice Bowls: The Returning Crisis? CABI Publishing.

Pingali, P.L., Rosegrant, M.W., 1994. Confronting the environmental consequences of the Green Revolution in Asia. Int. Food Policy Res. Inst.

Pingali, P.L., Shah, M., 2001. Policy re-directions for sustainable resource use: the rice-wheat cropping system of the Indo-gangetic plains. J. Crop. Prod. 3 (2), 103–118. https://doi.org/10.1300/J144v03n02_05.

Pingali, P.L., Rosegrant, M.W., 1994. Confronting the environmental consequences of the Green Revolution in Asia. https://hdl.handle.net/10568/156849.

Qaim, M., Traxler, G., 2005. Roundup ready soybeans in Argentina: farm level and aggregate welfare effects. Agric. Econ. 32 (1), 73–86.

Quisumbing, A., Meinzen-Dick, R., Malapit, H., 2022. Women's empowerment and gender equality in South Asian agriculture: measuring progress using the project-level Women's Empowerment in Agriculture Index (pro-WEAI) in Bangladesh and India. World Dev. 151, 105396. https://doi.org/10.1016/j.worlddev.2021.105396.

Qureshi, A.S., 2020. Groundwater governance in Pakistan: from colossal development to neglected management. Water 12 (11), 3017. https://doi.org/10.3390/w12113017.

Qureshi, H.U., Shah, S.M.H., Teo, F.Y., 2023. Trend assessment of changing climate patterns over the major agro-climatic zones of Sindh and Punjab. Front. Water 5, 1194540. https://doi.org/10.3389/frwa.2023.1194540.

Rahman, S., 2015. Green revolution in India: environmental degradation and impact on livestock. Asian J. Water, Environ. Pollut. 12 (1), 75–80.

Rahman, M.A., Lee, S., Ji, H.C., Kabir, A.H., Jones, C.S., Lee, K., 2018. Importance of mineral nutrition for mitigating aluminum toxicity in plants on acidic soils: current status and opportunities. Int. J. Mol. Sci. 19 (10), 3073.

Rahmann, G., Reza Ardakani, M., Bàrberi, P., Boehm, H., Canali, S., Chander, M., et al., 2017. Organic agriculture 3.0 is innovation with research. Org. Agric. 7 (3), 169–197.

Rajaram, S., Borlaug, N.E., Van Ginkel, M., 2002. CIMMYT International Wheat Breeding. Bread Wheat Improvement and Production. FAO, Rome, pp. 103–117.

Rajsic, P., 2025. Economic Incentives in Canada's Dairy Quota System: Implications for Market Entry and Policy Reform. Food Systems under Pressure-Local Practices, Risk Resilience, and Sustainability. IntechOpen, https://doi.org/10.5772/intechopen.1012757.

Raney, T., Anríquez, G., Croppenstedt, A., Gerosa, S., Lowder, S.K., Matuschke, I., et al., 2011. The role of women in agriculture. https://doi.org/10.22004/ag.econ.289018.

Rashid, S., 2019. Introduction [in The making of a blue revolution in Bangladesh]. https://hdl.handle.net/10568/147297.

Reganold, J.P., Wachter, J.M., 2016. Organic agriculture in the twenty-first century. Nat. plants 2 (2), 15221. https://doi.org/10.1038/nplants.2015.221.

Reijntjes, C., Haverkort, B., Waters-Bayer, A., 1992. Farming for the Future: An Introduction to Low-External-Input and Sustainable Agriculture. Macmillan Education Ltd., London.

Reveles, I.L.A., 2012. The limits and contradictions of agricultural technology in Latin America: lessons from Mexico and Argentina. Perspect. Glob. Dev. Technol. 11 (3), 386–400. https://doi.org/10.1163/156914912X651550.

Revenga Frauca, J.A., Soriano del Castillo, J.M., 2026. The new US food pyramid is scientifically questionable, and upends decades of trusted public health advice. The Conversation [online]. Available at: https://theconversation.com/the-new-us-food-pyramid-is-scientifically-questionable-and-upends-decades-of-trusted-public-health-advice-274082.

Reynolds, M.P., Borlaug, N.E., 2006a. Applying innovations and new technologies for international collaborative wheat improvement. J. Agric. Sci. 144 (2), 95–110. https://doi.org/10.1017/S0021859606005879.

Reynolds, M.P., Borlaug, N.E., 2006b. Impacts of breeding on international collaborative wheat improvement. J. Agric. Sci. 144 (1), 3–17. https://doi.org/10.1017/S0021859606005867.

Rizov, M., Pokrivcak, J., Ciaian, P., 2013. CAP subsidies and productivity of the EU farms. J. Agric. Econ. 64 (3), 537–557. https://doi.org/10.1111/1477-9552.12030.

Rockström, J., Falkenmark, M., Karlberg, L., Hoff, H., Rost, S., Gerten, D., 2009. Future water availability for global food production: the potential of green water for increasing resilience to global change. Water Resour. Res. 45 (7). https://doi.org/10.1029/2007WR006767.

Rogers, T.D., 2015. Agricultural transformations in sugarcane and labor in Brazil. Oxf. Res. Encycl. Lat. Am. History. https://doi.org/10.1093/acrefore/9780199366439.013.55.

Rose, M.T., Cavagnaro, T.R., Scanlan, C.A., Rose, T.J., Vancov, T., Kimber, S., et al., 2016. Impact of herbicides on soil biology and function. Adv. Agron. 136, 133–220.

Roy, D., Joshi, P.K., Chandra, R., 2017. Pulses for nutrition in India: changing patterns from farm to fork. Intl. Food Policy Res. Inst.

Sahu, G., 2020. Evergreen Revolution: a need to adopt in agricultural tradition in India. Pop. Kheti 8 (3), 109–113.

Sanchez, P.A., 2002. Soil fertility and hunger in Africa. Science 295 (5562), 2019–2020.

Sanchez, P.A., Shepherd, K.D., Soule, M.J., Place, F.M., Buresh, R.J., Izac, A.M.N., Mokwunye, A.U., Kwesiga, F.R., Ndiritu, C.G., Woomer, P.L., 1997. Soil fertility replenishment in Africa: An investment in natural resource capital. In: Buresh, R.J., Sanchez, P.A., Calhoun, F. (Eds.), Replenishing Soil Fertility in Africa. SSSA Special Publication No. 51, Madison, WI, pp. 1–46.

Sanginga, N., Woomer, P.L. (Eds.), 2009. Integrated Soil Fertility Management in Africa: Principles, Practices, and Developmental Process. CIAT.

Santillán-Urquiza, E., Ruiz-Espinosa, H., Angulo-Molina, A., Ruiz, J.F.V., Méndez-Rojas, M.A., 2017. Applications of nanomaterials in functional fortified dairy products: benefits and implications for human health. In: Grumezescu, A.M. (Ed.), Nutrient Delivery. Academic Press, pp. 293–328. https://doi.org/10.1016/B978-0-12-804304-2.00008-1.

Sarker, A., Kim, J.E., Islam, A.R.M.T., Bilal, M., Rakib, M.R.J., Nandi, R., et al., 2022. Heavy metals contamination and associated health risks in food webs—a review focuses on food safety and environmental sustainability in Bangladesh. Environ. Sci. Pollut. Res. 29 (3), 3230–3245.

Sarmah, D., Vinodh, S., Pandey, Y., Sagolsem, D., Pandey, A.K., 2025. Edible flowers and global food security: potential, challenges, and pathways forward. World Food Day 2025 28.

Schlenker, W., Lobell, D.B., 2010. Robust negative impacts of climate change on African agriculture. Environ. Res. Lett. 5 (1), 014010.

Schlett, A., 2015. The socialist-type process of innovation lessons of hungarian agrarian modernization between 1960 and 1990. The Hungarian historical review: new series of. Acta Historica Academiae Scientiarum Hungaricae 4 (4), 900–927.

Schrama, M., De Haan, J.J., Kroonen, M., Verstegen, H., Van der Putten, W.H., 2018. Crop yield gap and stability in organic and conventional farming systems. Agriculture, Ecosyst. & Environ. 256, 123–130.

Scott, R., 2000. Exported to death: the failture of agricultural deregulation. Minn. J. Glob. Trade 9, 87.

Scott, C.A., Sharma, B., 2009. Energy supply and the expansion of groundwater irrigation in the Indus-Ganges Basin. Int. J. River Basin Manag. 7 (2), 119–124. https://doi.org/10.1080/15715124.2009.9635374.

Scown, M.W., Nicholas, K.A., 2020. European agricultural policy requires a stronger performance framework to achieve the sustainable development goals. Glob. Sustain. 3, e11. https://doi.org/10.1017/sus.2020.5.

Sekhri, S., Shastry, G.K., 2024. The curse of plenty: the green revolution and the rise in chronic disease. Am. Economic Journal: Appl. Econ. 16 (4), 1–33. https://doi.org/10.1257/app.20220621.

Sevilla, R., 2006. Use of native and introduced maize diversity to improve cold tolerance in Andean maize. In: Reynolds, M.P., Ortiz-Monasterio, J.I., McNab, A. (Eds.), Enhancing the Use of Crop Genetic Diversity to Manage Abiotic Stress in Agricultural Production Systems. FAO, Rome, pp. 84–91.

Shah, T., 2007. The groundwater economy of South Asia: an assessment of size, significance and socio-ecological impacts. Agric. Groundw. Revolution: Opportunities Threat. Dev. 7–36.

Shah, K.K., Modi, B., Pandey, H.P., Subedi, A., Aryal, G., Pandey, M., et al., 2021. Diversified crop rotation: an approach for sustainable agriculture production. Adv. Agric. 2021 (1), 8924087. https://doi.org/10.1155/2021/8924087.

Shah, T., Roy, A.D., Qureshi, A.S., Wang, J., 2003. Sustaining Asia's groundwater boom: an overview of issues and evidence. Nat. Resour. Forum 27. Blackwell Publishing Ltd, Oxford, UK, pp. 130–141. https://doi.org/10.1111/1477-8947.00048.

Sharma, A., Jha, P., Reddy, G.V., 2018. Multidimensional relationships of herbicides with insect-crop food webs. Sci. Total. Environ. 643, 1522–1532.

Sibanda, M., Mushunje, A., Mutengwa, C.S., 2016. An evaluation on the profitability of growing improved maize open pollinated varieties in the Eastern Cape Province, South Africa. J. Dev. Agric. Econ. 8 (1), 1–13. https://doi.org/10.5897/JDAE2014.0713.

Singer, S.D., Laurie, J.D., Bilichak, A., Kumar, S., Singh, J., 2021. Genetic variation and unintended risk in the context of old and new breeding techniques. Crit. Rev. Plant. Sci. 40 (1), 68–108. https://doi.org/10.1080/07352689.2021.1883826.

Singh, R.B., 2000. Environmental consequences of agricultural development: a case study from the Green Revolution state of Haryana, India. Agric., Ecosyst. Environ. 82 (1-3), 97–103.

Smale, M., Byerlee, D., Jayne, T., 2012. Maize revolutions in sub-Saharan Africa. In: Otsuka, K., Larson, D.F. (Eds.), An African Green Revolution: Finding Ways to Boost Productivity on Small Farms. Springer Netherlands, Dordrecht, pp. 165–195.

Smale, M., Jayne, T.S., 2003. Maize in eastern and southern Africa:'Seeds' of success in retrospect. https://hdl.handle.net/10568/157009.

Smith, M.G., Urpelainen, J., 2016. Rural electrification and groundwater pumps in India: evidence from the 1982–1999 period. Resour. Energy Econ. 45, 31–45. https://doi.org/10.1016/j.reseneeco.2016.05.004.

Snowdon, R.J., Wittkop, B., Chen, T.W., Stahl, A., 2021. Crop adaptation to climate change as a consequence of long-term breeding. Theor. Appl. Genet. 134 (6), 1613–1623. https://doi.org/10.1007/s00122-020-03729-3.

Sobha, I., 2007. Green revolution: impact on gender. J. Hum. Ecol 22 (2), 107–113.
Somado, E.A., Guei, R.G, 2008. In: Keya, S.O. (Ed.), The New Rice for Africa—A Compendium. Africa Rice Center (WARDA), Cotonou, Benin.
Sotte, F., Vergamini, D., 2025. The green deal and the CAP 2023–2027. European Agricultural Policy: History and Analysis. Springer Nature Switzerland, Cham, pp. 219–235. https://doi.org/10.1007/978-3-031-83313-7_11.
Sponsler, D.B., Requier, F., Kallnik, K., Classen, A., Maihoff, F., Sieger, J., et al., 2022. Contrasting patterns of richness, abundance, and turnover in mountain bumble bees and their floral hosts. Ecology 103 (7), e3712. https://doi.org/10.1002/ecy.3712.
Stallwood, P., 2024, March 4. Desertification in Africa: causes, effects and solutions. Earth. Org. https://earth.org/desertification-in-africa/ [Accessed 01 March 2026].
Stevenson, J.R., Villoria, N., Byerlee, D., Kelley, T., Maredia, M., 2013. Green revolution research saved an estimated 18–27 million hectares from being brought into agricultural production. Proc. Natl. Acad. Sci. 110 (21), 8363–8368.
Sukhatme, P., 1961. The world's hunger and future needs in food supplies. J. R. Stat. Society: Ser. A (Gen). 124 (4), 463–508. https://doi.org/10.2307/2342930.
Swinnen, J.F., Mathijs, E., 2018. Agricultural privatisation, land reform and farm restructuring in Central and Eastern Europe: A comparative analysis. Agricultural Privatization, Land Reform and Farm Restructuring in Central and Eastern Europe. Routledge, pp. 333–371.
Swinnen, J.F., Van Herck, K., Vranken, L., 2009. Agricultural productivity in transition economies. Choices 24 (4).
Sylvain, Z.A., Wall, D.H., 2011. Linking soil biodiversity and vegetation: implications for a changing planet. Am. J. Botany 98 (3), 517–527.
Tabo, R., Bationo, A., Gerard, B., Ndjeunga, J., Marchal, D., Amadou, B., et al., 2007. Improving cereal productivity and farmers' income using a strategic application of fertilizers in West Africa. In: Bationo, A., Waswa, B., Kihara, J., Kimetu, J. (Eds.), Advances in Integrated Soil Fertility Management in Sub-Saharan Africa: Challenges and Opportunities. Springer Netherlands, Dordrecht, pp. 201–208.
Tadesse, W., Halila, H., Jamal, M., El-Hanafi, S., Assefa, S., Oweis, T., et al., 2017. Role of sustainable wheat production to ensure food security in the CWANA region. J. Exp. Biol. Agric. Sci. 5 (Spl-1–SAFSAW), S15–S32.
Tadesse, W., Sanchez-Garcia, M., Thabet, S.G., Tawkaz, S., Amri, A., El Bouhssini, M., 2019. In: Baum, M. (Ed.), Wheat Breeding Manual. International Center for Agricultural Research in the Dry Areas (ICARDA), Beirut, Lebanon.
Tadesse, W., Zegeye, H., Debele, T., Kassa, D., Shiferaw, W., Solomon, T., et al., 2022. Wheat production and breeding in Ethiopia: retrospect and prospects. Crop. Breeding, Genet. Genomics 4 (3). https://doi.org/10.20900/cbgg20220003.
Thiele-Bruhn, S., Bloem, J., de Vries, F.T., Kalbitz, K., Wagg, C., 2012. Linking soil biodiversity and agricultural soil management. Curr. Opin. Environ. Sustainability 4 (5), 523–528.
Tilman, D., 1998. The greening of the green revolution. Nature 396 (6708), 211–212. https://doi.org/10.1038/24254.
Tilman, D., Cassman, K.G., Matson, P.A., Naylor, R., Polasky, S., 2002. Agricultural sustainability and intensive production practices. Nature 418 (6898), 671–677.
Tittonell, P., Giller, K.E., 2013. When yield gaps are poverty traps: the paradigm of ecological intensification in African smallholder agriculture. Field Crop. Res. 143, 76–90. https://doi.org/10.1016/j.fcr.2012.10.007.
Tiwari, A.K., 2024. IPM essentials: combining biology, ecology, and agriculture for sustainable pest control. J. Adv. Biol. & Biotechnol. 27 (2), 39–47. https://doi.org/10.9734/jabb/2024/v27i2697.

Toenniessen, G., Adesina, A., DeVries, J., 2008. Building an alliance for a green revolution in Africa. Ann. N. Y. Acad. Sci. 1136 (1), 233–242. https://doi.org/10.1196/annals.1425.028.
Tracy, M., 1989. Government and agriculture in Western Europe 1880–1988.3rd ed. Harvester Wheatsheaf, New York.
Traxler, G., 2006. The GMO experience in North and South America. Int. J. Technol. Glob. 2 (1–2), 46–64. https://doi.org/10.1504/IJTG.2006.009126.
Trébuil, G., 1995. Pioneer agriculture, green revolution and environmental degradation in Thailand. In: Rigg, J. (Ed.), Counting the Costs: Economic Growth and Environmental Change in Thailand. Institute of Southeast Asian Studies (ISEAS), Singapore, pp. 67–89.
Tripathi, S., Srivastava, P., Devi, R.S., Bhadouria, R., 2020. Influence of synthetic fertilizers and pesticides on soil health and soil microbiology. In: Prasad, M.N.V. (Ed.), Agrochemicals Detection, Treat. Remediation. Butterworth-Heinemann, pp. 25–54.
Trostle, R., 2011. Why Have Food Commodity Prices Risen Again? Diane Publishing.
Tsige, M., Synnevåg, G., Aune, J.B., 2020. Gendered constraints for adopting climate-smart agriculture amongst smallholder Ethiopian women farmers. Sci. Afr. 7, e00250. https://doi.org/10.1016/j.sciaf.2019.e00250.
Uchimiya, M., Bannon, D., Nakanishi, H., McBride, M.B., Williams, M.A., Yoshihara, T., 2020. Chemical speciation, plant uptake, and toxicity of heavy metals in agricultural soils. J. Agric. Food Chem. 68 (46), 12856–12869.
Ullah, Z., Iqbal, J., Abbasi, B.A., Ijaz, S., Yaseen, T., Waqar, R., et al., 2024. The Green revolution: promoting environmental stewardship and plant growth. In: Fahad, S., Saud, S., Nawaz, T., Gu, L., Ahmad, M., Zhou, R. (Eds.), Environment, Climate, Plant and Vegetation Growth. Springer Nature Switzerland, Cham, pp. 425–469.
United States Department of Agriculture (USDA) Economic Research Service, 2025. Farm Labor [online]. Available at: https://www.ers.usda.gov/topics/farm-economy/farm-labor [Accessed 26 February 2026].
United States Environmental Protection Agency (EPA), 2025. DDT — Brief History and Status. Available at: https://www.epa.gov/ingredients-used-pesticide-products/ddt-brief-history-and-status [Accessed 26 February 2026].
Uphoff, N. (Ed.), 2013. Agroecological Innovations: Increasing Food Production with Participatory Development. Routledge.
van der Burg, M., 2019. "Change in the making": 1970s and 1980s building stones to gender integration in CGIAR agricultural research. In: Sachs, C.E. (Ed.), Gender, agriculture and agrarian transformations. Routledge, pp. 35–56.
van Etten, J., 2022. Revisiting the adequacy of the economic policy narrative underpinning the green revolution. Agriculture Hum. Values 39 (4), 1357–1372.
van Grinsven, H.J., Ten Berge, H.F.M., Dalgaard, T., Fraters, B., Durand, P., Hart, A., et al., 2012. Management, regulation and environmental impacts of nitrogen fertilization in northwestern Europe under the Nitrates Directive; a benchmark study. Biogeosciences 9 (12), 5143–5160.
Vanlauwe, B., Descheemaeker, K., Giller, K.E., Huising, J., Merckx, R., Nziguheba, G., et al., 2015. Integrated soil fertility management in sub-Saharan Africa: unravelling local adaptation. Soil 1 (1), 491–508.
Vargas, R., 2002. The anti-drug policy, aerial spraying of illicit crops and their social, environmental and political impacts in Colombia. J. Drug. Issues 32 (1), 11–60. https://doi.org/10.1177/002204260203200102.
Waal, A.D., 2002. What's new in the 'New Partnership for Africa's Development'? Int. Aff. 78 (3), 463–476. https://doi.org/10.1111/1468-2346.00261.
Williams, A.M., Couch, C.A., Emmerich, S.D., Ogburn, D.F., 2025. Ultra-processed food consumption in youth and adults: United States, August 2021–August 2023. NCHS Data Brief No. 536, August 2025. Centers for Disease Control and Prevention. Available at: https://www.cdc.gov/nchs/products/databriefs/db536.htm [Accessed 26 February 2026].

Williamson, H.F., 2024. Timothy Lorek: making the green revolution—agriculture and conflict in Colombia. Agric. Hum. Values. https://doi.org/10.1007/s10460-024-10677-x.
Winders, B., 2024. The Global Ascension of Corn and Soybeans: from managing supply to expanding production in the United States and the world. Agric. History 98 (3), 470–480. https://doi.org/10.1215/00021482-11225640.
Wise, T.A., 2020. Failing Africa's Farmers: An Impact Assessment of the Alliance for a Green Revolution in Africa (AGRA). Global Development and Environment Institute, Tufts University, Medford, MA.
Wise, T.A., 2021. Old Fertilizer in New Bottles: Selling the Past as Innovation in Africa's Green Revolution. Global Development and Environment Institute, Tufts University, Medford, MA.
World Bank, 2008. World Development Report 2008: Agriculture for Development. World Bank, Washington, DC [Accessed 26 February 2026]. https://openknowledge.worldbank.org/handle/10986/5990.
World Food Prize Foundation, 2026. About the Foundation. Available at: https://www.worldfoodprize.org/en/about_the_foundation/.
Wu, Z., Keizire, B.B., Fan, S., Feng, X., 2025. Reforming agricultural subsidy policies to facilitate the transformation of agrifood systems: insights from China and Africa. Front. Agric. Sci. & Eng. 12 (4). https://doi.org/10.15302/J-FASE-2025624.
Yang, T., Chandio, A.A., Zhang, A., Liu, Y., 2023. Do farm subsidies effectively increase grain production? Evidence from major grain-producing regions of China. Foods 12 (7), 1435. https://doi.org/10.3390/foods12071435.
Yeboah, J.O., 2021. Effect of heavy metal contamination on soil enzymes activities. J. Geosci. Environ. Prot.
Yousuf, S., Naqash, N., Singh, R., 2022. Nutrient cycling: an approach for environmental sustainability. Environ. Microbiol.: Adv. Res. Multidiscip. Appl. 77–104.
Zaidi, P.H., Das, R.R., Singamsetti, A., Seetharam, K., Fatma, S., Kuchanur, P.H., et al., 2026. Does breeding for abiotic stress tolerance mitigate G× E interactions under a changing climate? In: Mohammadi, R., Kamali, M.R.J., Amri, A., Kehel, Z., Sadeghzadeh, B. (Eds.), Genotype x Environment Interactions and its Implications for Plant Breeding. Springer Nature Singapore, Singapore, pp. 149–185. https://doi.org/10.1007/978-981-95-5664-9_6.
Zhang, W., 2024. The impact of agricultural machinery services on farmland transfer: a analytical perspective based on the profitability of grain production. Front. Sustain. Food Syst. 8, 1431005.
Zhao, F.J., Ma, J.F., Meharg, A.A., McGrath, S.P., 2009. Arsenic uptake and metabolism in plants. N. Phytol. 181 (4), 777–794.
Zhu, H., He, X., Wang, X., Long, P., 2024. Increasing hybrid rice yield, water productivity, and nitrogen use efficiency: optimization strategies for irrigation and fertilizer management. Plants 13 (12), 1717. https://doi.org/10.3390/plants13121717.
Zieliński, M., Gołębiewska, B., Jadczyszyn, J., Adamski, M., Tyburski, J., 2025. Importance of environmental measures under the CAP 2023–2027 on high nature value farmlands: evidence from Poland. Sustainability 17 (17), 7763. https://doi.org/10.3390/su17177763.
Zinati, G.M., 2002. Transition from conventional to organic farming systems: I. Challenges, recommendations, and guidelines for pest management. HortTechnology 12 (4), 606–610.
Zou, Y., Liu, Z., Chen, Y., Wang, Y., Feng, S., 2024. Crop rotation and diversification in China: enhancing sustainable agriculture and resilience. Agriculture 14 (9), 1465. https://doi.org/10.3390/agriculture14091465.
Zou, X., Xiao, X., Zhou, H., Chen, F., Zeng, J., Wang, W., et al., 2018. Effects of soil acidification on the toxicity of organophosphorus pesticide on Eisenia fetida and its mechanism. J. Hazard. Mater. 359, 365–372.

CHAPTER NINE

The nexus of biodiversity in food security and sustainability

Christian Joseph N. Ong[a,*], Jerico Bautista Ogaya[b,c,d], Rahmatullah Nazari[e], Dauda Ayuba[f], Oshibe Joseph Daberechi[g], Rich Milton R. Dulay[a,h], Kevin Smith P. Cabuhat[i], Oliver R. Alaijos[j], Raymundo F. Javier[j], Virginia P. Paitan[j], Joana May Nepomuceno[j], Debbie Ann Tuazon[j], Michelle Agustin[j], Razeale G. Resultay[k], Rose A. Arceño[l], Joan M. Recente[m], Richard Ian Mark T. Necosia[m], and Don Eliseo Lucero-Prisno III[n,o,p]

[a]Department of Biology, College of Science, De La Salle University, Manila, Philippines
[b]Department of Medical Technology, Institute of Health Sciences and Nursing, Far Eastern University, Manila, Philippines
[c]Center for University Research, University of Makati, Makati, Philippines
[d]University Research and Development Office, Guimaras State University, Buenavista, Guimaras, Philippines
[e]Faculty of Dentistry, Cheragh University of Medical Sciences, Kabul, Afghanistan
[f]Department of Local Government and Development Studies, Ahmadu Bello University, Zaria, Nigeria
[g]Department of Medical Laboratory Science, Ebonyi State University, Abakaliki, Nigeria
[h]Center for Tropical Mushroom Research and Development, Central Luzon State University, Nueva Ecija, Philippines
[i]Basic Education Department, La Consolacion University Philippines, Malolos, Bulacan, Philippines
[j]Science Department, College of Science, Bulacan State University, Malolos, Bulacan, Philippines
[k]Research Office, Pangasinan State University, Lingayen, Pangasinan, Philippines
[l]Research Services Office, Palompon Institute of Technology, Palompon, Leyte, Philippines
[m]Research Office, Bukidnon State University, Malaybalay, Bukidnon, Philippines
[n]Department of Global Health and Development, London School of Hygiene and Tropical Medicine, London, United Kingdom
[o]Research and Development Center, John B. Lacson Foundation Maritime University, Iloilo City, Philippines
[p]Office of Research and Extension, University of Camarines Norte, Daet, Camarines Norte, Philippines
*Corresponding author. e-mail address: Christian_joseph_ong@dlsu.edu.ph

Contents

Advances in Food Security and Sustainability, Volume 11
ISSN 2452-2635, https://doi.org/10.1016/bs.af2s.2026.06.008

Abstract

Biodiversity underpins global food security and sustainability by supporting essential ecosystem services, enhancing agricultural resilience, and sustaining nutritional diversity. However, accelerating biodiversity loss driven by land-use change, climate change, pollution, and intensive agricultural practices poses a significant threat to the stability of food systems worldwide. This chapter examines the nexus between biodiversity, food security, and sustainability, emphasizing the interdependence of ecological integrity and human well-being. It explores how genetic, species, and ecosystem diversity contribute to crop productivity, pest and disease regulation, soil fertility, pollination, and climate resilience. The role of agrobiodiversity in improving dietary diversity, livelihood security, and adaptive capacity—particularly in smallholder and resource-limited farming systems—is highlighted. The chapter further discusses the adverse impacts of modern high-input agriculture, monoculture expansion, and biodiversity homogenization on ecosystem functionality and long-term food system sustainability. Emerging challenges, including climate-induced shifts in species distribution and ecosystem services, are examined within the context of global environmental change. Integrated approaches such as agroecology, climate-smart agriculture, conservation agriculture, and landscape-level management are presented as viable pathways for reconciling food production with biodiversity conservation. Strengthening governance frameworks, promoting inclusive policies, and enhancing biodiversity-based solutions are identified as critical strategies for achieving sustainable food systems. Ultimately, safeguarding biodiversity is not only an environmental imperative but also a fundamental requirement for ensuring resilient, equitable, and sustainable food security in the face of growing global population pressures and climate uncertainty.

1. Introduction

Biodiversity forms the foundational pillar for ensuring global food security and promoting truly sustainable agricultural practices. Biodiversity, as defined by the Convention on Biological Diversity (CBD), encompasses the variety of life at the genetic, species, and ecosystem levels, including the ecological complexes of which they are part (CBD, 1992). This intricate and essential relationship between the diversity of life—from genes to ecosystems—and human food systems is crucial for maintaining the vital ecosystem services upon which agriculture depends (Convention on Biological Diversity, 2018). Specifically, a healthy level of biodiversity enhances agricultural productivity, bolsters the resilience of crops and livestock, and maintains ecological stability (World Health Organization, n.d.). The conservation and strategic utilization of this biological wealth are not merely environmental concerns; they are fundamental prerequisites for long-term food provision and planetary health. This broad ecological principle underscores that resilient food systems are those that work with nature, utilizing its inherent complexity to deliver food while regenerating the environment (Çakmakçı et al., 2023).

The importance of biodiversity in agriculture manifests primarily through essential ecosystem services (CBD, 2008). This includes the critical function of pollination by insects and other animals, the maintenance of soil fertility and structure through microbial and faunal activity, and natural pest control mediated by beneficial organisms (Frison et al., 2011; Tamburini et al., 2020). These biological processes are fundamental for sustainable practices and long-term food provision, effectively acting as nature's subsidies for the farmer. Furthermore, biodiversity provides the necessary genetic diversity within crops and livestock. This genetic library is indispensable for breeding programs aimed at improving yields and, more critically, developing crucial resilience to emerging pests, diseases, and the unpredictable nature of climate change (FAO, n.d.b). In essence, diverse farming systems are inherently more stable, reducing the environmental footprint and economic risks associated with external, often environmentally harmful, synthetic inputs.

Despite its clear importance, the nexus faces significant challenges and threats that jeopardize the sustainability of global food systems. Modern, intensive, high-input agricultural practices, particularly large-scale monocultures and intensive land use, have been identified as major drivers of dramatic biodiversity loss at genetic, species, and ecosystem levels, largely through habitat loss, synthetic agrochemical use, water depletion, and soil degradation (Pereira et al., 2025). Agricultural expansion and intensification fragment habitats and simplify landscapes, reducing species richness and ecological connectivity, which undermines the ecosystem services upon which food production depends (Jing et al., 2025). For example, repeated pesticide application and large monocultures reduce pollinator diversity and disrupt beneficial soil microbial communities, diminishing pollination services and soil health critical for crop productivity (Jing et al., 2025; Barros-Rodríguez et al., 2021). These processes make food systems more vulnerable to environmental stresses and instability. This vulnerability is severely compounded by the acceleration of climate change, which alters temperature regimes and increases the frequency of extreme weather events, further stressing biodiversity and ecosystem functions in agricultural landscapes (Muluneh, 2021). Addressing these intertwined threats requires a profound paradigm shift in how food is produced, moving away from systems that destroy biodiversity toward those that strategically integrate and enhance it to safeguard long-term food security (Raveloaritiana and Wanger, 2024).

This chapter examines the nexus of biodiversity, food security, and sustainability through a structured progression. Section 2 explores biodiversity's specific contributions to food security, including the roles of agrobiodiversity, microbial biodiversity, and genetic diversity. Section 3 analyzes the major challenges and trade-offs, focusing on agricultural intensification, environmental degradation, and climate change. Section 4 presents sustainable practices and solutions that enhance biodiversity, namely agroforestry systems, local seed systems with intercropping, and offset mechanisms. Section 5 discusses policy frameworks, collaborative efforts among stakeholders, and landscape-specific strategies for sustainable development. Finally, Section 6 concludes by synthesizing the chapter's key arguments and emphasizing the strategic necessity of biodiversity conservation for resilient food systems.

To successfully build resilient and sustainable food systems, a concerted effort focused on agroecology and strategic diversification is required. Agroecology involves integrating ecological principles into agricultural design and management, which naturally enhances biodiversity, improves soil health, and significantly reduces the need for synthetic agrochemicals. This practical approach includes techniques like intercropping (the simultaneous cultivation of two or more crops in the same field to enhance resource use efficiency and yield stability), integrated pest management (IPM), and conservation tillage (reduced soil disturbance that maintains crop residues on the soil surface to protect soil structure and biodiversity). IPM is a decision-based pest control strategy that prioritizes prevention, monitoring, and biological or cultural controls, thereby avoiding the use of synthetic pesticides under most circumstances; however, it permits their judicious application when pest populations exceed established economic thresholds and other methods prove insufficient. It is important to note that the effectiveness of IPM often depends on coordinated collective action across individual farms, particularly for mobile pests that disperse across property boundaries. A complementary measure to this is crop diversification, including the promotion of neglected and underutilized species, which serves to improve local dietary health, reduce agriculture's overall environmental footprint, and increase farmers' resilience to market fluctuations and climate risks. Ultimately, the successful conservation and advancement of biodiversity within agricultural landscapes require effective policies and governance, supported by Multi-Stakeholder Initiatives (MSIs) and Voluntary Sustainability Standards (VSS), to safeguard this critical biological wealth for the food security and planetary health of future generations.

2. Biodiversity's role in food security

Biodiversity plays a crucial role in ensuring food security by enhancing the resilience and productivity of agricultural systems. The diversity of species, genes, and ecosystems contributes to the stability and sustainability of food production. For instance, agricultural biodiversity, which includes a variety of crops and livestock, can increase productivity and resilience in different growing conditions, thereby enhancing food security (Frison et al., 2011). Diverse farming systems are more resilient to environmental perturbations, such as climate change, pests, and diseases, which can otherwise threaten food supply (Renard and Tilman, 2021). Additionally, microbial biodiversity is essential for promoting plant growth, improving soil fertility, and nutrient cycling, all of which are fundamental to sustainable food production (Chauhan and Chaurasia, 2025). The integration of biodiversity into agricultural practices, such as crop diversification and the use of indigenous species, can mitigate the risks associated with monocultures and enhance the nutritional value of diets (Zuza et al., 2024; Ulian et al., 2020). Entrenched perceptions that monocultures are essential for farm profitability have slowed the adoption of diversified systems. These perceptions are reinforced by supply chains that favor large-scale specialized producers, yet meta-analytical evidence shows that diversified farms are at least as profitable as simplified ones (Sánchez et al., 2022). Diversified systems may involve higher upfront costs, particularly for labor and inputs, although meta-analytical evidence shows that these are typically offset by higher gross incomes, resulting in profits that are at least equivalent to those of simplified systems (Sánchez et al., 2022). Addressing this trade-off requires supportive policy interventions. Mechanisms such as payments for ecosystem services (PES), whereby farmers receive direct financial compensation for adopting conservation practices that generate public environmental benefits, can help align farm-level economic incentives with broader biodiversity and sustainability goals (J-PAL Policy Insight, 2026).

Biodiversity serves as the foundation of food security by supporting its four key pillars: availability, access, utilization, and stability. It shows that diverse ecosystems enhance agricultural productivity, nutritional quality, and resilience against environmental and economic risks. Biodiversity also promotes cultural and dietary diversity, ensuring communities have access to safe and nutritious food. Overall, protecting biodiversity is essential for maintaining sustainable and resilient food systems that can withstand future challenges (see Fig. 1).

Moreover, biodiversity supports various ecosystem services that are vital for food security. These services include pollination, biological pest control, soil and water quality regulation, and carbon sequestration (Rehman et al., 2022). Sustainable agricultural practices that conserve and enhance

Fig. 1 The complex relationship between biodiversity and food security (Dannenberg et al., 2024).

biodiversity can provide these ecosystem services, leading to more stable and productive food systems (Thrupp, 2004; Raveloaritiana and Wanger, 2026). For example, traditional farming practices that incorporate local knowledge and biodiversity conservation can improve food security by maintaining ecosystem health and resilience (Allasi Canales et al., 2026). The conservation of forest genetic resources and the integration of forests into agricultural landscapes also contribute to food security by providing additional food sources and ecosystem services, including carbon sequestration (Sunderland, 2011; Laird et al., 2024). Therefore, the conservation and sustainable use of biodiversity are essential for achieving long-term food security and addressing the challenges posed by environmental degradation and climate change (Gupta and Kumar, 2022).

2.1 Agro-biodiversity in food security

Agrobiodiversity, encompassing the variety and variability of animals, plants, and microorganisms used in agriculture, plays a critical role in ensuring food security. It contributes to the resilience of food systems by enhancing ecosystem functions such as pollination, soil fertility, and pest control, which are essential for sustainable agricultural production (Hailu, 2025; Dhakad et al., 2022). The diversity of crops and livestock helps to increase nutritional diversity, providing a broader range of nutrients necessary for human health (Hailu, 2025). Moreover, agrobiodiversity supports the adaptability of food systems to climate change, reducing the vulnerability of subsistence-based economies by offering a variety of species that can thrive under different environmental conditions (Pascual et al., 2011; Bravo-Peña and Yoder, 2024).

The conservation of agrobiodiversity is vital for maintaining ecological balance and preserving socio-cultural norms, particularly in smallholder farmer food systems (Lokhandwala, 2022). These systems often coincide with agrobiodiversity hotspots, where traditional agricultural practices have evolved to maintain a high level of genetic diversity (Lokhandwala, 2022; Ba et al., 2023). For instance, the Rukai indigenous farmers in Taiwan practice swidden cultivation (also called "shifting cultivation" or "slash-and-burn agriculture"), which integrates mixed cropping systems and regular fallows to maintain soil fertility and reduce climate risks (Ba et al., 2023). When practiced with adequate fallow periods, swidden agriculture can function as an environmentally sustainable and biodiversity-supporting agroecological system. The Rukai case study demonstrates that such systems are highly knowledge intensive and labor intensive, relying on traditional ecological knowledge, mixed cropping, rotational cultivation, and customary land management institutions to maintain soil fertility and food security. The effectiveness of these practices depends on access to sufficient land for rotational fallow cycles and on social institutions that regulate land use, reciprocity, and crop diversity. However, demographic pressures, shortened fallow periods, and increasing commercialization may threaten the long-term sustainability of traditional swidden systems, highlighting the importance of secure land tenure and supportive governance frameworks (Ba et al., 2023). Similarly, the jhum agroforestry system in India enriches crop and dietary diversity, ensuring food and nutritional security for indigenous communities (Pandey et al., 2022). These traditional practices highlight the importance of preserving local knowledge and cultural practices to sustain agrobiodiversity and food security.

However, agrobiodiversity is under threat from various factors, including industrial agriculture, urbanization, climate change, and the loss of traditional knowledge (Hailu, 2025; Dhakad et al., 2022). The shift towards monocultures and the abandonment of biodiversity-based practices have led to a reduction in genetic variety, making food systems more vulnerable to pests, diseases, and environmental changes (Njeru et al., 2022). To combat these threats, integrated measures such as in situ conversation, which involves the protection and management of species within their natural habitats and ecosystems, and ex situ conservation, which involves the conservation of components of biological diversity outside their natural habitats, along with habitat management and the use of molecular techniques for genetic characterization and conservation, are essential (Dhakad et al., 2022). Additionally, policies that promote sustainable agricultural practices and support the economic viability of traditional farming systems are crucial for maintaining agrobiodiversity and ensuring long-term food security (Rasul et al., 2022; Liao et al., 2025).

2.2 Microbial biodiversity

Microbial biodiversity plays a crucial role in enhancing soil fertility and promoting plant growth, which are fundamental to achieving food security. Diverse microbial communities contribute to nutrient cycling, organic matter decomposition, and the suppression of soil-borne pathogens, thereby improving soil health and fertility (Chettri et al., 2022; Menon et al., 2024). For instance, plant growth-promoting bacteria (PGPB) enhance soil fertility and plant productivity through various mechanisms, including nitrogen fixation, phosphorus solubilization, and the production of phytohormones (Al-Tammar and Khalifa, 2022; Semwal et al., 2025). Beyond growth promotion, certain soil bacteria also contribute directly to pest management. *Bacillus thuringiensis* (Bt), for example, is a soil bacterium that produces crystal proteins with insecticidal properties, which have been widely adopted as a natural biopesticide in both conventional and organic agriculture (Li et al., 2024). These processes are essential for maintaining soil structure and nutrient availability, which directly impact crop yields and quality. Additionally, microbial inoculants and biofertilizers tailored to local soil conditions can significantly boost soil regeneration efforts, reducing the reliance on synthetic fertilizers and mitigating environmental pollution (Menon et al., 2024; Semwal et al., 2025). Thus, fostering a healthy soil microbiome, which comprises the diverse community of microorganisms including bacteria, fungi, archaea, viruses, and protozoa that inhabit the soil and interact with plants and each other, is integral to sustainable agricultural practices and long-term food security (Menon et al., 2024).

Microbial biodiversity also plays a pivotal role in pest and disease management, which is critical for maintaining crop health and productivity. Beneficial microbes, such as biocontrol agents, help manage pests and diseases by producing antimicrobial compounds and competing with pathogenic microorganisms (Chettri et al., 2022; Kumar et al., 2024). For example, certain soil bacteria and fungi can suppress plant pathogens, thereby reducing the incidence of diseases and minimizing crop losses (Toor et al., 2024; Singh et al., 2019). This biocontrol mechanism is particularly important in organic farming systems, where the use of synthetic pesticides is limited. Moreover, microbial diversity acts as a buffer against climate-induced disruptions, enhancing the resilience of agricultural ecosystems to environmental stresses (Toor et al., 2024). By integrating microbial approaches into agricultural practices, farmers can improve crop protection and reduce the dependency on synthetic agrochemicals, contributing to both food security and environmental sustainability.

Beyond the field, microbial biodiversity is essential in food processing and preservation, which are key links in the food supply chain. Microorganisms are involved in various stages of food production, from fermentation to the synthesis of vitamins and antimicrobial substances (Landa-Acuña et al., 2022; Panizzon et al., 2015). These processes not only enhance the nutritional value and safety of food products but also extend their shelf life, thereby reducing food waste. For instance, the use of probiotics, which are live microorganisms that confer a health benefit on the host when administered in adequate amounts, and prebiotics, which are non-digestible food ingredients that selectively stimulate the growth or activity of beneficial gut bacteria, in food products has gained popularity due to their health benefits and role in maintaining gut health (Panizzon et al., 2015; Valentino et al., 2024). Additionally, microbial biotechnologies enable the development of innovative food products and ecological technologies, such as protein-rich microbial cells, which can serve as sustainable food supplements (Chettri et al., 2022; Landa-Acuña et al., 2022). By leveraging the diverse capabilities of microorganisms, the food industry can address the challenges of food security and sustainability, ensuring the availability and accessibility of nutritious food for all populations.

2.3 Genetic diversity

Genetic diversity is fundamental to food security as it underpins the adaptability and resilience of agricultural systems. The genetic variation within plant species allows crops to adapt to various environmental conditions, which is crucial for overcoming challenges such as climate change, pest outbreaks, and declining biodiversity (Saini et al., 2020). This diversity is essential for developing new crop varieties that can withstand both biotic stresses (such as pests, diseases, and weeds) and abiotic stresses (such as drought, salinity, and temperature extremes) while also improving nutritional quality through biofortification, thereby ensuring stable and nutritious food production (Asdal, 2025; Kole et al., 2025; see also the chapter on biofortification in this volume). Advanced genetic techniques, such as next-generation sequencing, facilitate the exploration and utilization of genetic diversity, enabling the identification of valuable traits that can be harnessed for crop improvement (Onda and Mochida, 2016). The conservation and sustainable use of plant genetic resources are critical for maintaining this diversity and supporting global food security (Saini et al., 2020; Salgotra and Chauhan, 2023).

The erosion of genetic diversity poses a significant threat to food security. Industrial farming practices and the widespread replacement of diverse traditional landraces, which are locally adapted and genetically heterogeneous crop populations that have been selected and maintained by farmers over generations, with genetically uniform modern cultivars have narrowed the genetic base of many staple crops. While landraces have historically served as important sources of traits for developing modern varieties, their ongoing loss reduces the pool of genetic resources available for maintenance breeding and the continued development of new varieties capable of withstanding emerging environmental challenges and evolving pest and disease pressures (Gopi and Manjula, 2018; Gadissa et al., 2021). Conservation efforts employ both ex situ and in situ strategies. Ex situ approaches, such as the establishment of genebanks and international treaties like the International Treaty on Plant Genetic Resources for Food and Agriculture, aim to preserve genetic resources and make them available for breeding and research (Engels and Ebert, 2021). In situ conservation, which involves maintaining genetic resources in farmers' fields and natural habitats, is equally important as it allows crops to continue evolving in response to environmental pressures and farmers' selection (Salgotra and Chauhan, 2023). PES can support conservation in agricultural landscapes by providing farmers with direct financial compensation for adopting practices that protect or restore ecosystem services, such as soil and water conservation, afforestation, and biodiversity-friendly farming. By rewarding farmers for these practices, PES helps align farm-level economic incentives with broader environmental and sustainability goals (FAO, 2011). These combined efforts are vital for ensuring that genetic diversity is maintained and can be utilized to develop resilient crop varieties capable of meeting the food demands of a growing global population (Salgotra and Chauhan, 2023; Asdal, 2025).

Genetic diversity also plays a crucial role in sustainable agriculture by enhancing ecosystem resilience and supporting biodiversity. Diverse agricultural systems can better withstand environmental stresses and contribute to soil conservation, pest and disease control, and nutrient cycling (Loizou et al., 2025; Brooker et al., 2016). For example, the rich genetic diversity of rice varieties in Kerala allows adaptation to multiple agroecologies and provides resistance to various stresses, highlighting the importance of conserving traditional landraces (Gopi and Manjula, 2018). International cooperation and frameworks, such as the Nagoya Protocol, are essential for promoting the sustainable use of genetic resources and ensuring equitable benefit-sharing (Irie et al., 2025). This means that the benefits arising from the utilization of genetic resources, including monetary benefits such as royalties and licensing fees, as well as non-monetary benefits such as the sharing of research results, technology transfer, and capacity building, are fairly distributed to the countries and communities that have conserved and provided those resources (CBD, 2011). By fostering the conservation and utilization of genetic diversity, we can build more resilient agricultural systems and secure food production for future generations (Loizou et al., 2025; Irie et al., 2025).

3. Challenges and trade-offs

Agricultural intensification increases food production but usually reduces biodiversity and harms ecosystems. Environmental degradation weakens natural resources essential for sustainable food systems. Climate change adds unpredictable pressures on crop yields, ecosystems, and biodiversity resilience. Understanding these interconnected challenges and their trade-offs is necessary. This understanding helps design strategies that protect biodiversity while ensuring food security and environmental health.

3.1 Agricultural intensification

Agricultural intensification is essential to meet rising food demand while limiting farmland expansion. (Aihounton and Christiaensen, 2024) demonstrated contemporary intensification with some conventional inputs, such as improved seeds and fertilizers, precision tools, but also integrated pest management, and conservation practices. This hybrid approach can rapidly increase yields but introduces scientific, economic, and environmental trade-offs. Addressing these trade-offs requires targeted research and well-designed policies to achieve sustainable outcomes.

Lyu et al. (2021) demonstrated that the benefits of agricultural intensification are context-dependent. In fertile and well-served environments, targeted nutrient and seed packages significantly increase yields and farm income. However, in marginal areas, these packages usually underperform without complementary investments in irrigation, extension services (i.e., provision of technical advice to farmers), and market access. These findings shift the perspective on intensification from a single-technology solution to a system-level challenge. Successful implementation requires aligning inputs with local agroecological conditions and institutional capacity. Input dynamics in global agriculture have changed. (Vos et al., 2025) reported that fertilizer flows and prices had rebounded after recent disruptions. But in March 2026, the (FAO, n.d.) warned that disruption to the Strait of Hormuz, which typically carries up to 30 % of internationally traded fertilizers, had triggered a severe shock to global commodity flows, with urea prices rising by 19–28 % within the first week of the escalation (FAO, 2026). However, smallholder farmers still face major challenges in terms of affordability and access. Nutrient availability drives short-term yield gains, but excessive or poorly timed fertilizer use can degrade soils and pollute water, reducing long-term productivity. Research should identify optimal nutrient rates for different soils and cropping systems. It should also prioritize developing low-cost, field-ready soil health diagnostics to promote sustainable fertilizer use.

Technology integration plays a critical role in modern agriculture. (Adib Bin Rashid et al., 2025) stated that precision tools, such as satellite monitoring, variable-rate applications, and decision support systems, enhance input efficiency and mitigate environmental impacts. However, adoption is limited in regions with poor digital infrastructure, low literacy, and restricted financial resources. Addressing these challenges requires interdisciplinary collaboration. Agronomists can develop effective technical practices, economists can create suitable credit and insurance schemes, human-centered designers can improve usability for smallholder farmers, and extension advisers can work with farmers on uptake of the new approaches and applying them to specific contexts.

Policy and governance determine if agricultural intensification leads to sustainable outcomes. Factors such as secure property rights, subsidies, market incentives, extension services, and land-use planning shape farmers' decisions. (Pratzer et al., 2023) emphasize that empirical policy research should assess scalable governance models. Examples include performance-based payments for sequestration of carbon in the soil carbon, locally adapted input subsidies, and public–private extension partnerships. These strategies can align agricultural productivity with environmental stewardship. Agricultural intensification advances knowledge by shifting focus from merely increasing inputs to developing smarter, sustainable, context-specific systems. Future research should integrate field experiments, system modeling, and implementation studies. These approaches can identify interventions that increase yields, conserve natural capital, and enhance resilience among vulnerable producers (Aihounton and Christiaensen, 2024).

3.2 Environmental degradation

Mishra (2025) explained that environmental degradation undermines agricultural productivity, ecosystem services, and human well-being. It results from interconnected processes, including soil decline, water contamination, biodiversity loss, and landscape conversion. These processes occur at multiple scales and interact with climate change, increasing risks and reducing system resilience. (Kopittke et al., 2025) reported that global soil systems are deteriorating rapidly. Approximately one-third of the world's soils suffer moderate to severe degradation due to erosion, loss of organic carbon, salinization, compaction, and nutrient imbalance. This degradation lowers crop yields, decreases water retention, and impairs landscapes' capacity to absorb climatic shocks. Therefore, restoring soil organic matter and structure should be a key focus for sustainable land management.

Akinnawo (2023) explained that freshwater quality declines due to eutrophication. This problem arises from excessive nutrient application and poor waste management in rivers, lakes, and coastal zones. Nutrient pollution reduces oxygen levels, causing fishery collapse and increasing the costs of drinking water treatment. While national averages may appear moderate, local hotspots often experience severe ecological collapse. These spatial concentrations complicate regulation and remediation efforts. (Pfenning-Butterworth et al., 2024) noted that biodiversity loss weakens agroecosystems.

The Intergovernmental Science-Policy Platform on Biodiversity and Ecosystem Services (IPBES) estimates that about one million species face extinction risk. This threat results mainly from habitat conversion, overexploitation, pollution, and climate change. Loss of pollinators, natural pest enemies, and soil biota diminishes essential ecosystem services that support crop productivity and agroecological stability. Conservation that integrates protected areas with sustainable working landscapes can achieve better outcomes for both production and biodiversity. However, there can be significant trade-offs between the establishment of protected areas and the livelihoods of people who live in or near them, particularly in developing countries where communities depend directly on natural resources for food, fuel, and income (Wei et al., 2018). Restricting access to these resources without providing viable alternatives can exacerbate poverty and generate local opposition to conservation efforts (van Versendaal and Schickhoff, 2025). Addressing these trade-offs requires policies that balance conservation objectives with socioeconomic needs, including participatory governance frameworks that involve local communities in decision-making, benefit-sharing mechanisms such as payments for ecosystem services or revenue from ecotourism, and support for sustainable livelihood activities in buffer zones surrounding protected areas.

3.3 Climate change

Saleem et al. (2025) stated that climate change is now a primary and systemic threat to agrifood systems worldwide. Atmospheric warming has accelerated, with 2024 being the warmest year on record, briefly surpassing the 1.5 °C threshold above pre-industrial levels. The past decade includes the 11 warmest years in the observational record. These trends increase heat extremes, alter precipitation patterns, and increase the frequency and severity of droughts and floods. Such conditions directly reduce crop and livestock productivity. (Li et al., 2025) empirically and through model syntheses quantified the impact of warming on crop yields. Their recent meta-analyses estimated yield declines per degree Celsius of about 7.4 % for maize, 6 % for wheat, 3.2 % for rice, and 3.1 % for soybean. Without aggressive global mitigation of greenhouse gas emissions, climate change will continue to intensify, and in the absence of effective adaptation measures, the resulting crop yield declines pose substantial food security risks, particularly in vulnerable regions. Although projected yield losses vary by scenario and crop, most staple species show consistent negative effects when adaptation is limited (Kabato et al., 2025).

Lahlali et al. (2024) explained that climate change increases indirect risks. Warmer temperatures expand the range of pests and pathogens and disrupt phenology, which is the timing of recurring biological events in plants and animals, such as flowering, migration, and reproduction, that are influenced by seasonal and climatic conditions. This leads to greater exposure of crops and livestock to biological threats. Changes in rainfall patterns undermine the reliability of rainfed systems and strain groundwater and surface water resources. These direct and indirect effects together increase yearly yield volatility, which raises production risks, destabilizes farm income, and increases price fluctuations for consumers. (Akinyi et al., 2021) demonstrated that adaptation strategies provide clear benefits but also involve trade-offs. Improved crop varieties, adjusted sowing dates, and irrigation can restore local productivity. However, scaling these strategies requires capital, institutional capacity, collective action among affected farmers, and usually more energy or water use. These demands can increase greenhouse gas emissions or deplete scarce resources if not managed carefully. Therefore, adaptation should be integrated with mitigation efforts and governance reforms to prevent unintended ecological or socioeconomic harm. For instance, expanding irrigation can help farmers adapt to drought, but if the energy used to pump water comes from fossil fuels, it may increase greenhouse gas emissions and undermine mitigation goals. Similarly, promoting drought-tolerant crop varieties can enhance resilience, but if these varieties require higher inputs of synthetic fertilizers, they may contribute to nitrogen pollution. Integrating adaptation with mitigation means prioritizing strategies that deliver both objectives simultaneously, such as agroforestry systems that sequester carbon while providing shade and microclimate regulation for crops, or conservation agriculture that reduces emissions from tillage while improving soil water retention. Governance reforms are equally critical to ensure that adaptation policies do not disproportionately burden vulnerable groups, such as smallholder farmers, and that mitigation measures, such as carbon markets, provide equitable access to benefits rather than reinforcing existing inequalities (Dhar, 2025).

Shruti (Punjabi et al., 2025) synthesized three key insights for research and policy from a knowledge-contribution perspective. First, quantifying crop-specific sensitivity to warming (°C → % yield change) provides an operational benchmark for prioritizing adaptation investments at the crop level. Second, food security outcomes depend on the interaction between climate extremes and socio-institutional vulnerability, making multi-disciplinary studies that combine biophysical modeling with household and market analysis essential. Third, adaptation strategies should be evaluated against mitigation outcomes and resource constraints to identify interventions that are both effective and sustainable at scale. For example, solar-powered irrigation can help farmers adapt to drought while avoiding the greenhouse gas emissions associated with diesel or grid-powered pumping. However, if deployed without groundwater monitoring and usage limits, it may accelerate aquifer depletion, undermining long-term water security. Thus, an effective and sustainable intervention would pair solar pump deployment with groundwater governance measures, such as metered use, recharge requirements, or community-based water management (Dhar, 2025).

Wang et al. (2025) highlighted persistent evidence gaps in current research. Higher-resolution impact projections are required that integrate extreme events, pest dynamics, and compound stresses, such as heat combined with drought. There is also a lack of systematic evaluation of adaptation co-benefits for mitigation and biodiversity. Coordinated observational networks, targeted experiments, and transdisciplinary modeling frameworks are essential to address these gaps. These frameworks should link field-level interventions to regional food system outcomes.

Hultgren et al. (2025) demonstrated that accelerating warming has already reduced the margin for safe agrifood system operation. An effective response requires precise quantification of crop-level risks, integrated adaptation and mitigation planning, and governance that prioritizes equity and resilience. Without these measures, climate change will increasingly undermine progress on global food security.

Qin et al. (2024) reported that land-use conversion and intensive production create reinforcing feedback with the climate system. Between 2010 and 2019, agriculture, forestry, and other land uses (AFOLU) accounted for approximately 21 % of net human-caused greenhouse gas emissions. These emissions arise from deforestation, soil carbon loss, and livestock activities. They contribute to and intensify environmental degradation. Addressing these impacts requires mitigation and restoration strategies. (Nesshöver et al., 2017) emphasized that policy and practice responses should be multi-dimensional. Effective interventions include targeted soil restoration techniques, such as cover crops, reduced tillage, and organic amendments. They also include nutrient budgeting, landscape-scale water management, and incentives for biodiversity-friendly farming (Nesshöver et al., 2017). Use of feed methane-inhibiting feed additives and improved manure management can help reduce livestock-related emissions (Hegde et al., 2025). Monitoring should move beyond coarse national averages and employ spatially explicit indicators to detect degradation hotspots and underlying social drivers. Research should establish measurable links between restoration practices and crop yields under climatic stress. It should also assess the cost-effectiveness of interventions and governance mechanisms that align short-term farmer incentives with long-term public benefits. Environmental degradation is a linked socio-ecological syndrome, requiring the integration of science, spatial monitoring, and policies with restoration efforts, climate mitigation, and promotion of sustainable rural livelihoods.

Agricultural intensification should incorporate climate risk management. (Pereira et al., 2025) reported that climate change already reduces average crop yields and increases their variability. Adaptation strategies, including drought-tolerant crop varieties and water-saving irrigation, are vital complements to intensification. As already noted, however, some measures, such as expanded irrigation using fossil fuel pumps, may elevate greenhouse gas emissions. Therefore, integrated planning should balance immediate productivity gains with long-term mitigation goals. Research should prioritize crop modeling under plausible climate scenarios and evaluating low-carbon intensification pathways.

4. Sustainable practices and solutions enhancing biodiversity

Addressing the intertwined challenges of biodiversity loss and food insecurity requires a shift toward sustainable agricultural practices that actively conserve and enhance biodiversity while maintaining or improving productivity. The following subsections examine a range of such practices and solutions. Section 4.1 explores agroforestry systems (AFS), which integrate trees with crops and livestock to deliver multiple ecosystem services, including carbon sequestration, microclimate regulation, and habitat provision. Section 4.2 discusses local seed systems, intercropping, and other sustainable farming approaches that maintain on-farm crop diversity and strengthen resilience at the community level. Section 4.3 reviews offset mechanisms, which aim to compensate for unavoidable ecological harm through habitat protection or restoration, with attention to their design principles, social equity implications, and application in agricultural landscapes.

4.1 Agroforestry systems (AFS)

Agroforestry Systems (AFS) are a key nature-based solution for climate change adaptation, deliberately integrating trees into farming to build resilience deep within agricultural systems. By supporting the role of agroforestry, silvo-pastoralism, and trees on farms, these systems help mitigate the impact of heat waves, create better microclimates for crops and livestock, and contribute to better water management from plot to landscape (Cifor-Icraf, 2022). This makes AFS a vital strategy for securing food and building farms that can withstand a changing climate. Perhaps most importantly, selling products from trees generates cash for essentials like food and school fees, making the family more resilient to shocks. In this way, agroforestry does double duty: it softens the blow of environmental disasters and provides a vital economic cushion (Quandt, 2020).

The real proof of agroforestry's value comes from farmers themselves. In the drylands of Isiolo, Kenya, where droughts and floods are a constant threat, farmers explain how trees act as a natural shield. They talk about the relief of shade for both themselves and their crops during scorching heat, and how trees break the wind to stop dust storms and hold the soil together when the rains are heavy (Quandt, 2020). But the benefits go beyond just protection. For these farmers, trees are a living safety net. They provide fruit, fodder for animals, firewood, and building materials during tough times.

The way agroforestry looks on the ground is as diverse as the landscapes and cultures that practice it. The variety of agroforestry practices documented in Southern Tigray, Ethiopia, before the 2020–2022 civil war, illustrates how farmers developed systems suited to their needs: dense gardens around their homesteads (46.3 %), living fences (25.9 %), and fields dotted with valuable trees like papaya, mango, and native species like *Ziziphus spina-christi* (15 %) that provide food, fodder, and improve the soil (Gebru et al., 2019). The widespread adoption of these practices was strongly supported by federal, state, and local government policies during the 1990s and 2000s, which promoted agroforestry as part of broader environmental restoration and food security strategies (Balehegn et al., 2020). However, the conflict severely disrupted the region's agricultural systems and environmental restoration efforts, causing widespread deforestation, damage to soil and water conservation structures, and the collapse of local governance institutions (Birhane et al., 2025). Despite these setbacks, some community-based agroforestry practices, particularly homestead gardens, proved remarkably resilient, providing critical food and energy supplies during the crisis (Sharma et al., 2022; Haile et al., 2024). Whether a family adopts these practices depends on a mix of factors including their access to land, their education, and the roles of men and women, showing that successful agroforestry must be rooted in local social and economic realities (Gebru et al., 2019).

In places like Southern Tigray, where the climate is becoming hotter and drier, these trees are a direct response to a changing environment. Over a third of farmers see planting multipurpose trees as a core strategy for building resilience (Gebru et al., 2019). This local wisdom aligns perfectly with global initiatives like TreesAdapt, which champions trees as essential "green infrastructure" that must be adapted themselves to ensure they can keep protecting farms and landscapes for generations to come (Cifor-Icraf, 2022).

Farmers are astute observers of nature, and they choose trees based on key traits that ensure survival and usefulness. In Kenya, they prize trees with deep roots that can "travel far to get water" during droughts, and fast-growing species like papaya that produce food year-round, even with little rain (Quandt, 2020). This practical knowledge underscores a critical principle: success hinges on choosing the right tree for the right place and the right purpose, which in turn depends on having access to diverse, locally-suited seeds and seedlings (Cifor-Icraf, 2022).

Of course, simply planting trees is not a magic bullet. Their ability to strengthen a farm depends heavily on local conditions and management. A detailed study in the rocky, desert-prone karst regions of South China revealed a surprising insight: agroforestry systems in areas actively being restored were actually less vulnerable than those in seemingly stable, non-desertification areas. This shows that with the right interventions, resilience can be actively built. The study pinpointed that factors like how intensely the land is farmed, the steepness of the slope, the rockiness of the ground, and overall soil quality are major drivers of a system's vulnerability. When these are poorly managed, the farm's ability to adapt and recover from stress plummets (Xiao et al., 2026).

Naturally, there are hurdles. In Kenya, not all farmers have equal access to training programs (Hughes et al., 2020). In Ethiopia, a lack of water, limited land, and scarce quality seedlings are major barriers (Gebru et al., 2019). Overcoming these challenges requires more than just farmer effort; it needs a supportive policy environment. The Tigray region in Ethiopia provides a compelling example. Over three decades, the federal and state governments, with support from international organizations, strongly promoted soil and water conservation initiatives across the region, with agroforestry as an integral strategy. These coordinated efforts included establishing exclosures, constructing stone bunds and terraces, and promoting tree planting on farms and communal lands. The result was a dramatic increase in vegetation cover from less than 3 % in the 1980s to approximately 17 % by 2020, earning Tigray the 2017 Future Policy Gold Award from the World Future Council and the United Nations Convention to Combat Desertification (Birhane et al., 2025). However, as discussed above, this progress was severely disrupted by the 2020–2022 civil war, underscoring both the effectiveness of supportive policies and the vulnerability of such gains to conflict.

India provides another stellar example. Its 2014 National Agroforestry Policy broke down restrictive laws that made it difficult for farmers to cut and sell trees from their own land. It backed this up with a substantial budget and a nationwide mission to promote agroforestry, supported by strengthened research institutions and training for hundreds of thousands of farmers (Rizvi et al., 2024). This demonstrates the level of commitment required to scale up these practices.

4.2 Local seed systems, intercropping, and sustainable farming

4.2.1 Local seed systems and seed security

Local seed systems, often referred to as farmers' or informal seed systems, are fundamental to sustainable farming, particularly for an estimated 500 million smallholder farmers worldwide (Louwaars and Manicad, 2022). These systems are not merely sources of planting material; they are dynamic socioecological networks that sustain agrobiodiversity, strengthen food security, and enhance resilience to environmental and economic shocks.

Farmers' seed systems are characterized by the management of diverse portfolios of crops and varieties, including landraces, farmer-selected materials, improved varieties that have entered informal circulation, and semi-domesticated species (Louwaars and Manicad, 2022). Such diversity functions as an important risk-management strategy, enabling farmers to cope with climate variability, pest outbreaks, and fluctuating market conditions through the cultivation of crops with varying maturity periods and stress tolerances.

The importance of these systems for seed security has been further emphasized during periods of crisis. Sperling et al. (2020) highlighted during the COVID-19 pandemic that resilient seed systems depend on maintaining both formal and informal channels of seed access, rather than replacing one with the other. Their "seed system plurality" perspective emphasized that local markets, farmer exchange networks, and community-based seed systems provide critical redundancy during disruptions in commercial supply chains (Sperling et al., 2020). Similar findings emerged from Vermont, USA, where growers who relied on both formal and informal seed sources were better able to withstand pandemic-related shortages and transportation disruptions (Isbell et al., 2023). Informal seed systems also demonstrated substantial adaptive capacity during the pandemic through the rapid emergence of community initiatives such as online seed exchanges, neighborhood seed-sharing platforms, and locally organized seed catalogs (Isbell et al., 2023).

Smallholder farmers in many regions remain highly dependent on these systems. In Africa, approximately 90.2 % of seed used by smallholder farmers originates from informal sources, with local markets accounting for more than half of all seed sown (McGuire and Sperling, 2016). Community Seed Banks (CSBs) further strengthen seed security by conserving and distributing locally adapted crop genetic resources while supporting customary seed-sharing practices among farmers (Ramanna and Andersen, 2022). These institutions function as collective repositories of agrobiodiversity and play a crucial role in preserving varieties adapted to local climatic, nutritional, and cultural needs.

The organization of local seed networks is also deeply shaped by social and biocultural relationships. Research in south-eastern Senegal demonstrated that seed circulation occurs through multiple crop-specific exchange networks influenced by factors such as storability, cultural importance, and gendered management practices (Porcuna-Ferrer et al., 2023). Traditional crops such as sorghum and Bambara groundnut circulated primarily through kinship and neighborhood ties, reinforcing localized systems of exchange that maintain agrobiodiversity and support household resilience. Households occupying more central positions within these networks maintained greater on-farm varietal diversity, thereby enhancing their adaptive capacity under environmental and economic stress (Porcuna-Ferrer et al., 2023). However, the same study cautioned that external interventions by NGOs and development agencies can unintentionally centralize seed distribution systems, potentially weakening traditional sharing mechanisms and eroding the social relationships underpinning local resilience.

Consequently, strengthening local seed systems requires stewardship-oriented approaches that protect farmers' rights to save, exchange, use, and sell seed while reinforcing, rather than displacing, existing community-based structures and knowledge systems (Ramanna and Andersen, 2022). These practices are also recognized, albeit only weakly enforced, under the International Treaty on Plant Genetic Resources for Food and Agriculture, which affirms farmers' rights to conserve, use, exchange, and sell farm-saved seed as part of broader efforts to sustain agrobiodiversity and food security. However, recent analyses note that the treaty's implementation remains uneven and often constrained by intellectual property regimes and national regulatory frameworks, limiting the practical realization of these rights in many agricultural systems (Prip, 2024). Interventions such as improved seed storage and selection techniques, support for CSBs, and enhanced seed quality within local markets can amplify the resilience and sustainability benefits of informal systems (Louwaars and Manicad, 2022). Increasing evidence suggests that "integrated" or "pluralistic" seed systems, which combine the complementary strengths of formal and informal channels, offer one of the most promising pathways toward long-term agricultural resilience and food security (Isbell et al., 2023; Sperling et al., 2020).

4.2.2 Intercropping and sustainable farming systems

The agronomic importance of local seed systems is closely linked to diversified farming practices such as intercropping and crop rotation, which rely on access to diverse and locally adapted planting materials. Intercropping has consistently demonstrated substantial ecological, agronomic, and economic benefits across diverse farming systems. Multi-site and multi-year field experiments in semi-arid regions of China showed that oat/soybean strip intercropping achieved a mean Land Equivalent Ratio (LER) of 1.12, corresponding to a 12 % yield advantage compared with monocultures. The system also improved temporal yield stability by 5.5–62.6 % and increased net farm income by 5.2–9.4 %, directly linking crop diversification to enhanced food security and livelihood resilience (Feng et al., 2024).

Similar outcomes have been reported in Mediterranean farming systems. In Spain, broccoli/fava bean intercropping more than doubled the productivity of monoculture systems, with LER values consistently exceeding 1.5 while simultaneously improving soil fertility through increases in total nitrogen, available phosphorus, and exchangeable potassium, despite a 30 % reduction in fertilizer application (Marcos-Pérez et al., 2023). These findings align with a global meta-analysis demonstrating that intercropping systems generally produce 16–30 % higher yields per unit area than monocultures while reducing dependence on synthetic fertilizers and pesticides (Renard and Tilman, 2021). Importantly, crop diversity also acts as a natural insurance mechanism. Analysis of 50 years of agricultural data across 91 countries revealed that nations with higher effective crop diversity experienced severe harvest failures far less frequently, emphasizing the stabilizing effects of diversified production systems under environmental variability (Renard and Tilman, 2021).

Recent large-scale assessments further support the broad applicability of intercropping in climate-resilient agriculture. An evaluation of cereal-legume intercrops on commercial farms across Europe reported average yield gains of approximately 30 %, with benefits being especially pronounced in lower-yielding environments, underscoring the value of intercropping for risk reduction in resource-constrained systems (Brooker et al., 2024). These gains were achieved under varying climatic conditions and using organic nutrient sources, demonstrating the adaptability and sustainability of diversified systems. Beyond productivity, intercropping contributes to multiple ecosystem services. A review of smallholder systems in South Africa identified intercropping as a major strategy for enhancing soil fertility through nitrogen fixation and phosphorus mobilization, increasing carbon sequestration, suppressing weeds and pests naturally, and improving water and land-use efficiency (Mthembu et al., 2019).

4.3 Offset mechanisms

Offset mechanisms, the final step in the mitigation hierarchy, aim to compensate for unavoidable ecological harm to achieve No Net Loss (NNL) or a Net Gain of biodiversity (Arlidge et al., 2018). It is critical to distinguish true offsets from general compensation. As Pope et al. (2021) argue, only direct, measurable interventions like habitat protection or restoration qualify as offsets, unlike funding research or education, which constitute "other compensation" and risk diluting ecological credibility.

For offsets to be effective, they must adhere to core principles: additionality (creating new conservation gains), equivalency (matching lost biodiversity, preferably in-kind), and long-term security (Pope et al., 2021). Furthermore, calculating the required compensation is not a simple 1:1 swap. Credible offset ratios use multipliers (often 10:1 or higher) to account for restoration time lags, uncertainty, and risk of failure, ensuring NNL is genuinely achieved (Moilanen and Kotiaho, 2018). However, a significant implementation gap exists, particularly regarding social equity. Offsets can displace local communities and negatively impact livelihoods, especially in the Global South where people depend directly on natural resources for their livelihoods (Bidaud et al., 2017; Tupala et al., 2022). A case study in Madagascar revealed that costs (lost access to land) are immediate and widespread, while benefits (training, donations) are often delayed and captured by local elites, leaving communities as net losers (Bidaud et al., 2017). This underscores that a biologically successful offset can be socially unjust.

The institutional design is equally critical. Successful implementation requires a clear distribution of rights and responsibilities (Primmer et al., 2019). In practice, the focus often shifts to securing a developer's "license to operate" rather than safeguarding human or landowner rights. This is evident in how offset policies can profoundly restrict land use without adequately addressing the resultant loss of property rights (Primmer et al., 2019). Real-world application often falters. An analysis of England's mandatory Biodiversity Net Gain policy shows that a rigid preference for local offsets performs poorly, failing to create strategic ecological networks and representing an inefficient use of conservation resources (Mancini et al., 2024). This illustrates the pitfalls of poorly designed policy.

Applying this framework to agriculture requires tailoring. The mitigation hierarchy can be integrated by: avoiding conversion of high-value habitats, minimizing impacts through sustainable practices, and remediating on-site damage. For residual impacts, offsets should be strategically located within the bioregion for maximum gain, not just adjacent to the farm (Arlidge et al., 2018; Mancini et al., 2024). Equally important is ensuring that compensation mechanisms align with farmer priorities and socio-economic realities. Evidence suggests that farmers generally favor direct financial incentives, although long-term resilience and sustainability are more effectively supported when these are complemented by capacity-building measures such as technical training and institutional support (Li et al., 2020). Similarly, conservation programs that allow farmers to retain land ownership tend to achieve greater acceptance and participation than approaches involving land acquisition or displacement (Tupala et al., 2022).

Finally, for compensation to be truly effective, it needs to work across regional borders. A compelling model comes from China, where research tracks the hidden environmental footprint of food trade. Studies show that agricultural products—and the "virtual" land and water used to make them—flow from farming provinces to distant consuming cities. This creates a system of outsourced environmental costs. To address this, the concept of "Agricultural Horizontal Ecological Compensation" (AHEC) proposes that consuming regions should financially compensate producing regions for their ecological burden. This approach uses robust accounting to quantify the real cost of agriculture, ensuring the bill for environmental protection is paid by those who benefit from the products, not just those who produce them (Wang et al., 2022). Advanced models that integrate carbon and water footprints into a unified accounting system are now providing the scientific basis for such fair, inter-regional fiscal transfers (Pei et al., 2025). The effective offset governance requires SMART targets, that is, goals that are specific, measurable, ambitious, realistic, and timebound (Arlidge et al., 2018), and an explicit balance of rights and responsibilities (Primmer et al., 2019). It must combine strategic, landscape-scale planning (Mancini et al., 2024) with tailored farmer compensation (Li et al., 2020) and robust accounting for hidden ecological costs (Pei et al., 2025). When done correctly, offset mechanisms offer a structured pathway to balance agricultural production with essential ecological restoration.

5. Policy frameworks, collaborative efforts, and sustainable development

5.1 Integrated approaches: Agroforestry-bioenergy systems

In the face of climate change, continuous population growth, and depleting natural resources, it is very important to achieve a smooth relationship between food production and energy generation, and this comes with a lot of critical challenges (Peng et al., 2023). Ensuring affordable access to sustainable and reliable modern clean energy is one of the Sustainable Development Goals (SDG 7) and is very important towards achieving other SDGs, including promoting food and nutrition security (García-López et al., 2024). Integrating systems that combine agroforestry, multi-use cropping systems, and technological innovations offers a good opportunity to maximize bioenergy yields without compromising food security (Sharma et al., 2016). These approaches demonstrate how agriculture can evolve into a multi-faceted system providing food, feed, fiber, energy, and ecosystem services, all while conserving biodiversity (García-López et al., 2024).

Agroforestry plays a very important role in creating and maintaining this balance. It creates varied and strong landscapes where the combined benefits are greater than what each part could achieve alone (USDA, n.d.; FAO, n.d.a). Traditional agroforestry has been shown by research to be a promising way to make agriculture more productive and sustainable, helping to meet rising food needs while also supporting healthy ecosystems (Shukla and Dhyani, 2023; WAO, 2013). Trees in agroforestry systems provide biomass for sustainable energy in the form of firewood, charcoal, or processed biofuels, and offer shade that reduces water stress for crops (García-López et al., 2024). They enhance soil fertility through nitrogen-fixing species and organic matter deposition, which in turn boosts crop productivity. In addition, the organic matter from fallen leaves, roots, and decomposing plant residues enriches the soil with carbon, boosts its ability to hold water, and strengthens overall soil health This creates better conditions for sustained crop production and supports renewable energy efforts. This synergy helps farmers maintain or increase food yields while generating renewable energy, reducing the need for fossil fuels, and cutting greenhouse gas emissions (Yu et al., 2025; Barrios et al., 2023).

Multi-use cropping systems complement agroforestry by cultivating a mix of food crops, energy crops, and sometimes medicinal or fodder plants within the same land parcel or rotation cycle (Ogazie et al., 2022). This diversity not only reduces the risk of total crop failure but also maximizes land-use efficiency, thereby supporting farmers in boosting production to match what the market demands. For instance, intercropping fast-growing grasses suitable for bioenergy with staple food crops ensures that energy production does not encroach on food-growing land. Instead, it transforms agricultural landscapes into productive fusion that provides multiple outputs (Balandaitė et al., 2024; Wang et al., 2021). Furthermore, multi-use systems support beneficial insects and pollinators, which are essential for healthy crop yields and biodiversity conservation (Fijen et al., 2025).

Technological innovations further enhance these traditional practices by optimizing resource use and improving bioenergy production processes.

Recent advances include bioenergy gasification, pyrolysis to produce biochar, and precision agriculture tools (Li et al., 2023; EOS, 2025). Biochar, produced from biomass residues through pyrolysis, when applied to soils, improves fertility and water retention, which can lead to higher food yields and help sequester carbon in soils (American Biochar Institute, n.d.; Wei et al., 2023). Precision agriculture technologies, utilizing GPS, remote sensing, targeted spray systems, automated weeders, and data analytics, enable farmers to tailor input applications and monitor crop health in real time. This reduces wastage of water, fertilizers, and energy, ensuring environmental sustainability while boosting productivity (Singh, 2024; GAO, 2024).

Research has shown that integrated agroforestry-bioenergy systems can simultaneously address multiple sustainable development goals (García-López et al., 2024). By reusing biomass residues for energy, these systems reduce deforestation pressure and provide clean, affordable energy for rural households, empowering communities and improving health outcomes by reducing indoor air pollution (Okafor et al., 2022).

Sustainable forestry and using biomass are essential parts of Europe's climate action plans and tree-planting goals. For instance, Denmark, under the Green Tripartite Agreement, plans to plant 250,000 ha of new forest by 2045. This strong push for afforestation and carbon capture is mirrored by similar ambitious targets set by the EU and many individual member countries (DBDH, 2025).

Moreover, the increased productivity and diversified income streams foster economic resilience for smallholder farmers in the Global South, who are often women and/or members of marginalized groups, thus enabling poverty alleviation. One concrete example is the integration of nitrogen-fixing trees like *Gliricidia sepium* with food crops such as maize and energy crops like Napier grass in Kenya. The trees fix atmospheric nitrogen, enriching soil fertility, while Napier grass provides substantial biomass for bioenergy. These combinations help farmers maintain food production while generating sustainable energy, illustrating how agroecological knowledge combined with innovation yields multiple benefits (Gupta, 2025; Wanjira, 2019).

Globally, policy frameworks increasingly recognize the need for integrated land-use systems that address food and energy security alongside biodiversity conservation. Governments and international organizations promote agroforestry and bioenergy integration through incentives, technical assistance, and research funding (Mosquera-Losada et al., 2023).

However, challenges remain, including access to technologies, knowledge gaps, and sometimes conflicting land-use policies (Sustainability Dictionary, 2025). Addressing these requires participatory approaches that involve farmers, scientists, industry players, and policymakers to co-develop context-specific solutions. Policymakers must continue to foster enabling environments that support these integrative practices to build the resilient agricultural landscapes of the future (REFOREST, 2023; Jung and Vendrametto, 2025).

5.2 Stakeholder collaboration

In addressing complex global challenges such as biodiversity conservation, food security, and sustainability, stakeholder collaboration has emerged as a cornerstone for success (USAID, 2016; Ipara and Kipchumba, 2024). Diverse actors across scientific, industrial, and policy domains bring unique perspectives, resources, and capacities that, when integrated, create synergistic effects far greater than isolated efforts (USAID, 2016). Fostering collaboration among these stakeholders not only strengthens research capacity but also facilitates efficient data sharing, essential for evidence-based decision-making and adaptive management in food systems resilient to climate change and environmental degradation (Sjaf et al., 2022; Krug et al., 2020; Ponce Reyes et al., 2019). The scientific community plays a critical role in generating rigorous, empirical knowledge about ecosystem functions, agricultural technologies, and conservation strategies. However, the translation of this knowledge into practice hinges on effective collaboration with industrial stakeholders who are important in scaling innovations, and with policymakers who shape regulatory frameworks and resource allocation (MEA, 2005; FAO, 2019). Agricultural extension systems are crucial for making research results accessible and useful to farmers (Davis et al., 2021). Collaborative research initiatives encourage multidisciplinary and transdisciplinary approaches, recognizing food security and biodiversity as interconnected rather than isolated domains. For instance, joint research programs involving ecologists, agronomists, and socio-economists enable the holistic examination of agroecosystem resilience, bioenergy production potential, and socio-economic impacts on rural livelihoods (Ntihinyurwa and de Vries, 2021; FAO, 2017). It is crucial to integrate farmers' knowledge and perspectives into such research (Mustonen et al., 2022).

Industrial stakeholders contribute by developing and deploying sustainable agricultural technologies, bioenergy systems, and supply chain innovations that promote biodiversity-friendly practices (Krug et al., 2020). Their engagement ensures that research outcomes meet market demands and operational realities. By partnering with scientists, industries can adopt cutting-edge solutions such as precision agriculture, biochar technology, and integrated pest management, which increase productivity while minimizing environmental footprints (Potent, 2016; AGriNext Confrence, 2025). Moreover, industries often bring data analytics and digital tools that enhance real-time monitoring of biodiversity indicators and food production metrics. This technological interface is vital for optimizing resource use and ensuring traceability, transparency, and accountability across food and energy value chains (Hassoun et al., 2024; Dong et al., 2025).

Policymakers provide the enabling environment through policies, incentives, and regulatory mechanisms (FAO, 2019). Their collaboration with scientists and industries enhances the relevance and applicability of research findings in policy design and implementation (Ntihinyurwa and de Vries, 2021; Prajapati et al., 2025). Policy frameworks that encourage biodiversity conservation in agricultural landscapes, support investments in sustainable bioenergy, and facilitate market access for ecosystem services are evidence-based and adaptive to stakeholder feedback. Regular dialogue platforms, such as policy roundtables, public-private partnerships, and collaborative research networks, facilitate this exchange of knowledge and priorities.

Importantly, inclusive policies that engage smallholder farmers, indigenous communities, and local organizations promote social equity and contextualize conservation efforts within local cultural and economic realities. Data sharing is a critical outcome and enabler of stakeholder collaboration. Integrated databases combining ecological, agricultural, climatic, and socio-economic data provide comprehensive insights into the nexus of biodiversity, food security, and sustainability (Havinga et al., 2020). Open-access platforms and interoperable data standards ensure accessibility and usability by diverse stakeholders. For example, global initiatives like the Global Biodiversity Information Facility (GBIF) and national agricultural research centers offer repositories that facilitate cross-sectoral research and policy analysis (GBIF, 2016). Collaborative data efforts also include citizen science initiatives, which engage local communities in biodiversity monitoring and contribute valuable ground-level information that enriches scientific datasets (Krug et al., 2020).

In all, fostering collaboration among scientific, industrial, and policy communities is a fundamental strategy to strengthen research capacity and facilitate data sharing crucial for advancing biodiversity conservation in food security and sustainability. By leveraging diverse expertise and resources, these collaborative efforts enable integrated, adaptive, and equitable approaches that respond effectively to global environmental and food system challenges. Policymakers are encouraged to support institutional frameworks and incentives that cultivate such partnerships, while scientists and industry leaders should prioritize transparent communication and inclusive engagement. Together, they can build resilient food-energy-biodiversity systems necessary for a sustainable future (UNEP, 2021; IPBES, 2019).

5.3 Landscape-specific strategies

Promoting sustainable agriculture and biodiversity conservation demands strategies that recognize the unique ecological, social, and economic characteristics of each landscape. Landscape-specific approaches enable targeted, adaptive management that balances food production with ecosystem health, ensuring that interventions align with local conditions and stakeholder needs (Davis, 2026). In this endeavor, real-time monitoring tools combined with tailored solutions play an indispensable role, empowering decision-makers at all levels to respond swiftly and effectively to environmental changes, threats, and opportunities (Olawade et al., 2024). Advances in remote sensing, artificial intelligence (AI), and sensor technologies have revolutionized biodiversity and agricultural monitoring by shifting from periodic, labor-intensive surveys to continuous, scalable, and data-driven processes (Miller et al., 2025; Fu and Huang, 2025). Satellite imagery and drone surveillance provide landscape-wide overviews of vegetation patterns, species distributions, and habitat heterogeneity, essential indicators of biodiversity health and agricultural productivity (Gálvez-Salido et al., 2026). These technologies detect stress signals from crops, invasive species incursions, or habitat fragmentation early, enabling early interventions (Miller et al., 2025; Kondo et al., 2024). For instance, multispectral sensors capture changes in plant health and soil condition, while AI algorithms classify species and monitor their populations by analyzing images and sounds, reducing the need for extensive fieldwork and human bias (Kondo et al., 2024).

Platforms such as Deep Planet and Biotfy exemplify these innovations, integrating data from satellites, Internet of Things-enabled sensors, bioacoustics recorders, and environmental DNA (eDNA) samples into centralized dashboards (Biotify, 2025). These platforms provide decision-makers—from farmers to policymakers—with actionable insights through user-friendly interfaces and customizable alerts. Farmers receive real-time guidance on nutrient management, pest control, and water use efficiency, improving yields and reducing environmental impacts. Conservationists monitor threatened species and ecological dynamics more accurately, tailoring habitat restoration and protection efforts to the landscape's specific needs. Governments and industry can ensure compliance with environmental regulations more seamlessly, fostering transparency and accountability (DeepPlanet, 2025b).

Tailored solutions derive their strength from real-time data. Unlike one-size-fits-all approaches, landscape-specific strategies consider the diversity of soil types, climate zones, cropping systems, and socio-economic contexts (Miller and Dong, 2019; Tamene et al., 2022). For example, in drought-prone regions, satellite data combined with soil moisture sensors guide irrigation scheduling to optimize water use without stressing crops or ecosystems. In biodiversity hotspots, AI-assisted species identification tools help track the presence and behavior of pollinators and pest-controlling natural enemies, informing integrated pest management strategies that minimize synthetic inputs (Fontanet et al., 2018; EOS, 2025; Ihuoma et al., 2021; Xing and Wang, 2024; Miller and Dong, 2019). Additionally, these solutions support the design of multifunctional landscapes that allocate space for conservation corridors, agroforestry patches, and crop fields in ways that enhance connectivity, resilience, and carbon sequestration (Li et al., 2025; Tamene et al., 2022).

Beyond technical capacity, landscape-specific strategies must engage local communities and stakeholders in co-developing and co-owning monitoring and management processes. Participatory approaches ensure that data collection respects local knowledge and priorities while fostering trust and cooperation (Danielsen et al., 2021; Mugari et al., 2025). Citizen science initiatives, where farmers and community members contribute observations via mobile apps, enrich datasets and broaden monitoring coverage. This inclusive model enhances the relevance of tailored interventions and increases the likelihood of sustainable adoption (Davies et al., 2023).

Recent successes showcase the power of integrating real-time monitoring with landscape-specific adaptation (European Commission, 2020). The COALA project, for example, employed Sentinel-2 satellite data and a physics-based crop water-use model to deliver timely water and nutrient status information to farmers in Australia (European Commission, 2020; European Commission, 2024; Dongryeol et al., 2022). This capability enabled more efficient resource management and increased productivity under variable climatic conditions. Similarly, the MAMBO initiative harnessed remote sensing, drone-based LiDAR (Light Detection and Ranging), and AI-driven image analysis to map biodiversity and habitat conditions comprehensively, facilitating adaptive management plans that balance agriculture and conservation objectives at the landscape scale (MAMBO, n.d.).

Despite these advancements, challenges persist. Data heterogeneity, high costs of advanced technologies, and limited technical expertise can impede widespread adoption, especially in low-income regions (Sustainability Dictionary, 2025; Freitas and Gouveia, 2025). Data privacy, interoperability, and equitable access must be addressed to prevent digital divides and ensure that benefits reach marginalized groups. Additionally, translating complex environmental data into locally comprehensible and culturally appropriate information requires dedicated capacity-building and extension services (Freitas and Gouveia, 2025). Nevertheless, promoting sustainable agriculture and biodiversity conservation through landscape-specific strategies empowered by real-time monitoring tools offers a pragmatic yet innovative pathway forward (Freitas and Gouveia, 2025). By harnessing cutting-edge technologies to generate precise, dynamic, and context aware insights, these strategies enable adaptive management that respects the inherent complexity and diversity of agro-ecosystems (CGIAR, n.d.). Coupled with inclusive stakeholder engagement and tailored interventions, they enhance the resilience, productivity, and sustainability of food systems and the natural environments they depend upon (FAO, 2025).

6. Conclusion

The nexus between biodiversity, food security, and sustainability highlights the fundamental role that biological diversity plays in maintaining resilient and productive food systems. Across genetic, microbial, species, and ecosystem levels, biodiversity supports essential ecosystem services, including pollination, nutrient cycling, soil fertility, pest regulation, and climate resilience, that underpin sustainable agricultural productivity and nutritional diversity. However, the continued expansion of intensive agriculture, environmental degradation, and accelerating climate change threatens these ecological foundations, making food systems increasingly vulnerable to shocks and long-term instability. As discussed throughout this chapter, strengthening agrobiodiversity through practices such as agroforestry, crop diversification, local seed systems, and biodiversity-friendly farming can enhance both ecological integrity and livelihood security, particularly for smallholder and resource-limited communities. At the same time, effective policy frameworks; collaborative governance among scientific, industrial, and policy stakeholders; and the integration of landscape-specific technological monitoring tools are critical for balancing food production with biodiversity conservation. Ultimately, safeguarding biodiversity is not only an environmental priority but a strategic necessity for building resilient, equitable, and sustainable food systems capable of meeting the nutritional demands of a growing global population while preserving the ecological processes upon which agriculture and human well-being depend.

References

Adib Bin Rashid, A., Karim Kausik, A., Khandoker, A., Siddque, S.N., 2025. Integration of artificial intelligence and IoT with UAVs for precision agriculture. Hybrid. Adv. 10, 100458. https://doi.org/10.1016/j.hybadv.2025.100458.

AGriNext Confrence, 2025. The future of agriculture - banancing sustainability with productivity. https://agrinextcon.com/future-of-agriculture-balancing-sustainability/.

Aihounton, G., Christiaensen, L., 2024. Does agricultural intensification pay in the context of structural transformation? Food Policy 122, 102571. https://doi.org/10.1016/j.foodpol.2023.102571.

Akinnawo, S.O., 2023. Eutrophication: causes, consequences, physical, chemical and biological techniques for mitigation strategies. Environ. Chall. 12, 100733. https://doi.org/10.1016/j.envc.2023.100733.

Akinyi, D.P., Ng'ang'a, S.K., Girvetz, E.H., 2021. Trade-offs and synergies of climate change adaptation strategies among smallholder farmers in sub-Saharan Africa: a systematic review. Reg. Sustain. 2 (2), 130–143. https://doi.org/10.1016/j.regsus.2021.05.002.

Allasi Canales, N., Ryan, P., Jackson, A., Le, H.P., Borrell, J., Davis, A., et al., 2026. Past, present and future of local crop evolution. Plants, People, Planet 1–16. https://doi.org/10.1002/ppp3.70172.

Al-Tammar, F.K., Khalifa, A.Y.Z., 2022. Plant growth promoting bacteria drive food security. Braz. J. Biol. = Rev. brasleira de. biologia 82, e267257. https://doi.org/10.1590/1519-6984.267257.

American Biochar Institute (n.d.). Welcome to our new name, and new home. Retrieved July 14, 2026, from https://biochar.org/usbi/.

Arlidge, W.N.S., Bull, J.W., Addison, P.F.E., Burgass, M.J., Gianuca, D., Gorham, T.M., et al., 2018. A global mitigation hierarchy for nature conservation. BioScience 68 (5), 336–347. https://doi.org/10.1093/biosci/biy029.

Asdal, Å., 2025. Crop biodiversity and plant genetic resources. Acta Hortic (1421), 1–10. https://doi.org/10.17660/ActaHortic.2025.1421.1. https://doi.org/10.17660/ActaHortic.2025.1421.1.

Ba, Q.X., Wang, H.Z., Wang, M.H., 2023. Agrobiodiversity, social institutions, and indigenous farming practices: a case study of the Rukai in Wutai, Taiwan. Hum. Ecol. 51 (6), 1127–1140. https://doi.org/10.1007/s10745-023-00463-4.

Balandaitė, J., Romaneckas, K., Kimbirauskienė, R., Sinkevičienė, A., 2024. Comprehensive assessment of the effect of multi-cropping on agroecosystems. Plants 13 (10), 1372. https://doi.org/10.3390/plants13101372.

Balehegn, M., Haile, M., Fu, C., Liang, W., 2020. Ecosystem-based adaptation in Tigray, Northern Ethiopia: a systematic review of interventions, impacts, and challenges. In: Leal Filho, W. (Ed.), Handbook of Climate Change Resilience. Springer, Cham. https://doi.org/10.1007/978-3-319-93336-8_117.

Barrios, E., Coe, R., Place, F., Sileshi, G.W., Sinclair, F., 2023. Nurturing soil life through agroforestry. Biological Approaches to Regenerative Soil Systems. CRC Press, pp. 265–278. https://doi.org/10.1201/9781003093718-27.

Barros-Rodríguez, A., Rangseekaew, P., Lasudee, K., Pathom-aree, W., Manzanera, M., 2021. Impacts of agriculture on the environment and soil microbial biodiversity. Plants 10 (11), 2325. https://doi.org/10.3390/plants10112325.

Bidaud, C., Schreckenberg, K., Rabeharison, M., Ranjatson, P., Gibbons, J., Jones, J.P.G., et al., 2017. The sweet and the bitter: intertwined positive and negative social impacts of a biodiversity offset. Conserv. Soc. 15 (1), 1. https://doi.org/10.4103/0972-4923.196315.

Biotify, 2025. Environmental Intelligence for sustainability solutions. https://biotfy.com/.

Birhane, E., Siyum, Z.G., Shiferaw, H., Solomon, N., Berhe, M., Gebremedhin, M.A., et al., 2025. Armed conflict-driven environmental degradation and uneven socio-ecological resilience in Tigray, Ethiopia. Sustain. Futures 10, 101289. https://doi.org/10.1016/j.sftr.2025.101289.

Bravo-Peña, F., Yoder, L., 2024. Agrobiodiversity and smallholder resilience: a scoping review. J. Environ. Manag. 351, 119882. https://doi.org/10.1016/j.jenvman.2023.119882.

Brooker, R.W., Karley, A.J., Newton, A.C., Pakeman, R.J., Schöb, C., 2016. Facilitation and sustainable agriculture: a mechanistic approach to reconciling crop production and conservation. Funct. Ecol. 30 (1), 98–107. https://doi.org/10.1111/1365-2435.12496.

Brooker, R.W., Pakeman, R.J., Adam, E., Banfield-Zanin, J.A., Bertelsen, I., Bickler, C., et al., 2024. Positive effects of intercrop yields in farms from across Europe depend on rainfall, crop composition, and management. Agron. Sustain. Dev. 44 (4), 35. https://doi.org/10.1007/s13593-024-00968-2.

Çakmakçı, R., Salık, M.A., Çakmakçı, S., 2023. Assessment and principles of environmentally sustainable food and agriculture systems. Agriculture 13 (5), 1073. https://doi.org/10.3390/agriculture13051073.

CBD (Convention on Biological Diversity), 1992. Convention on biological diversity. United Nations. https://www.cbd.int/doc/legal/cbd-en.pdf.

CBD, 2008. Why is it important? https://www.cbd.int/agro/importance.shtml.

CBD, 2011. Nagoya protocol on access to genetic resources and the fair and equitable sharing of benefits arising from their utilization to the convention on biological diversity. United Nations. https://www.cbd.int/abs/doc/protocol/nagoya-protocol-en.pdf.

CGIAR (n.d.). PAS-AGRO-PAS—The making of fragile agro-ecosystems productive, adaptive and sustainable: Multifunctional agro-pastoralism. Monitoring, Evaluation & Learning (MEL) Platform. Retrieved July 14, 2026, from https://mel.cgiar.org/projects/1930.

Chauhan, P., Chaurasia, M., 2025. Food security and biodiversity. Biotechnological Innovations for Sustainable Biodiversity and Development. CRC Press, pp. 174–188. https://doi.org/10.1201/9781032697550-14.

Chettri, D., Verma, A.K., Verma, A.K., 2022. Microbes and their application in the food and agriculture industry. In: Verma, P. (Ed.), Industrial Microbiology and Biotechnology. Springer, Singapore. https://doi.org/10.1007/978-981-16-5214-1_21.

Cifor-Icraf, C.-I., 2022. TreesAdapt: Adaptation with Forests, Trees and Agroforestry for Agriculture, Landscapes and People. Center for International Forestry Research (CIFOR), https://doi.org/10.17528/cifor-icraf/008725.

Danielsen, F., Enghoff, M., Poulsen, M.K., Funder, M., Jensen, P.M., Burgess, N.D., 2021. The concept, practice, application, and results of locally based monitoring of the environment. BioScience 71 (5), 484–502. https://doi.org/10.1093/biosci/biab021.

Dannenberg, P., Braun, B., Greiner, C., Follmann, A., Haug, M., Semedi Hargo Yuwono, P., ... Kopriva, S., 2024. Eight arguments why biodiversity is important to safeguard food security. Plants People Planet 6 (3), 604–610. https://doi.org/10.1002/ppp3.10492.
Davies, M., Haklay, M., Kiprutto, T., Laws, M., Lewis, J., Lunn-Rockliffe, S., et al., 2023. Supporting the capacities and knowledge of smallholder farmers in Kenya for sustainable agricultural futures: a citizen science pilot project. UCL Open. Environ. 5. https://doi.org/10.14324/111.444/ucloe.000065.
Davis, J., 2026. Cultivating change: addressing shifts in knowledge and skills required for landscape-scale nature recovery. Ambio. https://doi.org/10.1007/s13280-026-02404-z.
Davis, K.E., Makhija, S., Spielman, D.J., 2021. Agricultural extension and rural advisory services: what have we learned? What's next? Int. Food Policy Res. Institute, Policies, Institutions, Mark. Synth. Brief. https://doi.org/10.2499/p15738coll2.134719.
DBDH, 2025. Biomass strengthens sustainable forestry. https://dbdh.org/biomass-strengthens-sustainable-forestry/.
DeepPlanet, 2025b. Why Biodiversity Monitoring in Agriculture Matters More Than Ever. https://www.deepplanet.ai/newsletters/biodiversity-monitoring-with-ai.
Dhar, A.R., 2025. Building climate-resilient food systems through the water–Energy–Food–Environment nexus. Environments 12 (5), 167. https://doi.org/10.3390/environments12050167.
Dhakad, A.K., Mohanapuria, P., Ranjan, R., Vaishnav, V., Yadav, S.K., 2022. Molecular approaches in conservation and restoration of agrobiodiversity. Agro-Biodiversity and Agri-Ecosystem Management. Springer Nature Singapore, Singapore, pp. 169–216. https://doi.org/10.1007/978-981-19-0928-3_10.
Dong, B., Zhang, M., Li, S., Xie, L., Xie, B., Chen, L., 2025. How digital intelligence integration boosts forestry ecological productivity: evidence from China. Forests 16 (8), 1343. https://doi.org/10.3390/f16081343.
Dongryeol R., Arash P., Danlu G., Andrew W.W., Oscar B., Guido D.'U., 2022. Sentinel-2 satellites map crop-water use over Aussie farms. https://www.coalaproject.eu/blog/sentinel-2-irrigated-farms/.
Engels, J.M.M., Ebert, A.W., 2021. A critical review of the current global Ex situ conservation system for plant agrobiodiversity. II. Strengths and weaknesses of the current system and recommendations for its improvement. Plants 10 (9), 1904. https://doi.org/10.3390/plants10091904.
EOS, 2025. Precision agriculture: technology to boost crop farming. https://eos.com/blog/precision-agriculture/.
European Commission, 2020. COpernicus applications and services for low impact agriculture in Australia. https://doi.org/10.3030/870518.
European Commission, (2024, April 30). COpernicus applications and services for low impact agriculture in Australia (COALA) (Grant Agreement No. 870518). CORDIS—EU Research Results. https://doi.org/10.3030/870518.
FAO (Food and Agriculture Organization of the United Nations), (n.d.). What is Agroforestry?. Retrieved February 20, 2026, from https://www.fao.org/agroforestry/about-agroforestry/overview/en.
FAO (n.d.). What is agricultural biodiversity? Retrieved February 20, 2026, from https://www.fao.org/agriculture/crops/thematic-sitemap/theme/compendium/tools-guidelines/what-is-agricultural-biodiversity/en/.
FAO, 2011. Payments for ecosystem services and food security. FAO. https://www.fao.org/4/i2100e/i2100e.pdf.
FAO, 2017. Sustainable forestry for food security and nutrition. https://openknowledge.fao.org/server/api/core/bitstreams/28b42702–5d06–4f81–8c40–67de0ccf0f02/content.

FAO, 2019. The state of the world's biodiversity for food and agriculture. https://openknowledge.fao.org/server/api/core/bitstreams/50b79369–9249-4486-ac07-9098d07df60a/content.

FAO, 2025. NSP - Adaptive management. https://www.fao.org/agriculture/crops/thematic-sitemap/theme/spi/soil-biodiversity/initiatives/adaptive-management/en/.

FAO (n.d.). Climate Change. Retrieved February 20, 2026, from https://www.fao.org/climate-change/what-we-do/policy-support/energy-and-food-nexus/en.

FAO, 2026. FAO chief economist warns of severe global food security risks from disruption to Strait of Hormuz trade corridor. FAO. https://www.fao.org/newsroom/detail/fao-chief-economist-warns-of-severe-global-food-security-risks-from-disruption-to-strait-of-hormuz-trade-corridor/en.

Feng, W., Ge, J., Rodríguez, A.R.S., Zhao, B., Wang, X., Peixoto, L., et al., 2024. Oat/soybean strip intercropping benefits crop yield and stability in semi-arid regions: A multi-site and multi-year assessment. Field Crop. Res. 318, 109560. https://doi.org/10.1016/j.fcr.2024.109560.

Fijen, T.P.M., Eeraerts, M., Osterman, J., Beyer, N., Hass, A., Lundin, O., et al., 2025. Crop diversification for pollinator conservation. Landsc. Ecol. 40 (1), 19. https://doi.org/10.1007/s10980-024-02027-3.

Fontanet, M., Fernàndez-Garcia, D., Ferrer, F., 2018. The value of satellite remote sensing soil moisture data and the DISPATCH algorithm in irrigation fields. Hydrol. Earth Syst. Sci. 22 (11), 5889–5900. https://doi.org/10.5194/hess-22-5889-2018.

Freitas, H., Gouveia, A.C., 2025. Biodiversity futures: digital approaches to knowledge and conservation of biological diversity. Web Ecol. 25 (1), 29–37. https://doi.org/10.5194/we-25-29-2025.

Frison, E.A., Cherfas, J., Hodgkin, T., 2011. Agricultural biodiversity is essential for a sustainable improvement in food and nutrition security. Sustainability 3 (1), 238–253. https://doi.org/10.3390/su3010238.

Fu, M., Huang, Z., 2025. Integrating remote sensing, IoT, and AI for biodiversity monitoring in hainan tropical rainforest National Park. Int. J. Agric. Environ. Inf. Syst. (IJAEIS) 16 (1), 1–18. https://doi.org/10.4018/IJAEIS.392505.

Gadissa, F., Abebe, M., Worku, B., 2021. Assessment on the current state of on-farm diversity and genetic erosion in barley (Hordeum vulgare L.) Landraces from Bale Highlands, Southeast Ethiopia. 12 pages, 2021. BioMed. Res. Int. 2021, 6677363. https://doi.org/10.1155/2021/6677363.

Gálvez-Salido, A., Robles, F., Gonçalves, R.J., Herrán, R., Ruiz Rejón, C., Navajas-Pérez, R., 2026. Analysis of biological images and quantitative monitoring using deep learning and computer vision. J. imaging 12 (2), 88. https://doi.org/10.3390/jimaging12020088.

GAO, 2024. Precision agriculture: benefits and challenges for technology adoption and use. https://www.gao.gov/products/gao-24–105962.

García-López, N., Bargués-Tobella, A., Goodman, R.C., Uwingabire, S., Sundberg, C., Boman, C., et al., 2024. An integrated agroforestry-bioenergy system for enhanced energy and food security in rural sub-Saharan Africa. Ambio 53 (10), 1492–1504. https://doi.org/10.1007/s13280-024-02037-0.

GBIF, 2016. Cooperation and Partnerships. https://www.cbd.int/cooperation/csp/gbif.shtml.

Gebru, B.M., Wang, S.W., Kim, S.J., Lee, W.-K., 2019. Socio-ecological niche and factors affecting agroforestry practice adoption in different agroecologies of southern Tigray, Ethiopia. Sustainability 11 (13), 3729. https://doi.org/10.3390/su11133729.

Gopi, G., Manjula, M., 2018. Speciality rice biodiversity of Kerala: need for incentivising conservation in the era of changing climate. Curr. Sci. 114 (5), 997–1006. http://www.jstor.org/stable/26495193.

Gupta, S., 2025. A review on the role of agroecology in achieving sustainable development goals. Br. J. Agroecology Environ. Stud. 2 (1), 38–48. https://doi.org/10.61424/bjaes.v2i1.433.
Gupta, S., Kumar, P., 2022. Declining biodiversity: a hanging sword to food security. Agriculture Issues Policies 99.
Haile, M., Gebremichael, D., Gebrekidan, H., Gebregziabher, D., Darcha, G., Gebreslassie, W., 2024. Contributions of homestead agroforestry during the war in Tigray, Ethiopia. Agrofor. Work. (Tropical For. Issues No) 62, 92–96. https://www.tropenbos.org/resources/contributions-of-homestead-agroforestry-during-the-war-in-tigray-ethiopia/.
Hailu, F., 2025. The role of agrobiodiversity and diverse causes of its losses and methods of conservation: a review. Food Humanity, 100500. https://doi.org/10.1016/j.foohum.2025.100500.
Hassoun, A., Alhaj Abdullah, N., Aït-Kaddour, A., Ghellam, M., Beşir, A., Zannou, O., et al., 2024. Food traceability 4.0 as part of the fourth industrial revolution: key enabling technologies. Crit. Rev. Food Sci. Nutr. 64 (3), 873–889. https://doi.org/10.1080/10408398.2022.2110033.
Havinga, I., Bogaart, P.W., Hein, L., Tuia, D., 2020. Defining and spatially modelling cultural ecosystem services using crowdsourced data. Ecosyst. Serv. 43, 101091. https://doi.org/10.1016/j.ecoser.2020.101091.
Hegde, S., Searchinger, T., Díaz, M.J., 2025. Methane Mitigation in Agriculture: Technological, Economic, and Regulatory ConsiderationsDC: World Resources Institute, Washington. https://www.wri.org/research/opportunities-methane-mitigation-agriculture-technological-economic-regulatory.
Hughes, K., Morgan, S., Baylis, K., Oduol, J., Smith-Dumont, E., Vågen, T.-G., et al., 2020. Assessing the downstream socioeconomic impacts of agroforestry in Kenya. World Dev. 128, 104835. https://doi.org/10.1016/j.worlddev.2019.104835.
Hultgren, A., Carleton, T., Delgado, M., et al., 2025. Impacts of climate change on global agriculture accounting for adaptation. Nature 642, 644–652. https://doi.org/10.1038/s41586-025-09085-w.
Ihuoma, S.O., Madramootoo, C.A., Kalacska, M., 2021. Integration of satellite imagery and in situ soil moisture data for estimating irrigation water requirements. Int. J. Appl. Earth Observation Geoinf 102, 102396. https://doi.org/10.1016/j.jag.2021.102396.
Ipara, M.O., Kipchumba, H.E., 2024. Effects of stakeholder collaboration on environmental conservation in Kanduyi constituency, Bungoma County, Kenya. World J. Adv. Res. Rev. 24 (2), 932–939. https://doi.org/10.30574/wjarr.2024.24.2.3403.
IPBES, 2019. The global assessment report on biodiversity and ecosystem services. https://files.ipbes.net/ipbes-web-prod-public-files/inline/files/ipbes_global_assessment_report_summary_for_policymakers.pdf.
Irie, K., Nagamine, T., Masuda, M., 2025. Role and perspectives of plant genetic resources for guaranteeing sustainable crop production in developing countries. In: Nagamine, T., Masuda, M., Irie, K. (Eds.), Plant Genetic Resources (PGR) for Sustainable Crop Production. Sustainable Development Goals Series. Springer, Singapore. https://doi.org/10.1007/978-981-96-7117-5_13.
Isbell, C., Tobin, D., Jones, K., Reynolds, T.W., 2023. Enhancing resilience through seed system plurality and diversity: challenges and barriers to seed sourcing during (and in spite of) a global pandemic. Agriculture Hum. Values 40 (4), 1399–1418. https://doi.org/10.1007/s10460-023-10456-0.
Rizvi, J., Dhyani, S.K., Arunachalam, A., Rouge, J.C., Singh, A., 2024. The success of a national agroforestry policy in India has become a model for other countries. Center for International Forestry Research and World Agroforestry (CIFOR-ICRAF), Bogor, Indonesia and Nairobi, Kenya. https://doi.org/10.17528/cifor-icraf/009240.
Jing, H., Liu, Y., Hou, J., 2025. Impacts of agricultural intensification on biodiversity: habitat loss, agrochemical use, water depletion, and soil degradation. J. Environ. Manag. 395, 128036. https://doi.org/10.1016/j.jenvman.2025.128036.

J-PAL Policy Insight, 2026. "Cash for conservation: Climate action on a budget." Cambridge, MA: Abdul Latif Jameel Poverty Action Lab (J-PAL). https://www.povertyactionlab.org/policy-insight/cash-conservation-climate-action-budget?lang=pt-br.

Jung, D.R., Vendrametto, O., 2025. Agroforestry for food security and public health: a comprehensive review. Int. J. Environ. Res. Public. Health 22 (4), 645. https://doi.org/10.3390/ijerph22040645.

Kabato, W., Getnet, G.T., Sinore, T., Nemeth, A., Molnár, Z., 2025. Towards climate-smart agriculture: strategies for sustainable agricultural production, food security, and greenhouse gas reduction. Agronomy 15 (3), 565. https://doi.org/10.3390/agronomy15030565.

Kole, C., Pandey, S., Yasin, J.K., Mamidi, S., Bohra, A., Bhattacharya, P., et al., 2025. Benefits, concerns, and sustainable alternatives to genetically modified crops from a global and Indian perspective. Plant. Genome 18, e70154. https://doi.org/10.1002/tpg2.70154.

Kondo, S., Yoshimoto, N., Nakayama, Y., 2024. Farm monitoring system with drones and optical camera communication. Sensors 24 (18), 6146. https://doi.org/10.3390/s24186146.

Kopittke, P.M., Harper, S.M., Asio, L.G., Asio, V.B., Batalon, J.T., Batuigas, A.M.T., et al., 2025. Soil degradation: an integrated model of the causes and drivers. Int. Soil. Water Conserv. Res. 13 (4), 744–755. https://doi.org/10.1016/j.iswcr.2025.07.010.

Krug, C.B., Sterling, E., Cadman, T., Geschke, J., Drummond de Castro, P.F., Schliep, R., et al., 2020. Stakeholder participation in IPBES: connecting local environmental work with global decision making. Ecosyst. People 16 (1), 197–211. https://doi.org/10.1080/26395916.2020.1788643.

Kumar, R., Kamboj, N., Mitra, D., 2024. Microbial innovations in agriculture: interdisciplinary approaches to leveraging microbes for food sustainability and security. Indian. J. Microbiol. Res. 11 (3), 129–139. https://doi.org/10.18231/j.ijmr.2024.026.

Lahlali, R., Taoussi, M., Laasli, S.-E., Gachara, G., Ezzouggari, R., Belabess, Z., et al., 2024. Effects of climate change on plant pathogens and host-pathogen interactions. Crop. Environ. 3 (3), 159–170. https://doi.org/10.1016/j.crope.2024.05.003.

Laird, S., Muir, G., Shanley, P., Pierce, A., López Binnquist, C., et al., 2024. Forests, trees and wild species inagrifood systems – optimizing benefits for biodiversity, climate and health. Rome, FAO. https://doi.org/10.4060/cd2883en.

Landa-Acuña, D., Solorzano-Acosta, A., Sánchez-Ortiz, V., Hualpa-Cutipa, E., Vargas-de-la-Cruz, C., Luis-Alaya, B., et al., 2022. Microbial food products: a sustainable solution to alleviate hunger. Relationship Between Microbes and the Environment for Sustainable Ecosystem Services, Volume 1. Elsevier, pp. 1–27. https://doi.org/10.1016/B978-0-323-89938-3.00001-3.

Li, C., Camac, J., Robinson, A., et al., 2025. Predicting changes in agricultural yields under climate change scenarios and their implications for global food security. Sci. Rep. 15, 2858. https://doi.org/10.1038/s41598-025-87047-y. Article.

Li, C., Su, Z., Nie, Y., Ye, Z., Li, J., Wang, J., et al., 2025. Scientific planning of dynamic crops in complex agricultural landscapes based on adaptive optimization hybrid SA-GA method. Sci. Rep. 15 (1), 28992. https://doi.org/10.1038/s41598-025-14188-5.

Li, K., Chen, M., Shi, J., Mao, T., 2024. An overview of the production and use of *Bacillus thuringiensis* toxin. Open. Life Sci. 19 (1), 20220902. https://doi.org/10.1515/biol-2022-0902.

Li, T., Wang, J., Chen, H., Li, W., Pan, P., Wu, L., et al., 2023. Performance analysis of an integrated biomass-to-energy system based on gasification and pyrolysis. Energy Convers. Manag. 287, 117085. https://doi.org/10.1016/j.enconman.2023.117085.

Li, X., Yan, Y., Yao, L., 2020. Get a Fish' vs. 'Get a Fishing Skill': farmers' preferred compensation methods to control agricultural nonpoint source pollution. Int. J. Environ. Res. Public. Health 17 (7), 2484. https://doi.org/10.3390/ijerph17072484.

Liao, M., Zhang, Z., Yan, R., Bai, K., 2025. Using the socio-ecological production landscapes and seascapes framework to assess the resilience of agroecosystems in three distinct agricultural production systems in China. Environ. Sustain. Indic, 100814. https://doi.org/10.1016/j.indic.2025.100814.

Loizou, E., Spinthiropoulos, K., Kalogiannidis, S., Chatzitheodoridis, F., Kalfas, D., Tzilantonis, G., 2025. Enhancing climate resilience and food security in greece through agricultural biodiversity. Land 14 (4), 838. https://doi.org/10.3390/land14040838.

Lokhandwala, Z., 2022. Peasants' rights as new human rights: promises and concerns for agrobiodiversity conservation. Asian J. Int. Law 12 (1), 105–120. https://doi.org/10.1017/S2044251322000042.

Louwaars, N.P., Manicad, G., 2022. Seed systems resilience—An overview. Seeds 1 (4), 340–356. https://doi.org/10.3390/seeds1040028.

Lyu, X., Peng, W., Yu, W., Xin, Z., Niu, S., Qu, Y., 2021. Sustainable intensification to coordinate agricultural efficiency and environmental protection: a systematic review based on metrological visualization. J. Land. Use Sci. 16 (3), 313–338. https://doi.org/10.1080/1747423X.2021.1922524.

MAMBO (2015). Integrated Agro-Business Strategies forImprovement of Food Safety and Security in theDemocratic Republic of Congo. https://web.archive.org/web/20200924195153/http:/www.mamboinitiatives.org/wp-content/uploads/2018/06/Mambo-Initiatives-Integrated-Agro-business-mega-project.pdf.

Mancini, M.C., Collins, R.M., Addicott, E.T., Balmford, B.J., Binner, A., Bull, J.W., et al., 2024. Biodiversity offsets perform poorly for both people and nature, but better approaches are available. One Earth 7 (12), 2165–2174. https://doi.org/10.1016/j.oneear.2024.10.002.

Marcos-Pérez, M., Sánchez-Navarro, V., Zornoza, R., 2023. Intercropping systems between broccoli and fava bean can enhance overall crop production and improve soil fertility. Sci. Horticulturae 312, 111834. https://doi.org/10.1016/j.scienta.2023.111834.

McGuire, S., Sperling, L., 2016. Seed systems smallholder farmers use. Food Security 8 (1), 179–195. https://doi.org/10.1007/s12571-015-0528-8.

MEA, 2005. Ecosystems and Human Well-being. https://www.millenniumassessment.org/documents/document.356.aspx.pdf.

Menon, K.G., Nithin, P., Boje Gowd, B.H., Reddy, K.V., 2024. Soil regeneration through soil microbiome: role in food security. In: Aransiola, S.A., Atta, H.I., Maddela, N.R. (Eds.), Soil Microbiome in Green Technology Sustainability. Springer, Cham. https://doi.org/10.1007/978-3-031-71844-1_14.

Miller, S., Dong, Y.S., 2019. Using soil moisture meters to complement irrigation scheduling. https://www.canr.msu.edu/news/using-soil-moisture-meters-to-compliment-irrigation-scheduling.

Mishra, H., 2025. Environmental degradation and impacts on agricultural production: a challenge to urban sustainability. In: Parray, J.A., Shameem, N., Haghi, A.K. (Eds.), Sustainable Urban Environment and Waste Management (Urban Sustainability). Springer. https://doi.org/10.1007/978-981-96-1140-9_3.

Miller, T., Mikiciuk, G., Durlik, I., Mikiciuk, M., Łobodzińska, A., Śnieg, M., 2025. The IoT and AI in agriculture: the time is now—A systematic review of smart sensing technologies. Sensors 25 (12), 3583. https://doi.org/10.3390/s25123583.

Moilanen, A., Kotiaho, J.S., 2018. Fifteen operationally important decisions in the planning of biodiversity offsets. Biol. Conserv. 227, 112–120. https://doi.org/10.1016/j.biocon.2018.09.002.

Mosquera-Losada, M.R., Santos, M.G.S., Gonçalves, B., Ferreiro-Domínguez, N., Castro, M., Rigueiro-Rodríguez, A., et al., 2023. Policy challenges for agroforestry implementation in Europe. Front. For. Glob. Change 6. https://doi.org/10.3389/ffgc.2023.1127601.

Mthembu, B.E., Everson, T.M., Everson, C.S., 2019. Intercropping for enhancement and provisioning of ecosystem services in smallholder, rural farming systems in KwaZulu-Natal Province, South Africa: a review. J. Crop. Improv. 33 (2), 145–176. https://doi.org/10.1080/15427528.2018.1547806.

Mugari, E., Nethengwe, N.S., Gumbo, A.D., 2025. A co-design approach for stakeholder engagement and knowledge integration in flood risk management in Vhembe district, South Africa. Front. Clim. 7. https://doi.org/10.3389/fclim.2025.1517837.

Muluneh, M.G., 2021. Impact of climate change on biodiversity and food security: a global perspective — a review article. Agric. Food Secur. 10, 36. https://doi.org/10.1186/s40066-021-00318-5.

Mustonen, T., Harper, S., Pecl, G., Castan Broto, V., Lansbury, N., Okem, A., et al., 2022. The role of indigenous knowledge and local knowledge in understanding and adapting to climate change. IPCC Clim. change 2713–2807.

Nesshöver, C., Assmuth, T., Irvine, K.N., Rusch, G.M., Waylen, K.A., Delbaere, B., et al., 2017. The science, policy and practice of nature-based solutions: an interdisciplinary perspective. Sci. Total. Environ. 579, 1215–1227. https://doi.org/10.1016/j.scitotenv.2016.11.106.

Njeru, E.M., Awino, R.O., Kirui, K.C., Koech, K., Jalloh, A.A., Muthini, M., 2022. Agrobiodiversity and perceived climatic change effect on family farming systems in semiarid tropics of Kenya. Open. Agric. 7 (1), 360–372. https://doi.org/10.1515/opag-2022-0099.

Ntihinyurwa, P.D., de Vries, W.T., 2021. Farmland fragmentation, farmland consolidation and food security: relationships, research lapses and future perspectives. Land 10 (2), 129. https://doi.org/10.3390/land10020129.

Ogazie, C.A., Ochekwu, E.B., Agbagwa, I.O., Ugiomoh, I.G., 2022. Multi-cropping practice: means to sustainable agriculture in the high humid rainforest agroecology of Southern Nigeria. Int. J. Sci. Technol. Res. Archive 3 (1), 066–090. https://doi.org/10.53771/ijstra.2022.3.1.0066.

Okafor, C.C., Nzekwe, C.A., Ajaero, C.C., Ibekwe, J.C., Otunomo, F.A., 2022. Biomass utilization for energy production in Nigeria: a review. Clean. Energy Syst. 3, 100043. https://doi.org/10.1016/j.cles.2022.100043.

Olawade, D.B., Wada, O.Z., Ige, A.O., Egbewole, B.I., Olojo, A., Oladapo, B.I., 2024. Artificial intelligence in environmental monitoring: advancements, challenges, and future directions. Hyg. Environ. Health Adv. 12, 100114. https://doi.org/10.1016/j.heha.2024.100114.

Onda, Y., Mochida, K., 2016. Exploring genetic diversity in plants using high-throughput sequencing techniques. Curr. Genomics 17 (4), 358–367. https://doi.org/10.2174/1389202917666160331202742.

Pandey, D.K., Momin, K.C., Dubey, S.K., Adhiguru, P., 2022. Biodiversity in agricultural and food systems of jhum landscape in the West Garo Hills, North-eastern India. Food security 14 (3), 791–804. https://doi.org/10.1007/s12571-021-01251-y.

Panizzon, J.P., Pilz, H.L., Knaak, N., Ramos, R.C., Ziegler, D.R., Fiuza, L.M., 2015. Microbial diversity: relevance and relationship between environmental conservation and human health. Braz. Arch. Biol. Technol. 58 (1), 137–145. https://doi.org/10.1590/S1516-8913201502821.

Pascual, U., Narloch, U., Nordhagen, S., Drucker, A.G., 2011. The economics of agrobiodiversity conservation for food security under climate change. Economía Agraria y. Recur. Naturales 11 (1), 191–200. https://doi.org/10.22004/ag.econ.117623.

Pei, B., Chen, S., Zhang, H., 2025. Integrating water footprint into cultivated land ecological compensation model: insights from a provincial-level analysis in mainland China. Agric. Water Manag. 322, 109967. https://doi.org/10.1016/j.agwat.2025.109967.

Peng, X., Chen, H., Zhong, H., Hong, J., Song, K., 2023. Editorial: curbing global warming with multi-scale and multi-sectoral water-energy-food nexus. Front. Earth Sci. 11. https://doi.org/10.3389/feart.2023.1271246.

Pereira, P., Inacio, M., Barcelo, D., Zhao, W., 2025. Impacts of agriculture intensification on biodiversity loss, climate change and ecosystem services. Curr. Opin. Environ. Sci. & Health 46, 100637. https://doi.org/10.1016/j.coesh.2025.100637.

Pfenning-Butterworth, A.P., Buckley, L.B., Drake, J.M., Farner, J.E., Farrell, M.J., Gehman, A.-L.M., et al., 2024. Interconnecting global threats: climate change, biodiversity loss, and infectious diseases. Lancet Planet. Health 8 (4), e270–e283. https://doi.org/10.1016/S2542-5196(24)00021-4[3].

Ponce Reyes, R., Firn, J., Nicol, S., Chadès, I., Stratford, D.S., Martin, T.G., et al., 2019. Building a stakeholder-led common vision increases the expected cost-effectiveness of biodiversity conservation. PLOS ONE 14 (6), e0218093. https://doi.org/10.1371/journal.pone.0218093.

Pope, J., Morrison-Saunders, A., Bond, A., Retief, F., 2021. When is an offset not an offset? A framework of necessary conditions for biodiversity offsets. Environ. Manag. 67 (2), 424–435. https://doi.org/10.1007/s00267-020-01415-0.

Porcuna-Ferrer, A., Labeyrie, V., Alvarez-Fernandez, S., Calvet-Mir, L., Faye, N.F., Ouadah, S., et al., 2023. Crop biocultural traits shape seed networks: implications for social-ecological resilience in south eastern Senegal. Agric. Syst. 211, 103750. https://doi.org/10.1016/j.agsy.2023.103750.

Potent, J., 2016. Sustainable agriculture through innovation and collaboration. https://news.climate.columbia.edu/2016/11/16/sustainable-agriculture-through-innovation-and-collaboration/.

Prajapati, C.S., Priya, N.K., Bishnoi, S., Vishwakarma, S.K., Buvaneswari, K., Shastri, S., et al., 2025. The role of participatory approaches in modern agricultural extension: bridging knowledge gaps for sustainable farming practices. J. Exp. Agric. Int. 47 (2), 204–222. https://doi.org/10.9734/jeai/2025/v47i23281.

Pratzer, M., Fernández-Llamazares, Á., Meyfroidt, P., et al., 2023. Agricultural intensification, Indigenous stewardship and land sparing in tropical dry forests. Nat. Sustain. 6 (6), 671–682. https://doi.org/10.1038/s41893-023-01073-0.

Primmer, E., Varumo, L., Kotilainen, J.M., Raitanen, E., Kattainen, M., Pekkonen, M., et al., 2019. Institutions for governing biodiversity offsetting: an analysis of rights and responsibilities. Land. Use Policy 81, 776–784. https://doi.org/10.1016/j.landusepol.2018.11.040.

Prip, C., 2024. Farmers' rights in the plant treaty: interrelations and recent interactions with other international regimes and processes. Dev. Stud. Res. 11 (1). https://doi.org/10.1080/21665095.2024.2357095.

Punjabi, S., Misra, S., Rippy, M.A., Grant, S.B., Galappaththi, E., Lim, T., et al., 2025. A conceptual framework for knowledge integration in cross-disciplinary collaborations. Environ. Sci. & Policy 172, 104197. https://doi.org/10.1016/j.envsci.2025.104197. Article.

Qin, Z., Zhu, Y., Canadell, J.G., Chen, M., Li, T., Mishra, U., et al., 2024. Global spatially explicit carbon emissions from land-use change over the past six decades (1961–2020). One Earth 7 (5), 835–847. https://doi.org/10.1016/j.oneear.2024.04.002.

Quandt, A., 2020. Contribution of agroforestry trees for climate change adaptation: narratives from smallholder farmers in Isiolo, Kenya. Agrofor. Syst. 94 (6), 2125–2136. https://doi.org/10.1007/s10457-020-00535-0.

Ramanna, A., Andersen, R., 2022. Stewardship or ownership in India: options for community seed banks in managing crop genetic resources in relation to intellectual property rights. J. World Intellect. Property 25 (2), 432–459. https://doi.org/10.1111/jwip.12233.

Rasul, G., Hussain, A., Adhikari, L., Molden, D.J., 2022. Conserving agrobiodiversity for sustainable food systems in the Hindu Kush Himalaya. Int. J. Agric. Sustain. 20 (6), 1117–1135. https://doi.org/10.1080/14735903.2022.2057642.

Raveloaritiana, E., Wanger, T.C., 2024. Decades matter: agricultural diversification increases financial profitability, biodiversity, and ecosystem services over time (arXiv:2403.05599). arXiv. https://doi.org/10.48550/arXiv.2403.05599.

Raveloaritiana, E., Wanger, T.C., 2026. Long-term agricultural diversification increases financial profitability, biodiversity, and ecosystem services: a second-order meta-analysis. Nat. Commun. 17, 1016. https://doi.org/10.1038/s41467-025-67757-7.

REFOREST, 2023. Agroforestry sector in the EU: Towards novel policies and financing frameworks. https://agroreforest.eu/wp-content/uploads/2023/10/Reforest-Policy-Session-Report.pdf.

Rehman, A., Farooq, M., Lee, D.J., Siddique, K.H.M., 2022. Sustainable agricultural practices for food security and ecosystem services. Environ. Sci. Pollut. Res. Int. 29 (56), 84076–84095. https://doi.org/10.1007/s11356-022-23635-z.

Renard, D., Tilman, D., 2021. Cultivate biodiversity to harvest food security and sustainability. Curr. Biol. 31 (19), R1154–R1158. https://doi.org/10.1016/j.cub.2021.06.082.

Saini, P., Saini, P., Kaur, J.J., Francies, R.M., Gani, M., Rajendra, A.A., et al., 2020. Molecular approaches for harvesting natural diversity for crop improvement. Rediscovery of genetic and genomic resources for future food security. Springer Singapore, Singapore, pp. 67–169. https://doi.org/10.1007/978-981-15-0156-2_3.

Saleem, A., Anwar, S., Nawaz, T., et al., 2025. Securing a sustainable future: the climate change threat to agriculture, food security, and sustainable development goals. J. Umm Al-Qura Univ. Appl. Sci. 11 (4), 595–611. https://doi.org/10.1007/s43994-024-00177-3.

Salgotra, R.K., Chauhan, B.S., 2023. Genetic diversity, conservation, and utilization of plant genetic resources. Genes 14 (1), 174. https://doi.org/10.3390/genes14010174.

Sánchez, A.C., Kamau, H.N., Grazioli, F., Jones, S.K., 2022. Financial profitability of diversified farming systems: a global meta-analysis. Ecol. Econ. 201, 107595. https://doi.org/10.1016/j.ecolecon.2022.107595.

Semwal, P., Dave, A., Israr, J., Misra, S., Kumar, M., Paul, D., 2025. Exploring microbial ecosystem services for environmental stress amelioration: a review. Int. J. Mol. Sci. 26 (10), 4515. https://doi.org/10.3390/ijms26104515.

Singh, V., 2024. Advances in precision agriculture technologies for sustainable crop production. J. Sci. Res. Rep. 30 (2), 61–71. https://doi.org/10.9734/jsrr/2024/v30i21844.

Singh, D.P., Prabha, R., Gupta, V.K., 2019. Microbial inoculants for sustainable crop management. In: Singh, D., Gupta, V., Prabha, R. (Eds.), Microbial interventions in agriculture and environment. Springer, Singapore. https://doi.org/10.1007/978-981-13-8383-0_1.

Sharma, N., Bohra, B., Pragya, N., Ciannella, R., Dobie, P., Lehmann, S., 2016. Bioenergy from agroforestry can lead to improved food security, climate change, soil quality, and rural development. Food Energy Security 5 (3), 165–183. https://doi.org/10.1002/fes3.87.

Sharma, R., Mina, U., Kumar, B.M., 2022. Homegarden agroforestry systems in achievement of sustainable development goals. A review. Agron. Sustain. Dev. 42 (2022), 44. https://doi.org/10.1007/s13593-022-00781-9.

Shukla, J., Dhyani, S., 2023. Potential of traditional agroforestry systems for intensification of agriculture and meeting global goals: lessons from Asia and Africa (pp. 665–683). https://doi.org/10.1007/978–981-19–4602-8_20.

Sjaf, S., Arsyad, A.A., Mahardika, A.R., Gandi, R., Elson, L., Hakim, L., et al., 2022. Partnership 4.0: smallholder farmer partnership solutions. Heliyon 8 (12), e12012. https://doi.org/10.1016/j.heliyon.2022.e12012.

Sperling, L., Louwaars, N., de Ponti, O., Smale, M., Baributsa, D., van Etten, J., 2020. Viewpoint: COVID-19 and seed security response now and beyond. Food Policy 97, 102000. https://doi.org/10.1016/j.foodpol.2020.102000.

Sunderland, T.C., 2011. Food security: why is biodiversity important? Int. Forestry Rev. 13 (3), 265–274.

Sustainability Dictionary, 2025. Digital divide in agroecological AI adoption. https://prism.sustainability-directory.com/scenario/digital-divide-in-agroecological-ai-adoption/.

Tamburini, G., Bommarco, R., Wanger, T.C., Kremen, C., van der Heijden, M.G.A., Liebman, M., et al., 2020. Agricultural diversification promotes multiple ecosystem services without compromising yield. Sci. Adv. 6 (45), eaba1715. https://doi.org/10.1126/sciadv.aba1715.

Tamene, L., Abera, W., Bendito, E., Erkossa, T., Tariku, M., Sewnet, H., et al., 2022. Data-driven similar response units for agricultural technology targeting: an example from Ethiopia. Exp. Agriculture 58, e27. https://doi.org/10.1017/S0014479722000126.

Thrupp, L.A., 2004. The importance of biodiversity in agroecosystems. J. Crop. Improvement 12 (1-2), 315–337. https://doi.org/10.1300/J411v12n01_03.

Toor, M.D., Ur Rehman, M., Abid, J., et al., 2024. Microbial ecosystems as guardians of food security and water resources in the era of climate change. Water Air Soil. Pollut. 235, 741. https://doi.org/10.1007/s11270-024-07533-3.

Tupala, A.-K., Huttunen, S., Halme, P., 2022. Social impacts of biodiversity offsetting: A review. Biol. Conserv. 267, 109431. https://doi.org/10.1016/j.biocon.2021.109431.

Ulian, T., Diazgranados, M., Pironon, S., Padulosi, S., Liu, U., Davies, L., et al., 2020. Unlocking plant resources to support food security and promote sustainable agriculture. Plants, People, Planet 2 (5), 421–445. https://doi.org/10.1002/ppp3.10145.

UNEP., 2021. Making peace with nature: A scientific blueprint to tackle the climate, biodiversity and pollution emergencies. https://wedocs.unep.org/handle/20.500.11822/34948.

USAID., 2016. Measuring impact stakeholder engagement for biodiversity conservation goals assessing the status of the evidence. https://citieswithnature.org/wp-content/uploads/2019/02/USAID_Stakeholder-Engagement-for-Biodiversity-Conservation-Goals.pdf.

USDA (n.d.). Agroforestry practices. Retrieved February 20, 2026, from https://www.fs.usda.gov/nac/practices/.

Valentino, V., Magliulo, R., Farsi, D., Cotter, P.D., O'Sullivan, O., Ercolini, D., et al., 2024. Fermented foods, their microbiome and its potential in boosting human health. Microb. Biotechnol. 17, e14428. https://doi.org/10.1111/1751-7915.14428. Available from:.

van Versendaal, L., Schickhoff, U., 2025. Viability of nature conservation in the midst of crises. A review of community perspectives during the COVID-19 pandemic in Madagascar. Trees, For. People 21, 100893. https://doi.org/10.1016/j.tfp.2025.100893.

Vos, R., Glauber, J., Hebebrand, C., Rice, B., 2025. Global shocks to fertilizer markets: impacts on prices, demand and farm profitability. Food Policy 133, 102790. https://doi.org/10.1016/j.foodpol.2024.102790.

Wang, L., Zhu, J., Wang, D., 2025. Comparative analysis of high-resolution CMIP6 GCM and CMIP5 RCM: unveiling biases and advancements in simulating compound extreme events in China. Clim. Dyn. 63 (1), 91. https://doi.org/10.1007/s00382-024-07571-x.

Wang, J., Lu, X., Zhang, J., Wei, H., Li, M., Lan, N., et al., 2021. Intercropping perennial aquatic plants with rice improved paddy field soil microbial biomass, biomass carbon and biomass nitrogen to facilitate soil sustainability. Soil. Tillage Res. 208, 104908. https://doi.org/10.1016/j.still.2020.104908.

Wang, T., Mao, D., Jiang, Z., 2022. Quantitative assessment of agricultural horizontal ecological compensation in China, from the perspective of virtual land and virtual water. Environ. Sci. Pollut. Res. 30 (9), 22885–22899. https://doi.org/10.1007/s11356-022-23220-4.

Wanjira, E. O. (2019). Gliricidia sepium factsheet: Establishment, management and benefits. World Agroforestry. https://www.regreeningafrica.org/wp-content/uploads/2019/10/Gliricidia-sepium-Factsheet.pdf.

WAO, 2013. Agroforestry for sustainable intensification of agriculture. https://worldagroforestry.org/news/agroforestry-sustainable-intensification-agriculture.

Wei, F., Wang, S., Fu, B., Zhang, L., Fu, C., Kanga, E.M., 2018. Balancing community livelihoods and biodiversity conservation of protected areas in East Africa. Curr. Opin. Environ. Sustainability 33, 26–33. https://doi.org/10.1016/j.cosust.2018.03.013.

Wei, B., Peng, Y., Lin, L., Zhang, D., Ma, L., Jiang, L., et al., 2023. Drivers of biochar-mediated improvement of soil water retention capacity based on soil texture: a meta-analysis. Geoderma 437, 116591. https://doi.org/10.1016/j.geoderma.2023.116591.

World Health Organization (n.d.). Biodiversity. Retrieved July 14, 2026, from https://www.who.int/teams/environment-climate-change-and-health/climate-change-and-health/biodiversity.

Xiao, J., Yang, Y., Xiong, K., Jiang, S., Huang, Y., 2026. Enhancing agroforestry system resilience: harnessing environmental vulnerability data to map restoration and pave the way for karst desertification control. Agric. Syst. 231, 104538. https://doi.org/10.1016/j.agsy.2025.104538.

Xing, Y., Wang, X., 2024. Precision agriculture and water conservation strategies for sustainable crop production in arid regions. Plants 13 (22), 3184. https://doi.org/10.3390/plants13223184.

Yu, B., Liu, X., Bi, X., Sun, H., Buysse, J., 2025. Agricultural resource management strategies for greenhouse gas mitigation: the land-energy-food-waste nexus based on system dynamics model. Environ. Impact Assess. Rev. 110, 107647. https://doi.org/10.1016/j.foodpol.2020.102000.

Zuza, E.J., Lambert, K., Macmillan, T., Chiyemura, F., Araya, Y., Bowskill, V., et al., 2024. Crop species diversity: a key strategy for sustainable food system transformation and climate resilience. Food Energy Security 13 (3), e558. https://doi.org/10.1002/fes3.558.

CHAPTER TEN

Leveraging AI for coral reef assessment and food systems resilience: A framework for coastal nutrition security in the pacific blue economy

Jannie Fleur V. Oraño[a,*], Larmie S. Feliscuzo[a], Chris Jordan G. Aliac[a], and Jerome Jack O. Napala[b]

[a]Cebu Institute of Technology – University,Cebu City, Philippines
[b]Southern Leyte State University, Sogod, Southern Leyte, Philippines
*Corresponding author. e-mail address: janniefleur.orano@cit.edu

Contents

Abstract

Coral reef ecosystems are integral to food security across the Pacific and Southeast Asia, where many coastal communities depend on reef fisheries for nutrition and livelihoods. However, widespread reef degradation driven by climate change, pollution, and destructive fishing practices threatens these ecosystem services, underscoring the need for efficient and scalable monitoring methods. This chapter presents an AI-powered framework for coral cover assessment that integrates automated image preprocessing, YOLO-based coral detection and segmentation, and GIS

Advances in Food Security and Sustainability, Volume 11
ISSN 2452-2635, https://doi.org/10.1016/bs.af2s.2026.06.005

mapping to produce spatially explicit indicators of reef health. A case study in Sogod Bay, Southern Leyte, Philippines demonstrates the practical application of the framework using expert annotated quadrat images. Through transfer learning, the model effectively detected and segmented diverse coral morphologies, including Acropora branching, massive, and mushroom types, enabling detailed estimation of coral cover and benthic composition. By automating coral cover estimation and linking outputs to future geospatial analysis, the proposed framework offers a scalable tool to support reef monitoring. The integration of artificial intelligence and geospatial analysis supports evidence based decision making in fisheries management, marine protected area planning, and coastal governance. Ultimately, the framework offers a practical tool for strengthening food security and advancing sustainable blue economy initiatives by enhancing reef conservation, supporting adaptive management, and improving resilience in coral dependent communities.

1. Introduction

Food security in the Pacific islands and coastal regions of Southeast Asia is deeply intertwined with the condition and the sustained ecosystem stability and recovery of coral reef ecosystems (Bell et al., 2018). These ecosystems sustain food webs, livelihoods, economies, shoreline protection, and cultural identities, serving as foundational pillars for low income communities, as well as for developing nations and small island states with strong reliance on marine ecosystems (Chaijaroen, 2022). Although coral reefs occupy less than 0.2 % of the global ocean floor, they harbor approximately 38 % of marine biodiversity, recognizing as one of the planet's most biologically diverse and productive systems (Jury et al., 2024). This extraordinary biodiversity directly contributes to food availability, particularly in island nations and archipelagic regions where terrestrial agriculture is limited. In many Pacific Island communities, reef fisheries provide 50–90 % of total animal protein intake (Bell et al., 2018), highlighting their irreplaceable role in sustaining daily nutritional requirements. Reef-derived foods with fish being a key source of high quality protein and essential micronutrients, such as omega 3 fatty acids, iodine, iron, zinc, selenium, and vitamin A, necessary for human well-being. Consequently, the degradation of coral reef can compromise the availability of these nutrient-rich food sources, increasing the risk of food insecurity and contributing to malnutrition in vulnerable coastal populations (Rabbitt et al., 2023).

The ecological structure of coral reefs fundamentally drives the productivity of reef-derived food systems. Hard corals, including branching, tabulate, massive, encrusting, and foliose morphologies, form three-dimensional frameworks that sustain diverse fish communities and support essential ecological processes. These structural habitats provide refuge from predation and competition, create nursery grounds for juvenile fishes, and buffer organisms from environmental stressors (Beese, 2024). Structurally complex reefs are important for maintaining a wide range of fish functional groups, thereby supporting diverse assemblages of reef organisms and enhancing overall ecosystem productivity (Helder et al., 2022). Empirical evidence shows that coral cover is positively correlated not only with herbivorous invertebrates (e.g., sea urchins) but also with fish family richness, demonstrating how habitat structure directly influences species diversity and trophic functioning (Ditzel et al., 2022). These ecological interactions underpin small-scale fisheries that supply households with nutrient-rich seafood, reinforce traditional food-sharing networks, and significantly contribute to local income, market stability, and the broader resilience of coastal communities.

However, the ongoing decline of coral reef ecosystems constitutes a significant threat to food security. Coral reefs are becoming increasingly susceptible to the compounding effects of climate change, particularly ocean warming, mass bleaching events, sedimentation, pollution, destructive fishing practices, and coastal development (Eddy et al., 2021). Alongside these environmental stressors, periodic outbreaks of coral predators, particularly the crown of thorns starfish (Acanthaster spp.), contribute substantially to coral mortality by rapidly consuming live coral tissue and reducing reef building capacity (Castro-Sanguino et al., 2023; Fontoura et al., 2020) demonstrate that climate driven shifts in coral morphological structure are strongly associated with declines in juvenile reef fish populations, with changes in coral morphology explaining a substantial proportion of the observed decrease in juvenile abundance. Their findings further indicate that alterations in the composition of coral growth forms can disrupt habitat quality and reduce population replenishment, suggesting that continued declines in coral cover and structural complexity will negatively affect the future dynamics of reef fish communities.

Fish and aquatic products are vital sources of nutrition and health throughout Pacific Island societies, where seafood consumption rates are among the highest globally and provide essential micronutrients and lean protein that complement carbohydrate rich staple diets (O'Meara et al., 2023). Marine fishes and invertebrates serve as an irreplaceable and affordable source of key micronutrients, such as iron and zinc, particularly in low income countries where seafood supports physical and cognitive development (Cheung et al., 2023). Despite their nutritional importance, the sustainability of these marine food resources is increasingly under threat, as global marine fisheries production has declined since its peak in the 1990s and a substantial proportion of fish stocks are now classified as overexploited or fully exploited (MRAG Asia Pacific, 2022). These pressures are further intensified by climate change, which threatens survival of coral reefs and alters species distribution and productivity, thereby reducing the availability and accessibility of nutrient rich seafood and increasing the risk of food insecurity among populations that rely heavily on marine resources (Chomariyah, 2025). Within this context, coral reef degradation further amplifies these challenges by undermining the ecological foundation of coastal fisheries, as declines in reef structure and productivity reduce fish abundance and diversity, ultimately creating widespread ecological and social vulnerabilities, particularly for poor fishing dependent households whose incomes and diets rely directly on reef productivity (MRAG Asia Pacific, 2022).

The contribution of coral reefs to food security can be better understood within the broader context of the blue economy, which promotes the sustainable use of ocean resources to support economic growth, enhance livelihoods, and maintain ecosystem health. Coral reef ecosystems are key drivers of blue economy sectors, particularly fisheries, tourism, and coastal protection, all of which are critical to the economies of Pacific Island countries and developing nations (Chomariyah, 2025). Sustainable reef fisheries not only provide food and nutrition but also generate income and employment for millions of coastal households, reinforcing their importance in inclusive economic development (Fauzi and Rahman, 2025). However, the degradation of coral reefs undermines these benefits by reducing fisheries productivity, weakening coastal resilience, and limiting economic opportunities. As such, maintaining coral reef health is essential for advancing blue economy objectives, ensuring that marine resources continue to support both ecological sustainability and human well-being (Fauzi and Rahman, 2025).

The relationship between coral reefs and food security therefore reflects a tightly interconnected system in which ecological health, nutritional well-being, and socioeconomic resilience are mutually reinforcing. Without healthy corals, reef-based food systems cannot sustain adequate fish production, dietary diversity diminishes, and coastal populations become increasingly vulnerable to hunger, malnutrition, and climate-related shocks (Charlton et al., 2016). Protecting coral ecosystems is therefore not only an environmental imperative but also a critical strategy for ensuring food security. Achieving this, however, requires continuous and systematic coral reef monitoring capable of detecting ecological change, quantifying coral cover trends, and identifying emerging threats before they lead to irreversible losses (Rivera et al., 2020). Accurate and timely monitoring using machine learning and artificial intelligence provides the scientific basis needed to guide fisheries management, restoration initiatives, environmental monitoring, climate adaptation planning, and evidence based decision making and conservation strategies (Wahsha and Wahsheh, 2025), making coral assessment a central component of present and future food security strategies in reef dependent regions.

2. Coral monitoring and coral cover estimation

Systematic monitoring of coral reefs is fundamental to detecting ecological trends, assessing shifts in reef condition, and identifying early signs of degradation. Long term monitoring and integrated ecosystem observations generate essential data that enable coastal communities and resource managers to understand reef health and recognize their dependence on these ecosystems for human well-being, while also supporting the assessment of ecological, economic, and social impacts of reef decline (Obura et al., 2019). Standardized coral reef monitoring protocols have been developed and implemented by international organizations such as Global Coral Reef Monitoring Network, Reef Check Foundation, and NOAA Coral Reef Conservation Program, which define key indicators including live coral cover, benthic composition, reef fish biomass, coral recruitment, and incidence of bleaching and disease. These protocols ensure consistency, comparability, and reliability of data across spatial and temporal scales (Obura et al., 2019; Towle et al., 2022; Rivera-Sosa et al., 2025).

Traditional coral monitoring programs aim to quantify benthic composition, detect bleaching, assess disease prevalence, and evaluate long term changes in coral community structure using established survey methods such as line intercept transects, photo quadrats, and belt transects. Consistent monitoring enables researchers and resource managers to assess recovery trajectories following disturbances such as storms or bleaching events, while also providing the empirical foundation for fisheries management and coastal planning (Eddy et al., 2018). Because coral decline has immediate implications for food security in reef dependent communities (Rogers et al., 2018), monitoring the condition of coral ecosystems has become an indispensable tool for understanding and strengthening socioecological resilience.

Among the key metrics used in coral assessments, coral cover estimation remains one of the most widely adopted indicators of reef health (Obura et al., 2019). Coral cover, measured as the proportion of reef substrate occupied by live scleractinian corals, serves as a key indicator of reef condition and ecosystem function. Higher coral cover is associated with greater habitat complexity and supports higher diversity and abundance of reef-dependent fish and invertebrates (Eddy et al., 2018). High live coral cover is typically associated with abundant fish biomass, strong recruitment, and balanced trophic interactions. Conversely, declining coral cover often signals shifts toward algal dominance, reduced species richness, and weakened fishery productivity (Koester et al., 2023; Fahim et al., 2024). As such, coral cover measurements serve not only as ecological indicators but also as proxies for predicting changes in food availability and nutritional outcomes for coastal populations.

Traditional coral monitoring techniques rely heavily on in situ transect surveys and manual visual assessment. The Line Intercept Transect (LIT) method involves laying a measuring tape directly across the reef substrate and recording every benthic category intercepted by the line, producing high-resolution estimates of live coral cover but requiring substantial field time and experienced observers (Rivera-Sosa et al., 2025; Facon et al., 2016). The Point Intercept Transect (PIT) method applies a similar transect approach but records benthic categories only at predefined points along the tape, making it faster and more feasible for long-term or large-area surveys, albeit with lower taxonomic specificity compared to LIT. Meanwhile, photo-quadrat approaches capture standardized photographs of quadrats placed along transects, providing a permanent visual record that can be re-examined, compared across years, and processed digitally for improved consistency and annotation accuracy (Rivera-Sosa et al., 2025; Hill and Wilkinson, 2004) (Fig. 1).

Although these traditional methods have produced invaluable long-term datasets, they depend heavily on manual labor and expert judgment. Manual annotation of hundreds or thousands of images is time-consuming and can significantly delay data availability during critical events such as bleaching episodes (Williams et al., 2019). These limitations underscore the growing need for automated, scalable, and rapid coral cover assessment methods, especially in regions where declining coral ecosystems threaten fisheries and the food security of coastal communities (Fig. 2).

Fig. 1 Point intercept transect (PIT) method for coral reef monitoring, showing the transect tape laid along the reef substrate, fixed sampling points at regular intervals, and in situ recording of benthic categories by a diver using an underwater data slate.

Fig. 2 Photo quadrat method for coral reef monitoring. A quadrat frame is placed along a transect to capture standardized images of benthic composition, providing a permanent visual record for comparison and digital analysis.

3. Advancements in AI for coral cover assessment

Recent developments in artificial intelligence (AI) and deep learning have transformed coral reef monitoring, enabling applications that include species classification, semantic segmentation, bleaching detection, and long-term ecosystem assessment. These innovations directly address the limitations of traditional survey techniques, particularly the need for scalable, efficient, and high-resolution methods capable of producing consistent coral cover estimates across diverse and dynamic reef environments.

A rapidly expanding body of research has focused on improving semantic segmentation, a critical step for automated coral cover quantification. Li et al. (2024) provided one of the most comprehensive surveys of deep learning–based coral segmentation approaches, introducing a densely annotated benchmark dataset and evaluating multiple state-of-the-art models to expose performance gaps in manual assessment workflows. Building on this, Zhang et al. (2024) developed CNet, a three-branch encoder architecture that integrates RGB, depth, and shape inputs, achieving an mIoU of 81.83 % and demonstrating improved detection of minority coral classes.

Significant progress has also been made in addressing annotation bottlenecks. Raine et al. (2024) introduced a human-in-the-loop annotation pipeline that combines DINOv2 feature extraction with a K-Nearest Neighbors (KNN) classifier, achieving substantial gains in pixel-level accuracy even with sparse expert labels. Similarly, enhancements to segmentation architectures such as PSPNet, specifically through the incorporation of spatial attention and pixel shuffle modules, have yielded high performance metrics, including a mean Intersection over Union of 89.51 % and pixel accuracy of 94.47 %, while maintaining computational efficiency for field deployment (Li et al., 2024). Hybrid methodologies combining 3D photogrammetry and AI segmentation are also emerging; Kopecky et al. (2023), for instance, demonstrated that 3D surface models better capture structural complexity and reveal size-dependent bleaching mortality compared to traditional 2D imagery.

AI-enabled coral classification has likewise advanced, supporting biodiversity assessments and targeted conservation planning. Banala et al. (2024) proposed RL-BoHVF, a hybrid classifier that combines AlexNet with handcrafted visual features to mitigate underwater imaging distortions and detect bleached corals with 96.2 % accuracy. YOLO-based approaches have also gained traction for field-ready classification and detection workflows. For instance, Oraño et al. (2023) developed a YOLOv5 classifier that achieved 89.29 % accuracy across seven coral lifeforms, demonstrating rapid performance suitable for real-time applications. Multi-view approaches such as TwinNet and nViewNet-8 (King et al., 2019) integrate stereoscopic imagery to enhance taxonomic classification in visually complex environments.

More recently, YOLO architectures have been further adapted for fast and reliable coral detection under variable underwater conditions. Younes et al. (2023) trained a YOLOv5 model on 580 annotated images, achieving robust detection performance across different lighting and habitat types, thereby reinforcing the suitability of YOLO models for in situ monitoring. Beyond detection, AI continues to expand into ecological forecasting and restoration science. Macadam et al. (2021) developed a machine learning pipeline for coral juvenile assessment that reduced annotation time thirtyfold while maintaining over 90 % accuracy. Similarly, long-term monitoring studies such as Gapper et al. (2024) have used support vector machines (SVMs) applied to nearly two decades of Landsat imagery to detect decadal shifts in coral cover in the Red Sea.

Ensuring model robustness and generalizability remains a critical priority, given the high variability of underwater imaging environments. Wyatt et al. (2025) evaluated multiple out-of-distribution (OOD) detection methods for coral imagery and identified KNN and ViM as the most effective for maintaining stable performance when exposed to unfamiliar conditions. At the same time, human expertise continues to play an essential role in the data preparation pipeline. Pavoni et al. (2022) introduced TagLab, a human-in-the-loop annotation platform that integrates AI-assisted pre-labeling with expert refinement, reducing annotation time by 90 % without compromising reliability.

Collectively, these advances illustrate how architectural innovations, improved annotation strategies, and multimodal data integration are transforming coral reef monitoring into a more accurate, efficient, and scalable process. The rapid evolution of YOLO models, transformer architectures, and hybrid annotation tools positions AI as a critical foundation for the future of coral cover estimation and reef health assessment, particularly in regions where reef decline directly affects fisheries productivity and food security.

4. A proposed framework: AI-powered coral assessment for food security

This figure presents an iterative and adaptive framework that connects artificial intelligence based coral monitoring with ecological assessment, participatory spatial analysis, and governance decision making to support reef derived food security in coastal and island communities. The framework is intentionally designed as a cycle, emphasizing continuous learning, feedback, and refinement across monitoring, management, and policy stages (Fig. 3).

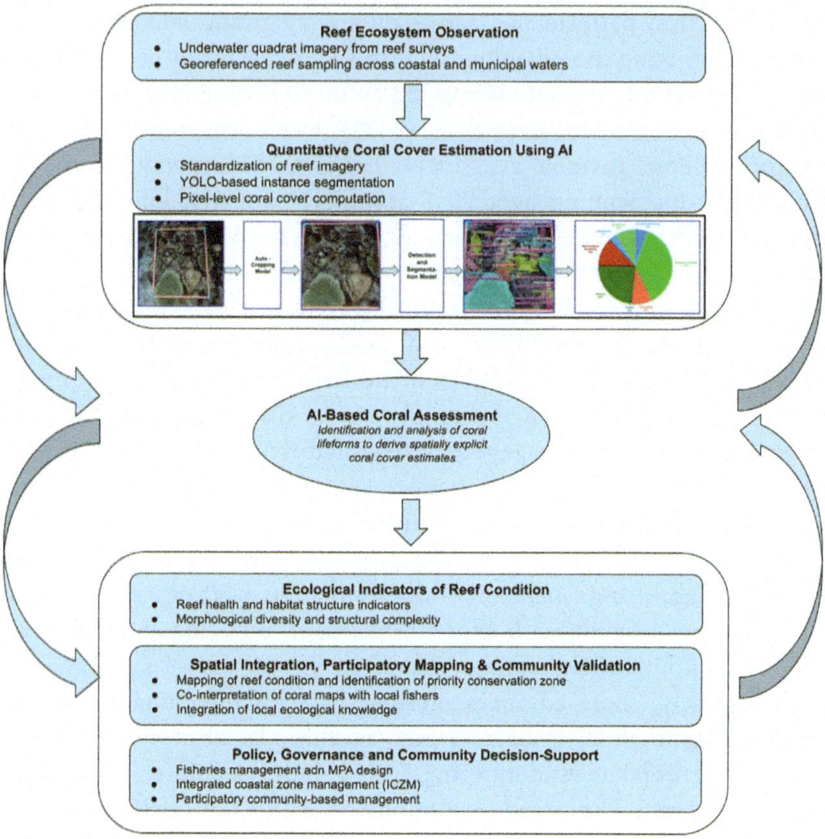

Fig. 3 Framework for AI-powered coral cover assessment and decision support for reef-derived food security.

The process begins with reef ecosystem observation, where georeferenced underwater quadrat imagery is collected through systematic reef surveys across coastal and municipal waters. These images represent standardized sampling units using quadrat that capture benthic composition and coral morphology at fine spatial resolution. The collected imagery undergoes automated preprocessing, including image cropping to isolate ecologically relevant regions, reduction of peripheral noise, and brightness adjustment and data augmentation to enhance image quality and improve model robustness under variable underwater conditions.

Preprocessed images are subsequently analyzed through quantitative coral cover estimation using artificial intelligence, implemented via a YOLO based instance segmentation model optimized for underwater environments. At this stage, the model detects and segments multiple hard coral morphologies, including branching, tabulate, massive, foliose, submassive, non Acropora branching, and mushroom corals, at the pixel level. The segmentation outputs are translated into proportional coral cover estimates for each morphological class and are further summarized through visual analytics to support interpretation at site and transect scales. These outputs collectively constitute the AI based coral assessment, which provides spatially explicit and quantitative insights into reef condition.

The outputs of the AI driven segmentation process can be translated into ecological indicators of reef condition, including habitat structure, morphological diversity, and structural complexity These indicators are strongly linked to reef fish abundance, fisheries productivity, and overall ecosystem resilience. As such, they directly contribute to food availability by supporting the productivity of reef based fisheries and sustaining marine biodiversity. Within the proposed framework, these ecological indicators are integrated into a spatial decision support environment through GIS based analysis, where coral cover metrics are overlaid with spatial layers such as fishing grounds, management boundaries, and other locally relevant ecological features.

The framework further incorporates participatory mapping and community validation, enabling the co interpretation of AI derived outputs with fishers, local stakeholders, and resource managers. Through these participatory processes, coral cover estimates are contextualized using local ecological knowledge, thereby enhancing the relevance and applicability of scientific outputs. This component supports both food utilization and access, as it ensures that resource management strategies align with local practices, livelihood needs, and equitable access to marine resources.

The interpreted and validated outputs are intended to inform policy, governance, and community decision-support mechanisms, including fisheries management, marine protected area (MPA) design, integrated coastal zone management (ICZM), land-use planning, and community-based resource management initiatives. By grounding governance actions in quantitative and spatially explicit coral cover data, the framework strengthens alignment between reef condition, food availability, and livelihood sustainability.

Crucially, the framework is cyclical rather than linear. Governance decisions, management interventions, and community feedback are designed to feed back into subsequent monitoring cycles, informing future survey design, model refinement, and priority-setting. This feedback loop enables adaptive management, supports long-term reef monitoring, and ensures that AI-powered coral assessment evolves in response to ecological change, stakeholder needs, and food-security objectives. Through this integrated and adaptive structure, coral cover estimation is positioned as a dynamic decision support system that not only sustains reef ecosystems but also reinforces the four pillars of food security namely availability, access, utilization, and stability (Guiné et al., 2021), in reef dependent regions.

5. Case insights from the Philippines (Sogod Bay)

Located within the Coral Triangle, the Philippines represents one of the most ecologically significant marine regions in the world., with coastal communities that are heavily reliant on coral reef ecosystems for fisheries production and food security (Muallil et al., 2020). Fisheries contribute substantially to local livelihoods, and Sogod Bay itself is recognized as one of the major fishing grounds in Southern Leyte, supporting food supply and income for multiple coastal municipalities (Major products, province of southern leyte, n.d.). The bay sustains diverse marine resources, including pelagic and reef associated species that are harvested for both subsistence and commercial purposes. According to Ratilla et al. (2016), the estimated average total annual volume of frigate tuna traded in Sogod Bay was 77,781.50 kg, indicating the economic importance of marine resources and their direct role in sustaining local food systems and livelihoods. However, increasing fishing pressure and environmental changes pose risks to the long term sustainability of these resources.

Within this context, Sogod Bay in Southern Leyte offers a particularly informative setting in which to examine the practical implementation of AI-based coral assessment. The bay supports extensive fringing and patch reef formations and is noted for its rich assemblages and high abundance of hard corals (Pascual et al., 2022). These characteristics make Sogod Bay an ideal location for evaluating how automated coral cover estimation can support both reef-health assessment and food-security planning.

To operationalize the coral cover estimation, the study applied a YOLO-based segmentation model to high-resolution underwater quadrat imagery collected across multiple reef sites in the bay. The approach integrates image acquisition, annotation, segmentation, coral-cover calculation, and geospatial visualization—providing a systematic example of how AI-driven coral assessment can be employed within real coastal monitoring programs (Fig. 4).

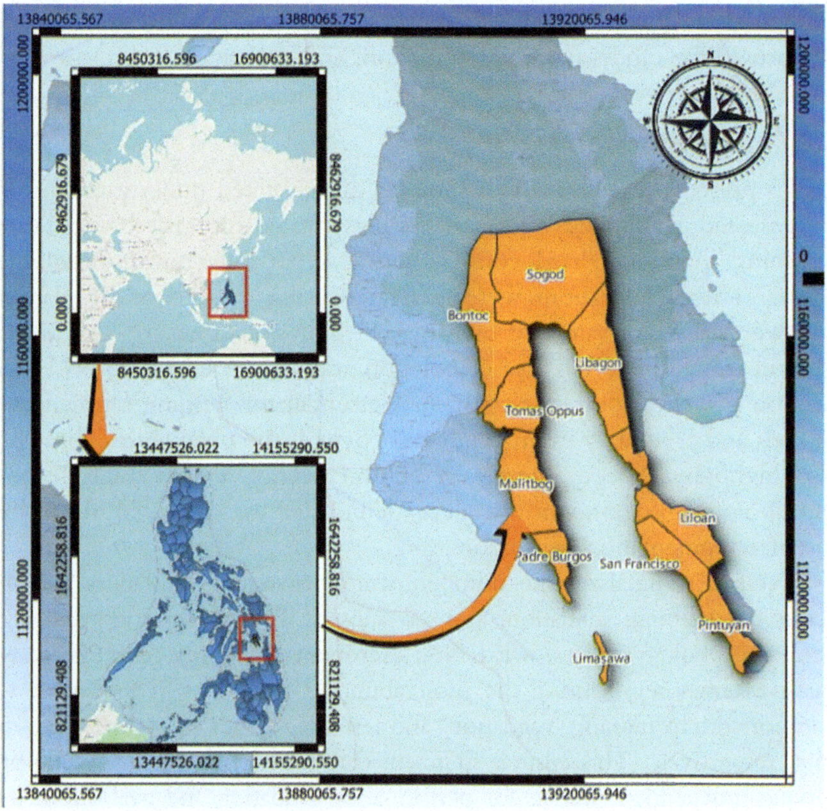

Fig. 4 Map of Sogod Bay, Southern Leyte, Philippines.

5.1 Dataset acquisition

Each image in the dataset represents a fixed sampling unit delineated by a 1 m × 1 m PVC quadrat frame, systematically or randomly positioned along predetermined reef transects. The dataset is derived from primary data collected through in situ underwater surveys conducted in Sogod Bay, Southern Leyte, Philippines. Photographs were taken using a high-resolution underwater camera oriented perpendicular to the reef substrate and maintained at a consistent height to ensure uniform coverage and reduce parallax distortion. The dataset captures a wide variety of scleractinian coral morphologies, including Acropora (Branching and Tabulate), Encrusting, Foliose, Massive, Non-Acropora branching, and Submassive forms. Non-coral substrates, such as algae, sand, and rubble, are also represented, enhancing the robustness of both classification and segmentation tasks by exposing the model to diverse benthic conditions. All images were annotated by trained coral expert, producing high-quality ground-truth labels used for model training, validation, and evaluation.

5.2 Image annotation

Fig. 5 presents representative examples of annotated underwater quadrat images used in this study. The images illustrate polygon based annotations outlining individual hard coral colonies across different morphological classes. These annotations were generated using the Roboflow platform and were carefully validated by a marine biologist to ensure high quality ground truth labels. The use of polygon annotations enables precise delineation of coral boundaries, which is essential for training segmentation models and accurately estimating coral cover at the pixel level. The figure also highlights the complexity of benthic scenes, where multiple coral morphologies coexist and overlap, demonstrating the need for robust instance segmentation techniques.

Table 1 summarizes the number of annotated coral instances per class across the training, validation, and test subsets. The dataset comprises eight coral morphology classes, with Non Acropora branching, encrusting, and massive corals representing the most abundant categories. The dataset was partitioned into training, validation, and test sets using a 70%, 15%, and 15% split, respectively. This ensures sufficient data for model learning while preserving independent subsets for performance validation and evaluation. The presence of all coral classes across each subset supports model generalization; however, the uneven distribution of samples may influence classification and segmentation performance, particularly for less represented classes.

Fig. 5 Sample annotated quadrat images showing bounding polygons around hard coral instances.

Table 1 Distribution of annotated coral instances across training, validation, and test datasets.

Class	Train	Validation	Test
Non-Acropora branching	721	171	145
Encrusting	657	142	144
Massive	608	114	131
Mushroom	275	64	73
Acropora branching	242	42	53
Submassive	198	28	29
Foliose	162	27	24
Acropora Tabulate	96	18	21

5.3 Training and validation

The training process employed the YOLOv8 segmentation architecture, selected for its real time segmentation capability, detection mechanism, and improved feature representation compared to earlier models (Swathi and Challa, 2024). These characteristics enable more accurate delineation of complex and irregular coral morphologies, which is essential for precise coral cover estimation in underwater environments. Pretrained weights from the Ultralytics repository were fine-tuned through transfer learning using the curated coral quadrat dataset. All images were auto oriented based on metadata and were uniformly resized to 640 by 640 pixels to ensure consistency across the dataset. To address data quality issues and class imbalance, only annotated images were retained, and underrepresented classes were excluded from training.

Multiple data augmentation strategies were applied to improve the model's ability to generalize under varying underwater imaging conditions. These included horizontal flipping, random rotations within a range of 15°, brightness adjustments ranging from 20 % darker to 20 % brighter, application of Gaussian blur with a two pixel kernel, and mosaic augmentation. These techniques resulted in five synthetic variants for each original training image.

6. Model evaluation

The performance of the YOLOv8-Segmentation (YOLOv8-seg) model in detecting Scleractinian corals from quadrat images was assessed using standard object detection and segmentation metrics. The primary indicators included the Mean Average Precision (mAP) at both mAP@ 0.5 and mAP@ 0.5:0.95, which evaluate the model's ability to accurately localize and segment coral instances across varying Intersection over Union (IoU) thresholds.

Additionally, class-specific evaluation metrics—Precision, Recall, F1-Score, and Specificity—were calculated to provide a detailed assessment of detection performance for each coral morphology. These metrics offered critical insights into the model's ability to distinguish between different coral types while minimizing false positives and false negatives. The combination of pixel-level segmentation analysis and morphological classification ensured a comprehensive and robust evaluation, demonstrating the model's applicability for real-world coral reef monitoring and management.

6.1 Calculating the coral cover

To calculate the percentage cover of each coral lifeform within a quadrat image, the YOLOv8 segmentation model was used to produce pixel level instance masks for all detected classes. Each coral instance was classified into one of the predefined morphological categories, including Acropora branching, Acropora Tabulate, Encrusting, Foliose, Massive, Non Acropora branching, Submassive, and Mushroom.

First, the percentage cover for each individual coral lifeform class was computed by dividing the number of pixels belonging to that class by the total number of pixels in the quadrat image, then multiplying by 100. This is expressed as:Mathematically, this is defined as formula (1):

$$Coral\ Cover_i\ (\%) = \left(\frac{Coral\ Pixels_i}{Total\ Pixels_{quadrat}} \right) x\ 100 \tag{1}$$

Where:

- Coral $Cover_i$ (%): Percentage cover of the i^{th}coral lifeform class (e.g., *Acropora branching, Massive,* etc).
- Coral $Pixels_i$: Number of segmented pixels belonging to the i^{th}coral class
- Total $Pixels_{quadrat}$: Total number of pixels in the quadrat image (e.g., $640 \times 640 = 409{,}600$ pixels if resized).

After computing the cover for each class, the overall total coral cover was obtained by summing percentage cover of all coral lifeform classes per quadrat.

$$\textbf{Total Coral Cover}(\%) = \left(\sum_{i=1}^{n} Coral\ Cover_i \right) \tag{2}$$

Where:

- n Total number of distinct scleractinian coral lifeform classes in the quadrat.

All masks were thresholded to binary values and overlaid onto the cropped quadrat region to ensure accurate spatial correspondence. Due to the application of an autocropping technique, the total number of pixels considered for each image may vary depending on the detected region of interest. Consequently, pixel counts per coral class are computed relative to the cropped area rather than a fixed image size. To maintain consistency in coral cover estimation, all computations were normalized using the total number of pixels within the cropped quadrat region. Images that did not contain any valid coral detections were excluded from cover computation. The resulting coral cover estimates provide a quantifiable measure of benthic composition, enabling reliable comparisons across sites and supporting integration into long term coral reef monitoring programs.

6.2 Results and findings

Accurate coral segmentation is a cornerstone of modern reef monitoring, especially as coastal ecosystems face increasing pressures from climate change, overfishing, and pollution. Segmenting coral morphologies at the pixel level facilitates detailed ecological assessments, enabling researchers and resource managers to quantify reef composition, monitor structural changes, and evaluate reef health over time. In this study, the YOLOv8-segmentation (YOLOv8-seg) model demonstrated strong performance in detecting and segmenting various scleractinian coral morphologies from quadrat images. As illustrated in Fig. 6, the model achieved a peak mean Average Precision (mAP)@ 0.5 of 0.8652 and a mAP@ 0.5:0.95 of around 0.6329 at epoch 172. This gradual improvement reflects the model's growing ability to accurately segment coral morphologies under more challenging conditions.

Table 2 presents the performance of the YOLOv8 segmentation model across eight coral morphology classes, evaluated using precision, recall, and

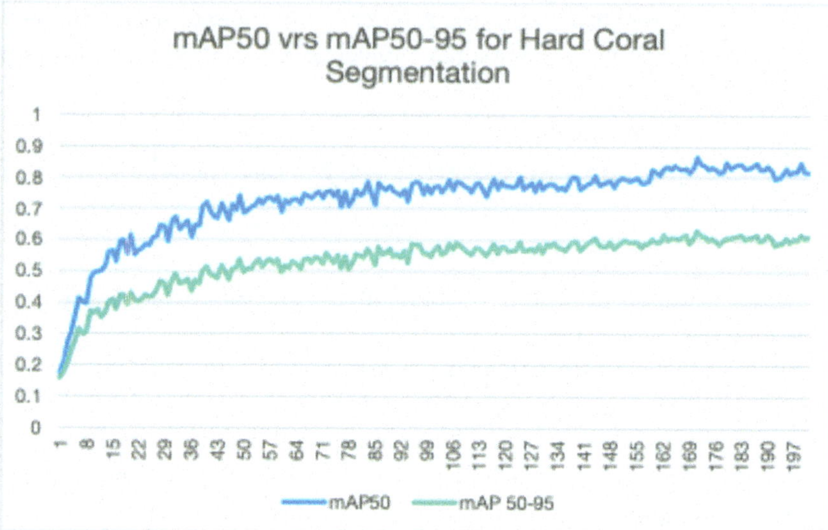

Fig. 6 YOLOv8-seg model mAP@ 0.5 and mAP@ 0.5:0.95 over 200 training epochs.

mean Average Precision for both bounding box detection and segmentation mask outputs. Overall, the model demonstrates strong detection and segmentation capability, with consistently high mAP@ 0.5 values across most classes, indicating reliable identification of coral instances.

Among the evaluated classes, Acropora tabulate and submassive corals exhibit the highest overall performance. Acropora tabulate achieved the highest detection accuracy, with a box mAP@ 0.5 of 0.878 and a mask mAP@ 0.5:0.95 of 0.739, reflecting the model's ability to accurately detect and segment morphologies with distinct structural patterns. Similarly, submassive corals achieved strong performance across both detection and segmentation metrics, including a mask mAP@ 0.5:0.95 of 0.632, indicating robust boundary delineation and consistent classification. Massive and non Acropora branching corals also demonstrated high mAP values, further confirming the model's effectiveness for well represented and structurally defined classes.

In contrast, comparatively lower performance was observed in Acropora branching and foliose corals, particularly in terms of recall and mask mAP@ 0.5:0.95. These classes exhibit more complex and irregular structures, which can lead to partial detections or segmentation inaccuracies. The relatively lower recall values suggest that some coral instances may not be fully detected, while the reduced mask accuracy indicates challenges in precisely delineating fragmented or overlapping boundaries.

Table 2 Performance metrics of the YOLOv8 segmentation model across different coral morphology classes.

Class	Box P	Box R	Box mAP50	Box mAP50–95	Mask P	Mask R	Mask mAP50	Mask mAP50–95
Acropora branching	0.536	0.697	0.597	0.404	0.546	0.710	0.621	0.405
Acropora Tabulate	0.930	0.751	0.878	0.754	0.940	0.745	0.878	0.739
Encrusting	0.740	0.648	0.698	0.523	0.754	0.669	0.725	0.486
Foliose	0.793	0.568	0.654	0.534	0.781	0.556	0.627	0.470
Massive	0.718	0.772	0.786	0.607	0.712	0.763	0.780	0.572
Non-Acropora branching	0.727	0.766	0.808	0.579	0.707	0.743	0.769	0.499
Submassive	0.835	0.821	0.832	0.691	0.815	0.785	0.784	0.632
Mushroom	0.805	0.688	0.738	0.554	0.808	0.688	0.734	0.541

Notably, differences between bounding box and mask performance across all classes highlight that while object detection remains strong, pixel level segmentation is more challenging, especially in heterogeneous reef environments with overlapping coral colonies and variable textures. This gap underscores the importance of further improving segmentation models and training data diversity to enhance boundary precision.

To further evaluate the model's classification reliability, a confusion matrix was generated from the test dataset (Fig. 7). The results show strong true positive predictions for dominant coral morphologies, particularly Non-Acropora branching (143), Massive (94), and Encrusting (102), indicating the model's effectiveness in identifying prevalent coral types. However, several misclassifications were noted, especially among visually similar or structurally overlapping categories. For instance, Acropora branching exhibits notable confusion with the Background class (22 instances), indicating difficulty in distinguishing fine branching structures from surrounding substrates. Likewise, Encrusting (30), Massive (29), and Non Acropora branching (50) are frequently misclassified as background, suggesting that low contrast, partial

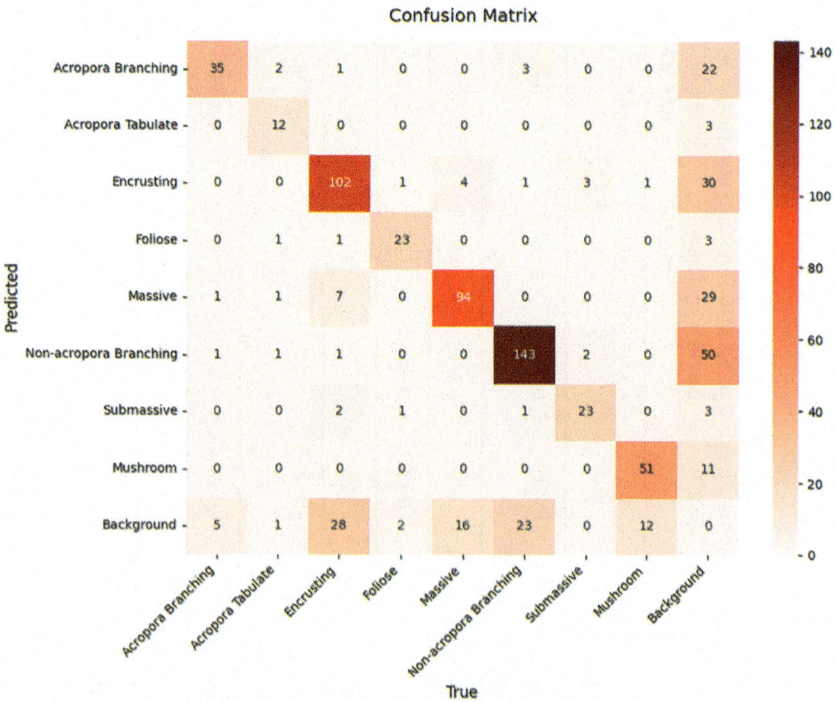

Fig. 7 Confusion matrix for YOLOv8 coral classification results.

visibility, or segmentation uncertainty may affect classification accuracy. Confusion is also evident between structurally related coral types. For example, Massive corals are occasionally misclassified as Encrusting (7), reflecting similarities in surface texture and growth form. These misclassifications highlight potential areas for improvement, such as refining class definitions or expanding the training dataset to better capture intra-class variability. Nonetheless, the overall pattern affirms the YOLOv8 model's utility for detailed coral segmentation, supporting its application in ecological monitoring and coral reef health assessment.

A visual inspection of the segmentation results (see Fig. 8) confirmed the model's real-world application. The predicted bounding boxes and class labels, rendered with confidence scores, illustrate the model's ability to diffentiate between multiple coral morphologies within a complex benthic scene. Notably, *Mushroom* coral instances were detected with high confidence (up to 0.96), alongside accurate predictions for *Massive*, *Submassive*, and *Acropora Tabulate.*

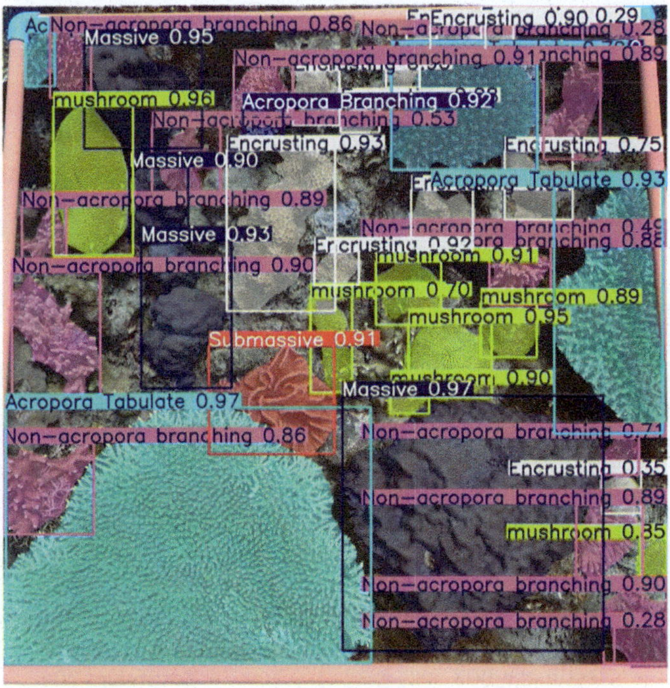

Fig. 8 Sample YOLOv8-segmentation output showing predicted coral classes with confidence scores.

Using the segmented masks generated by the YOLOv8-segmentation model, coral cover estimates were computed across selected test images. The visual outputs shown in Fig. 9 illustrate how the model effectively identified and labeled various coral lifeforms—such as Mushroom, Massive, Acropora Tabulate, and Non-Acropora branching—with corresponding confidence scores. These detections reflect the model's real-world application in analyzing benthic images with complex coral compositions.

Quantitative results derived from these segmentations are summarized in Table 3, which presents the computed percent coral cover for each coral lifeform per test image. Among the four samples, TestImage_004 displayed the highest overall coral cover at 92.02 %, dominated by Non-Acropora branching (60.16 %) and Acropora Tabulate (21.39 %). This suggests a densely colonized reef area, with high structural complexity that supports greater ecological functions. Meanwhile, TestImage_001 recorded the

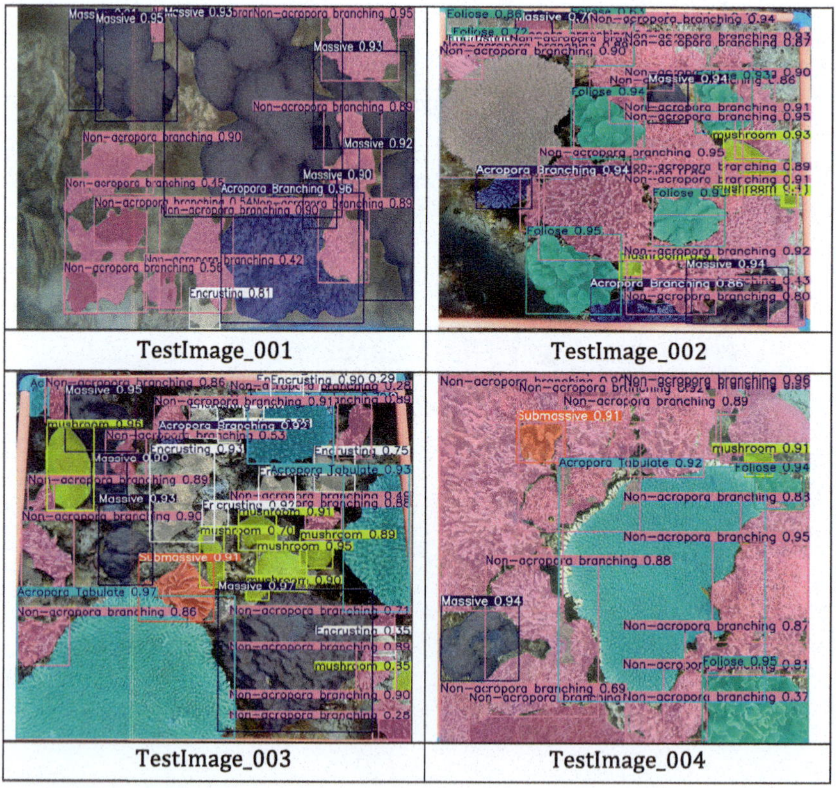

Fig. 9 Sample YOLOv8 segmentation outputs on test images showing detected coral morphologies with confidence scores.

Table 3 Percent coral cover per lifeform in four test images as predicted by the YOLOv8-segmentation model.

Image	Acropora branching (%)	Acropora tabulate (%)	Encrusting (%)	Foliose (%)	Massive (%)	Non-Acropora branching (%)	Submassive (%)	Mushroom (%)	Total coral cover (%)
TestImage_001.jpg	9.00	0.00	0.52	0.00	32.42	19.64	0.00	0.00	61.59
TestImage_002.jpg	2.78	10.43	0.66	12.77	4.85	43.53	0.00	2.35	77.38
TestImage_003.jpg	3.22	23.50	5.70	0.00	14.78	7.92	2.23	5.43	62.79
TestImage_004.jpg	0.00	21.39	0.00	6.05	2.95	60.16	1.06	0.41	92.02

lowest coral cover at 61.59 %, primarily consisting of Massive (32.42 %) and Non-Acropora branching (19.64 %) colonies, indicating a simpler benthic composition. Intermediate values were observed in TestImage_002 (77.38 %) and TestImage_003 (62.79 %), with substantial contributions from Acropora and Foliose types. For instance, TestImage_002 featured a significant portion of Foliose (12.77 %) and Non-Acropora branching (43.53 %), while TestImage_003 had notable amounts of Acropora Tabulate (23.50 %), Massive (14.78 %), and Mushroom (5.43 %).

Coral cover estimates derived from segmentation models such as YOLOv8 not only support ecological monitoring but also serve as proxy indicators of reef-related food resource availability, particularly fish biomass. This link is crucial in the context of food security for coastal communities. Numerous studies have demonstrated a strong positive relationship exists between live coral cover and species diversity and abundance of reef-associated fish (Seraphim et al., 2020). Different coral morphologies contribute to habitat complexity, a key determinant of reef fish richness and productivity. For instance, branching corals offer refuge and nursery space for juvenile fish, while massive and tabulate forms provide stable foraging and spawning grounds (Zhang et al., 2025; Tropical Marine Ecology, n.d.).

Fisher (2023) highlighted that fish density and biomass increased with coral colony size and structural intricacy, emphasizing the foundational role of live coral in sustaining fish populations. Similarly, Huang et al. (2024) reported that losses in coral cover and morphological diversity lead to pronounced reductions in reef fish abundance by diminishing habitat complexity. Additional evidence by Fahim et al. (2024) showed that microhabitat degradation, especially from coral loss, leads to measurable declines in fish species reliant on live coral structures. Supporting this, a meta-review by Samoilys et al. (2025) affirmed the long-standing ecological connection between coral health and reef fish biomass.

The image-specific results in Fig. 9 and Table 3 support these observations. TestImage_004, for instance, recorded the highest coral cover at 92.02 %, predominantly composed of Non-Acropora branching and Acropora Tabulate, suggesting a structurally diverse reef likely to support greater fish density. Conversely, TestImage_001 exhibited the lowest cover at 61.59 %, with fewer branching corals—possibly indicating a less complex habitat. These findings underscore the model's ability to quantify coral cover at the image level, offering a valuable tool for evaluating reef health and food provisioning potential.

Despite its strengths, several challenges emerged. The confusion matrix (Fig. 7) revealed misclassifications between visually or morphologically similar coral types, such as Mushroom and Background or Acropora branching and Massive. These errors may lead to under- or over-estimations in coral cover, potentially affecting downstream assessments of reef condition. Furthermore, real-world underwater imagery presents technical challenges like variable lighting, occlusions, and image blur due to currents or poor visibility.

To mitigate these issues, the model's performance can be further improved through several strategies: (1) expanding and diversifying training data from multiple reef locations to enhance generalization; (2) refining class definitions and annotations to improve morphological discrimination; and (3) employing advanced ensemble methods or multimodal inputs, such as depth or texture data, to assist in disambiguating visually similar features. Enhancing annotation consistency and leveraging expert feedback loops may also help fine-tune predictions in borderline cases.

7. Policy, governance, and community integration

Artificial intelligence has become an increasingly important enabler for advancing the monitoring, management, and conservation of marine resources by improving decision-making processes, optimizing resource allocation, and strengthening environmental governance systems (Alam et al., 2024). In coral reef management, AI-enabled monitoring systems and GIS technologies provide timely and spatially explicit ecological information that is critical for supporting strategies that facilitate early threat detection and predictive risk assessment in marine ecosystems (Chowdhury et al., 2024). These capabilities highly important in data-limited coastal regions, where conventional monitoring is often constrained by time, cost, and human expertise. Findings from the Coral Reef Health Assessment Study in the Gulf of Aqaba illustrate how the integration of advanced imaging techniques with machine-learning algorithms enhances the precision, efficiency, and reliability of coral reef assessments, ultimately strengthening the scientific basis for conservation and management interventions (Wahsha and Wahsheh, 2025).

The application of AI techniques has the potential to transform multiple stages of marine habitat mapping, from data acquisition to the production and interpretation of analytical outputs, including automated image analysis.

The automation of biodiversity recognition through the analysis of underwater photographs and videos offers substantial opportunities for scaling up biodiversity monitoring programs and reducing reliance on time-intensive manual methods (Fraschetti et al., 2024). In this context, site-level coral cover trends derived from AI-based analysis can be used to assess the ecological impacts of coastal development, regulate fishing pressure, and prioritize reef restoration investments in areas that directly support food security and coastal livelihoods. Conservation planning increasingly emphasizes the importance of ecosystem condition, rather than mere spatial presence, in ensuring the long-term persistence of biodiversity. As highlighted by Vercammen et al. (2019), failing to account for reef condition can lead to the designation of marine protected areas (MPAs) in ecologically degraded or unproductive locations, thereby limiting their effectiveness in sustaining biodiversity and fisheries productivity, thereby limiting their effectiveness. Incorporating AI-generated coral cover data into spatial planning processes therefore enhances MPA adequacy by aligning conservation objectives with ecosystem productivity, fish biomass, and habitat quality.

This governance approach is especially critical in the Coral Triangle, a region at the western edge of the Pacific Ocean recognized as a global center of marine biodiversity (Heck, 2022). As a "center of the center," this Pacific hub harbors the world's most extensive reef systems, currently prioritized under the CTI-CFF Regional Plan of Action 2.0 to ensure climate resilience for the 120 million people who depend on them (CTI-CFF, 2019). Because municipal waters across the Pacific support subsistence fishing and coastal economies highly sensitive to reef condition (Govan et al., 2009), AI-enhanced frameworks integrated with GIS tools enable decision-makers to detect patterns of decline and implement adaptive management across multiple scales. Furthermore, automated historical baselines provide the context necessary for scientists to set realistic recovery targets amidst cumulative human pressures (Eddy et al., 2018).

Community integration remains a cornerstone of effective coral reef governance, especially in reef-dependent regions where local stewardship plays a central role in resource management. Many coastal households actively participate in informal monitoring, seasonal fishing restrictions, and local enforcement practices that complement formal governance structures. Locally Managed Marine Areas (LMMAs), which are widespread across the Pacific, have proven to be effective mechanisms for conserving coral reefs and sustaining fisheries when supported by sound scientific information and strong community engagement (Govan et al., 2009). Empirical evidence from Kenya demonstrates that well-managed LMMAs have led to measurable increases in live hard coral cover as well as reef-associated fish biomass, underscoring the effectiveness of community-led approaches when underpinned by reliable ecological data (Newell et al., 2019). Integrating AI-generated coral cover assessments into LMMA frameworks enhances transparency, facilitates shared decision-making, and empowers communities with accessible and objective indicators of reef health. Participatory mapping, citizen science initiatives, and community involvement in data interpretation and the establishment of local data repositories further strengthen governance systems, improve ecological understanding, and foster trust between scientific institutions and local stakeholders (Fletcher et al., 2015).

Finally, coral cover estimation plays an increasingly vital role in land-use and coastal planning, particularly in countries such as the Philippines, where municipalities are required to prepare Comprehensive Land Use Plans (CLUPs) on a regular basis in accordance with the Clup Guidebook (n.d.). Reef health is inextricably linked to terrestrial activities, as sedimentation and nutrient runoff resulting from watershed degradation directly impair reef resilience (Carlson et al., 2019). In this context, coral monitoring serves not only as an indicator of marine ecosystem condition but also as a signal of watershed management effectiveness. Integrating AI-derived coral cover data into land-use planning processes reinforces importance of Integrated Coastal Zone Management (ICZM), a policy framework that seeks to achieve sustainable development (Ivčević et al., 2025) by explicitly recognizing the interconnectedness of terrestrial and marine systems under conditions of limited space and resources.

8. Future directions

Future research should prioritize expanding spatial and temporal coverage to enable continuous, long term monitoring of coral reefs across broader geographic scales. Integrating time series coral cover estimates derived from repeated field surveys or satellite imagery would facilitate trend analysis, early detection of reef degradation, and improved forecasting of recovery or collapse under climate driven stressors. Advancements in model architecture also present important opportunities. Emerging transformer based vision models, multimodal learning frameworks, and lightweight attention mechanisms can further enhance segmentation accuracy, robustness, and computational efficiency under challenging underwater conditions. In addition, incorporating complementary data streams, such as depth information from photogrammetry, water quality indicators, temperature records, and hydrodynamic variables, would support more comprehensive ecosystem assessments that move beyond coral cover alone to include habitat complexity, bleaching severity, and reef structural integrity. These advancements directly contribute to food availability by improving the understanding of reef productivity and sustaining fisheries resources.

Significant research opportunities also exist in developing larger, regionally representative training datasets that capture the diversity of reef conditions across the Coral Triangle and the South Pacific. Expanding the range and diversity of annotated coral imagery would improve model generalization and reduce bias, particularly for visually similar coral morphologies. Greater involvement of marine biologists and trained annotators can further enhance ground-truth quality, while citizen science platforms offer a promising avenue for expanding dataset collection and accelerating AI deployment in data-limited and rural coastal communities.

Future work should also explore stronger integration between AI-derived coral metrics and socio-economic datasets. Linking coral cover estimates with fisheries catch data, household nutrition surveys, and market indicators would enable more direct quantification of the impacts of reef condition on food availability, livelihoods, and community resilience. Such integrative approaches would strengthen the role of coral monitoring as a decision-support tool for fisheries management, disaster risk reduction, and climate adaptation planning at both local and national scales.

As automation becomes increasingly feasible, governance frameworks must evolve to incorporate AI-derived coral cover metrics into mainstream resource management. Future efforts should focus on developing user-friendly dashboards, mobile applications, and open-access platforms that allow local governments, marine managers, and communities to interpret coral cover data without requiring advanced technical expertise. Embedding AI-generated outputs into existing planning instruments, such as Marine Protected Area (MPA) management plans, Comprehensive Land Use Plans (CLUPs), and Integrated Coastal Zone Management (ICZM) frameworks, will be essential for translating scientific insights into actionable policy. Achieving this integration will require new data-sharing mechanisms, institutional capacity-building initiatives, and targeted training programs to support local governments, fisheries officers, and community groups in interpreting and applying AI-generated ecological information.

Finally, strengthening community participation and capacity-building will be critical to the long-term sustainability of AI-based coral monitoring. Future directions should emphasize participatory mapping, citizen science initiatives, and co-designed monitoring programs that integrate local ecological knowledge with automated analytics. Training local stakeholders to collect imagery, validate AI outputs, and interpret spatial results can enhance trust, promote local ownership, and ensure data continuity, ensuring that technological advances contribute meaningfully to inclusive, adaptive, and resilient reef governance.

9. Conclusion

Coral reefs are foundational to food security in the Pacific and Southeast Asia, supporting coastal livelihoods, small-scale fisheries, and the daily nutritional requirements of millions of people. Yet continued degradation threatens not only ecological resilience but also household nutrition, economic stability, and cultural continuity across reef-dependent communities. Addressing these vulnerabilities requires monitoring systems capable of detecting ecological change with sufficient spatial detail and temporal frequency to guide policy responses. Artificial intelligence offers a transformative opportunity to scale coral monitoring, automate coral cover estimation, and provide decision-ready ecological information for climate-vulnerable coastal nations.

The case study from Sogod Bay demonstrates that AI-based segmentation models such as YOLOv8 can accurately detect and quantify multiple coral morphologies while producing fine-scale coral cover estimates that are immediately relevant to food-security planning. When integrated with GIS platforms, these models enable spatial analysis at ecological and governance scales—highlighting sites requiring urgent intervention while supporting evidence-based resource management. Nevertheless, technical refinement, policy integration, and community participation remain essential to ensure that AI adoption strengthens rather than marginalizes local knowledge systems and food-security strategies.

References

Alam, M., Khan, I.R., Siddiqui, F., Alam, M.A., 2024. Artificial intelligence as key enabler for safeguarding the marine resources. In: De, D., Sengupta, D., Tran, T.A. (Eds.), Artificial Intelligence and Edge Computing for Sustainable Ocean Health. The Springer Series in Applied Machine Learning. Springer, Cham. https://doi.org/10.1007/978-3-031-64642-3_18.

Banala, S., Gladson, V.B., Kittappa, T., Balasubramanian, P., Parida, P.K., 2024. An effective AI system for accurate segmentation of the hybrid visual character analysis of a coral bleaching domain. In 2024 International Conference on Advances in Computing, Communication and Materials (ICACCM) (pp. 1–6). IEEE.
Beese, C., 2024. Modelling coral reef futures: exploring the role of structural complexity in sustaining ecosystem services (Doctoral dissertation, Open Access Te Herenga Waka-Victoria University of Wellington).
Bell, J.D., Cisneros-Montemayor, A., Hanich, Q., Johnson, J.E., Lehodey, P., Moore, B.R., et al., 2018. Adaptations to maintain the contributions of small-scale fisheries to food security in the Pacific Islands. Mar. Policy 88, 303–314.
Carlson, R.R., Foo, S.A., Asner, G.P., 2019. Land use impacts on coral reef health: A ridge-to-reef perspective. Front. Mar. Sci. 6, 562.
Castro-Sanguino, C., Bozec, Y.M., Condie, S.A., Fletcher, C.S., Hock, K., Roelfsema, C., et al., 2023. Control efforts of crown-of-thorns starfish outbreaks to limit future coral decline across the great barrier reef. Ecosphere 14 (6), e4580.
Chaijaroen, P., 2022. Coral reef deterioration and livelihoods of coastal communities: an economics perspective. Corals-Habitat Formers in the Anthropocene. IntechOpen.
Charlton, K.E., Russell, J., Gorman, E., Hanich, Q., Delisle, A., Campbell, B., et al., 2016. Fish, food security and health in Pacific Island countries and territories: a systematic literature review. BMC Public. Health 16 (1), 285.
Cheung, W.W., Maire, E., Oyinlola, M.A., Robinson, J.P., Graham, N.A., Lam, V.W., et al., 2023. Climate change exacerbates nutrient disparities from seafood. Nat. Clim. Change 13 (11), 1242–1249.
Chowdhury, A., Jahan, M., Kaisar, S., Khoda, M.E., Rajin, S.A.K., Naha, R., 2024. Coral reef surveillance with machine learning: a review of datasets, techniques, and challenges. Electronics 13 (24), 5027.
Clup Guidebook. (n.d.-a). https://dhsud.gov.ph/wp-content/uploads/Publication/Guidebooks/HLURB_CLUP_Vol_1.pdf.
Chomariyah, 2025. Strengthening national legal frameworks for coral reef protection: toward a sustainable blue economy. In IOP Conference Series: Earth and Environmental Science (Vol. 1473, No. 1, p. 012063). IOP Publishing.
Ditzel, P., König, S., Musembi, P., Peters, M.K., 2022. Correlation between coral reef condition and the diversity and abundance of fishes and sea urchins on an East African coral reef. In: Oceans. 3. MDPI, pp. 1–14 1.
Eddy, T.D., Cheung, W.W.L., Bruno, J.F., 2018. Historical baselines of coral cover on tropical reefs as estimated by expert opinion. PeerJ 6, e4308. https://doi.org/10.7717/peerj.4308.
Eddy, T.D., Lam, V.W., Reygondeau, G., Cisneros-Montemayor, A.M., Greer, K., Palomares, M.L.D., et al., 2021. Global decline in capacity of coral reefs to provide ecosystem services. One Earth 4 (9), 1278–1285.
Facon, M., Pinault, M., Obura, D., Pioch, S., Pothin, K., Bigot, L., et al., 2016. A comparative study of the accuracy and effectiveness of Line and Point Intercept Transect methods for coral reef monitoring in the southwestern Indian Ocean islands. Ecol. Indic. 60, 1045–1055.
Fahim, H., Naaykens, T., D'Aloia, C.C., 2024. Habitat quality effects on the abundance of a coral-dwelling fish across spatial scales. Ecol. Evolution 14 (9), e70322.
Fauzi, A., Rahman, N.A., 2025. Sustainable marine fisheries and their contribution to food security and coastal economic development. J. Mar. Fish 1 (2), 64–71.
Fletcher, P.J., Spranger, M., Hendee, J.C., Li, Y., Clark, M., Kiker, G.A., 2015. Decision tools for coral reef managers: using participatory decision support to integrate potential climate impacts and informed decision making. Glob. Ecol. Conserv. 4, 491–504.

Fisher, W.S., 2023. Relating fish populations to coral colony size and complexity. Ecol. Indic. 148, 110117.

Fontoura, L., Zawada, K.J., D'agata, S., Álvarez-Noriega, M., Baird, A.H., Boutros, N., et al., 2020. Climate-driven shift in coral morphological structure predicts decline of juvenile reef fishes. Glob. Change Biol. 26 (2), 557–567.

Fraschetti, S., Strong, J., Buhl-Mortensen, L., Foglini, F., Gonçalves, J., Goncalez-Irusta, J.M., et al., 2024. EMB Future Science Brief, 11 European Marine Board, Ostend, Belgium, 76. https://doi.org/10.5281/zenodo.11203127.

Gapper, J.J., Maharjan, S., Li, W., Linstead, E., Tiwari, S.P., Qurban, M.A., et al., 2024. A generalized machine learning model for long-term coral reef monitoring in the Red Sea. Heliyon 10, 18.

Govan, H., Tawake, A., Tabunakawai, K., Jenkins, A., Lasgorceix, A., Schwarz, A.M., et al., 2009. Status and potential of locally-managed marine areas in the South Pacific: meeting nature conservation and sustainable livelihood targets through wide-spread implementation of LMMAs: Study Report.

Guiné, R.D.P.F., Pato, M.L.D.J., Costa, C.A.D., Costa, D.D.V.T.A.D., Silva, P.B.C.D., Martinho, V.J.P.D., 2021. Food security and sustainability: discussing the four pillars to encompass other dimensions. Foods 10 (11), 2732.

Heck, S.A., 2022. Ocean governance in the coral triangle: a multi-level regulatory governance structure. Politics Gov. 10 (3), 70–79.

Helder, N.K., Burns, J.H., Green, S.J., 2022. Intra-habitat structural complexity drives the distribution of fish trait groups on coral reefs. Ecol. Indic. 142, 109266.

Hill, J., Wilkinson, C.L.I.V.E., 2004. Methods for ecological monitoring of coral reefs. Australian Institute of Marine Science, Townsville, pp. 117.

Huang, L., McWilliam, M., Liu, C., Yu, X., Jiang, L., Zhang, C., et al., 2024. Loss of coral trait diversity and impacts on reef fish assemblages on recovering reefs. Ecol. Evolution 14 (11), e70510.

Ivčević, A., Škugor, D.P., Snoussi, M., Karner, M., Kervyn, M., Huge, J., 2025. Identifying success factors for integrated coastal zone management: development of a regional coastal plan in Morocco. Environ. Sci. & Policy 167, 104027.

Jury, C.P., Bahr, K.D., Cros, A., Dobson, K.L., Freel, E.B., Graham, A.T., et al., 2024. Experimental coral reef communities transform yet persist under mitigated future ocean warming and acidification. Proc. Natl. Acad. Sci. 121 (45) e2407112121.

King, A., Bhandarkar, S.M.., Hopkinson, B.M., 2019. Deep learning for semantic segmentation of coral reef images using multi-view information. In Proceedings of the IEEE/CVF conference on computer vision and pattern recognition workshops (pp. 1–10).

Koester, A., Gordó−Vilaseca, C., Bunbury, N., Ferse, S.C., Ford, A., Haupt, P., et al., 2023. Impacts of coral bleaching on reef fish abundance, biomass and assemblage structure at remote Aldabra Atoll, Seychelles: insights from two survey methods. Front. Mar. Sci, 1230717.

Kopecky, K.L., Pavoni, G., Nocerino, E., Brooks, A.J., Corsini, M., Menna, F., et al., 2023. Quantifying the loss of coral from a bleaching event using underwater photogrammetry and AI-assisted image segmentation. Remote. Sens. 15 (16), 4077.

Li, Z., Zhao, S., Lu, Y., Song, C., Huang, R., Yu, K., 2024. Deep learning-based automatic estimation of live coral cover from underwater video for coral reef health monitoring. J. Mar. Sci. Eng. 12 (11), 1980.

Macadam, A., Nowell, C.J., Quigley, K., 2021. Machine learning for the fast and accurate assessment of fitness in coral early life history. Remote. Sens. 13 (16), 3173.

Major products | province of southern leyte. (n.d.-b). https://southernleyte.gov.ph/major-products/.

MRAG Asia Pacific, 2022. The State of Fish in Nutrition Systems in the Philippines. 87 p.

Muallil, R.N., Tambihasan, A.M., Enojario, M.J., Ong, Y.N., Nañola Jr, C.L., 2020. Inventory of commercially important coral reef fishes in Tawi-Tawi Islands, Southern Philippines: the Heart of the Coral Triangle. Fish. Res. 230, 105640.
Newell, S.L., Nagabhatla, N., Doubleday, N.C., Bloecker, A., 2019. The potential for locally managed marine area (LMMAs) as a participatory strategy for coastal and marine ecosystems-the global commons. OIDA Int. J. Sustain. Dev. 12 (04), 47–62.
O'Meara, L., Cohen, P.J., I'ilu Kafa, R., Wate, J.T., Albert, J., Bogard, J., et al. 2023. Pacific food systems – The role of fish and other aquatic foods for nutrition and health. Apia. https://doi.org/10.4060/cc5796en.
Obura, D.O., Aeby, G., Amornthammarong, N., Appeltans, W., Bax, N., Bishop, J., et al., 2019. Coral reef monitoring, reef assessment technologies, and ecosystem-based management. Front. Mar. Sci. 6, 580.
Oraño, J.F.V., Napala, J.J.O., Maaghop, J.F.O., Elecito, J.C., 2023. Automated coral lifeform classification using YOLOv5: a deep learning approach. Novel & Intelligent Digital Systems Conferences. Springer Nature Switzerland, Cham, pp. 13–22.
Pascual, G.L., Napala, J.J., Tomol, R., Oyong, L.J., Alfante, J.M., Lumain, J.P.L., 2022. Crown-of-Thorns starfish (COTS) outbreak in Sogod Bay, Southern Leyte, Philippines. J. Ecosyst. Sci. Eco-Governance 4 (1), 24–30.
Pavoni, G., Corsini, M., Ponchio, F., Muntoni, A., Edwards, C., Pedersen, N., et al., 2022. TagLab: AI-assisted annotation for the fast and accurate semantic segmentation of coral reef orthoimages. J. field Robot 39 (3), 246–262.
Rabbitt, S., Lilley, I., Albert, S., Albert, J., Tibbetts, I., 2023. The importance of nutrition-sensitive fisheries management: Women's dietary diversity in Marovo Lagoon, Solomon Islands. J. Agriculture, Food Systems, Community Dev. 13 (1), 197–213.
Raine, S., Marchant, R., Kusy, B., Maire, F., Sunderhauf, N., Fischer, T., 2024. Human-in-the-loop segmentation of multi-species coral imagery. In Proceedings of the IEEE/CVF Conference on Computer Vision and Pattern Recognition (pp. 2723–2732).
Ratilla, M., Abamo, A., Tampus, D., 2016. Value chain analysis for frigate tuna (Auxis thazard) in selected aquatic and agricultural systems communities along the Sogod bay area, Southern Leyte, Philippines. J. Educ. Hum. Resour. Dev. (JEHRD) 4, 72–87.
Rivera, H.E., Chan, A.N., Luu, V., 2020. Coral reefs are critical for our food supply, tourism, and ocean health. We can protect them from climate change. MIT. Sci. Policy Rev. 1.
Rivera-Sosa, A., Muñiz-Castillo, A.I., Charo, B., Asner, G.P., Roelfsema, C.M., Donner, S.D., et al., 2025. Six decades of global coral bleaching monitoring: a review of methods and call for enhanced standardization and coordination. Front. Mar. Sci. 12, 1547870.
Rogers, A., Blanchard, J.L., Mumby, P.J., 2018. Fisheries productivity under progressive coral reef degradation. J. Appl. Ecol. 55 (3), 1041–1049.
Samoilys, M., Roche, R.C., Gudka, M., Thouless, C., Obura, D., Rowlands, S., et al., 2025. Fish as indicators of coral reef ecosystem status. Ecol. Indic.
Seraphim, M.J., Sloman, K.A., Alexander, M.E., Janetski, N., Jompa, J., Ambo-Rappe, R., et al., 2020. Interactions between coral restoration and fish assemblages: implications for reef management. J. Fish. Biol. 97 (3), 633–655.
Swathi, Y., Challa, M., 2024. YOLOv8: advancements and innovations in object detection. International Conference on Smart Computing and Communication. Springer Nature Singapore, Singapore, pp. 1–13.
Towle, E.K., Donovan, E.C., Kelsey, H., Allen, M.E., Barkley, H., Blondeau, J., et al., 2022. A national status report on United States coral reefs based on 2012–2018 data from National Oceanic and Atmospheric Administration's National Coral Reef Monitoring Program. Front. Mar. Sci. 8, 812216.
Tropical Marine Ecology. (n.d.). https://marine-ecology.nl/.

Vercammen, A., McGowan, J., Knight, A.T., Pardede, S., Muttaqin, E., Harris, J., et al., 2019. Evaluating the impact of accounting for coral cover in large-scale marine conservation prioritizations. Diversity Distrib. 25 (10), 1564–1574.

Wahsha, M., Wahsheh, H., 2025. From data to action toward sustainable marine conservation: ai-based coral health assessment. In: Kumar, A., Swaroop, A., Shukla, P. (Eds.), Proceedings of fourth international conference on computing and communication networks. ICCCN 2024. Lecture Notes in Networks and Systems, vol 1291. Springer, Singapore. https://doi.org/10.1007/978-981-96-3244-2_16.

Williams, I.D., Couch, C.S., Beijbom, O., Oliver, T.A., Vargas-Angel, B., Schumacher, B.D., et al., 2019. Leveraging automated image analysis tools to transform our capacity to assess status and trends of coral reefs. Front. Mar. Sci. 6, 222.

Wyatt, M., Hickey, S., Radford, B., Gonzalez-Rivero, M., Boutros, N., Callow, N., et al., 2025. Safe AI for coral reefs: benchmarking out-of-distribution detection algorithms for coral reef image surveys. Ecol. Inform, 103207.

Younes, O., Jihad, Z., Noël, C., Mohsen, K., Philippe, A.M., Eric, C., et al. (2023). Automatic coral detection with yolo: a deep learning approach for efficient and accurate coral reef monitoring. In: European Conference on Artificial Intelligence (pp. 170–177). Cham: Springer Nature Switzerland.

Zhang, H., Li, M., Zhong, J., Qin, J., 2024. CNet: a novel seabed coral reef image segmentation approach based on deep learning. In: Proceedings of the IEEE/CVF winter conference on applications of computer vision (pp. 767–775).

Zhang, J., Ke, J., Liu, X., Zhao, H., Zhu, W., Rinkevich, B., et al., 2025. Survival and growth of "branches of opportunity" from ten coral species outplanted on framed reef modules. Ecol. Eng. 212, 107529.

CHAPTER ELEVEN

Revisiting the policy environment of food security in the Philippines: Pathways to sustainability and reform

Ederson de los Trino Tapia[a,b,*], Sonia Fontanilla-Pimentel[a], Rhene C. Tabajen[a], Gary G. Ador Dionisio[c], and Roselle A. Jardin[d]
[a]University of Makati, Philippines
[b]University of the Philippines, Diliman, Philippines
[c]De La Salle College of Saint Benilde, Philippines
[d]Cebu Normal University, Philippines
*Corresponding author. e-mail address: ederson.tapia@umak.edu.ph;

Contents

Advances in Food Security and Sustainability, Volume 11
ISSN 2452-2635, https://doi.org/10.1016/bs.af2s.2026.06.003

Abstract

Food security in the Philippines is a complex problem, with dimensions that cut across climate change, high levels of economic inequality, governance inefficiencies, and excessive dependence on food imports. In this chapter, a detailed analysis of the policy environment of food security in the country is presented with a great level of scrutiny of the existing frameworks and governance architecture, and the multifaceted challenges that are hampering the development of a sustainable and resilient food system.

Key policy milestones, such as the Rice Tariffication Law (RA 11203), the Mandanas-Garcia decision, and the Philippine Development Plan (PDP), are put under critical scrutiny to determine their effectiveness in improving food supply, accessibility, and affordability to provide a strong base for further scholarly inquiry.

This chapter employs a policy analysis framework that incorporates governance, economic, and sustainability perspectives to identify significant deficiencies in food security policies, such as insufficient enforcement, disjointed inter-agency coordination, inadequate investment in agricultural innovation, and vulnerabilities related to climate change. It examines the role of digital technology, the empowerment of smallholder farmers, and public-private partnerships in tackling these difficulties. It further advocates for a holistic, multi-sectoral strategy to attain sustainable food security in the Philippines.

1. Introduction

Food security functions as a critical national development base because it provides people with continuous access to healthy food needed for healthy development (FAO, 2009). Food security, as defined by the Food and Agriculture Organization (FAO, 2009), represents a state where all people receive guaranteed access to sufficient, safe food to maintain their health throughout their entire lives (Pinstrup-Andersen, 2009; Mwarage, 2016). Food security issues continue to affect the Philippines because economic factors collaborate with natural influences and governmental guidelines (Israel and Briones, 2013). Studies demonstrate how the Philippines has boosted agricultural productivity by implementing policy reforms, but the continued risk to food security appears from climate change, the growing population along with economic inequalities, and poor governance systems (Habito, 2019).

Food security is widely understood as involving availability, access, utilization, and stability over time. In the Philippine context, these dimensions are shaped by climate vulnerability, trade dependence, land-use pressures, and uneven institutional capacity.

Natural disasters, including typhoons, floods, and droughts, damage Philippine agricultural output because the country faces significant exposure to such events (Yumul et al., 2011). Rural farming communities experience detrimental weather effects that reduce agricultural output and the financial stability of their agricultural workforce (Jaranilla-Sanchez et al., 2007). Food imports of staple crops, including rice, make the Philippines vulnerable to external shocks from global supply chain disruptions and price volatilities alongside geopolitical tensions (Briones, 2020). Rapid industrial growth, together with urban expansion, has generated various land disputes that have led to reduced availability of national agricultural land (David et al., 2012). OECD (2021) states the need for food security to have a sustainable and resilient framework that upholds national development goals along with global sustainability targets.

National food security challenges in the Philippines receive attention through government-initiated policy measures and programs. Since its establishment, the National Food Authority (NFA) maintained control over rice imports alongside buffer stock management until the implementation of the Rice Tariffication Law (RA 11203) to achieve rice trade liberalization and lower rice prices for consumers (Tolentino, 2021). The Department of Agriculture runs programs that support local farmers through higher production levels and lower food prices, based on a Philippine Institute for Development Studies [PIDS] (2021) report. The food security targets of the national policy framework Philippines Development Plan (PDP) combine with social and economic development strategies (NEDA, 2021). Davila (2018) shows that political strategies achieve limited progress in enhancing program results and administration.

A comprehensive food security plan is essential for the country because it needs to strictly link agricultural production initiatives to food distribution networks and consumer systems (Briones and Galang, 2022; Golloso-Gubat et al., 2024). Social security measures remain ineffective because organization leaders implement policies inconsistently, and agencies fail to establish adequate coordination among themselves (Dai & Guo, 2024; Muliati et al., 2021). Local government units in food security leadership gained control over essential operations through the Mandanas-Garcia ruling, but its implementation creates operational challenges (Franklin and Ebdon, 2020; Diokno-Sicat et al., 2022). Foreign aid and the influence of international institutions increasingly dictate the formulation of policies on food security in the Philippines, while simultaneously creating barriers to both sustainable development and the presidential policy agenda (World Bank, 2020).

This chapter performs a detailed analysis of the policy framework that supports food security in the Philippine context, critically examining existing practices and emphasizing strongly on some possibilities for systemic improvement and transformative interventions. It uses a policy analysis framework structured with governance perspectives along with economic considerations and sustainability principles to evaluate food security strategies (Anderson et al., 2012; Kingdon, 2011). An analysis of existing policy frameworks helps this research in recommending approaches to create a more integrated food security system with relevant evidence as a basis for future strategies.

A sustainable food security future for the Philippines must be achieved through a comprehensive partnership between public entities with local communities and both private sector representatives and international supporters (OECD, 2021). Long-term solutions for strengthening agricultural resilience involve three key components: sustainable farming practices combined with digital food security monitoring systems and improvements to supply chain operations (FAO, 2022). The findings emphasize critical agricultural reforms that ensure resilience and an equitable food system while employing combined policies to secure food security status. The Philippines needs to introduce reforms for building food systems that ensure population health and advance national goals (United Nations Development Programme [UNDP], (2021)).

In this chapter, food security in the Philippines is discussed as a complex governance challenge that spans the dimensions of availability, access, utilization, and stability, and is embedded in the context of climate vulnerability, dependence on trade, skewed distribution of local capacity, and a strong fragmentation of institutions. Instead of considering theoretical perspectives as separate, isolated explanatory frameworks, this analysis draws on institutional theory, policy process analysis, and systems-oriented perspectives to explain the design, implementation, and contested dynamics of food security reforms at both the national and local levels.

2. Philippine food security landscape: Issues and challenges

The Philippines faces multifaceted challenges, ranging from government concerns to ecological damage to fiscal issues that gradually affect the availability, accessibility and stability of food resources (Aldaba, 2009; Lapada, 2019). The country faces an array of challenges to food security, from excessive dependence on imported foods to environmental hazards, territorial conflicts over land, poor governance, disruption to food supply chains, and food insecurity caused by poverty. Addressing these issues requires a deep understanding of the balance between the market and the environment, as well as the efficient application of institutional capacity.

2.1 Reliance on food imports and external supply risks

Philippine food security remains structurally exposed to external shocks because domestic policy has not sufficiently reduced dependence on imported staples. As an agricultural society, the Philippines demands large food imports that include substantial amounts of rice and wheat, together with corn. Rice consumption requires that the Philippines import large quantities from Vietnam, Thailand and India (Agricultural Indicators System: Agricultural Exports and Imports, 2024; Lapada, 2019; Briones, 2020). Foreign food dependence makes the Philippines vulnerable to market price instability, as well as commercial trade controls and supply chain issues. The COVID-19 outbreak revealed additional risks because the pandemic led to supply disruptions of imported food commodities between nations (Tolentino, 2021).

RA 11203, also known as the Rice Tariffication Law of 2019, was enacted to enhance food security by shifting from rice import quantity limitations to a tariff system. Rice tariffication initially had positive effects for consumers by moderating prices. However, its side effects negatively affected Philippine rice growers, leading to reduced sales and reduced competitiveness vis- a vis imports of rice from overseas (Alcuitas and Petralba, 2023). The policy has been criticized due to its dependence on external market forces, thus making the nation vulnerable to fluctuations that are beyond its control. Such a dependency may be untenable in light of volatile geopolitical tensions (World Bank, 2022).

A balanced method must exist to provide reasonably priced food products for consumers without neglecting the improvement of domestic agricultural production systems. Implementing domestic rice sufficiency programs combined with upgrading post-harvest facilities and bettering procurement systems from local sources can establish sustainable approaches for decreasing external food dependence.

2.2 Climate change and environmental degradation

Climate vulnerability is not a peripheral risk but a central determinant of food insecurity outcomes in the Philippines. It is a country highly vulnerable to climate change, with extreme weather events such as typhoons, droughts, and heavy rainfall significantly affecting agricultural productivity (Jaranilla-Sanchez et al., 2007). Variability in temperature and precipitation patterns disrupts traditional farming cycles, leading to lower yields and increased production costs (David et al., 2012). Rice-producing regions in Central Luzon and Mindanao have been particularly affected by droughts linked to the El Niño phenomenon, which has resulted in declining water availability for irrigation (Perez et al., 2022; Stuecker et al., 2018; Briones, 2020).

Environmental damage also presents a lasting danger that jeopardizes food security. Agricultural land productivity drops because poor land management methods alongside insufficient forest retention and elevated reliance on chemical fertilizers cause soil fertility to deteriorate (OECD, 2021). The fusion of excessive fishing activities with marine pollution has triggered fish population depletion, thus restricting seafood resources for local fishing communities (Ngatia et al., 2019; Israel and Briones, 2013).

The Climate Resilient Agriculture Program is a key driver behind the systematic introduction of climate adaptation techniques in farm management regimes. These initiatives function in relative isolation and have notable spatial irregularities across various regions of implementation. Achieving food security over the longer term requires enhanced commitment to climate-smart agronomic practices and ecosystem-centered resource management strategies.

2.3 Land-use conflicts and urbanization

Land conversion has emerged as a governance issue affecting long-term domestic production capacity. Modern urban development and industrial growth have caused extensive conversion of agricultural territories to residential areas and commercial and infrastructure facilities. During the period from 1990 to 2020, the nation experienced a 30% decline in useful farmland because of urban development and land reclassification processes (NEDA, 2021). Less available arable land limits Philippine food production domestically, thus forcing the country to depend on imported food products.

The Comprehensive Agrarian Reform Program (CARP) served as the foundation to redistribute land resources for smallholder farmers in pursuance of agricultural development equity. Agricultural land reclassification has occurred due to weak enforcement and regulatory loopholes in land conversion rules, according to Vista et al. (2011) and Davila (2018). Agribusiness expansion, which mostly focuses on exports of bananas and pineapples, has consistently prioritized commercial goals at the expense of food security (Briones, 2020).

A strategic land-use policy needs implementation to maintain a proper balance between urban development needs and food production land preservation. A combination of strict enforcement of land protection laws with financial rewards for sustainable land management approaches should reduce food security effects due to land-use conflicts.

2.4 Inequitable food access and persistent poverty

Food insecurity in the Philippines reflects distributive inequalities as much as production constraints. Increased imports, together with agricultural growth, have boosted food availability in recent years, yet poor families encounter extreme barriers when buying healthful food. Economic resource limitations stop poor populations from purchasing dietary food that benefits their health. According to Southeast Asia statistics, the Philippines stands as one of the countries with the highest rates of childhood stunting, with chronic malnutrition affecting 30 % of children below five years old (UNICEF, 2022). Malnutrition appears as a symptom of wider social problems, which include expensive food coupled with inadequate variety in dietary choices and deficient access to necessary nutrients.

World Bank (2020) reports that farming and fishing communities face extreme food insecurity because poor incomes and market price instability create unpredictable food access. The *Pantawid Pamilyang Pilipino Program (4Ps)* and other social protection initiatives face reduced impact from bureaucratic problems along with limited financial support, according to Bustos et al. (2023).

Food security policies need to combine nutritional goals with current production and distribution measures for enhanced effectiveness. An expanded network of community-based feeding programs alongside improved food fortification and strengthened livelihood support systems forms the basis for a food security approach that addresses all populations effectively.

2.5 Weak agricultural policies and governance structures

Fragmented mandates and uneven administrative capacity continue to undermine food security policy implementation. Food security governance throughout the Philippines suffers from multiple policy divisions and institutional performance issues together with weak enforcement capabilities (Barrion et al., 2023; Building Food Security and Managing Risk in Southeast Asia, 2017).

The implementation of agricultural subsidy and credit assistance schemes has experienced several challenges. The problems with corruption combined with inadequate distribution of funds undermine farmer support, which worsens food-producing constraints. No substantial investments in agricultural research and development prevent farmers from adopting advanced farming technologies, which restricts both their productivity levels and sectoral innovation.

The establishment of a unified governance structure will help national and local food security plans work together (Rosegrant et al., 2018). More effective food security governance emerges when institutions collaborate strongly, when subsidies become transparent, and when we expand agricultural research funding. To establish clear lines of authority, the government needs to define agency responsibilities better while implementing rigorous evaluation systems alongside public and private sector alliances for resource usage. Local government units, along with community-based organizations, should receive empowerment to increase the effectiveness and responsiveness of food security programs in remote communities.

2.6 COVID-19 and supply chain vulnerabilities

The COVID-19 pandemic exposed significant weaknesses in the country's food supply chains. The implementation of lockdowns combined with transportation restrictions and logistical problems forced the food market to suffer from shortages of essentials while prices rose steeply (OECD, 2021). The inability of farmers and fisherfolks to reach markets, together with restricted consumer mobility, resulted in decreased food availability (Tolentino, 2021).

Emergency response programs from government authorities supplied temporary food assistance and monetary benefits to the population, yet highlighted that the food systems required improvement. Food distribution efficiency will increase through expanded cold storage capacity together with digital supply chain developments and enhanced farm-to-market linkages.

3. The Philippine food security policy environment and reform dynamics

Food security policies in the Philippines experienced major changes through various economic and environmental elements as well as governance concerns. The Philippines implements its food security approach through a series of important policy instruments as well as institutional structures. This section analyzes both the present operations of the National Food Authority (NFA) and the effects of the Rice Tariffication Law (RA 11203) on food security while also discussing the Department of Agriculture's (DA) governance responsibilities and the strategic objectives of the Philippine Development Plan (PDP) and local government and international partnerships supporting policy creation.

3.1 The national food authority (NFA) and its evolving role in food security policy

The evolution of the NFA reflects a broader transition from state-centered market intervention to liberalized food governance. The National Food Authority (NFA) was established in 1972 to maintain price stability while guaranteeing food availability, specifically for rice, which is the country's main dietary staple. The NFA started operations in 1972 through Presidential Decree No. 4 to oversee rice production alongside importation and distribution and stabilize prices (Briones, 2020). The agency maintained a central position in government-executed initiatives while leading efforts to sustain buffer stock reserves and distribute discounted rice products.

Reforms emerged throughout decades targeting NFA inefficiency alongside financial constraints and market distortion concerns to minimize government control of rice markets. RA 11203, passed as the Rice Tariffication Law in 2019, drastically reduced NFA's market influence by confining its activities to emergency food storage and buffer stock management (Araza et al., 2025; Bartelet et al., 2025). The government's move to dismantle NFA regulatory powers was intended to enhance market efficiency, but critics maintain that it reduced the ability to control rice prices and shield local farmers from market price fluctuations (PIDS, 2021).

The primary focus of the National Food Authority centers on disaster preparedness and food safety through its role in maintaining strategic rice reserves for use during calamities and emergencies. The debate continues about whether the NFA should take a proactive approach to stabilizing food market prices while facing external supply uncertainties.

3.2 The rice tariffication law (RA 11203) and its impact on local farmers and consumers

The Rice Tariffication Law (RA 11203) marked a fundamental shift in Philippine food security policy by replacing quantitative restrictions on rice imports with a tariff-based system. RA 11203 granted private importers the right to import rice when they deposited rice tariffs for use in RCEF programs that supported the Philippine rice industry (Tolentino, 2021).

The main directive of RA 11203 established a dual purpose to lower rice prices while making rice affordable for low-income consumers. The rice price decrease after RA 11203 went into effect offered price relief for urban consumers who dedicate most of their budget to purchasing food (Briones and Galang, 2022).

Rice farmers throughout the country faced economic difficulties due to inferior market prices compared to cheaper international rice imports. The decrease in *palay* (unmilled rice) terminal prices at farm gates caused financial difficulties for rice farmers operating on small scales (Agcaoili et al., 2023). The RCEF aimed to deliver mechanization support with credit access and seed distribution but encountered implementation delays, which, combined with administrative inefficiencies, decreased its effect on target beneficiaries (Diokno-Sicat et al., 2022).

The Rice Tariffication Law persists as an issue of public debate despite its initially logical economic basis. Supporters think the law boosts market efficiency through better food prices, yet opponents say it damages rural systems and the domestic growth of agriculture. The ongoing policy debate concerning the Rice Tariffication Law centers around the request for various modifications that focus on supporting struggling farmers through subsidies while strengthening local procurement systems alongside yearly reviews of import taxes. The reform embodies a distributive trade-off between consumer price stabilization and producer protection.

3.3 The role of the Department of Agriculture (DA) in food security planning and implementation

The Department of Agriculture (DA) serves as the lead agency responsible for agricultural development, food security, and rural livelihood programs. The organization works together with other entities to execute policies, as well as the NFA with the Bureau of Plant Industry and local government units (LGUs) (NEDA, 2021).

The DA implements various programs, including the National Rice Program, the Agricultural Credit Policy Council, and Agri-Agra Credit Programs, to build farm productivity while improving rural employment and farm financing accessibility. Fundamental obstacles exist in executing programs and funding restrictions together with LGU coordination, thus producing inconsistent policy outcomes throughout various regions (Briones, 2020).

One critical issue about the food security governance role of the DA rests in its ability to execute evidence-based climate-resilient agricultural policies. Because climate change, along with environmental degradation, has become increasingly dangerous for food security, the agency needs to embed sustainability and adaptation strategies into nationwide food security plans. The development of digital agriculture and precision farming and effective risk mitigation measures stand essential for building sustainable agricultural resilience. Despite its central mandate, the DA's effectiveness remains constrained by coordination challenges and uneven local implementation.

3.4 The Philippine Development Plan (PDP) and agricultural modernization

The Philippine Development Plan (PDP) functions as the national blueprint for government economic and social development by establishing strategic agricultural and food security targets. Within the PDP 2023–2028 framework, these main reforms stand out: environmentally sustainable agricultural methods for food safety while fighting climate change, investments into rural infrastructure along with irrigation systems and road and post-harvest facilities, and development of agro-industry value-added products for improving market competitiveness (NEDA, 2021).

The PDP establishes clear policies but needs proper funding, effective implementation, and strong cooperation between agencies to produce successful outcomes. Positive results from policy targets will require effective governance systems alongside institutional bottleneck resolutions. The effectiveness of PDP food security priorities depends less on strategic clarity than on administrative capacity and fiscal alignment.

3.5 Local government initiatives in food security under the Mandanas-Garcia ruling

With the 2022 Mandanas-Garcia ruling, LGUs secured more national tax revenue to independently plan and execute food security programs (Decentralization, Digitalization, and Development: Strengthening Local Governance for Crisis Response, Recovery, Resilience, and the Sustainable Development Goals, 2024; Diokno-Sicat et al., 2022). The decision redistributed various agricultural operations with social welfare duties, which until then operated under national authority to local administrative units.

The new reform provides LGUs with the power to develop localized food security approaches, yet it introduces concerns about differences in local governance capability, limited funding, and inconsistent administration. The lack of expertise and inadequate administrative systems in LGUs make it difficult for them to manage extensive agricultural programs successfully. The lack of clear national supervision over local governing bodies might lead to substantial food security differences between regions because of varied local governance practices.

Food security governance will experience maximum benefits through improved capacity development for LGUs combined with local government collaboration and standardized national policies. Decentralization has expanded local discretion while exposing asymmetries in institutional capability across LGUs.

3.6 The role of international organizations and foreign aid in Philippine food security

International organizations, in partnership with foreign aid, offer significant support to Philippine food security policies through programs focusing on climate adaptation, disaster response, and rural development initiatives. The

Food and Agriculture Organization (FAO, 2022) and World Bank, together with the Asian Development Bank (ADB) and United Nations World Food Programme (WFP), serve food security initiatives by offering technical support, financial backing, and policy direction (World Bank, 2020). International support helps fund short-term food security programs, but it generates questions about how dependence on these funds impacts local government planning autonomy. For sustainable food security to be achieved, it is crucial to verify that international funding supports national objectives and promotes lasting agricultural autonomy. External support strengthens short-term resilience but raises long-term questions about domestic policy autonomy.

4. Policy gaps and areas for reform

The reform priorities presented in this section are the direct outcomes of the weaknesses of the institution and policies described above. The discussion shows that food insecurity in the Philippines does not persist only due to the restrictions of production, but also due to the disjointed governance, lack of homogeneous local capacity, insufficient agricultural innovation systems, poor integration of climate concerns, and persistent inequities in access. The suggested reforms are, therefore, seen as specific interventions to correct these structural and administrative gaps.

Despite the presence of various food security policies in the Philippines, several structural, governance, and implementation gaps persist, undermining the effectiveness of national and local initiatives. RA 11203 and the Mandanas-Garcia ruling introduced changes to the food security governance structure, yet multiple barriers prevent a coherent framework and sustainable operation of policy governance. Government and private sector partners need to work systematically using multi-sectoral methods that improve policy enforcement and agricultural development and climate-friendly strategies to resolve current challenges.

4.1 The need for a comprehensive national food security strategy

The absence of integrated planning across production, trade, nutrition, and climate policy has contributed to fragmented implementation outcomes. The Philippines requires a unified national food security strategy that unites food production with distribution and consumption to create an organized framework. Multiple national policies exist, but these policies operate independently, which leads to disjointed implementation across the country. The Philippines struggles to address current and future food security risks because it lacks sustained integrated planning (Briones and Galang, 2022).

A successful food security strategy requires thorough attention to production efficiency through farming support systems and distribution infrastructure development with storage facilities and market connectivity for minimizing post-harvest losses and consumption and nutrition policies that integrate food security and public health standards to ensure proper nutrition and food safety.

A complete food security plan based on research-driven policy development and coordinated agency collaboration would boost current programs. A National Food Security Council operating as a unified body to merge DA and NNC and other related agency policies represents an essential reform for developing coherent food security governance.

4.2 Weak enforcement and regional disparities in policy implementation

Decentralization has intensified implementation disparities due to uneven technical and fiscal capacity among LGUs. The challenge of inconsistent agricultural policy enforcement negatively affects Philippine food security as different regions apply different levels of enforcement. The development of food security policies at the national level through the Department of Agriculture depends heavily on local government units for their execution after the Mandanas-Garcia ruling extended LGUs' power in agricultural development (Diokno-Sicat et al., 2022).

Local governance implementation becomes inconsistent due to technical capability differences and funding and institutional priority variations across local government units. Local government units demonstrate successful food security program implementations, yet many encounter limitations from insufficient budgets and governance weaknesses plus bureaucratic inefficiencies (Lahoti et al., 2023; Van et al., 2022).

The national government needs to establish three measures for supporting LGUs: first, delivering training that elevates food security planning abilities; second, implementing reward systems based on results for pushing local agricultural advancements; and third, developing evaluation and inspection systems that keep authorities accountable to national food security targets. Persistent food access and agricultural productivity gaps between regions will continue unless extensive work is done to harmonize national and local food security programs.

4.3 Insufficient investment in agricultural research and innovation

Chronic underinvestment in research and extension has limited productivity gains and adaptation capacity. The Philippines lags behind its regional counterparts in agricultural research, development, and technology adoption. Research and development spending on farming by the Philippines falls below the investments of neighboring nations Thailand and Vietnam since the government funds little of its agricultural budget (Briones, 2020).

The country faces difficulties in boosting farm productivity because scarce research funding does not support institutions and services required for technology adoption programs. Traditional low-yield farming techniques persist among farmers because they have restricted availability of climate-resistant seeds and advanced agricultural instruments (PIDS, 2021).

The resolution of this gap necessitates three key strategies: additional funds for agricultural research in climate-resistant crops and soil protection and smart irrigation systems, stronger agricultural innovation alliances between public and private entities for quick technology implementation and usage expansion, and widened outreach services for small-scale farmers to obtain contemporary agricultural knowledge. Investing actively in agricultural innovation will enhance food security resilience and boost farm productivity while decreasing food imports for the Philippines.

4.4 Strengthening climate adaptation and disaster risk reduction in food security planning

The increasing frequency of extreme weather events, droughts, and typhoons poses a significant risk to Philippine food security, yet climate adaptation remains weakly integrated into national food security policies (Jaranilla-Sanchez et al., 2007). Rice fields and livestock farms, together with fisheries, experience recurring damages from typhoon-induced climate-related disasters. Most farmers remain unable to acquire climate adaptation tools that include drought-resistant seeds, together with early warning systems and insurance mechanisms (NEDA, 2021).

Food security improvement demands three critical elements: the inclusion of climate resilience throughout agricultural planning as well as the assessment of farming risk for different regions, an expansion of insurance schemes based on weather data to prevent farmers' income loss during disasters, and an expansion of climate-smart agriculture programs implementing sustainable land-use approaches together with water conservation methods.

The adaptation of climate change strategies into food security policies creates essential conditions for protecting agricultural sustainability with environmental resilience over extended timescales.

4.5 Policy fragmentation and the need for improved inter-agency coordination

Overlapping mandates and siloed governance structures continue to weaken policy coherence. Philippine food security governance is characterized by policy fragmentation, with multiple government agencies implementing overlapping and sometimes contradictory programs. Poor coordination between the DA, NFA, NNC, Department of Social Welfare and Development (DSWD), and LGUs creates inefficiencies along with redundancies in food production and distribution efforts (Briones and Galang, 2022).

The DA works to improve agricultural autonomy, yet the Department of Trade and Industry (DTI) makes import freedom its top priority, resulting in harm to domestic farmers (Golloso-Gubat et al., 2024). The social welfare program known as 4Ps provides food access, yet its contributions to sustainable nutritional security are uncertain because of inconsistent policy connections between agricultural and health departments.

Enhancing inter-agency coordination requires establishing a National Food Security Council to consolidate and harmonize food security policies across agencies, strengthening data-sharing mechanisms to improve planning and decision-making, ensuring policy coherence between agriculture, trade, and social protection strategies to balance economic efficiency with food security objectives; and work for a more integrated governance approach, necessary to create synergistic policies that effectively address food security challenges.

4.6 Enhancing participatory governance and stakeholder engagement

Limited stakeholder engagement has reduced policy responsiveness and legitimacy at the local level. Efficient food security governance depends on full participation between farmers' cooperatives along with private sector actors while civil society organizations and academic institutions join the process. In the Philippines, traditional food security policymaking proceeded from the top level of government while offering few openings for community members to take part (Diokno-Sicat et al., 2022).

Food security decision-making becomes more successful when stakeholders gain the power to influence choices. Participatory governance increases its strength through two main pathways: the establishment of formal consultative structures for farmer organizations and cooperatives and the support of community-driven food safety initiatives along with collaboration between businesses to advance supply chain markets and financing options. Food security policies strengthen their effectiveness through participatory governance, which makes them inclusive and sustains their ability to respond to the needs of their stakeholders.

An integrated solution for addressing Philippine food security policy gaps will unite strategic national planning with local governance and community participation from all stakeholders. A sustainable food system will function more inclusively and resiliently through the creation of a well-planned national food security strategy in combination with improved climate resilience, strengthened inter-agency coordination, and enhanced agricultural innovation.

5. Pathways for sustainability and reform

Based on the gaps in governance and implementation outlined above, the next phase of the reform agenda should be guided by the need to strengthen the domestic production resilience, improve institutional coordination, increase technological capacity, and attain fair access to our food systems.

5.1 Strengthening local food production through climate-resilient agricultural practices

This responds directly to the risks associated with import dependence and climate stress discussed earlier. Food import dependency for the Philippines has made the country more exposed to international natural and political factors like supply disruption, price turbulence, and geopolitical conflicts. Prime domestic farming must take priority for sustainable food security because present agricultural constraints stem from old farming practices, poor irrigation systems, and worsening climate change effects (Briones, 2020).

The need to shift toward resilient, sustainable agricultural practices stands vital for both productivity growth and environmental preservation purposes. Laudable sustainable farming methods, including integrated pest management and agroforestry with organic farming together with crop diversification, improve soil fertility, lower dependence on chemicals, and increase enduring agricultural productivity (Jaranilla-Sanchez et al., 2007). Farmers' water efficiency will increase when they dedicate resources to modern irrigation systems together with rainwater harvesting programs.

The government should focus its programs on developing farmer capacities through training that provides knowledge about climate-smart agricultural technology solutions, which include drought-tolerant crops, biofertilizer use, and conservation agricultural practices. To achieve sustainable food security, the policymaking process should prioritize both financial and technical support that helps farmers implement resilient agricultural techniques.

5.2 Enhancing the role of cooperatives and smallholder farmers in food security

Reform must address the distributive consequences of liberalization policies for small producers. Despite being the main pillar of Philippine agriculture, smallholder farmers encounter insufficient resources such as credit services market outlets and modern agricultural technology because of their limited access (Galang, 2022). Even with the presence of agricultural cooperatives, many struggle due to poor governance, inadequate financial support, and inefficient supply chain integration. The food security initiatives require enhanced participation from smallholders through strengthening both cooperative institutions and farmer organizations.

Cooperatives provide members with advantages through their collective bargaining strength in addition to resource-sharing capabilities, enhanced market networks, and larger operating size benefits. Cooperatives and smallholder farmer associations will build their institutional capabilities by receiving financial incentives and by implementing governance reforms and training programs, which will strengthen their ability to ensure food stability and fair market entry (Briones and Galang, 2022).

Government support through subsidized credit and logistical assistance combined with storage facilities will help smallholder farmers become more resistant to price changes and production uncertainty. The establishment of farmer-led processing facilities with value-adding enterprises helps farmers create additional revenue streams and maintain agricultural sustainability.

5.3 Leveraging digital technologies and data analytics for food security planning

Digitalization should be understood as a governance tool for monitoring, targeting, and supply-chain transparency. Digital technologies offer innovative solutions for addressing food security challenges, particularly in data-driven decision-making, market efficiency, and precision farming. The use of remote sensing, geographic information systems (GIS), and artificial intelligence (AI) in crop monitoring, weather forecasting, and soil analysis can enhance productivity and climate adaptation (PIDS, 2021).

One of the most promising applications of digital technology in food security is e-agriculture, which involves real-time agricultural data collection to monitor crop yields and soil conditions; mobile-based platforms providing farmers with weather updates, market prices, and agronomic advice; and blockchain technology to ensure transparency and efficiency in food supply chains.

Moreover, digital traceability systems can help track food production, distribution, and storage to reduce post-harvest losses and ensure food safety (World Bank, 2020). Strengthening public and private investments in agricultural digitization will be key to modernizing Philippine agriculture and improving long-term food security outcomes.

5.4 Policy recommendations for improving food accessibility, affordability, and distribution

Improving affordability addresses the access inequalities identified in the structural pressures section. While food production plays a crucial role in food security, improving food accessibility and affordability is equally essential. High food prices and inefficiencies in food distribution systems remain major barriers to equitable access (Grote et al., 2021). Improving food supply chain efficiency can help reduce logistical bottlenecks and price markups that make food unaffordable for low-income consumers. Investing in cold storage, transportation networks, and regional food hubs can help stabilize food prices and reduce post-harvest losses (Briones, 2020).

Additionally, strengthening social protection programs, such as conditional cash transfers and targeted food subsidies, can ensure that vulnerable populations have consistent access to nutritious food. Expanding community-based feeding programs and school meal initiatives can also address malnutrition and improve long-term food security at the household level.

5.5 The role of public-private partnerships in advancing food security initiatives

Partnerships are most effective where they compensate for state capacity limitations. Public-private partnerships (PPPs) have emerged as an effective mechanism for advancing food security initiatives through investment mobilization, infrastructure development, and supply chain improvements. The private sector plays a crucial role in agri-business innovations, market expansion, and technology adoption (Briones and Galang, 2022).

Collaborations between government agencies, private enterprises, research institutions, and farmer cooperatives can enhance resource-sharing and expertise exchange, particularly in agricultural R&D, value chain development, and financial inclusion programs. PPP models in food security may include: agri-tech partnerships that introduce precision agriculture and smart irrigation; infrastructure investments in cold storage, farm-to-market roads, and food processing centers; and microfinancing schemes that provide accessible credit to smallholder farmers.

Ensuring transparent regulatory frameworks and mutual accountability in PPP engagements will be crucial in maximizing the benefits of private sector participation in food security governance.

5.6 Long-term policy reforms for a resilient and self-sufficient food security system

These reforms collectively aim to build a coordinated, climate-sensitive, and socially inclusive food governance framework. Achieving long-term food security sustainability requires structural policy reforms that align agricultural productivity, social protection mechanisms, and environmental sustainability. Policymakers must adopt a forward-looking approach that anticipates future food security challenges, including climate change, population growth, and trade disruptions.

Key policy reforms should focus on the following: revising trade and tariff policies to balance domestic farmer protection and consumer affordability; expanding agricultural financing to support smallholder farmers and climate-resilient farming projects; and institutionalizing a National Food Security Strategy that integrates nutrition, health, trade, and environmental considerations (NEDA, 2021).

By ensuring that food security policies are adaptive, inclusive, and forward-looking, the Philippines can build a resilient food system that supports long-term economic and social stability.

Sustainable food security in the Philippines requires a multifaceted approach that strengthens local food production, empowers smallholder farmers, integrates digital technologies, and fosters public-private partnerships. Addressing challenges in food affordability, accessibility, and governance will be essential to achieving a resilient and self-sufficient food system. Moving forward, long-term policy reforms that emphasize climate adaptation, agricultural modernization, and inter-agency coordination will play a crucial role in ensuring that future generations have stable access to sufficient, safe, and nutritious food.

6. Conclusion

Climate vulnerability, reliance on imports, institutional fragmentation, and endemic inequalities in access to sufficient nutrition still dictate food security in the Philippines. Despite reforms like the Rice Tariffication Law and the decentralizing implications of the Mandanas-Garcia ruling, the results of these reforms have been unequal, due in part to differences in administrative capacity and coordination. Sustainable food security, thus, requires a combined approach to governance, which balances agricultural output, trade policy, social protection, climate adaptation, and institutional empowerment at the local level. Conceptualized this way, food security is no longer a sectoral concern but a general development issue that is core to long-term resilience and equity.

References

Agricultural Indicators System: Agricultural Exports and Imports, 2024. Philippine statistics authority. Retrieved from https://psa.gov.ph/content/agricultural-indicators-system-agricultural-exports-and-imports-0.

Agcaoili, D.C., Cabral, J.S., Nuñez, C.L.R., Castro, M.M.E.D., Manuben, J.J.P., Herrera, M.N.Q., Labios, R.J.D., 2023. Initiatives towards inclusive trading practices of rice value chain in Pila, Laguna, Philippines. E3S Web Conf. 444, 2033. https://doi.org/10.1051/e3sconf/202344402033.

Alcuitas, A.B., Petralba, J., 2023. The sectoral impact of the rice tariffication law on Filipino rice supply: a time-series analysis. Recoletos Multidiscip. Res. J. 11 (2), 85–95.

Aldaba, F.T., 2009. Poverty Philippines: causes, constraints, opportunities. Retrieved from https://archium.ateneo.edu/economics-faculty-pubs/65/.

Anderson, K., Jha, S., Nelgen, S., Strutt, A., 2012. Re-examining policies for food security in Asia. Food Security 5 (2), 195–210. https://doi.org/10.1007/s12571-012-0237-5. Springer Science+Business Media.

Araza, A.B., Gagarin, W., Corales, Osorio, C.P., Mendoza, M.D., Ancog, R., 2025. Land-use conversion from agricultural production areas to built-up areas in the Philippines for decades 2000–2020: Spatial analysis and policy implications. Land Use Policy 162, 107874. https://doi.org/10.1016/j.landusepol.2025.107874.

Barrion, A.S.A., Calayag, J.A.S., Nguyen-Orca, M.F., Melo-Rijik, M.N., 2023. Food loss and waste in the Philippines: a literature review. Food Res. 7 (6), 278. https://doi.org/10.26656/fr.2017.7(6).127. Rynnye Lyan Resources.

Briones, R., 2020. Rice trade liberalization and food security in the Philippines. Philippine Institute for Development Studies (PIDS) Discussion Paper. https://pids.gov.ph.

Bartelet, H.A., Prodigalidad, A.J.D., Dy, J.S., Manzano, J., 2025. Understanding rice production stagnation in the Philippines: regional evidence and development implications. PLoS One 20 (10). https://doi.org/10.1371/journal.pone.0335344.

Briones, R., Galang, I., 2022. Food security policies and governance challenges in the Philippines. Philippine J. Public. Policy 45 (2), 12–36.

Building Food Security and Managing Risk in Southeast Asia, 2017. OECD eBooks. Organization for Economic Cooperation and Development. https://doi.org/10.1787/9789264272392-en.

Bustos, M., Lau, L., Kirkpatrick, S.I., Dubin, J.A., Manguerra, H., Dodd, W., 2023. Examining the association between household enrollment in the Pantawid Pamilyang Pilipino Program (4Ps) and wasting and stunting status among children experiencing poverty in the Philippines: A cross-sectional study. Asia Pac. J. Public Health 420–434. https://doi.org/10.1177/10105395231189570.

Dai, L., Guo, Y., 2024. The problem and countermeasures of the failure of the rural social security system to reduce poverty. Discov. Sustain. 5 (1). https://doi.org/10.1007/s43621-024-00244-z.

David, C.C., Intal, P.S., Balisacan, A.M., 2012. Agricultural Policy Reforms and Food Security in the Philippines. Ateneo de Manila University Press.

Decentralization, Digitalization, and Development: Strengthening Local Governance for Crisis Response, Recovery, Resilience, and the Sustainable Development Goals, 2024. United Nations Development Programme (UNDP). Retrieved from https://www.undp.org/philippines/publications/decentralization-digitalization-and-development-strengthening-local-governance-crisis-response-recovery-resilience-and.

Diokno-Sicat, C., Capuno, J., Jandoc, K., 2022. Decentralization and fiscal transfers: The impact of the Mandanas ruling. National Economic and Development Authority (NEDA) Working Paper.

Davila, F., 2018. Human ecology and food systems: insights from the Philippines. Hum. Ecol. Rev. 24 (1). https://doi.org/10.22459/her.24.01.2018.02.

Food and Agriculture Organization, 2009. The State of Food Insecurity in the World 2009. FAO Publications.

Food and Agriculture Organization, 2022. Digital Agriculture for Food Security in Southeast Asia. FAO Regional Office for Asia and the Pacific.

Franklin, A.L., Ebdon, C., 2020. Participatory budgeting in the Philippines. Chin. Public. Adm. Rev. 11 (1), 60–78. https://doi.org/10.22140/cpar.v11i1.250.

Galang, I.M. Is Food Supply Accessible, Affordable, and Stable? The State of Food Security in the Philippines. https://doi.org/10.62986/dp2022.21.

Golloso-Gubat, M.J., Felix, A., Tandang, N.A., Acuin, C.C.S., Gordoncillo, P.U., Duante, C., 2024. Sustainability of the Philippine food system. Front. Sustain. Food Syst 8. https://doi.org/10.3389/fsufs.2024.1278891. Article 1278891. Frontiers Media.

Grote, U., Faße, A., Nguyễn, T.T., Erenstein, O., 2021. Food security and the dynamics of wheat and maize value chains in Africa and Asia. Front. Sustain. Food Syst. 4. https://doi.org/10.3389/fsufs.2020.617009.

Habito, C., 2019. Agriculture and Food Security in the Philippines: Trends and Prospects. Asian Development Bank.

Israel, D.C., Briones, R.M., 2013. Climate change and agriculture: Implications for food security in the Philippines. Philippine Institute for Development Studies (PIDS) Policy Notes.

Jaranilla-Sanchez, P.A., Lasco, R.D., Villamor, G.B., Gerpacio, R., Nilo, G.P., Villegas, K.L., 2007. A Primer on climate change adaptation in the Philippines. World Agroforestry Centre.
Kingdon, J., 2011. Agendas, Alternatives, and Public Policies, 2nd ed. Pearson.
Lahoti, S.A., Withaningsih, S., Lomente, L., Kamiyama, C., Luna, C.D., Sahle, M., Malik, A.D., Parikesit, P., Pulhin, J.M., Hashimoto, S., Saito, O., 2023. Exploring bioproduction systems in socio-ecological production landscapes and seascapes in Asia through solution scanning using the nature futures framework. Sustain. Sci. https://doi.org/10.1007/s11625-023-01338-9.
Lapada, A.A., 2019. Rice production monitoring system in the Philippines. Indian. J. Sci. Technol. 12 (2), 1–10. https://doi.org/10.17485/ijst/2019/v12i2/138388. Indian Society for Education and Environment.
Muliati, I., Putro, U.S., Anggoro, Y., 2021. Ensuring social security for all: key considerations for policy options in Indonesia. Policy Gov. Rev. 5 (3), 240. https://doi.org/10.30589/pgr.v5i3.456.
Mwarage, E., 2016. Research Guides: Food Security and Nutrition - A Global Issue. United Nations Retrieved from. https://research.un.org/en/foodsecurity.
National Economic and Development Authority, 2021. Philippine Development Plan 2023–2028. NEDA Retrieved from. https://neda.gov.ph.
Ngatia, L.W., Grace, J.M., Moriasi, D.N., Taylor, R.W., 2019. Nitrogen and phosphorus eutrophication in marine ecosystems. IntechOpen eBooks. IntechOpen. https://doi.org/10.5772/intechopen.81869.
Organisation for Economic Co-operation and Development, 2021. Sustainable Food Systems in Southeast Asia: Policy Recommendations for Resilience and Food Security. OECD Publishing.
Perez, G.J., Enricuso, O., Manauis, K.A., Valete, M.A., 2022. Characterizing the drought development in the Philippines using multiple drought indices during the 2019 El Niño. ISPRS Ann. Photogrammetry, Remote. Sens. Spat. Inf. Sci. 5 (3), 463–472. https://doi.org/10.5194/isprs-annals-v-3-2022-463-2022. Copernicus Publications.
Philippine Institute for Development Studies, 2021. Assessing the impact of the rice Tariffication Law on farmers and consumers. PIDS Discussion Paper.
Pinstrup-Andersen, P., 2009. Food security: definition and measurement. Food Security 1 (1), 5–7.
Reyes, C.M., Mina, C.D., Gloria, R.T, 2020. Food security and poverty in the Philippines: The role of government interventions. PIDS Discussion Paper.
Rosegrant, M.W., Magalhães, E., Valmonte-Santos, R., Mason-D'Croz, D., 2018. Returns to investment in reducing postharvest food losses and increasing agricultural productivity growth. Cambridge University Press eBooks. Cambridge University Press, pp. 322. https://doi.org/10.1017/9781108233767.020.
Stuecker, M.F., Tigchelaar, M., Kantar, M.B., 2018. Climate variability impacts on rice production in the Philippines. PLoS ONE 13 (8). https://doi.org/10.1371/journal.pone.0201426. Public Library of Science.
Tolentino, B., 2021. The rice tariffication law: economic implications and future directions. Philippine Rev. Econ 58 (3), 22–41.
UNDP, 2021. Food Systems Transformation and Sustainable Development Goals. UNDP Report.
United Nations Children's Fund, 2022. Philippines Nutrition Report: Addressing malnutrition and child stunting. UNICEF. Retrieved from https://www.unicef.org.
Van, V.T.S., Siguin, C.P., Lacsina, A.C., Yao, L.F., Sales, Z.G., Gordoncillo, N.P., Advincula-Lopez, L., Sescon, J.T., Miro, E.D., 2022. A community-led central kitchen model for school feeding programs in the philippines: learnings for multisectoral action for health. Glob. Health Sci. Pract. 10 (6). https://doi.org/10.9745/ghsp-d-21-00391.

Vista, B.M., Nel, E., Binns, T., 2011. Land, landlords and sustainable livelihoods: the impact of agrarian reform on a coconut hacienda in the Philippines. Land. Use Policy 29 (1), 154. https://doi.org/10.1016/j.landusepol.2011.06.002. Elsevier BV.

World Bank, 2020. Agricultural Resilience and Food Security Policies in Developing Economies. World Bank Publications.

World Bank, 2022. Global Food Crisis: Impacts and Policy Responses. World Bank Group.

Yumul, G., Cruz, N., Servando, N., 2011. Extreme weather events and disaster resilience in the Philippines. Clim. Res. J. 20 (4), 56–72.

CHAPTER TWELVE

The duality of food production and consumption in the United States: Science, health, and practices

Deborah Oluwaseun Shomuyiwa[a,*], Sedoten Oluwaseun Ogun[b], Oluwatosin Seyide Alabi[c], Nsikakabasi Samuel George[d], Kehinde Precious Fadele[e], Clement Ogar Elam[f], Olabode Ekerin[c], and Don Eliseo Lucero-Prisno III[g,h,i]

[a]Department of Health Promotion & Behavior, College of Public Health, University of Georgia, Athens, GA, United States
[b]Department of Food Science and Technology, University of Nebraska-Lincoln, Nebraska, United States
[c]Department of Child, Youth & Family Studies, College of Education and Human Sciences, University of Nebraska-Lincoln, Nebraska, NE, United States
[d]Faculty of Health Sciences, Institute of Public Health, Jagiellonian University Medical College, Krakow, Poland
[e]Department of Medicine and Surgery, College of Medicine, University of Nigeria, Enugu, Nigeria
[f]Department of Medicine and Surgery, Faculty of Clinical Sciences, College of Medicine, University of Ibadan, Ibadan, Nigeria
[g]Department of Global Health and Development, London School of Hygiene and Tropical Medicine, London, United Kingdom
[h]Research Office, Palompon Institute of Technology, Palompon, Leyte, Philippines
[i]Office for Research, Extension and Innovations, Bukidnon State University, Malaybalay City, Bukidnon, Philippines
*Corresponding author: e-mail address: deborahshomuyiwa@gmail.com

Contents

Advances in Food Security and Sustainability, Volume 11
ISSN 2452-2635, https://doi.org/10.1016/bs.af2s.2026.06.001

Abstract

The United States food system is a paradox of abundance and inequity, characterized by high agricultural productivity and innovation on one hand, and growing public health, environmental, and socio-economic challenges on the other. This chapter examines the duality of food production and consumption in the U.S., tracing how scientific advances, industrial-scale farming, and market dynamics have shaped food availability, quality, and accessibility. While innovations such as biofortification, precision agriculture, and controlled environment farming have improved yield and efficiency, they often come at the cost of nutritional value, environmental sustainability, and equity. The chapter highlights how marginalized communities face disproportionate exposure to food insecurity, ultra-processed diets, and environmental health risks, while farmers themselves contend with economic pressures, mental health challenges, and shrinking autonomy. Community-based food systems, including Community Supported Agriculture (CSA), urban agriculture, food co-ops, and grassroots food sovereignty movements, offer promising models for a healthier, more equitable future. Policy recommendations emphasize the need for holistic, cross-sectoral approaches that realign agricultural priorities with public health goals, support farmer livelihoods, and ensure access to nutritious food for all. By addressing the intersections of health, sustainability, and justice, this chapter contributes to a broader reimagining of food systems that center people, planet, and shared prosperity.

1. Introduction

The global food system is a vast, interconnected network that sustains over 8 billion people, with the United States playing a pivotal role as both a leading producer and consumer of food. The U.S. food system's influence extends across global food security, public health, and environmental sustainability, but it is marked by a paradox: while it excels in producing abundant, affordable food, it also contributes to significant public health and environmental challenges, and a significant number of people in the United States remain food insecure. The United States is among the world's top agricultural producers, with total farm output nearly tripling between 1948 and 2021, an average annual growth rate of 1.46 % (USDA, 2019). Remarkably, this increase in output occurred even as total input use (land, labor, capital, and intermediate goods) declined slightly, at an average annual rate of −0.03 % over the same period. This efficiency is largely attributed to technological advancements, including mechanization, improved crop genetics, and the adoption of synthetic fertilizers and pesticides (Wang et al., 2025). For example, labor and land inputs declined by 76 % and 28 %, respectively, while intermediate inputs such as energy and agricultural chemicals grew by 109 % (Wang et al., 2025). Specialty crops, fruits, vegetables, tree nuts, and horticultural products accounted for $64.7 billion in farmgate sales in 2017, representing one-third of total U.S. crop receipts (Astill et al., 2020). From 2008 to 2018, the U.S. Department of Agriculture (USDA) invested $287.7 million in 213 projects to develop automation and mechanization for specialty crop production, further enhancing productivity and efficiency (Astill et al., 2020).

Despite its productivity, the U.S. food system faces a public health crisis. Poor diet is the leading cause of mortality in the United States, directly linked to malnutrition and chronic diseases such as type 2 diabetes (T2D), cardiovascular disease (CVD), obesity, and certain cancers (Matthews and Kurnat-Thoma, 2024). Nearly one million Americans die annually from diet-related chronic diseases, and these numbers are rising (Matthews and Kurnat-Thoma, 2024). As of 2022, 38 % of U.S. adults have pre-diabetes, and 11.3 % have diabetes, with the majority of cases being type 2 diabetes (T2D). An estimated 126.9 million Americans aged 20 and older, approximately 37 % of the population, have some form of CVD. CVD remains the leading cause of death in the United States, accounting for 695,000 deaths in 2021. In the same year, diabetes was the eighth leading cause of death, resulting in 103,294 deaths. Diet-related chronic diseases disproportionately affect underserved communities, particularly those in areas of higher poverty with limited access to affordable, healthy foods, and facing so-called "food deserts" and "food swamps" (Matthews and Kurnat-Thoma, 2024). These communities often rely on ultra-processed foods (UPFs), which are widely available, aggressively marketed, and less expensive, but contribute to poor diet quality and chronic disease.

The U.S. agricultural sector's reliance on synthetic inputs has environmental and health impacts. Nearly 90 % of pesticides in the U.S. are used in agriculture, exposing farmers, farmworkers, and their families to harmful chemicals (Donley et al., 2022; Curl et al., 2020). Farmworkers can have pesticide levels in their bodies up to 400 times higher than the national average, leading to an estimated 13,000–15,000 acute illnesses annually, potentially as high as 300,000 when accounting for undiagnosed cases (Donley et al., 2022). These exposures also affect children and communities near agricultural fields and manufacturing plants (Curl et al., 2020; Prado et al., 2017; Leskovac and Petrović, 2023). Addressing these challenges requires comprehensive policies and community-driven approaches that center local food systems, food sovereignty, equity, and sustainability to create a healthier, more just U.S. food system. This chapter explores these complexities by examining how scientific innovations, agricultural practices, consumer behaviors, and socio-economic factors shape the dual realities of food production and consumption in the United States. By elucidating these dynamics, the chapter seeks to identify pathways toward a healthier, more sustainable, and equitable food future.

2. Food insecurity in the United States

In 2024, the last year for which complete data are available, food insecurity in the United States reached levels not seen since the aftermath of the 2007–2009 recession (USDA Economic Research Service, 2026). According to the final USDA Household Food Security report, released in October 2025, 48 million Americans, accounting for 13.7 % of the population, lived in food-insecure households in 2024, including 16.5 million people experiencing very low food security, characterized by disrupted eating patterns and reduced food intake rather than anxiety alone (Rabbitt et al., 2025). The number of people living in food insecurity increased each year since 2022 (Gao, 2024).

2.1 Defining and measuring food security in the United States

Food security is defined by the Food and Agriculture Organization of the United Nations (FAO) as a condition in which "all people, at all times, have physical and economic access to sufficient, safe and nutritious food to meet their dietary needs and food preferences for an active and healthy life" (Fao et al., 2022). This broad framework emphasizes four pillars, availability, access, utilization, and stability, that together shape global nutrition policy. In the United States, this concept has been adapted into a more operational classification system by the U.S. Department of Agriculture's Economic Research Service (ERS). The ERS identifies four levels of food security: high, marginal, low, and very low (USDA Economic Research Service, 2026a). These categories distinguish food insecurity from hunger, an individual physiological response to prolonged deprivation. Importantly, food insecurity also differs from malnutrition, which encompasses deficiencies or excesses of nutrients and can occur even in the absence of caloric insufficiency.

The measurement of food security within the United States has evolved through decades of policy development and empirical refinement (USDA Economic Research Service, 2026a). The National Nutrition Monitoring and Related Research (NNMRR) Act of 1990 mandated the creation of standardized tools to assess food accessibility at the national level, leading to the development of the Household Food Security Survey Module (HFSSM) by the USDA's Food and Nutrition Service and the Department of Health and Human Services' National Center for Health Statistics (USDA Economic Research Service, 2026b). First administered in 1995 as a supplement to the U.S. Census Bureau's Current Population Survey, the HFSSM became the basis for the annual *Household Food Security in the United States* reports, providing consistent national estimates of food insecurity between 1999 and 2025 (USDA, 2019; USDA Economic Research Service, 2026a; USDA Economic Research Service, 2026b). The HFSSM's 18-item design, augmented by shorter versions for space-limited surveys, has also been incorporated into major federal studies, such as the National Health and Nutrition Examination Survey (NHANES), the National Health Interview Survey (NHIS), the Panel Study of Income Dynamics (PSID), and the Early Childhood Longitudinal Study (ECLS) (USDA, 2019; USDA Economic Research Service, 2026a; USDA Economic Research Service, 2026b).

2.2 Prevalence of food insecurity in the U.S

In 2024, 5.4 % of Americans regularly skipped meals or went hungry. The food-insecure population included more than 14 million children under the age of 18, accounting for 20 % of all children (Rabbitt et al., 2025; USDA Economic Research Service, 2026a; Rabbitt et al., 2024). This was the highest rate of child food insecurity since 2014 (Rabbitt et al., 2025; USDA Economic Research Service, 2026a; Rabbitt et al., 2024). Food insecurity has trended upward for several years: 47.4 million people (13.5 % of households) were food insecure in 2023, an increase of 3.2 million from 2022, when food insecurity had temporarily declined due to enhanced pandemic-era nutrition assistance, expanded benefits under the Supplemental Nutrition Assistance Program (SNAP, formerly known as Food Stamps, the largest federal government nutrition assistance program), and an augmented child tax credit (Rabbitt et al., 2025; USDA Economic Research Service, 2026a). The upward trajectory from 2022 to 2024 corresponds with the expiration of those emergency supports (Rabbitt et al., 2024).

Food insecurity remains deeply patterned along lines of race, household structure, geography, and age. In 2023, 23.3 % of Black households and 21.9 % of Latino(a) households experienced food insecurity, compared to 9.9 % of White non-Latino(a) households, meaning Black individuals were approximately 2.5 times and Latino(a) individuals roughly twice as likely to face hunger as their White counterparts (Rabbitt et al., 2024). Indigenous communities also experience disproportionately high rates of food insecurity, reflecting the combined legacy of land dispossession, underinvestment in infrastructure, and structural exclusion from decision-making in food and land governance (Gao, 2024). Single-parent households headed by women are particularly vulnerable, with 34.7 % experiencing food insecurity, and the Southern region continues to record the highest rates nationally at 14.7 % (Rabbitt et al., 2025, 2024). Rural areas exhibit higher food insecurity than urban areas, reflecting the combined effects of lower incomes, limited employment opportunities, and greater geographic remoteness from retail food stores (Rabbitt et al., 2025, 2024). Older adults face growing risk: in 2024, more than one million additional senior households became food insecure compared to 2023, pushing the senior food insecurity rate to 11 %, reflecting the unique challenges posed by fixed incomes, medical diet requirements, mobility limitations, and social isolation (Rabbitt et al., 2025). Historically, national food security data have revealed cyclical patterns tied to macroeconomic shocks, rising during recessions, easing in periods of expansion, and spiking during the COVID-19 pandemic before temporarily improving with emergency interventions, but the termination of the annual USDA Household Food Security Report in 2025 will impede longitudinal tracking of these dynamics at a particularly critical juncture (Rabbitt et al., 2025).

2.3 Structural drivers of food insecurity

Understanding why food insecurity is concentrated in particular populations requires moving beyond individual behaviors to the structural and historical forces that shape exposure to economic and material hardship (Drewnowski, 2022). Food insecurity in the United States is not fundamentally a problem of insufficient food production, as domestic output far exceeds domestic need, but rather a manifestation of the inequitable distribution of wealth and power and the historic and ongoing inequities (Odoms-Young and Bruce, 2018). Systemic racism is central: the Home Owners' Loan Corporation (HOLC) "redlining" maps of the 1930s directed capital away from Black, Brown, and immigrant neighborhoods for decades, and contemporary research shows that census tracts with higher proportions of racial and ethnic minority residents continue to have reduced food access (Shaker et al., 2023; Singleton et al., 2023). "Supermarket redlining," in which full-service grocery retailers avoid or withdraw from minority neighborhoods, reinforces these patterns of disinvestment, while gentrification displaces long-term residents from both social networks and emerging food retail, further destabilizing access (Drewnowski, 2022; Shaker et al., 2023).

Economic structures interact with racialized disadvantage to produce and maintain food insecurity. Approximately 33.5 % of households with incomes below 185 % of the federal poverty threshold experience food insecurity, reflecting the combined effects of poverty, unstable employment, wages that lag behind the cost of living, unaffordable housing, and inadequate healthcare, conditions that disproportionately affect communities of color (Drewnowski, 2022). These economic forces are embedded in the built environment: so-called food deserts do not arise naturally but from zoning decisions, transportation infrastructure, and real estate investment patterns that leave many low-income neighborhoods, particularly Black and Brown communities, without proximate full-service grocery stores, meaning households may have some financial resources yet still lack reliable physical access to healthy food (Weiler et al., 2015). Intergenerational trauma adds a further, often overlooked layer, as research shows that caregivers with high levels of Adverse Childhood Experiences and depressive symptoms are more likely to report low and very low food security even long after material access has improved (Chilton et al., 2017).

3. The science of food production in the U.S

The food production system in the U.S. has evolved over time, driven by agricultural research, innovations in animal and crop genetics, new technologies, and scientific advancement. This has led to maximized efficiency and output, evident in reduced labor, land use decline, and increased productivity over the years (Alston and Pardey, 2022; Chávez-Dulanto et al., 2020; USDA, 2019). According to ERS, the total farm output nearly tripled between 1948 and 2021 (USDA, 2019). Likewise, the labor and land inputs declined by 76 % and 28 %, respectively, within this period (Wang et al., 2025). According to data from the Bureau of Economic Analysis at the U.S. Department of Commerce, agriculture and related industries contributed 5.5 % of the U.S. gross domestic product (GDP) in 2023 (Kaufman, 2025). The advancement in the science of food production has helped the U.S. meet its food demands, allowing for surplus food locally with enough for exportation. The U.S. is a major player in global staples production (USDA, 2019), as illustrated by the interstate food flow network in Fig. 1, which highlights the Midwest's dominant role as a net exporter to key distribution hubs along the Gulf and Pacific coasts (Lin et al., 2014).

However, the mass production of food has been linked to a decline in the nutritional quality of foods (Bhardwaj et al., 2024). A typical example in the U.S. is the selection of maize for its increased grain yield and competitiveness against weeds due to its larger seed size. This larger seed size can give rise to larger and more persistent seedlings that are able to compete better with weeds. This favorable trait selection by plant breeders has been associated with an increase in starch content and a decrease in protein content (Floros et al., 2010). This trend is also seen in other food varieties, as traditional breeding techniques, which involve selection of specific genetic traits accompanied by uncontrolled genetic mutations, often lead to underlying changes in other traits, which could lead to negative changes as seen in the case of maize (Floros et al., 2010; Hamdan et al., 2022). Hence, there is a need to shift agricultural practices to create sustainable, nutritionally balanced food.

3.1 Advances in agricultural technology and innovation for healthier crops

Agricultural technology is continuously evolving in the U.S. to ensure that crops are not only high yielding but also nutritious and resilient to environmental and climate changes, as shown in Fig. 2. Traditional initiatives taken towards an increase in food production have been based on selective breeding for favorable traits, mechanization of farming processes, irrigation, and use of fertilizers to boost growth and yield, as well as the use of synthetic pesticides to prevent crop damage or diseases. This package of practices is commonly described as the

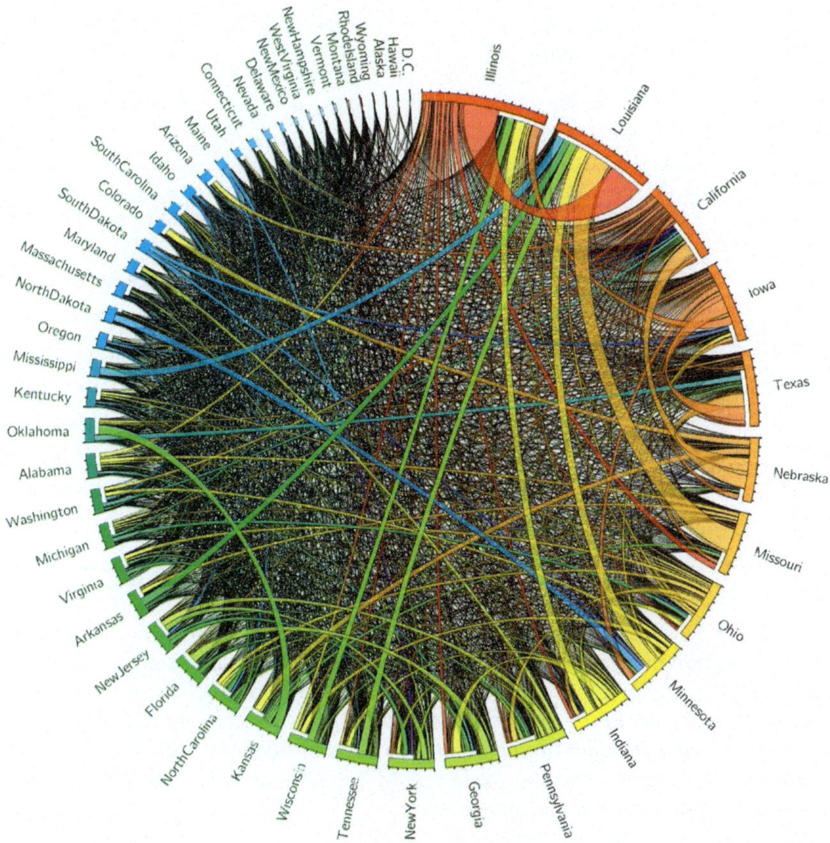

Fig. 1 Network representation of food flows between the 50 USA states. *The states are ranked according to the total food trade volume and plotted clockwise in descending order. The size of the outer bar indicates the total trade volume of each state as a percentage of total USA trade. Export volume is indicated with links emanating from the outer bar of the same color. Import volume is indicated with a white area separating the outer bar from links of a different color. Note that the largest link in the USA food flow network is from Illinois to Louisiana. Midwestern states are shown as major exporters of food to the key ports in Louisiana, California, and Texas. The total trade volume of food represented by this graph is 414.7×106 tons* (Lin et al., 2014).

Green Revolution (see the Green Revolution chapter in this volume for further discussion) (Chávez-Dulanto et al., 2020; Hamdan et al., 2022).

However, decades of reliance on broad-spectrum synthetic pesticides have contributed to the development of resistance in many target pest populations, which can lead to repeated applications and, in some cases, the use of more potent products to maintain control (Chávez-Dulanto et al.,

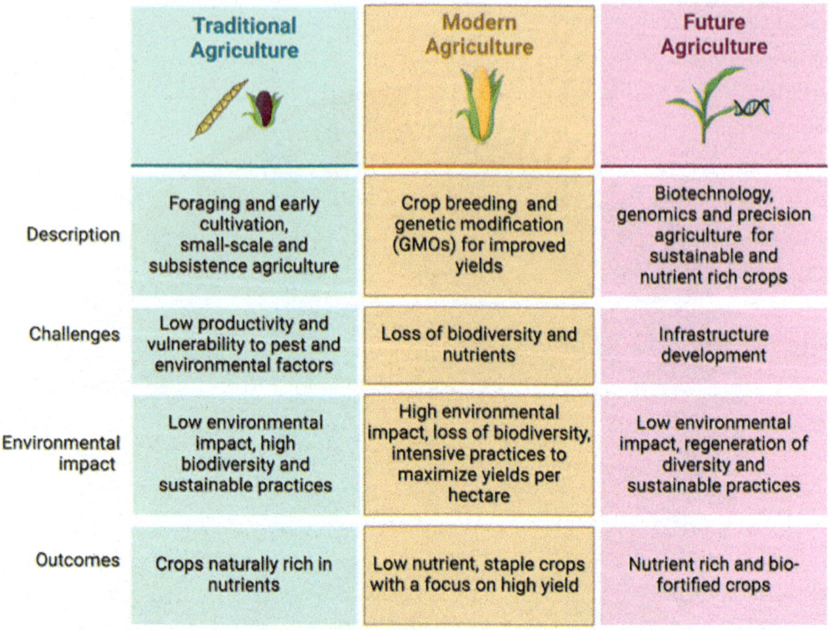

Fig. 2 Evolution of agriculture and its impact on nutrition and sustainability. *This figure describes the progression of agriculture through traditional, modern, and future agriculture, highlighting the challenges, environmental impacts, and outcomes at each phase. It highlights the shift from nutrient-rich crops in traditional systems to high-yield approaches in modern agriculture, and finally, to sustainable, bio-fortified solutions in future farming. Source: Authors' analysis.*

2020). At the same time, pesticide chemistry and delivery systems continue to evolve, with more precise, targeted application technologies being developed to reduce off-target impacts, improve efficiency, and lower overall chemical use (Chávez-Dulanto et al., 2020; Wallinga et al., 2022). Nevertheless, residues from some pesticides can accumulate in foods and, when exposures are repeated over time, may contribute to chronic health conditions such as kidney impairment, hypertension, endocrine disruption, and certain cancers (Chávez-Dulanto et al., 2020; Swanson et al., 2014; Mnif et al., 2011; Nicolopoulou-Stamati et al., 2016).

Crop breeding has historically prioritized traits associated with high yield and uniformity, often with comparatively less emphasis on traits such as pest and drought resistance or nutrient density, even though stress tolerance and micronutrient content can themselves contribute to stable yields and long-term crop performance (Ofori et al., 2022). Plant domestication has also led to a reduction in quality and genetic diversity within most crops

(Chávez-Dulanto et al., 2020). A typical example is the wild tomato *Solanum pimpinellifolium,* which contains a high level of the antioxidant lycopene, 270 mg/kg compared to its domesticated counterpart *(Solanum lycopersicum)* which had 60–120 mg/kg (Zsogon et al., 2018). Lycopene has been shown to have anti-inflammatory properties, with its intake associated with a reduction of cardiovascular and cancer risks (Zsogon et al., 2018). Similarly, in wheat, domestication and the introduction of semidwarfing genes have increased yield but have been linked to declines in protein content and micronutrients such as zinc, iron, copper, and magnesium (Chávez-Dulanto et al., 2020).

Diet has been associated with the onset of chronic and non-communicable diseases leading to a high mortality rate (Qiao et al., 2022). The modernization of U.S. agriculture and food processing has altered the nutrient composition of foods (Komarnytsky et al., 2022), contributing to the nutrient deficiencies and imbalances that are commonly observed. To address these challenges, innovations like biofortification, controlled environment agriculture, microbial inoculation, genetic resource conservation, and digital and precision agriculture have been explored to bridge the gap between yield and healthy crops.

3.1.1 Biofortification

Biofortification involves the increase in the nutrient density of crops through plant breeding, agronomic practices, or biotechnology without sacrificing any characteristics important to farmers or consumers (see the Biofortification chapter in this volume for more detail) (Talsma and Pachón, 2017). Unlike conventional food fortification, which adds nutrients to foods after harvest, biofortification is a sustainable strategy that enhances the quantity and bioaccessibility of nutrients during the crop's growth phase (Ofori et al., 2022; Kumar et al., 2023). Approaches to biofortification include agronomic biofortification, plant breeding, and genetic engineering. Agronomic biofortification involves the application of mineral fertilizers enriched with micronutrients such as zinc, iron, or selenium to soils or crops to increase their concentration in edible plant tissues (Ofori et al., 2022; Sheoran et al., 2022). This approach is relatively simple to implement and can be integrated into existing fertilizer regimes, but the higher cost of micronutrient-enriched fertilizers can be a major barrier in lower-income settings. Biofortification through plant breeding focuses on developing crop varieties with enhanced capacity to take up, transport, and redistribute key vitamins and minerals within plant tissues.

Conventional breeding involves crossbreeding varieties of the same species; mutation breeding involves the use of chemicals or irradiation to introduce genetic variations to develop nutrient-rich varieties; and molecular breeding involves the identification of gene positions associated with the desired traits and utilizing genetic markers to incorporate these traits efficiently (18, 20). All three approaches can be used to enhance the nutritional quality of crops. Biofortification through plant breeding is sustainable and less costly; however, it involves a lot of investment during the research and developmental stages (Ofori et al., 2022).

Genetic engineering is another approach to biofortification. It is used to transfer and overexpress a desirable trait from an unrelated plant species or organism to another crop to boost nutrients (Ofori et al., 2022; Sheoran et al., 2022).

HarvestPlus, an organization based in Washington, DC, focuses on the biofortification of foods and has released over 13 types of biofortified staple crops, including high-iron beans, cassava rich in provitamin A, and high-zinc maize, amongst others, in over 128 countries (Sheoran et al., 2022). It seeks to biofortify varieties that also have agronomically attractive traits (high yield, drought and pest resistance).

3.1.2 Controlled environment agriculture

Controlled Environment Agriculture (CEA) utilizes the optimization of environmental factors including light, temperature, humidity, and nutrients to improve crop productivity while also enhancing nutritional quality (Dsouza et al., 2023). Cultivation is mainly soilless with the use of hydroponic (growing plants in nutrient-enriched water without soil), aeroponic (suspending plant roots in air and misting them with nutrients), and smart fertigation (automated, sensor-guided delivery of fertilizer through irrigation water) systems (Dsouza et al., 2023; Cowan et al., 2022). CEA implementation consists of various methods to optimize nutrient concentration and plant quality, such as adjusting nutrient solutions like nitrogen and potassium, biofortification of minerals, continuous manipulation of light wavelength and intensity, and application of plant growth regulators and hormones. Optimizing these processes across different plant species enhances crop quality and yield; improves nutrient content, color, and taste; boosts disease resistance; and prevents post-harvest spoilage. An example is the pre-harvest continuous LED light and nitrogen depletion for 3 days, resulting in an increase in soluble sugar, soluble protein, anthocyanins (plant pigments with antioxidant properties), and phenolic compounds (plant-derived compounds with antioxidant and anti-inflammatory activity) in lettuce under hydroponic conditions (Dsouza et al., 2023; Yang et al., 2023). CEA systems are able to limit exposure to external factors due to their enclosure and separation from the environment, thereby potentially reducing the need for pesticide application (Cowan et al., 2022).

3.1.3 Genetic resource conservation

Plant genetic diversity is crucial for ensuring a continuous supply of diverse crops as well as for enhancing the nutritional quality. Focused breeding efforts over the years aimed at increasing yield and other beneficial agronomic traits have resulted in a streamlined genetic base in almost all major crop cultivars (Fassil et al., 2000). Preserving and utilizing the genetic diversity of crops through traditional knowledge from local farmers, seed banks, and conservation programs allows breeders to access a wide range of varieties and traits, including those related to nutritional quality. This genetic diversity is crucial for developing new, nutrient-rich crop varieties (Fassil et al., 2000). In the U.S., there is a shift towards healthy eating and utilization of previously underutilized healthy crops, creating a market for crop diversification (Mustafa et al., 2019). For example, millet is gaining popularity as it is gluten-free and high in protein and amino acid (IFT, 2024).

3.1.4 Digital and precision agriculture

Digital agriculture utilizes information technologies, which include the internet, the internet of things, predictive analytics, and digital devices to enhance the collection and analysis of digital content within the agricultural sector (McFadden et al., 2024). Precision agriculture is an element of digital agriculture that uses the geographic positioning system (GPS) and geographic information systems (GIS) through advanced software and equipment to spatially and temporally apply agricultural inputs such as fertilizer and irrigation for optimal productivity (McFadden et al., 2024; Ammal and Rajakumar, 2023). This approach ensures that crops receive the precise amount of nutrients, water, and care needed, leading to improved growth and nutrient content. Precision nutrient management tools include optical sensors like multispectral and hyperspectral sensors that can assess spectral responses from plants (i.e., how they use light) (Ammal and Rajakumar, 2023), chlorophyll meters, leaf color charts, aerial imaging, and nutrient management models, among others, for facilitating need-based nutrient applications. Overall, precision nutrition management focuses on harmonizing the demand and supply of nutrients to achieve better crop quality (Ammal and Rajakumar, 2023).

3.2 Sustainable farming practices and their effects on food quality, nutrition, and the environment

Sustainable agriculture aims to meet rising global food demand while upholding environmental integrity (Chandel et al., 2024). Given the significant contributions of the agricultural sector to the U.S. economy, there is a need for continuous improvement and adjustment to the changing climatic conditions. Climate change has introduced variability in weather patterns, resulting in unpredictable rainfall, droughts, storms, floods, and heat waves (Thornton et al., 2014). Inadequate water supply and drought stress are threats to agricultural productivity (Nicolopoulou-Stamati et al., 2016). Sustainable farming practices like sensor-based irrigation, plant breeding, and biotechnology, as well as drought-resistant microbial inoculants, can help mitigate the impacts of water scarcity on crop yields and quality (Chandel et al., 2024; Alori et al., 2017). Various agricultural practices are targeted towards sustainable food production and quality management, as shown in Fig. 3.

3.2.1 Agronomic biofortification

Agronomic biofortification adds micronutrient-rich fertilizer to plant leaves or the soil. It is an easy approach for farmers to adopt, but has a number of

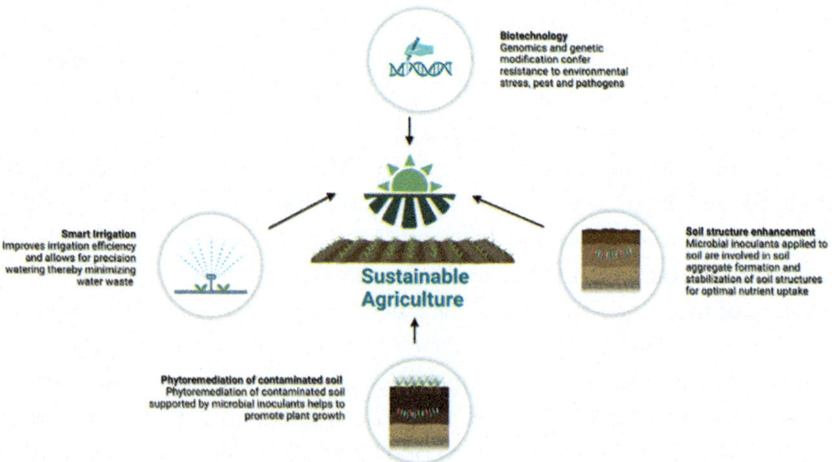

Fig. 3 Sustainable agricultural practices. *Source: Authors' analysis.*

significant drawbacks. The fertilizer can be easily washed off, thereby contaminating neighboring water bodies. Fertilizers can also accumulate in the soil over time, causing toxicity if not applied in the right amounts (de Valença et al., 2017). Repeated fertilizer use can lead to soil nutrient imbalances, and certain micronutrients (e.g., selenium) may accumulate to toxic levels if improperly managed (Bhardwaj et al., 2022). Other bio-fortification strategies like conventional breeding and genetic engineering ensure reduced pollution and contamination. Also, agronomic methods require continuous input, making them less cost-effective for resource-limited farmers (Ofori et al., 2022; Teklu et al., 2023).

3.2.2 Microbial inoculation and soil remdiation

Microbial inoculation involves the use of formulations of microorganisms that, when added to the soil, improve the availability of nutrients to plants, thereby promoting plant growth (Alori et al., 2017). These microbial inoculants are not nutrients themselves but increase nutrient availability through their activities in the soil, including solubilizing and mobilizing minerals to plant roots. Key examples include phosphate-solubilizing bacteria and fungi such as *Pseudomonas*, *Azospirillum*, *Burkholderia*, *Enterobacter*, *Rhizobium*, *mycorrhiza*, and *Streptomyces* (Owen et al., 2015). Microbial inoculants can be applied to seeds, plant surfaces, or soil to increase nutrient supply. Microbial biofertilizers may offer cost advantages over reliance on animal and plant manures alone, and have been reported to increase yields in crops like soybeans when co-inoculated with rhizobium and fungal inoculants (Alori et al., 2017; N'cho et al., 2013). Beyond nutrient provision, microbial inoculants added to contaminated soil can help transform pollutants into harmless substances. This phytoremediation function is supported by the production of growth hormones such as indole acetic acid, nitrogen-fixing siderophore activity, and nutrient solubilization (Alori et al., 2017). The use of microbial inoculants is an environmentally friendly and cost-effective approach to phytoremediation that helps to reduce the cost of contamination removal in the soil. It also enhances soil nutrient availability to plants, helping promote plant growth and quality (Alori et al., 2017).

3.2.3 Irrigation technology

Advancements in irrigation technology have become central to sustainable agriculture, particularly in the face of increasing water scarcity and climate variability (Chandel et al., 2024; Su and Singh, 2024). Modern irrigation systems, such as drip and micro-irrigation, deliver water directly to the plant root zone, significantly reducing water loss due to evaporation and runoff (Su and Singh, 2024). Studies have shown that drip irrigation can reduce water use by up to 50 % compared to traditional flood irrigation, while simultaneously increasing crop yields by 20–30 % (Chu, 2017). Precision irrigation systems, which utilize soil moisture sensors, weather data, and automated controls, further optimize water application by adjusting schedules and amounts based on real-time crop needs. For example, the adoption of sensor-based irrigation scheduling in California almond orchards led to a 33 % reduction in water use without compromising yield (Palmgren and Shabala, 2024). Additionally, solar-powered irrigation pumps are increasingly used in regions with limited access to electricity, reducing operational costs by 30–50 % and supporting sustainable water management (Mishra et al., 2024). These technologies not only conserve water but also minimize nutrient leaching and soil erosion, contributing to healthier crops and improved food quality (Chandel et al., 2024; Su and Singh, 2024).

3.2.4 Biocontrol of pests and pathogens

Some microbes can produce metabolites like antifungals (2, 4-diacetylphloroglucinol) and enzymes (lytic enzymes, proteases, and chitinase) that act or protect against plan pathogens, thereby offering sustainable and ecofriendly controls (Bonaterra et al., 2022; Malik et al., 2023; Xu et al., 2021). Microbial inoculants that act as biocontrol agents can also prevent pathogens by competitive colonization of plant roots or biofilm formation (N'cho et al., 2013). Typical examples of these biocontrols include *Streptomyces* that prevents damping off in tomatoes (a condition in which fungi kill seedlings); *Bacillus pumilus*, *Pseudomonas alcaligenes*, and *Rhizobium sp* that prevent wilt disease in lentils; and *Bacillus subtilis* that prevents *Fusarium* wilt in maize (N'cho et al., 2013; Bonaterra et al., 2022; Malik et al., 2023; Xu et al., 2021). Biocontrol agents also include biological pest and disease control methods like the use of biopesticides and integrated pest management systems. These are used instead of the synthetic pesticides that have caused considerable damage to the environment (Nega, 2014).

3.2.5 Soil structure enhancement

Enhancing soil structure is fundamental for improving water retention, nutrient availability, and overall crop resilience, especially under drought conditions. The application of organic amendments, such as compost and manure, has been shown to increase soil porosity by 12–18 % and water-holding capacity by up to 20 %, which is critical for plant growth during dry periods (Franzluebbers, 2020; Xuan et al., 2023). In arid and semi-arid regions, the use of super absorbent polymers (SAPs) has proven particularly effective, increasing saturated hydraulic conductivity by 35 % and boosting maize yields by 22 % (Xuan et al., 2023; Chang et al., 2021). Recent innovations include nanobubble technology, which infuses irrigation water with high concentrations of oxygen, reducing soil compaction by 20 % and enhancing root development and nutrient uptake by 15–25 % (Yang et al., 2022). These soil enhancements not only improve crop productivity but also support long-term soil health, reducing the need for mineral fertilizers, and mitigate the impacts of extreme weather events (Alori et al., 2017).

3.2.6 Crop tolerance to drought stress

Developing crops with enhanced tolerance to drought stress is a key strategy for ensuring food security in the face of climate change. One approach involves the use of microbial inoculants, such as phosphate-solubilizing bacteria and mycorrhizal fungi, which improve nutrient availability and water uptake under stress conditions. Field trials have demonstrated that co-inoculation of soybeans with rhizobium and fungal inoculants can increase yields by 18–25 % in drought-prone areas (Bashan et al., 2014; Aloo et al., 2022). Genetic improvement is another critical avenue, with plant breeders utilizing both traditional and molecular techniques to develop drought-resistant varieties. For instance, the reintroduction of underutilized crops like millet, which is naturally drought-tolerant, is gaining traction in the U.S. as a climate-resilient alternative (Li et al., 2023). CEA also plays a role, as hydroponic and aeroponic systems can precisely manage water and nutrient delivery, reducing water use by up to 90% compared to conventional farming while maintaining or improving crop quality (Benke and Tomkins, 2017). These combined strategies, microbial, genetic, and technological, are essential for building resilient agricultural systems capable of withstanding future droughts and ensuring a stable food supply.

3.3 Challenges in ensuring the nutritional value of mass-produced foods

As global populations soar and the world becomes increasingly urbanized, there is an increased demand for particular foods like wheat, rice, and maize, which have undergone selective breeding over the years with major losses in nutritional contents (Bhardwaj et al., 2024). It is therefore crucial at this time that consumers and farmers embrace traditional foods high in nutrient content, creating increased diversity in the food supply chain. Though we see advances in technology for the production of nutrient-dense crops, the effectiveness of these technologies is dependent on compliance and implementation. There are several challenges faced in the implementation of these technologies, including the adherence to fortification standards by farmers and food producers, inadequate monitoring and enforcement of fortification standards, and lack of awareness of standards by farmers and producers (Rowe, 2020).

The United States is a major exporter of grains and soybeans, with grains accounting for an estimated 64.7 % of total U.S. agricultural export volume between 2013 and 2022, underscoring the importance of nutrient-rich grains for supporting global nutrition (Kaufman, 2025). Post-harvest handling and storage practices further influence the nutritional value of mass-produced foods. The nutrient availability for consumers depends on the harvest time and method, type of crop, type of nutrients, post-harvest management, and storage conditions. Most vegetables and fruits require low temperature and high relative humidity for nutrient optimization post-harvest, as high temperatures destroy vitamins (Bhardwaj et al., 2024). Furthermore, optimizing the food processing method is essential for various foods and the nutrients they contain. Water-soluble vitamins like vitamins B and C, as well as polyphenols, are susceptible to loss during thermal treatment of vegetables (Bhardwaj et al., 2024).

3.4 Role of biotechnology and its potential health impacts (GMOs, pesticides use)

Genetic engineering is the backbone of genetically modified organism (GMO) production, using recombinant DNA techniques and, more recently, genome-editing tools such as CRISPR-Cas9 to introduce, remove, or modify specific genes in plants and other organisms (Haris et al., 2023). These technologies allow breeders to confer traits such as nutrient enhancement, pest resistance, and stress tolerance with much greater precision than conventional breeding. For example, Golden Rice has been genetically engineered to produce beta-carotene, with the goal of reducing vitamin A deficiency in populations with limited dietary diversity (Tang et al., 2009).

In the United States, GMOs constitute a large share of major commodity crops such as soybeans, corn, and canola. In 2020, an estimated 94 % of soybeans and 92 % of corn planted were genetically engineered varieties, with most of the harvest used for animal feed, biofuel production (e.g., corn-based ethanol, soy-based biodiesel), and highly processed ingredients such as refined oils, starches, and sweeteners rather than for direct consumption as whole foods (Fda, 2024). Oversight of GMOs in the U.S. is shared across multiple federal agencies, including the Food and Drug Administration (FDA), Environmental Protection Agency (EPA), and Department of Agriculture (USDA), which evaluate food and feed safety, environmental impacts, and plant health risks before commercialization (Tsatsakis et al., 2017; USDA, 2025). While the advancement of agricultural biotechnology has contributed to farm efficiency and crop protection, ongoing post-market surveillance and risk assessment remain important components of responsible GMO governance.

GMOs hold potential benefits for food security, climate resilience, and yield stability, but public concern about their safety and broader social implications remains substantial (Haris et al., 2023; Fda, 2024; Kole et al., 2025). Health-related debates often focus on potential allergenicity, particularly when genes from allergen-containing organisms are introduced into previously non-allergenic crops, and on possible toxic effects of traits conferring pest resistance (Haris et al., 2023; Giraldo et al., 2019). At the same time, GM crops approved for food use in the U.S. and other high-income countries have been consumed for several decades without clear evidence of population-level adverse health effects, and non-GMO or certified organic products remain widely available for consumers who wish to avoid GM ingredients (Uconn Extension, 2025; Goodman, 2024). Many of the concerns raised in public debates do not apply uniformly across all GM traits and therefore need to be assessed on a case-by-case basis (Uconn Extension, 2025; Goodman, 2024).

Many GMO crops have been engineered not only to increase yield but also to tolerate specific herbicides and resist insect pests (Fda, 2024; Kole et al., 2025). One widely used approach is the introduction of genes from the soil bacterium *Bacillus thuringiensis* (Bt), enabling plants to produce Bt proteins that are toxic to certain insect pests and thereby reducing the need for conventional synthetic insecticides (Haris et al., 2023). Although the presence of Bt proteins in plant tissues has raised questions for some stakeholders, it is notable that Bt formulations derived directly from the bacterium have long been used as biopesticides in organic agriculture, and regulatory assessments have concluded that Bt traits in approved crops do not pose known risks to human health when consumed at typical dietary levels (Haris et al., 2023; Tian et al., 2015). The more complex and ongoing concern lies in ecological and agronomic effects, such as pest resistance evolution and non-target impacts, which underscore the need to integrate biotechnology with broader integrated pest management and sustainable agriculture frameworks, rather than treating it as a standalone solution (Kole et al., 2025).

4. Community-based approaches to food systems in the U.S

Community-based approaches to food systems in the U.S. are gaining momentum as innovative, localized responses to economic, environmental, and nutritional challenges. At the forefront is Community Supported Agriculture (CSA), a model in which consumers subscribe to regular shares of a farm's harvest. This experiential system fosters stronger relationships between farmers and customers, enabling producers to recapture a greater share of the food dollar while providing consumers with fresh, locally grown food. By centering consumer values, these models help food system stakeholders better understand and address community priorities. More broadly, community food systems aim to integrate sustainable food production, processing, distribution, and consumption to enhance local environmental health, economic opportunity, and food equity (Rossi and Woods, 2024). Across the country, such systems are being designed to increase access to nutritious, affordable food; sustain family farms that use low environmental impact methods; strengthen farmer–consumer connections; reduce transportation-related resource use; generate local employment; and shape supportive food and agricultural policies. These systems not only boost local economies but also empower communities through direct marketing, value-added activities, and reinvestment of capital.

4.1 The role of local food systems: Farmers' markets, food co-ops, and CSA

Local food systems, typically defined as regionally bounded networks of production and consumption, have gained prominence for their potential to support health, sustainability, and community development (Markow et al., 2014). As public concern over health, sustainability, and food access grows, consumer preferences have shifted toward models that emphasize transparency, locality, and nutrition. CSA has emerged as a particularly successful direct-to-consumer model in the U.S., allowing customers to subscribe to a farm's seasonal harvest (Rossi and Woods, 2024). This model offers farmers upfront capital to plan production and diversify markets while fostering deeper consumer engagement through food festivals, farmers markets, and health-oriented programs. CSA networks not only strengthen local food systems through civic collaboration and knowledge-sharing but also serve as a gateway for consumers to access other localized food markets and sustainable food initiatives (Rossi and Woods, 2024).

Complementing CSAs are food cooperatives, solidarity grocery stores, and farmers' markets, each playing a distinct role in building equitable, sustainable food systems (FoodSecureCanada, 2024). Food co-ops and solidarity stores promote democratic governance, affordability, and access in underserved neighborhoods, often operating as member-owned or nonprofit entities. These initiatives empower communities by redistributing control over food sourcing and promoting food sovereignty. Examples such as the Mandela Grocery Cooperative in Oakland, California, the Weaver Street Market in North Carolina, or the Daily Groceries Co-Op in Athens, Georgia, illustrate how locally driven, culturally responsive models can reduce inequality and support small-scale producers (Mandela Grocery, 2026; Weaver Street Market, 2026; Daily Groceries, 2026). Farmers' markets further advance these goals by shortening supply chains, ensuring fair pricing, and fostering environmentally conscious practices like heirloom crop cultivation (Chandel et al., 2024; Giller et al., 2021). They also educate consumers and strengthen trust in food systems through direct interactions with growers (Chandel et al., 2024). Collectively, these localized approaches not only improve access to healthy food but also reinvest in communities, reduce environmental impacts, and contribute to a more resilient and just food economy (Chandel et al., 2024; Giller et al., 2021).

A central question for food security, however, is whether these models reach households experiencing food insecurity or remain primarily accessible to higher-income consumers with greater financial and transportation resources. Evidence suggests that when intentional inclusion strategies are implemented, local food systems can meaningfully serve food-insecure households (Garrity et al., 2024; Hammon and Goralnik, 2024). The expansion of SNAP/Electronic Benefit Transfer acceptance at thousands of farmers' markets, often supported by wireless payment infrastructure, has increased low-income shoppers' access to fresh produce, and incentive programs such as Double Up Food Bucks effectively double SNAP purchasing power for fruits and vegetables, with documented improvements in diet quality and food security among participants (Goddeeris et al., 2017). In addition, the WIC (Special Supplemental Nutrition Program for

Women, Infants, and Children) Farmers Market Nutrition Program allows low-income pregnant and postpartum women and young children to use federal government coupons to purchase fresh, locally grown foods at participating farmers' markets and roadside stands, further integrating federal nutrition assistance with local food systems (USDA, 2026). CSA programs that offer subsidized or sliding-scale shares, sometimes coupled with nutrition education and cooking support, as in Garden City Harvest's model in Montana, along with food co-ops in low-income neighborhoods that accept SNAP and prioritize affordability, demonstrate that local food initiatives need not function as exclusive spaces for the well resourced (Haris et al., 2023; Tsatsakis et al., 2017). When deliberately designed with equity in mind, local food systems can complement federal nutrition assistance and contribute to a more just distribution of healthy food access rather than merely reflecting existing social and economic divides (Markow et al., 2014; Giller et al., 2021).

4.2 Equity and access in community-based food programs

Food insecurity remains a pressing global and national issue and is directly linked to multiple Sustainable Development Goals, including ending poverty (SDG 1), achieving zero hunger (SDG 2), and promoting good health and well-being (SDG 3). It is commonly defined along four key dimensions: availability, access, utilization, and stability, and affects more than 2 billion people globally, with over 700 million experiencing severe food insecurity or extreme hunger (Doustmohammadian et al., 2022). Community-Based Participatory Research (CBPR) has emerged as a promising strategy to address these disparities by involving local communities in identifying food challenges, co-developing solutions, and implementing context-sensitive interventions. CBPR enhances the relevance and impact of food programs, builds trust through shared leadership, and fosters university-community partnerships (Doustmohammadian et al., 2022). While effective, it faces barriers such as limited funding, time constraints, and insufficient evidence for standardized models, underscoring the need for more research into sustainable, participatory strategies for food security.

Alongside CBPR, community-driven initiatives such as food banks and urban agriculture play a critical role in expanding food access. Food banks partner with supply chain actors to rescue and redistribute surplus food, ensuring consistent, safe, and culturally appropriate nutrition for vulnerable populations (The Global Food Banking Network Promoting Food

Security, no date). Urban agriculture, ranging from backyard gardens to large-scale city farms, has re-emerged as a vital tool for food justice in diverse settings across North America and globally, particularly in dense urban areas where land use patterns shape food behaviors (Horst et al., 2024). These efforts often support both household subsistence and local markets while advancing ecologically oriented practices, including approaches such as vertical farming, which can reduce water use and pesticide reliance but may also entail substantial energy demands.

Yet, sustainable food access also depends on addressing systemic challenges such as climate change, supply chain consolidation, and rising food prices (Dailey et al., 2015). The COVID-19 pandemic revealed the fragility of centralized food systems and the need for resilient, localized alternatives. Programs such as farmers' markets, produce prescriptions, and CSA models improve diet quality and access, particularly when paired with financial incentives, nutrition education, and culturally relevant outreach (Lynch et al., 2019). Strengthening equity in food systems therefore requires a holistic approach that integrates environmental resilience, policy reform, and deep community engagement, alongside community-led innovations in food production and distribution.

4.3 Food sovereignty and grassroots efforts to reclaim local food systems

Food sovereignty, first introduced by the international peasant movement La Vía Campesina at the 1996 United Nations World Food Summit, advocates for the rights of people, not corporations or market institutions, to define and control their own food systems (Sampson et al., 2021). The movement outlines seven core principles that position food as a basic human right, calling for agrarian reform, environmental protection, and the restructuring of food trade to favor local production over global profit-driven models. Central to this vision is the belief that communities should have access to healthy, culturally appropriate food produced through sustainable and ecologically sound methods. Unlike food security, which primarily emphasizes food availability and access, food sovereignty shifts the focus toward autonomy, justice, and democratic participation in how food is grown, distributed, and consumed, while recognizing that these two perspectives can be complementary rather than mutually exclusive (SanDiego Foundation, 2020; Lusk and McCluskey, 2018).

In practice, food sovereignty seeks to dismantle systemic inequalities that allow affluent areas to access organic, locally grown food while underserved communities remain trapped in cycles of food insecurity. It promotes local control of land, natural resources, and agricultural practices, reducing reliance on harmful chemicals and external inputs. Grassroots efforts rooted in food sovereignty take shape in the form of community food projects that strengthen local economies, urban gardens that provide nutritional and social benefits, farmers' markets that offer affordable produce while boosting farmer incomes, and food banks that supply culturally appropriate foods (Weiler et al., 2015; Sampson et al., 2021). These initiatives not only enhance nutrition and economic development but also advance racial and economic justice by centering the needs of communities over corporate interests. By empowering local producers and consumers, food sovereignty fosters resilience, self-determination, and long-term sustainability, particularly in marginalized areas.

4.4 Ideals of food consumption in the U.S

Technological advancements and globalization drastically changed farming and food production in the United States. Critics have arisen to emphasize the power of mega-agribusiness over the food system and its role in limiting farmers' freedom to operate and promoting fast food culture in the U.S., thereby threatening the sustainability of the food system that was previously trusted to produce healthy food for communities (Lyson, 2014). This has developed insecurity and concerns about knowing where food comes from and the relationships the local farmers once encouraged. Regardless of this, the change in food supply is global, and the study of alternative food movements is being championed by scholars, especially in the U.S. context. The last ten years witnessed this advocacy via books, articles, and public documentaries. The perception the movement runs with is how the United States food system has created unhealthy food and people (Lyson, 2014).

4.5 Shifting consumer preferences and dietary trends

For the past 50 years, the U.S. food system has faced mounting pressures from climate change, economic instability, environmental degradation, and persistent food insecurity, heightening public concern about sustainability and health (Feenstra and Ohmart, 2012). These pressures have prompted solution-driven conversations centered on food equity, labor rights, and long-term sustainability worldwide (Feenstra and Ohmart, 2012). In response, consumer preferences are shifting away from cheap, commercially produced food toward fresh, minimally processed, and health-promoting alternatives, driven by growing awareness of nutrition, sourcing, and environmental impact (Garrity et al., 2024). Fresh fruit and vegetable consumption has risen progressively, and parents have become more deliberate about food selection, scrutinizing ingredients, sourcing, and nutritional content, especially for their children (Lusk and McCluskey, 2018). While this shift has placed competitive pressure on the market for unhealthy foods, fast food and processed products remain dominant, underscoring that convenience, affordability, and aggressive marketing continue to shape American dietary habits (Lusk and McCluskey, 2018).

According to Thilmany et al., consumers desire the most benefits from the food purchased by selecting their optimal choice of fresh produce (Thilmany et al., 2008). Food sourcing and products reflect qualities that are of personal preference to consumers, like cleanliness, benefits, and additional transport expenses to get it, and others can be public-related and have broader societal benefits, like knowing whether the products are sourced locally or are environmentally acceptable. These food attributes give consumers additional options when deciding what to purchase, weighing out between personal gains and general benefits (Thilmany et al., 2008). With the food advocacy movement calling for changes in the food system, many people are looking for ways to care for their well-being, both practically and traditionally. These include growing and cooking their own food, and preserving food at home. Also, obtaining edible wild plants and resources for food, beverages, and medicine in nature and sourcing food from non-mainstream sources have been embraced tremendously (Lyson, 2014). Cultural and structural factors shape people's perspectives on food and health greatly. The advancement of sustainable food advocacy as an active social movement involving grassroots community activism has positively impacted how families and communities can nurture themselves (Lyson, 2014).

Theoretical models seek to understand consumers' motivations for choosing a specific food source (Thilmany et al., 2008). These models consider influencing factors like health and nutritional attributes, socio-economic concerns, and intrinsic qualities, raising answers to questions on motivation such as safety, food product features, available variety of food products, and how purchasing from a source was beneficial to local businesses and farmers. Incorporating these elements, along with the demographics and the purchasing habits of consumers, gives a better understanding of the varied motivations moderating the relationship between consumers' food source and selection (Thilmany et al., 2008).

Thilmany et al. also examined where consumers prefer to shop, distinguishing between primary, secondary, and seasonal purchasing patterns. Their findings showed that 30 % of consumers made their primary food purchases directly from producers, while secondary and seasonal shopping occurred predominantly at farmers' markets (Thilmany et al., 2008). Since the late 1990s, the School Food and Farm-to-School movement has gained influence. It was launched to address food concerns arising in the U.S. food system. The movement has received consistently favorably evaluations, attracting support from both USDA and the Kellog Foundation, leading to improved policies and programs, as well as positive effects on consumption habits, food choices, and health (Feenstra and Ohmart, 2012).

In the context of heightened awareness of food sourcing and efforts to ensure that children have access to fresh, healthy food, farmers seek to educate children about agriculture and farm practices. They have created a platform for building connections and strengthening their brands in the food chain system (Feenstra and Ohmart, 2012). As the Farm-to-School movement expands and grows, its interrelationships with other international food industries to support food security are likely to deepen as well (Feenstra and Ohmart, 2012). The resultant effect still lies in the accessibility to fresh and healthy foods, wrapped up around resources, environment, cash flow, and consumer awareness (Lyson, 2014). A recurring question in this food activism is the power of food choice (Lyson, 2014).

4.6 Factors influencing food choices and the impact of advertising

Dietary patterns have undergone significant transformation in recent decades, shaped by evolving lifestyle preferences, cultural traditions, and media influence (Kalpe et al., 2023). Contemporary trends include plant-based, ketogenic, DASH (Dietary Approaches to Stop Hypertension), intermittent fasting, paleo, Mediterranean, Sirtfood, flexitarian, vegan, and vegetarian diets. These fluctuate in popularity but are generally adopted in pursuit of eating patterns that align with personal health goals, values, and identities (Kalpe et al., 2023). A common feature across these approaches is the selective inclusion or exclusion of particular food groups or nutrients, distinguishing them from more conventional mixed diets and underscoring the importance of understanding overall dietary patterns rather than single nutrients in isolation (Kalpe et al., 2023).

Terms such as "plant-based," "vegan," and "vegetarian" are distinct dietary categories that are nonetheless frequently conflated in popular discourse, with meaningful implications for both health and environmental impact (Thelwell, 2022). Plant-based diets primarily emphasize foods derived from plants, legumes, nuts, fruits, vegetables, whole grains, and seeds, with varying inclusion of animal products such as dairy and eggs, whereas vegan diets exclude all animal products entirely, and vegetarian diets eliminate meat, poultry, and seafood but may allow other animal by-products (King et al., 2020; Hemler and Hu, 2019). Evidence indicates that appropriately planned plant-based, vegetarian, and vegan diets can reduce body mass index, improve glycemic control, and lower the risk of type 2 diabetes, cardiovascular disease, and some cancers, largely through higher fiber intake, lower saturated fat, and improved lipid profiles (King et al., 2020; Hemler and Hu, 2019; Vitale et al., 2023). These patterns can also contribute to reduced greenhouse gas emissions by lowering demand for animal-source foods, thus aligning individual health benefits with environmental co-benefits (King et al., 2020; Hemler and Hu, 2019). At the same time, strict avoidance of animal products can increase the risk of inadequate intake of nutrients such as protein, iron, and zinc, making attention to diet quality and, in some cases, supplementation important for maintaining nutritional adequacy across the life course (Thelwell, 2022).

By contrast, fad diets, often highly restrictive, heavily marketed, and weakly supported by evidence, are widely used for weight loss but carry significant health risks. Estimates suggest that about half of the roughly 45 million Americans attempting to lose weight in a given year try at least one fad diet, despite limited scientific support for their safety or long-term effectiveness (Kalpe et al., 2023). Health organizations have cautioned that such diets may be particularly harmful for people living with obesity or CVD and generally advise against their use in favor of sustainable, balanced dietary patterns (Kalpe et al., 2023). Perceptions of organic food illustrate another dimension of consumer decision-making: surveys from the Organic Trade Association indicate that many people view organic products as healthier, citing beliefs about higher nutritional value and lower contamination risk as key reasons for their choices, even though evidence on nutritional differences remains mixed (Rehm et al., 2016).

4.7 Challenges of food deserts and chronic diseases

Chronic diseases such as obesity, diabetes, and hypertension are leading causes of morbidity and mortality in the United States, and their burden falls disproportionately on communities with limited access to healthy foods (Copilot.microsoft.com and Imran, 2024). Health care and management costs for these conditions exceed hundreds of billions of dollars annually and are especially concentrated in low-income neighborhoods that are often characterized as food deserts, where residents have limited access to affordable fresh fruits and vegetables (Copilot.microsoft.com and Imran, 2024; Oh et al., 2024). Food deserts are commonly defined as urban or rural areas in which residents have constrained physical and economic access to full-service food retailers offering nutritious options, and a growing body of evidence links these environments to higher rates of obesity, diabetes, CVD, and other diet-related illnesses (Copilot.microsoft.com and Imran, 2024; Oh et al., 2024; Karpyn et al., 2011; Taylor and Ard, 2015; Gordon et al., 2011). Spatial analyses indicate that food deserts and elevated burdens of obesity and diabetes cluster in many Midwestern and Southern states, underscoring the importance of place-based interventions that expand access to supermarkets and other healthy food outlets as part of chronic disease prevention strategies (Copilot.microsoft.com and Imran, 2024; Oh et al., 2024; Taylor and Ard, 2015).

The health effects of food deserts intersect with cultural norms, family practices, and targeted marketing to shape patterns of unhealthy food consumption. In many communities, particularly among racial and ethnic minorities, beliefs that heavier children are healthier; the use of food, often sugary snacks and fast food, as a reward; and the perception of energy-dense, branded products as "status foods" can normalize diets high in fat, sugar, and refined carbohydrates (Kumanyika, 2008).

These norms develop within broader media and commercial environments that heavily promote energy-dense, nutrient-poor products; food advertising frequently associates fast food and sugary snacks with happiness, social belonging, and success, influencing what children and adolescents view as desirable and typical foods (Beaudoin, 2014; Hughto et al., 2022). Young people are exposed to a high volume of such marketing through television and digital media, and those who are less able to recognize persuasive intent are more susceptible to forming preferences for advertised products, increasing their intake of unhealthy foods and elevating their risk for obesity and related conditions (Beaudoin, 2014; Hughto et al., 2022). This convergence of structural constraints, cultural meanings, and aggressive marketing has led public health experts to call for stronger regulation of unhealthy food advertising to children and adolescents, alongside investments in healthy retail access and culturally grounded nutrition education, as part of a comprehensive response to diet-related chronic disease (Karpyn et al., 2011; Batada and Borzekowski, 2008).

4.8 Focus on farmers in the U.S

Agriculture is one of the most dangerous jobs, with many people suffering from occupational injuries and illnesses. The misuse of harmful substances, frequently due to insufficient protective measures, raises concerns about both individual and environmental safety. However, the arduous tasks performed by agricultural laborers, especially those related to pesticide exposure, influence farmers' and farmworkers' health and well-being (Khode et al., 2024). Farmers and ranchers in the United States confront difficult occupations in addition to economic uncertainties, unpredictable weather, and social isolation. Rural populations' access to healthcare and mental health services is limited, making it difficult for them to seek help for stress, worry, or depression. Addressing these issues is critical for their overall well-being and ability to cope with daily stress. Economic troubles in recent years have been compared to the Farm Crisis of the 1980s, with escalating financial difficulty contributing to a rise in farmer suicides. In response, a number of federal, state, and nonprofit efforts are attempting to provide mental health support in rural communities (RHIhub, 2024).

4.9 Farmer livelihoods and how economic pressures affect the production of healthy, nutritious food

U.S. farmers face intensifying economic pressures that threaten both their livelihoods and the nation's ability to produce healthy, nutritious food. Rising production costs and declining profit margins create severe financial instability. The U.S. agricultural sector painted a paradoxical picture in 2025, with projected net farm income rebounding to \$180.1 billion– a 29.5 % (\$41 billion) increase from 2024 – yet this growth stemmed almost entirely from government interventions rather than market fundamentals (U.S. Department of Agriculture, E. R. S, 2025). This artificial recovery masks deeper structural challenges: while direct government payments are forecast to surge 354.5 % to \$42.4 billion (including \$35.7 billion in ad hoc disaster relief), farm receipts continue declining with corn prices down 20 % and soybeans falling 14.6 % from recent highs (U.S. Department of Agriculture, E. R. S, 2025; AGWEB, 2025). Production costs remain elevated at 11 % above pre-pandemic levels, squeezing operating margins to their tightest in a decade despite a projected 4.3 % decrease in total farm expenses (U.S. Department of Agriculture, E. R. S, 2025; AGWEB, 2025). The sector's growing dependence on federal support – government payments will comprise 23.5 % of net farm income in 2025 versus just 5 % in

2024 - raises serious questions about long-term sustainability (AGWEB, 2025; MarketIntel, 2025).

These economic pressures fundamentally alter production decisions and food quality outcomes. Contract farming now dominates 40 % of U.S. agricultural output (Hu, 2013), offering risk reduction but constraining farmer autonomy; only 6 % of small farms (< $250 K sales) participate compared to 60 % of large operations, and for those who do participate, decisions about what to grow, when, and how are largely predetermined by the contracting agribusiness (Hu, 2013; Li and Wang, 2024). The system creates concerning incentives: USDA data show contract growers prioritize yield and consistency over nutritional quality, with corn and soybean producers under contracts achieving 12–18 % higher volumes but lower protein content (Hu, 2013). Meanwhile, the average farmer age reaches 58, as young entrants face prohibitive barriers; – farm debt hit $542.5 billion in 2024 while prime cropland requires 22 years to recoup investment at current returns (U.S. Department of Agriculture, E. R. S, 2025). Labor shortages compound these issues, with 78 % of operations reporting reduced production capacity due to workforce gaps (Hu, 2013). Many workers are seasonal and undocumented immigrants, leaving them vulnerable to exploitation, wage theft, unsafe working conditions, and deportation (Nesheim et al., 2015a). Union representation remains limited, although groups like the United Farm Workers of America continue to fight for fair labor practices (Nesheim et al., 2015a). These structural constraints threaten both farmer livelihoods and the nation's capacity to produce diverse, nutrient-dense foods, suggesting the 2025 farm income rebound may represent temporary relief rather than meaningful recovery.

Creative public policies can tackle economic insecurity in the farm sector while simultaneously expanding food access for low-income families. The National School Lunch Act of 1946 is a landmark example: passed in the aftermath of World War II, the legislation was galvanized in part by the alarming discovery that large numbers of young men had been rejected for military service due to malnutrition — a failure that revealed the nation's food system was not adequately nourishing its own population (USDA, 2024; ILSNA History of School Lunch, no.date). Congress responded by creating the National School Lunch Program, which channeled federal funding to farmers to supply schools with nutritious meals while ensuring those meals were free or subsidized for low-income children (Hammon and Goralnik, 2024). Now in its eighth decade, the program remains a foundational example of how federal policy can align farmer support with public nutrition goals, though its scope, funding, and eligibility criteria have evolved considerably since its inception.

4.10 The health of farmers: Occupational hazards, access to healthcare, and mental health issues in agricultural communities

The ongoing crisis in U.S. farming communities mirrors the devastating 1980s farm crisis, with suicide rates among farmers remaining alarmingly high. Research from the U.S. Centers for Disease Control and Prevention and the National Rural Health Association (NRHA) reveals that male farmers and ranchers had a suicide rate of 43.2 per 100,000 in 2016, 57 % higher than the 27.4 per 100,000 rate for working-age men in other occupations (Eisenreich, Pollari, no date). This disparity persists today, with farmers' suicide rates 3.5 times higher than the general population (Scheyett et al., 2024). Financial instability, including mounting debt, volatile commodity prices, and labor shortages, exacerbates mental health struggles. Changing government regulations, role overload, and poor housing conditions also take a toll, as do environmental stresses. Shifting rainfall patterns and climate-related disasters, such as droughts, floods, and wildfires, further destabilize farm incomes and make agricultural work more unpredictable and physically demanding (Scheyett et al., 2024). Exposure to excessive heat, pesticide drift, and long hours in isolated environments can lead to physical illnesses, cognitive fatigue, and emotional strain. Chronic stress related to adaptation challenges and environmental degradation can erode farmers' connection to the land, often referred to as topophilia, impacting their identity and psychological resilience, as well as their livelihoods (MarketIntel, 2025; Hu, 2013). The COVID-19 pandemic intensified these pressures, disrupting supply chains and isolating rural communities (RHIhub, 2024; Norrod et al., 2023). Older farmers are particularly vulnerable: 45 % of suicides occur among those aged 65 or older, often linked to chronic health issues and economic despair (Scheyett et al., 2024; Norrod et al., 2023).

Compounding these issues, a cultural norm of stoicism in farming communities, along with limited access to mental health resources, often discourages individuals from seeking help, further diminishing social cohesion and support networks (Ecopsychepedia, 2025). Rural and agricultural communities also face structural challenges that increase mental health risks. While strong familial and community ties can be protective, many rural areas contend with aging populations, social isolation, and limited access to healthcare. Global studies have linked these factors, as well as pesticide exposure and previous injuries, to heightened risks of depression, anxiety, and burnout (FarmFoundation, 2025). Despite their reputation for resilience and independence, the many challenges facing farmers call into question the long-term viability of sustainable agriculture.

Efforts to mitigate this crisis are growing, with organizations like the American Farm Bureau Federation launching initiatives such as the Farm State of Mind program, which connects farmers to mental health resources (RHIhub, 2024). However, systemic barriers remain, including limited access to rural healthcare and the stigma surrounding mental illness. Studies show 70 % of farmer suicides involve firearms, highlighting the need for culturally tailored prevention strategies (Norrod et al., 2023). Advocacy groups emphasize peer support networks and financial counseling as critical tools, yet long-term solutions require policy reforms to stabilize farm economies and expand mental health services in rural areas (RHIhub, 2024; Scheyett et al., 2024). Without intervention, the compounding stressors of trade disputes, climate volatility, and aging farmer demographics threaten to deepen this public health emergency.

4.11 The impact of farming practices on the health of consumers

Modern farming practices, particularly the widespread use of antibiotics in livestock production, pose significant health risks to consumers. Antibiotics such as tetracyclines, β-lactams, and macrolides are frequently used to prevent or treat bacterial infections in animals, but their overuse contributes to the emergence of antibiotic-resistant bacteria. This is particularly concerning given that many of these antibiotics are also used in human medicine, making cross-resistance a serious threat to public health. In the United States, approximately 80 % of all antibiotics sold are used in livestock production, and the FDA's largely voluntary guidelines for phasing out growth-promotion uses have had limited impact on overall consumption to date (Wallinga et al., 2022). Antibiotic-resistant infections affect an estimated 2.8 million Americans annually and result in more than 35,000 deaths, with foodborne transmission recognized as one important pathway for the spread of resistant bacteria (CDC, 2025). In countries like China, where antibiotic production and usage are highest, residues have been found to pollute water and soil, further spreading resistant genes and bacteria (Manyi-Loh et al., 2018). These residues in food can increase the risk of drug toxicity, allergic reactions, and treatment-resistant infections. As a result, the World Health Organization's Global Action Plan on antimicrobial resistance frames the issue as a "One Health" challenge, calling for coordinated global efforts, though low-resource countries still face major surveillance and enforcement gaps (Manyi-Loh et al., 2018).

Similarly, pesticide use in agriculture has raised alarms about its environmental and health impacts. While pesticides play an important role in protecting crops and boosting yields, many are linked to long-term health effects, including cancer, neurological disorders, and reproductive issues. Widespread use of organochlorines, organophosphates, carbamates, and pyrethroids has led to contamination of food, water, and soil, especially in nations like China, Brazil, and the United States where industrial farming dominates (Zhou et al., 2025). The World Health Organization and UN Environment Programme estimate that pesticide poisoning affects three million people annually, resulting in 200,000 deaths, primarily in developing countries (Ahmad et al., 2024). These chemicals can produce reactive oxygen species that cause oxidative stress and cellular damage, contributing to diseases such as Alzheimer's, Parkinson's, leukemia, and diabetes. Despite these dangers, pesticides remain essential for crop protection and disease vector control. A more sustainable path forward lies in integrated pest management and other responsible practices that balance productivity with human and ecological health (Ahmad et al., 2024).

5. The duality of production and consumption: Implications for public health

The relationship between food production and consumption involves interconnected economic, environmental, and public health concerns (Reisch et al., 2017). In the United States, the food system is highly efficient, producing abundant, relatively affordable food, yet it also promotes widespread availability of unhealthy options that contribute to major public health problems (Vitale et al., 2023; Nesheim et al., 2015b; Monteiro et al., 2018). Food production directly impacts consumer health through the availability and affordability of nutritious foods (Wood et al., 2023). However, the emphasis on large-scale production has often prioritized high-calorie, nutrient-poor foods. For example, the U.S. food system provides many low-cost, calorie-dense foods, which leads to an abundance of calories in the food supply but also to an increased likelihood of excessive calorie consumption, overweight, obesity, diabetes, and CVDs (Nesheim et al., 2015b).

Policies, legislation, and economic incentives have played a key role in creating this situation. There is a connection between food and agricultural policies and other systems, such as the health system and the environmental system (Fao, Ifad, Unicef, & Who, 2022). These policies have an impact on the decisions that consumers and the food industry make because they have an impact on the availability and affordability of food throughout the entire value chain, from the initial production to final consumption (Fao, Ifad, Unicef, & Who, 2022). Federal subsidies for production of particular crops have resulted in overproduction of processed, energy-dense foods as some experts have claimed (Franck et al., 2013). The US Farm Bill, renewed approximately every five years, is one example of such legislation that includes subsidies, crop insurance, and price supports for major commodity crops. Among other policies are the Direct and Counter-Cyclical Payment Programs (DCP) and Price Loss Coverage (PLC), which provide payments to farmers when market prices for staple commodities fall below a certain threshold; Crop Insurance Subsidies; Marketing Assistance Loans (MALs); and Loan Deficiency Payments (LDPs) (Dimitri and Effland, 2020). These policies encourage less diversity of foods and have created a market environment where unhealthy foods are cheaper to produce and purchase, while healthier options remain relatively expensive (Franck et al., 2013; Dimitri and Effland, 2020).

The availability of inexpensive, ultra-processed foods has shifted dietary patterns in the U.S., with implications for a range of public health issues. In a prospective longitudinal analysis, a positive association was observed between higher daily consumption of ultra-processed foods and an increased incidence of systemic lupus erythematosus, an autoimmune disease, among U.S. registered nurses, which was particularly strong in younger women (Rossato et al., 2025). Similarly, reviews of prospective cohort studies and epidemiological meta-analyses revealed a consistently positive association between high ultra-processed foods intake and increased risk of developing CVDs and obesity (Vitale et al., 2023; Lane et al., 2024). Food production practices further distort dietary patterns in the U.S. by shaping what foods are cheapest and most available, reinforcing consumer demand for processed over whole foods (Lynch et al., 2019; Nesheim et al., 2015b; Fernqvist et al., 2024; Martinho, 2020).

Meanwhile, it is worth noting that the food industry has, in the last four decades, responded to consumer preferences for convenience by expanding the production of processed and ultra-processed foods, which are designed for long shelf lives, easy preparation, and appealing taste profiles. But this has come at the cost of public health (Hazen, 2023). The rise of fast food, ready-to-eat meals, and packaged snacks has reshaped the American diet, often at the expense of nutritional quality. Dietary habits in the United States are shaped by consumer preferences and demands, which are influenced by factors such as culture, affordability, flavor, and convenience (Nesheim et al., 2015b). It is also worthy of note that this demand for convenience foods will probably increase in the U.S. considering that consumers choose to spend less time on food preparation in light of globalization and fast-paced societies (Brunner et al., 2010; Rahkovsky et al., 2018). However, in recent years, there is a dramatic increase in the demand and consumption of fresh foods, particularly fruits and vegetables, as well as organic and minimally processed foods, which is because of growing awareness of nutrition and sustainability (Vigar et al., 2020; Gomes et al., 2023). Industry has responded in kind, making healthier options increasingly accessible through innovations like prepackaged fresh produce and ready-to-eat whole food products. The higher costs associated with these products can create an accessibility gap, reinforcing existing health disparities (Kern et al., 2017; Lewis et al., 2023; Russell et al., 2022).

Moreover, food prices and quality also vary within community food environments, with patterns of inequity occurring along lines of racial and socio-economic composition of communities (Fanzo et al., 2022). Health disparities are perpetuated by production practices that prioritize the cultivation of commodity crops over fruits and vegetables, as well as by distribution systems that favor affluent areas with higher purchasing power (Nesheim et al., 2015b; Agurs-Collins et al., 2024; Neff et al., 2009). An analysis of a nationally representative survey in the U.S. highlights a significant relationship between household food expenditure at various retail outlets and diet quality, revealing that households allocating larger portions of their food budgets to convenience stores and fast-food restaurants tend to have poorer diet quality and higher body mass index levels, with this effect being particularly pronounced among low-income households (Dhakal and Khadka, 2021).

Individuals living in environments dominated by unhealthy food retailers, such as fast-food chains, face food insecurity and are more likely to consume high-energy, nutrient-poor foods, leading to increased rates of obesity, diabetes, and hypertension (Testa, 2019). After adjusting for socio-demographic factors in a nationally representative survey of 19,058 US adults aged 65 years or older, all physical and mental health conditions were significantly associated with low or very low food security (Cai and Bidulescu, 2023).

The severity of food insecurity further amplified these associations, as individuals with diabetes, for instance, had a 49% higher likelihood of experiencing low food security and a 74% higher likelihood of very low food security compared to those without diabetes (Cai and Bidulescu, 2023). Conversely, increased neighborhood access to food retailers offering healthier options is associated with a reduced risk of food insecurity (Mayer et al., 2014). Individuals residing in such areas generally exhibit healthier dietary patterns and a lower prevalence of obesity and other diet-related illnesses (Testa, 2019). These findings underscore the critical role of food access and affordability in shaping dietary habits and overall health.

Food production practices, while meeting the energy needs of the teeming U.S. population, also result in negative environmental impacts (Nesheim et al., 2015b). The global transition and industrialization in agriculture has increased the use of pesticides, fertilizers, antibiotics, and GMOs, which raises concerns about antimicrobial resistance and chemical residues in food, as well as the long-term health effects these may have on consumers (Leskovac and Petrović, 2023; Manyi-Loh et al., 2018; Buiatti et al., 2013). Despite the anticipated risks posed by global warming and the widespread extreme heat events projected to intensify through 2050, about 500,000 tons of pesticides, 12 million tons of nitrogen fertilizer, and four million tons of phosphorus fertilizer are applied annually to crops in the continental United States according to the EPA (Merem et al., 2022; EPA, 2025a). These substances, which can leach into water sources through rainfall, irrigation, and infiltration, contribute to algal blooms, hypoxic conditions (oxygen-depleted "dead zones" where aquatic life cannot survive), sedimentation, and bacterial contamination, affecting aquatic life, recreational waters, and drinking water supplies (Merem et al., 2022; EPA, 2025a; Clark et al., 2019). Groundwater pollution from agricultural chemicals poses health risks, particularly for households relying on private wells. The extent of these effects varies with farm practices, landscape conditions, and climate factors (Merem et al., 2022; EPA, 2025a). Furthermore, these impacts disproportionately affect marginalized communities, which are often located near industrial farming operations and are more vulnerable to the health effects of pollution, thus exacerbating existing health disparities, and creating a cycle of inequality that is difficult to break.

As the U.S. population continues to grow, ensuring a sustainable and healthy food system is crucial. Achieving a sustainable food system requires collaboration and multi-sectoral engagement, involving not only traditional stakeholders but also unconventional actors including farmers, consumers, researchers, government agencies, and advocacy groups to shape food safety policies and sustainable production practices, and transition from theoretical frameworks to measurable outcomes (Fao, Ifad, Unicef, & Who, 2022). Local and regional Food Policy Councils (FPCs) have emerged as key players in fostering community-driven initiatives that enhance food accessibility (Lange et al., 2021; Oza-Frank et al., 2025). Municipalities with active FPCs have been shown to be more likely to implement policies supporting grocery stores, farmers' markets, and transportation to healthy food retailers (Lange et al., 2021; Oza-Frank et al., 2025). These councils help mobilize stakeholders, prioritize community needs, and drive systemic change that promotes equitable food systems, economic growth, and resilience.

To further expand access, structured dialogues between vendors and low-income consumers should be established to address barriers to food affordability and availability (Andress and Fitch, 2016). Modifying policies to encourage the availability of healthy foods in retail outlets serving low-income communities while limiting the prominence of unhealthy food options can shape purchasing behaviors. Supermarkets can further support this shift by adjusting in-store marketing strategies to emphasize nutrient-dense foods over processed, high-calorie alternatives. Also, investments in food quality over sheer quantity can yield long-term benefits for both public health and farming communities (Franck et al., 2013). To this regards, technological advancements, such as precision agriculture, offer promising solutions by optimizing crop production through data-driven techniques, increasing efficiency, and reducing environmental impact. In addition, farmers can adopt nutrient management techniques by applying fertilizers and manure at optimal times and locations to enhance soil health and crop yields (EPA, 2025b).

Considering that improving food systems extends beyond production, a critical challenge lies in balancing consumer demand for convenience with the promotion of health-conscious food systems. One of the key strategies for striking this balance is improving the nutritional quality of processed foods through food reformulation and fortification (Fao, Ifad, Unicef, & Who, 2022; Komarnytsky et al., 2022). Governments and food manufacturers can collaborate to reduce harmful ingredients, such as trans fats and excessive sodium, while incorporating healthier alternatives like whole grains, fiber, and plant-based proteins (Fao, Ifad, Unicef, & Who, 2022; Komarnytsky et al., 2022). Mandatory food labeling policies, such as front-of-pack nutrition labels and warning labels on unhealthy products, can also help consumers make more informed choices (Golan and Unnevehr, 2008). Beyond reformulation and labeling, enhancing consumer awareness through nutrition education and public health campaigns can play a role in shifting preferences. While nutrition education campaigns provide some benefits, policy-driven interventions, regulatory policies, and financial incentives have proven more effective in shifting consumption patterns (Temple, 2022). Examples include banning ultra-processed foods from school meals, taxing sugar-sweetened beverages (SSBs), and shifting agricultural subsidies towards fruits and vegetables that could lower healthcare costs by reducing diet-related chronic

diseases (Temple, 2022; Ramesh et al., 2021). Additionally, state-issued incentives can encourage investment in rural areas to develop full-service grocery stores or informal transportation networks that connect residents with healthier food options. To this end, understanding the intersection of food production and consumption is essential for maintaining economic viability while prioritizing public health and creating an environment where healthier choices become the default.

6. Conclusion

While scientific innovations such as biofortification, precision agriculture, and controlled environment farming have improved crop yields and resilience, they often fail to reach smallholder farmers and marginalized communities due to systemic inequities. At the same time, the overreliance on ultra-processed foods, shaped by advertising, affordability, and food policy, has contributed to rising rates of chronic diseases, particularly among low-income populations. To create a more just and resilient food system, it is essential to adopt a holistic approach, one that integrates health, environmental sustainability, and social justice at every level of the food chain. This includes shifting agricultural subsidies toward nutrient-dense crops, supporting farmer mental health, expanding access to healthy food in underserved areas, and regulating the marketing of unhealthy products. Only through coordinated, multi-sectoral action can the U.S. food system move beyond its current paradox and chart a path toward a future where food fosters health, dignity, and ecological balance for all.

References

Agurs-Collins, T., Alvidrez, J., ElShourbagy Ferreira, S., Evans, M., Gibbs, K., Kowtha, B., et al., 2024. Perspective: nutrition health disparities framework: a model to advance health equity. Adv. Nutr. 15 (4), 100194. https://doi.org/10.1016/j.advnut.2024.100194.

AGWEB (2025). Government payments drive U.S. Farm income surge in 2025. https://www.agweb.com/news/policy/ag-economy/government-payments-drive-u-s-2025-farm-income-surge.

Ahmad, M.F., Ahmad, F.A., Alsayegh, A.A., Zeyaullah, M., AlShahrani, A.M., Muzammil, K., et al., 2024. Pesticides impacts on human health and the environment with their mechanisms of action and possible countermeasures. Heliyon 10 (7), e29128. https://doi.org/10.1016/j.heliyon.2024.e29128.

Aloo, B.N., Tripathi, V., Makumba, B.A., Mbega, E.R., 2022. Plant growth-promoting rhizobacterial biofertilizers for crop production: the past, present, and future. Front. Plant Sci. 13, 1002448. https://doi.org/10.3389/fpls.2022.1002448.

Alori, E.T., Babalola, O.O., 2018. Microbial inoculants for improving crop quality and human health in Africa. Front. Microbiol. 9. https://doi.org/10.3389/fmicb.2018.02213.

Alston, J.M., Pardey, P.G., 2022. Innovation, Growth, and Structural Change in American Agriculture. National Bureau of Economic Research.

Ammal, U.B., Rajakumar, R., 2023. Precision nutrient management in agriculture for sustaining crop productivity. J. Nat. Resour. Conserv. Manag. 4 (2), 147–160. https://doi.org/10.51396/anrcm.4.2.2023.147-160.

Andress, L., Fitch, C., 2016. Juggling the five dimensions of food access: perceptions of rural low income residents. Appetite 105, 151–155. https://doi.org/10.1016/j.appet.2016.05.013.

Astill, G., Perez, A., Thornsbury, S., 2020. Developing automation and mechanization for specialty crops: a review of U.S. Department of agriculture programs a report to congress. https://ers.usda.gov/sites/default/files/_laserfiche/publications/95828/AP-082.pdf?v=70966.

Bashan, Y., de-Bashan, L.E., Prabhu, S.R., Hernandez, J.-P., 2014. Advances in plant growth-promoting bacterial inoculant technology: formulations and practical perspectives (1998–2013). Plant Soil 378 (1), 1–33. https://doi.org/10.1007/s11104-013-1956-x.

Batada, A., Borzekowski, D., 2008. Snap! Crackle! What? J. Child. Media 2, 19–36. https://doi.org/10.1080/17482790701733179.

Beaudoin, C., 2014. The mass media and adolescent socialization. J. Mass Commun. Q. 91, 544–561. https://doi.org/10.1177/1077699014538829.

Benke, K., Tomkins, B., 2017. Future food-production systems: vertical farming and controlled-environment agriculture. Sustain.: Sci. Pract. Policy 13 (1), 13–26. https://doi.org/10.1080/15487733.2017.1394054.

Bhardwaj, A.K., Chejara, S., Malik, K., Kumar, R., Kumar, A., Yadav, R.K., 2022. Agronomic biofortification of food crops: an emerging opportunity for global food and nutritional security. Front. Plant Sci. 13, 1055278. https://doi.org/10.3389/fpls.2022.1055278.

Bhardwaj, R.L., Parashar, A., Parewa, H.P., Vyas, L., 2024. An alarming decline in the nutritional quality of foods: the biggest challenge for future generations health. Foods 13 (6). https://doi.org/10.3390/foods13060877.

Bonaterra, A., Badosa, E., Daranas, N., Frances, J., Rosello, G., Montesinos, E., 2022. Bacteria as biological control agents of plant diseases. Microorganisms 10 (9). https://doi.org/10.3390/microorganisms10091759.

Brunner, T.A., van der Horst, K., Siegrist, M., 2010. Convenience food products. Drivers for consumption. Appetite 55 (3), 498–506. https://doi.org/10.1016/j.appet.2010.08.017.

Buiatti, M., Christou, P., Pastore, G., 2013. The application of GMOs in agriculture and in food production for a better nutrition: two different scientific points of view. Genes & Nutrition 8 (3), 255–270. https://doi.org/10.1007/s12263-012-0316-4.

Cai, J., Bidulescu, A., 2023. The association between chronic conditions, COVID-19 infection, and food insecurity among the older US adults: findings from the 2020–2021 National Health Interview Survey. BMC Public Health 23 (1), 179. https://doi.org/10.1186/s12889-023-15061-8.

CDC, 2025. Antimicrobial resistance facts and stats. https://www.cdc.gov/antimicrobial-resistance/data-research/facts-stats/index.html.

Chandel, N., Kumar, A., Kumar, R., 2024. Towards sustainable agriculture: integrating agronomic practices, environmental physiology and plant nutrition. Int. J. Plant Soil Sci. 36 (6), 492–503. https://doi.org/10.9734/ijpss/2024/v36i64651.

Chang, L., Xu, L., Liu, Y., Qiu, D., 2021. Superabsorbent polymers used for agricultural water retention. Poly. Test. 94, 107021. https://doi.org/10.1016/j.polymertesting.2020.107021.

Chávez-Dulanto, P.N., Thiry, A.A.A., Glorio-Paulet, P., Vögler, O., Carvalho, F.P., 2020. Increasing the impact of science and technology to provide more people with healthier and safer food. Food Energy Secur. 10 (1). https://doi.org/10.1002/fes3.259.

Chilton, M., Knowles, M., Bloom, S.L., 2017. The intergenerational circumstances of household food insecurity and adversity. J Hunger Environ. Nutr. 12 (2), 269–297. https://doi.org/10.1080/19320248.2016.1146195.

Chu, J., 2017. Watering the world. https://news.mit.edu/2017/design-cuts-costs-energy-drip-irrigation-0420.

Clark, M.A., Springmann, M., Hill, J., Tilman, D., 2019. Multiple health and environmental impacts of foods. Proc. Natl. Acad. Sci. U S A 116 (46), 23357–23362. https://doi.org/10.1073/pnas.1906908116.

Copilot.microsoft.com, Imran, S., 2024. Exploring the relationship between food deserts and chronic disease prevalence. Non Human Journal 1, 15–19. https://doi.org/10.70008/nhj.v1i01.15.

Cowan, N., Ferrier, L., Spears, B., Drewer, J., Reay, D., Skiba, U., 2022. CEA systems: the means to achieve future food security and environmental sustainability? Front. Sustain. Food Syst. 6. https://doi.org/10.3389/fsufs.2022.891256.

Curl, C.L., Spivak, M., Phinney, R., Montrose, L., 2020. Synthetic pesticides and health in vulnerable populations: agricultural workers. Curr. Environ. Health Rep. 7 (1), 13–29. https://doi.org/10.1007/s40572-020-00266-5.

Dailey, A.B., Hess, A., Horton, C., Constantian, E., Monani, S., Wargo, B., et al., 2015. Healthy options: a community-based program to address food insecurity. J. Prev. Interv. Commun. 43 (2), 83–94. https://doi.org/10.1080/10852352.2015.973248.

Daily Groceries (2026). Daily groceries co-Op. https://www.dailygroceriescoop.com/.

de Valença, A.W., Bake, A., Brouwer, I.D., Giller, K.E., 2017. Agronomic biofortification of crops to fight hidden hunger in sub-Saharan Africa. Global Food Security 12, 8–14. https://doi.org/10.1016/j.gfs.2016.12.001.

Dhakal, C.K., Khadka, S., 2021. Heterogeneities in consumer diet quality and health outcomes of consumers by store choice and income. Nutrients 13 (4). https://doi.org/10.3390/nu13041046.

Dimitri, C., Effland, A., 2020. From farming to food systems: the evolution of US agricultural production and policy into the 21st century. Renew. Agric. Food Syst. 35 (4), 391–406. https://doi.org/10.1017/S1742170518000522.

Donley, N., Bullard, R.D., Economos, J., Figueroa, I., Lee, J., Liebman, A.K., et al., 2022. Pesticides and environmental injustice in the USA: root causes, current regulatory reinforcement and a path forward. BMC Public Health 22 (1), 708. https://doi.org/10.1186/s12889-022-13057-4.

Doustmohammadian, A., Mohammadi-Nasrabadi, F., Keshavarz-Mohammadi, N., Hajjar, M., Alibeyk, S., Hajigholam-Saryazdi, M., 2022. Community-based participatory interventions to improve food security: a systematic review. Front. Nutr. 9, 1028394. https://doi.org/10.3389/fnut.2022.1028394.

Drewnowski, A., 2022. Food insecurity has economic root causes. Nature Food 3 (8), 555–556. https://doi.org/10.1038/s43016-022-00577-w.

Dsouza, A., Newman, L., Graham, T., Fraser, E.D.G., 2023. Exploring the landscape of controlled environment agriculture research: a systematic scoping review of trends and topics. Agric. Syst. 209, 103673. https://doi.org/10.1016/j.agsy.2023.103673.

Ecopsychepedia (2025). Farmers and climate mental health. https://ecopsychepedia.org/glossary/farmers-and-climate-mental-health/.

Eisenreich, R., Pollari, C., Addressing higher risk of suicide among farmers in rural America. (no.date).

EPA, 2025a. Nonpoint Source: Agriculture. https://www.epa.gov/nps/nonpoint-source-agriculture.

EPA, 2025b. Sources and solutions: agriculture. https://www.epa.gov/nutrientpollution/sources-and-solutions-agriculture.

Fanzo, J., Rudie, C., Sigman, I., Grinspoon, S., Benton, T.G., Brown, M.E., et al., 2022. Sustainable food systems and nutrition in the 21st century: a report from the 22nd annual harvard nutrition obesity symposium. Am. J. Clin. Nutr. 115 (1), 18–33. https://doi.org/10.1093/ajcn/nqab315.

Fao, Ifad, Unicef, & Who, 2022. The State of Food Security and Nutrition in the World. FAO, Rome.

FarmFoundation, 2025. Perspective: farmer mental and physical health as components of sustainability. https://www.farmfoundation.org/2022/07/20/perspective-farmer-mental-and-physical-health-as-components-of-sustainability/.

Fassil, H., Guarino, L., Sharrock, S., BhagMal, Hodgkin, T., Iwanaga, M., 2000. Diversity for food security: improving human nutrition through better evaluation, management, and use of plant genetic resources. Food Nutr. Bull. 21 (4), 497–502. https://doi.org/10.1177/156482650002100429.

Fda, 2024. GMO crops, animal food, and beyond. https://www.fda.gov/food/agricultural-biotechnology/gmo-crops-animal-food-and-beyond.

Feenstra, G., Ohmart, J., 2012. The evolution of the school food and farm to school movement in the United States: connecting childhood health, farms, and communities. Child. Obes. 8 (4), 280–289. https://doi.org/10.1089/chi.2012.0023.

Fernqvist, F., Spendrup, S., Tellström, R., 2024. Understanding food choice: a systematic review of reviews. Heliyon 10 (12), e32492. https://doi.org/10.1016/j.heliyon.2024.e32492.

Floros, J.D., Newsome, R., Fisher, W., Barbosa-Cánovas, G.V., Chen, H., Dunne, C.P., et al., 2010. Feeding the world today and tomorrow: the importance of food science and technology. Compr. Rev. Food Sci. Food Saf. 9 (5), 572–599. https://doi.org/10.1111/j.1541-4337.2010.00127.x.

FoodSecureCanada, 2024. Toward fair and sustainable food systems: The role of food cooperatives and solidarity grocery stores. https://foodsecurecanada.org/2024/03/28/toward-fair-and-sustainable-food-systems-the-role-of-food-cooperatives-and-solidarity-grocery-stores/.

Franck, C., Grandi, S.M., Eisenberg, M.J., 2013. Agricultural subsidies and the American obesity epidemic. Am. J. Prev. Med. 45 (3), 327–333. https://doi.org/10.1016/j.amepre.2013.04.010.

Franzluebbers, A.J., 2020. Holding water with capacity to target porosity. Agric. Environ. Lett. 5 (1), e20029. https://doi.org/10.1002/ael2.20029.

Gao, 2024. Tribal food security: Opportunities exist to address challenges in federal nutrition programs. https://www.gao.gov/products/gao-24–106218.

Garrity, K., Krzyzanowski Guerra, K., Hart, H., Al-Muhanna, K., Kunkler, E.C., Braun, A., et al., 2024. Local food system approaches to address food and nutrition security among low-income populations: a systematic review. Adv. Nutr. 15 (4), 100156. https://doi.org/10.1016/j.advnut.2023.100156.

Giller, K.E., Delaune, T., Silva, J.V., Descheemaeker, K., van de Ven, G., Schut, A.G.T., et al., 2021. The future of farming: who will produce our food? Food Secur. 13 (5), 1073–1099. https://doi.org/10.1007/s12571-021-01184-6.

Giraldo, P.A., Shinozuka, H., Spangenberg, G.C., Cogan, N.O.I., Smith, K.F., 2019. Safety assessment of genetically modified feed: is there any difference from food? Front. Plant Sci. 10, 1592. https://doi.org/10.3389/fpls.2019.01592.

Goddeeris, L., Mann, J., O'Hara, J., Miller, S., Trumbull, E., Pirog, R., 2017. Analysis of healthy food incentive programs' impact on farmers market vendors in Michigan. Michigan State University, Center for Regional Systems: Michigan State University, Center for Regional Systems.

Golan, E., Unnevehr, L., 2008. Food product composition, consumer health, and public policy: introduction and overview of special section. Food Policy 33 (6), 465–469. https://doi.org/10.1016/j.foodpol.2008.05.004.

Gomes, B.A.F., Alexandre, A.C.S., de Andrade, G.A.V., Zanzini, A.P., de Barros, H.E.A., Ferraz e Silva, L.M. d S., et al., 2023. Recent advances in processing and preservation of minimally processed fruits and vegetables: a review – Part 2: physical methods and global market outlook. Food Chem. Adv. 2, 100304. https://doi.org/10.1016/j.focha.2023.100304.

Goodman, R.E., 2024. Twenty-eight years of GM food and feed without harm: why not accept them? GM Crops Food 15 (1), 40–50. https://doi.org/10.1080/21645698.2024.2305944.

Gordon, C., Purciel-Hill, M., Ghai, N.R., Kaufman, L., Graham, R., Van Wye, G., 2011. Measuring food deserts in New York City's low-income neighborhoods. Health & Place 17 (2), 696–700. https://doi.org/10.1016/j.healthplace.2010.12.012.

Hamdan, M.F., Mohd Noor, S.N., Abd-Aziz, N., Pua, T.L., Tan, B.C., 2022. Green revolution to gene revolution: technological advances in agriculture to feed the world. Plants (Basel) 11 (10). https://doi.org/10.3390/plants11101297.

Hammon, A., Goralnik, L., 2024. Barriers and bridges: contributions of mobile farmers markets to fresh food access for low-income and minority consumers. Local Environ. 30 (4), 447–462. https://doi.org/10.1080/13549839.2024.2436024.

Haris, M., Ajmal, H.M.N., Bashir, H., 2023. Impacts of genetically modified organisms (GMOs) on environment and agriculture: a comprehensive review. Trends Anim. Plant Sci. 2 (16), 9. https://doi.org/10.62324/taps/2023.012.

Hazen, C., 2023. Consumers are willing to pay more for ultra-processed foods if they are healthier. SupplySide Food Beverage J.

Hemler, E.C., Hu, F.B., 2019. Plant-based diets for personal, population, and planetary health. Adv. Nutr. 10, S275–S283. https://doi.org/10.1093/advances/nmy117.

Horst, M., McClintock, N., Hoey, L., 2024. The intersection of planning, urban agriculture, and food justice: a review of the literature. In: Raja, S., Caton Campbell, M., Judelsohn, A., Born, B., Morales, A. (Eds.), Planning for Equitable Urban Agriculture in the United States: Future Directions for a New Ethic in City Building. Springer International Publishing, Cham, pp. 89–120.

Hu, W.-Y., 2013. Effect of contract farming on the U.S. crop farmers' average return. Agric. Econ. (Zemědělská ekonomika) 59 (5), 195–201. https://doi.org/10.17221/100/2012-agricecon.

Hughto, J.M.W., Clark, K.A., Daken, K., Brömdal, A., Mullens, A.B., Sanders, T., et al., 2022. Victimization within and beyond the prison walls: a latent profile analysis of transgender and gender diverse adults. J. Interpers. Violence 37 (23-24), Np23075–np23106. https://doi.org/10.1177/08862605211073102.

IFT, 2024. The promise of underutilized and biofortified crops. https://www.ift.org/news-and-publications/blog/2024/the-promise-of-underutilized-and-biofortified-crops.

ILSNA History of School Lunch. (no.date) https://www.ilsna.net/resources/schoolnutrition/historyschoollunch.

Kalpe, S., Mathur, A., Kharat, P.P., 2023. How fad diets May Jeopardize your oral well-being: the hidden consequences. Hum. Nutr. Metab.

Karpyn, A., Young, C., Weiss, S., 2011. Reestablishing healthy food retail: changing the landscape of food deserts. Child. Obes. 8 (1), 28–30. https://doi.org/10.1089/chi.2011.0113.

Kaufman, J., 2025. U.S. Agricultural Trade - U.S. Agricultural Trade at a Glance.

Kern, D.M., Auchincloss, A.H., Stehr, M.F., Diez Roux, A.V., Moore, L.V., Kanter, G.P., et al., 2017. Neighborhood prices of healthier and unhealthier foods and associations with diet quality: evidence from the multi-ethnic study of atherosclerosis. Int. J. Environ. Res. Public Health 14 (11). https://doi.org/10.3390/ijerph14111394.

Khode, D., Hepat, A., Mudey, A., Joshi, A., 2024. Health-related challenges and programs among agriculture workers: a narrative review. Cureus 16 (3), e57222. https://doi.org/10.7759/cureus.57222.

King, K.M., Pierce, G., Folsom, L.J., Cessna, W., Jaggers, J.R., Wintergerst, K.A., 2020. Championing whole-food and plant-based diets: the role of the health and fitness professional. ACSM's Health Fit. J. 24, 50–54.

Kole, C., Pandey, S., Yasin, J.K., Mamidi, S., Bohra, A., Bhattacharya, P., et al., 2025. Benefits, concerns, and sustainable alternatives to genetically modified crops from a global and Indian perspective. Plant Genome 18 (4), e70154. https://doi.org/10.1002/tpg2.70154.

Komarnytsky, S., Retchin, S., Vong, C.I., Lila, M.A., 2022. Gains and losses of agricultural food production: implications for the twenty-first century. Annu. Rev. Food Sci. Technol. 13, 239–261. https://doi.org/10.1146/annurev-food-082421-114831.

Kumanyika, S.K., 2008. Environmental influences on childhood obesity: ethnic and cultural influences in context. Physiol. Behav. 94, 61–70.

Kumar, S., DePauw, R.M., Kumar, S., Kumar, J., Kumar, S., Pandey, M.P., 2023. Breeding and adoption of biofortified crops and their nutritional impact on human health. Ann. N. Y. Acad. Sci. 1520 (1), 5–19. https://doi.org/10.1111/nyas.14936.

Lane, M.M., Gamage, E., Du, S., Ashtree, D.N., McGuinness, A.J., Gauci, S., et al., 2024. Ultra-processed food exposure and adverse health outcomes: umbrella review of epidemiological meta-analyses. BMJ (Clinical Research ed.) 384, e077310. https://doi.org/10.1136/bmj-2023-077310.

Lange, S.J., Calancie, L., Onufrak, S.J., Reddy, K.T., Palmer, A., Lowry Warnock, A., 2021. Associations between food policy councils and policies that support healthy food access: a national survey of community policy supports. Nutrients 13 (2). https://doi.org/10.3390/nu13020683.

Leskovac, A., Petrović, S., 2023. Pesticide use and degradation strategies: food safety, challenges and perspectives. Foods 12 (14). https://doi.org/10.3390/foods12142709.

Lewis, M., Herron, L.-M., Chatfield, M.D., Tan, R.C., Dale, A., Nash, S., et al., 2023. Healthy food prices increased more than the prices of unhealthy options during the COVID-19 pandemic and concurrent challenges to the food system. Int. J. Environ. Res. Public Health 20 (4). https://doi.org/10.3390/ijerph20043146.

Li, J., Wang, J., Liu, H., Macdonald, C.A., Singh, B.K., 2023. Microbial inoculants with higher capacity to colonize soils improved wheat drought tolerance. Microb Biotechnol. 16 (11), 2131–2144. https://doi.org/10.1111/1751-7915.14350.

Li, Q., Wang, Z., 2024. Impact of contract farming on green technological efficiency of farmers: a comparative study of two contract organizational models. Front Sustain Food Syst. 8. https://doi.org/10.3389/fsufs.2024.1368997.

Lin, X., Dang, Q., Konar, M., 2014. A network analysis of food flows within the United States of America. Environ. Sci. Technol. 48 (10), 5439–5447. https://doi.org/10.1021/es500471d.

Lusk, J.L., McCluskey, J., 2018. Understanding the impacts of food consumer choice and food policy outcomes. Appl. Econ. Perspect. Policy 40 (1), 5–21. https://doi.org/10.1093/aepp/ppx054.

Lynch, H., Berardy, A., Wharton, C., 2019. Chapter 6 - Food production and dietary patterns. In: Sabaté, J. (Ed.), Environmental Nutrition. Academic Press, pp. 101–122.

Lyson, M.C., 2014. The class politics of alternative food: informing public health policy and remedying health inequality. Sociol. Compass 8 (10), 1216–1228. https://doi.org/10.1111/soc4.12202.

Malik, M.S., Rehman, A., Khan, I.U., Khan, T.A., Jamil, M., Rha, E.S., et al., 2023. Thermo-neutrophilic cellulases and chitinases characterized from a novel putative antifungal biocontrol agent: Bacillus subtilis TD11. PLoS One 18 (1), e0281102. https://doi.org/10.1371/journal.pone.0281102.

Mandela Grocery, 2026. Mandela Grocery Worker-Owned Cooperative. https://www.mandelagrocery.coop/.

Manyi-Loh, C., Mamphweli, S., Meyer, E., Okoh, A., 2018. Antibiotic use in agriculture and its consequential resistance in environmental sources: potential public health implications. Mol. (Basel, Switzerland) 23 (4). https://doi.org/10.3390/molecules23040795.

MarketIntel, 2025. Disaster assistance fuels 2025's farm income rebound. https://www.fb.org/market-intel/disaster-assistance-fuels-2025s-farm-income-rebound.

Markow, K., Booth, S., Savio, S., Coveney, J., 2014. Improving access to community-based food systems: comparing perspectives of low socioeconomic individuals and food system representatives. Nutr. Diet. 73 (1), 19–27. https://doi.org/10.1111/1747-0080.12153.

Martinho, V.J., 2020. Food marketing as a special ingredient in consumer choices: the main insights from existing literature. Foods 9 (11). https://doi.org/10.3390/foods9111651.

Matthews, E.D., Kurnat-Thoma, E.L., 2024. U.S. food policy to address diet-related chronic disease. Front Public Health 12, 1339859. https://doi.org/10.3389/fpubh.2024.1339859.

Mayer, V.L., Hillier, A., Bachhuber, M.A., Long, J.A., 2014. Food insecurity, neighborhood food access, and food assistance in Philadelphia. J. Urban Health 91 (6), 1087–1097. https://doi.org/10.1007/s11524-014-9887-2.

McFadden, J., Njuki, E., Griffin, T., (2024). Precision agriculture in the digital era: recent adoption on U.S. Farms. https://www.ers.usda.gov/publications/105893.

Merem, E.C., Twumasi, Y., Wesley, J., Olagbegi, D., Crisler, M., Romorno, C., et al., 2022. Assessing the environmental impacts of farm subsidy in north Texas region. Int. J. Agric. For. 12 (1), 9–28. https://doi.org/10.5923/j.ijaf.20221201.02.

Mishra, A.K., Sudarsan, J.S., Suribabu, C.R., Murali, G., 2024. Cost-benefit analysis of implementing a solar powered water pumping system – A case study. Energy Nexus 16, 100323. https://doi.org/10.1016/j.nexus.2024.100323.

Mnif, W., Hassine, A.I., Bouaziz, A., Bartegi, A., Thomas, O., Roig, B., 2011. Effect of endocrine disruptor pesticides: a review. Int. J. Environ. Res. Public Health 8 (6), 2265–2303. https://doi.org/10.3390/ijerph8062265.

Monteiro, C., Cannon, G., Levy, R., Moubarac, J.-C., Louzada, M., Rauber, F., et al., 2018. Ultra-processed foods: what they are and how to identify them. Public Health Nutr.

Mustafa, M.A., Mayes, S., Massawe, F.J., 2019. Crop diversification through a wider use of underutilised crops: a strategy to ensure food and nutrition security in the face of climate change. Sustain. Solutions Food Secur.

N'cho, C., Yusuf, A.A., Ama-Abina, J., Jemo, M., Abaidoo, R.C., Savane, I., 2013. Effects of commercial microbial inoculants and foliar fertilizers on soybean nodulation and yield in northern Guinea savannah of Nigeria. Int. J. Adv. Agric. Res. 66–73. https://cgspace.cgiar.org/items/b0c74ef9-2214-493c-bec2-cf99de501c7a.

Neff, R.A., Palmer, A.M., McKenzie, S.E., Lawrence, R.S., 2009. Food systems and public health disparities. J. Hunger Environ. Nutr. 4 (3-4), 282–314. https://doi.org/10.1080/19320240903337041.

Nega, A., 2014. Review on concepts in biological control of plant pathogens. 4.

Nesheim, M.C., Oria, M., Yih, P.T.C.O.A.F.F.A.T.H., Environmental, Social Effects of the Food, S., Food, Nutrition, B., et al., 2015a. In: Nesheim, M.C., Oria, M., Yih, P.T. (Eds.), A Framework for Assessing Effects of the Food System. National Academies Press (US), Washington (DC) Copyright 2015 by the National Academy of Sciences. All rights reserved.

Nesheim, M., Oria, M., Yih, P.T., Resources, B., System, C., 2015b. A Framework for Assessing Effects of the Food System (Environmental Effects of the U.S. Food System). The National Academies Press.

Nicolopoulou-Stamati, P., Maipas, S., Kotampasi, C., Stamatis, P., Hens, L., 2016. Chemical pesticides and human health: the urgent need for a new concept in agriculture. Front. Public Health 4, 148. https://doi.org/10.3389/fpubh.2016.00148.

Norrod, P.E., Sanderson, W.T., Abner, E.L., Seals, J., Browning, S., 2023. Farmer suicides among states reporting violent deaths, 2003 - 2017. Rural Ment. Health 47 (3), 139–151. https://doi.org/10.1037/rmh0000232.

Odoms-Young, A., Bruce, M.A., 2018. Examining the impact of structural racism on food insecurity: implications for addressing racial/ethnic disparities. fam community health, 41 Suppl 2 Suppl, Food Insecurity and Obesity(Suppl 2 FOOD INSECURITY AND OBESITY), S3–S6, https://doi.org/10.1097/FCH.0000000000000183.

Ofori, K.F., Antoniello, S., English, M.M., Aryee, A.N.A., 2022. Improving nutrition through biofortification-a systematic review. Front. Nutr. 9, 1043655. https://doi.org/10.3389/fnut.2022.1043655.

Oh, J.I., Lee, K.J., Hipp, A., 2024. Food deserts exposure, density of fast-food restaurants, and park access: exploring the association of food and recreation environments with obesity and diabetes using global and local regression models. PLoS One 19 (4), e0301121. https://doi.org/10.1371/journal.pone.0301121.

Owen, D., Williams, A.P., Griffith, G.W., Withers, P.J.A., 2015. Use of commercial bio-inoculants to increase agricultural production through improved phosphrous acquisition. Appl. Soil Ecol. 86, 41–54. https://doi.org/10.1016/j.apsoil.2014.09.012.

Oza-Frank, R., Warnock, A.L., Calancie, L., Bassarab, K., Palmer, A., Cooksey Stowers, K., et al., 2025. Food policy councils and healthy food access policies: a 2021 national survey of community policy supports. Prev. Chronic Dis. 22. https://doi.org/10.5888/pcd22.240335. E03.

Palmgren, M., Shabala, S., 2024. Adapting crops for climate change: regaining lost abiotic stress tolerance in crops. Front. Sci. 2. https://doi.org/10.3389/fsci.2024.1416023.

Prado, J.B., Mulay, P.R., Kasner, E.J., Bojes, H.K., Calvert, G.M., 2017. Acute pesticide-related illness among farmworkers: barriers to reporting to public health authorities. J. Agromedicine 22 (4), 395–405. https://doi.org/10.1080/1059924X.2017.1353936.

Qiao, J., Lin, X., Wu, Y., Huang, X., Pan, X., Xu, J., et al., 2022. Global burden of non-communicable diseases attributable to dietary risks in 1990–2019. J. Hum. Nutr. Diet. 35 (1), 202–213. https://doi.org/10.1111/jhn.12904.

Rabbitt, M.P., Reed-Jones, M., Hales, L.J., Suttles, S., Burke, M.P., United States. Department of Agriculture. Economic Research Service, S.B., 2025. Household food security in the United States in 2024. Washington, D.C: Economic Research Service, U.S. Department of Agriculture.

Rabbitt, M., Reed-Jones, M., Hales, L., Burke, M., (2024). Household Food Security in the United States in 2023. https://www.ers.usda.gov/publications/109895.

Rahkovsky, I., Jo, Y., Carlson, A., 2018. What drives consumers to purchase convenience foods? https://www.usda.gov/about-usda/news/blog/what-drives-consumers-purchase-convenience-foods.

Ramesh, G., Belardo, D., Gulati, M., Ostfeld, R.J., Michos, E.D., 2021. Agricultural policy and societal factors influence patients' ability to follow a healthy diet. Am. J. Prev. Cardiol. 8, 100285. https://doi.org/10.1016/j.ajpc.2021.100285.

Rehm, C.D., Peñalvo, J.L., Afshin, A., Mozaffarian, D., 2016. Dietary intake among US adults, 1999-2012. JAMA: J. Am. Med. Assoc. 315 (23), 2542–2553. https://doi.org/10.1001/jama.2016.7491.

Reisch, L., Eberle, U., Lorek, S., 2017. Sustainable food consumption: an overview of contemporary issues and policies. Sustain. Sci. Pract. Policy 9 (2), 7–25. https://doi.org/10.1080/15487733.2013.11908111.

RHIhub, 2024. Rural Response to Farmer Mental Health and Suicide Prevention. https://www.ruralhealthinfo.org/topics/farmer-mental-health.

Rossato, S., Oakes, E.G., Barbhaiya, M., Sparks, J.A., Malspeis, S., Willett, W.C., et al., 2025. Ultraprocessed food intake and risk of systemic lupus erythematosus among women observed in the nurses' health study Cohorts. Arthritis Care Res. 77 (1), 50–60. https://doi.org/10.1002/acr.25395.

Rossi, J., Woods, T., 2024. Placing community supported agriculture in local food systems. Int. J. Agric. Sustain. 22 (1). https://doi.org/10.1080/14735903.2024.2318936.

Rowe, L.A., 2020. Addressing the fortification quality gap: a proposed way forward. Nutrients 12 (12). https://doi.org/10.3390/nu12123899.

Russell, C., Whelan, J., Love, P., 2022. Assessing the cost of healthy and unhealthy diets: a systematic review of methods. Curr. Nutr. Rep. 11 (4), 600–617. https://doi.org/10.1007/s13668-022-00428-x.

Sampson, D., Cely-Santos, M., Gemmill-Herren, B., Babin, N., Bernhart, A., Bezner Kerr, R., et al., 2021. Food Sovereignty and rights-based approaches strengthen food security and nutrition across the globe: a systematic review. Front. Sustain. Food Syst. 5. https://doi.org/10.3389/fsufs.2021.686492.

SanDiego Foundation, 2020. What is food sovereignty? https://www.sdfoundation.org/news-events/sdf-news/what-is-food-sovereignty/.

Scheyett, A., Marburger, I.L., Scarrow, A., Hollifield, S.M., Dunn, J.W., 2024. What do farmers need for suicide prevention: considerations for a hard-to-reach population. Neuropsychiatr. Dis. Treat. 20, 341–352. https://doi.org/10.2147/ndt.S452561.

Shaker, Y., Grineski, S.E., Collins, T.W., Flores, A.B., 2023. Redlining, racism and food access in US urban cores. Agric. Hum. Values 40 (1), 101–112. https://doi.org/10.1007/s10460-022-10340-3.

Sheoran, S., Kumar, S., Ramtekey, V., Kar, P., Meena, R.S., Jangir, C.K., 2022. Current status and potential of biofortification to enhance crop nutritional quality: an overview. Sustainability 14 (6). https://doi.org/10.3390/su14063301.

Singleton, C.R., Wright, L.A., McDonald, M., Archer, I.G., Bell, C.N., McLoughlin, G.M., et al., 2023. Structural racism and geographic access to food retailers in the

United States: a scoping review. Health & Place 83, 103089. https://doi.org/10.1016/j.healthplace.2023.103089.
Su, Q., Singh, V.P., 2024. Advancing irrigation management: integrating technology and sustainability to address global food security. Environ. Monit. Assess. 196 (11), 1018. https://doi.org/10.1007/s10661-024-13145-5.
Swanson, N.L., Leu, A., Abrahamson, J., Wallet, B., 2014. Genetically engineered crops, glyphosate and the deterioration of health in the United States of America. J Org Syst 9 (2), 6–37.
Talsma, E.F., Pachón, H., 2017. Biofortification of crops with minerals and vitamins. https://www.who.int/tools/elena/bbc/biofortification.
Tang, G., Qin, J., Dolnikowski, G.G., Russell, R.M., Grusak, M.A., 2009. Golden rice is an effective source of vitamin A. Am. J. Clin. Nutr. 89 (6), 1776–1783. https://doi.org/10.3945/ajcn.2008.27119.
Taylor, D.E., Ard, K., (2015). Detroit's food justice and food systems. https://www.irp.wisc.edu/resource/detroits-food-justice-and-food-systems/.
Teklu, D., Gashu, D., Joy, E.J.M., Amede, T., Broadley, M.R., 2023. Effectiveness of agronomic biofortification strategy in fighting against hidden hunger. Agronomy 13 (8). https://doi.org/10.3390/agronomy13082173.
Temple, N.J., 2022. The origins of the obesity epidemic in the USA–lessons for today. Nutrients 14 (20). https://doi.org/10.3390/nu14204253.
Testa, A.M., 2019. Access to healthy food retailers among formerly incarcerated individuals. Public Health Nutr. 22 (4), 672–680. https://doi.org/10.1017/S1368980018002720.
The global food banking network promoting food security. (no.date) https://www.foodbanking.org/promoting-food-security/.
Thelwell, L., 2022. Planting health: analyzing the role and accessibility of plant-based diets in the U.S. SSRN Electr. J. https://doi.org/10.2139/ssrn.4184000.
Thilmany, D., Bond, C.A., Bond, J.K., 2008. Going local: exploring consumer behavior and motivations for direct food purchases. Am. J. Agric. Econ. 90 (5), 1303–1309. https://doi.org/10.1111/j.1467-8276.2008.01221.x.
Thornton, P.K., Ericksen, P.J., Herrero, M., Challinor, A.J., 2014. Climate variability and vulnerability to climate change: a review. Glob. Chang. Biol. 20 (11), 3313–3328. https://doi.org/10.1111/gcb.12581.
Tian, J.C., Yao, J., Long, L.P., Romeis, J., Shelton, A.M., 2015. Bt crops benefit natural enemies to control non-target pests. Sci. Rep. 5, 16636. https://doi.org/10.1038/srep16636.
Tsatsakis, A.M., Nawaz, M.A., Kouretas, D., Balias, G., Savolainen, K., Tutelyan, V.A., et al., 2017. Environmental impacts of genetically modified plants: a review. Environmental Research 156, 818–833. https://doi.org/10.1016/j.envres.2017.03.011.
U.S. Department of Agriculture, E. R. S., 2025. Farm sector income & finances: Highlights from the farm income forecast. https://www.ers.usda.gov/topics/farm-economy/farm-sector-income-finances/highlights-from-the-farm-income-forecast.
Uconn Extension, 2025. GMOs and Food Safety. https://gmo.extension.uconn.edu/about-gmos/food-safety/.
USDA Economic Research Service, 2026. Food Security Status of U.S. Households in 2024. https://www.ers.usda.gov/topics/food-nutrition-assistance/food-security-in-the-us/key-statistics-graphics.
USDA Economic Research Service (2026a). Food security in the U.S. - definitions of food security. https://www.ers.usda.gov/topics/food-nutrition-assistance/food-security-in-the-us/definitions-of-food-security.
USDA Economic Research Service (2026b). Food security in the U.S. - survey tools. https://www.ers.usda.gov/topics/food-nutrition-assistance/food-security-in-the-us/survey-tools.

USDA, 2019. Science Breakthroughs to Advance Food and Agricultural Research by 2030. U.S. Department of Agriculture (USDA), Washington, DC.
USDA (2024). History of the National School Lunch Program.
USDA (2025). Regulation of Biotech Plants.
USDA (2026). WIC Farmers Market Nutrition Program.
Vigar, V., Myers, S., Oliver, C., Arellano, J., Robinson, S., Leifert, C., 2020. A systematic review of organic versus conventional food consumption: is there a measurable benefit on human health? Nutrients 12 (1). https://doi.org/10.3390/nu12010007.
Vitale, M., Costabile, G., Testa, R., D'Abbronzo, G., Nettore, I., Macchia, P., et al., 2023. Ultra-processed foods and human health: a systematic review and meta-analysis of prospective cohort studies. Adv. Nutr. 15. https://doi.org/10.1016/j.advnut.2023.09.009.
Wallinga, D., Smit, L.A.M., Davis, M.F., Casey, J.A., Nachman, K.E., 2022. A review of the effectiveness of current US policies on antimicrobial use in meat and poultry production. Curr. Environ. Health Rep. 9 (2), 339–354. https://doi.org/10.1007/s40572-022-00351-x.
Wang, S.L., Njuki, E., Mosheim, R., Nehring, R., 2025. Agricultural productivity in the United States - productivity growth in U.S. agriculture. https://www.ers.usda.gov/data-products/agricultural-productivity-in-the-united-states/productivity-growth-in-us-agriculture#:~:text=With%20aggregat%20inputs%20(including%20land,percent%20above%20its%201948%20level.
Weaver Street Market, 2026. Weaver Street Market. https://www.weaverstreetmarket.coop/.
Weiler, A.M., Hergesheimer, C., Brisbois, B., Wittman, H., Yassi, A., Spiegel, J.M., 2015. Food sovereignty, food security and health equity: a meta-narrative mapping exercise. Health Policy Planning 30 (8), 1078–1092. https://doi.org/10.1093/heapol/czu109.
Wood, B., Robinson, E., Baker, P., Paraje, G., Mialon, M., van Tulleken, C., et al., 2023. What is the purpose of ultra-processed food? An exploratory analysis of the financialisation of ultra-processed food corporations and implications for public health. Global. Health 19 (1), 85. https://doi.org/10.1186/s12992-023-00990-1.
Xu, K., Li, X.Q., Zhao, D.L., Zhang, P., 2021. Antifungal secondary metabolites produced by the fungal endophytes: chemical diversity and potential use in the development of biopesticides. Front. Microbiol. 12, 689527. https://doi.org/10.3389/fmicb.2021.689527.
Xuan, K., Li, X., Zhang, J., Jiang, Y., Ma, B., Liu, J., 2023. Effects of organic amendments on soil pore structure under waterlogging stress. Agronomy 13 (2). https://doi.org/10.3390/agronomy13020289.
Yang, X., Hu, J., Wang, Z., Huang, T., Xiang, Y., Zhang, L., et al., 2023. Pre-harvest nitrogen limitation and continuous lighting improve the quality and flavor of lettuce (Lactuca sativa L.) under hydroponic conditions in greenhouse. J. Agric. Food Chem. 71 (1), 710–720. https://doi.org/10.1021/acs.jafc.2c07420.
Yang, Y., Wu, J., Du, Y.-L., Gao, C., Tang, D.W.S., van der Ploeg, M., 2022. Effect on soil properties and crop yields to long-term application of superabsorbent polymer and manure. Front. Environ. Sci. 10. https://doi.org/10.3389/fenvs.2022.859434.
Zhou, W., Li, M., Achal, V., 2025. A comprehensive review on environmental and human health impacts of chemical pesticide usage. Emerg. Contam. 11 (1), 100410. https://doi.org/10.1016/j.emcon.2024.100410.
Zsogon, A., Cermak, T., Naves, E.R., Notini, M.M., Edel, K.H., Weinl, S., et al., 2018. De novo domestication of wild tomato using genome editing. Nat. Biotechnol. https://doi.org/10.1038/nbt.4272.

Printed and bound by CPI Group (UK) Ltd, Croydon, CR0 4YY
17/09/2026
02218532-0012